Modern Techniques in High-Resolution FT-NMR

Narayanan Chandrakumar
Sankaran Subramanian

Modern Techniques in High-Resolution FT-NMR

With 259 Figures

Springer-Verlag
New York Berlin Heidelberg London Paris Tokyo

Narayanan Chandrakumar
Scientist-in-Charge
FT-NMR Laboratory
Central Leather Research Institute
Adayaru, Madras-600 020
Tamil Nadu
India

Sankaran Subramanian
Professor of Chemistry
Regional Sophisticated
Instrumentation Center
Indian Institute of Technology
Madras-600 036
Tamil Nadu
India

Library of Congress Cataloging in Publication Data
Chandrakumar, N., 1951-
Modern techniques in high resolution FT-NMR.
Bibliography: p.
Includes index.
1. Liquids--Spectra. 2. Solids--Spectra.
3. Nuclear magnetic resonance spectroscopy.
4. Fourier transform optics. I. Subramanian, S.,
1942- . II. Title.
QC145.4.06C48 1986 543'.0877 86-6583

Typeset by Asco Trade Typesetting, Ltd., Hong Kong.
Printed and bound by R.R. Donnelley & Sons, Harrisonburg, Virginia.
Printed in the United States of America.

9 8 7 6 5 4 3 2 1

ISBN 0-387-96327-8 Springer-Verlag New York Berlin Heidelberg
ISBN 3-540-96327-8 Springer-Verlag Berlin Heidelberg New York

கல்வி கரையில கற்பவர் நாள்சில . . .

Kalvi kaṟayila karpavar nāḻsila . . .
Nāladiyār 2:14:135

Learning is endless, (and) life so short . . .
Excerpt from Tamil Anthology of Quartets,
Nāladiyār, circa 50 B.C.- 150 A.D.

Preface

The magnetism of nuclear spin systems has proved an amazingly fertile ground for the creativity of researchers. This happy circumstance results from the triple benediction that nature appears to have bestowed on nuclear spins: they are sporting spies—being infinitely manipulable (one is even tempted to say malleable), not unduly coy in revealing their secrets, and having a whole treasure house of secrets to reveal in the first place.

Since spin dynamics are now orchestrated by the NMR researcher with ever more subtle scores, it is important to be able to tune into the proceedings with precision, if one is to make sense of it at all. Fortunately, it is not terribly difficult to do so, since in many ways spin dynamics are the theoretician's dream come true: they are often finite dimensional and quite tractable with basic quantum mechanics, frequently allowing near-exact treatments and readily testable predictions.

This book was conceived two years ago, with the objective of providing a simple, consistent introduction to the description of the spin dynamics that one encounters in modern NMR experiments. We believed it was a good time to attempt this, since it was possible by then to give sufficiently general descriptions of powerful classes of new NMR experiments. The choice of experiments we discuss in detail is necessarily subjective, although we hope to have given a flavor of most of the important classes of pulse sequences, including some surface coil imaging applications. Except for brief treatments of the nuclear Overhauser effect, 2D NOESY and exchange spectroscopy, and cross-relaxation dynamics in the solid state, however, we have limited the exposition essentially to *coherent* spin dynamics. In particular, therefore, classical relaxation measurements and their relation to *molecular* dynamics are not treated. We have attempted to make the book self-contained within this scope and have formatted it in a manner we hope will encourage self-teaching. We have included the relevant algebra in substantial quantity throughout the book, because we believe: (a) it is simple enough and deserves to be commonly used, and (b) one pays a heavy price in precision and detail in choosing to ignore it.

We record our deep appreciation and gratitude to Professor Dr. E. Fluck, who displayed great faith in our project and encouraged us to go ahead. We were also egged on to wrap up the work without undue delay, haunted as we were by the spectre of a book rendered unmanageable by having to include the ever newer NMR experiments that continue to be invented with every passing month!

A large number of examples have been used from the literature in illustrating the new techniques we discuss, and we have received unstinted cooperation from researchers and publishers around the world in kindly permitting us the use of their work. As is customary, we acknowledge these permissions individually in the relevant figure captions. In particular, we have employed a number of illustrations from works of research groups of Professors R.R. Ernst, R. Freeman and A. Pines, and we acknowledge here their kind, prompt courtesy in according us their permission to do so.

We would like to thank the Directors of our respective Institutes, Dr. G. Thyagarajan (CLRI) and Dr. L.S. Srinath (IIT Madras), for their support and encouragement. One of us (N.C) also thanks Dr. D. Ramaswamy, Assistant Director, CLRI, for his kind encouragement. G.V. Visalakshi and D. Srinivas cheerfully presided over the metamorphosis of a perfectly illegible manuscript through a confused typescript, into a final, usable version. They deserve our thanks; we are afraid we must take all the credit for the remaining errors, however! We also thank the editors at Springer-Verlag for their friendly cooperation in seeing this work through the throes of final production.

Finally, our primary debt of gratitude is to our long-suffering better halves: Parvathy Chandrakumar and Rajalakshmi Subramanian who were quite willing to have us work at the book even when, as was frequently the case, we were less willing to do so. They let us know, in unmistakable terms, when the occasion demanded advice!

August, 1986

N.C
S.S

Contents

CHAPTER 1

Introduction and General Theory

Nuclear magnetic resonance (NMR), which originated some 40 years ago primarily as a potentially accurate method for measuring nuclear magnetogyric ratios, turned out to be something of an embarrassment in that application when it transpired that the rf magnetic susceptibility it measured could be a quite complicated function, exhibiting many sharp, close-lying resonances. When it was realized however, that this complexity rather subtly reflected exceedingly fine characteristics of the electronic environment in which the nuclei were embedded, NMR began being developed as a high-resolution (HR) spectroscopic technique for the elucidation of molecular structure, dynamics, and, most recently, distribution (i.e., NMR imaging). Here again it soon became apparent that HR-NMR spectra were generally too complicated to admit of straightforward, unambiguous interpretations of molecular structures. Major effort, since then, has been spent on developing ever more powerful methods to help produce and interpret HR-NMR spectra. On the experimental side, this has, on the one hand, led to attempts (a) to develop NMR as a truly multinuclear technique, and (b) to improve the sensitivity or signal-to-noise ratio of NMR spectra as well as their resolution. On the other, people have sought to devise NMR experiments that can generate unambiguous, clearly recognizable features in the spectra by various means of selectively monitoring different kinds of nuclear magnetic interactions while suppressing others as required. Pulse Fourier transform (FT) NMR has emerged as the method of choice, allowing the spectroscopist maximum flexibility in the pursuit of practically any combination of these objectives. In conjunction with the rapid advances in the commercially available instrumentation, this situation has led to an explosion in the development of new techniques in NMR that shows no signs of letting up.

We attempt in this book to capture the spirit of NMR as it is practiced today, by presenting a coherent view of the fundamentals on which it rests. We make in the process no claims to exhaustiveness in cataloging the emerging techniques and their applications.

Introduction

A large number of nuclear isotopes have a nonzero spin angular momentum $Ih/2\pi$, where h is the Planck's constant. In accordance with the principles of quantum mechanics, I can take on only the values 1/2, 1, 3/2, 2, 5/2, etc. Associated with this spin quantum number I is the magnetic dipole moment μ, of the isotope in question:

$$\boldsymbol{\mu} = \gamma h \mathbf{I}/2\pi \tag{1}$$

γ being the magnetogyric ratio, which is a nuclear property. In an external magnetic field $\mathbf{B}_0$, the magnetic moment can take up one of $(2I + 1)$ allowed orientations, each with its characteristic energy corresponding to the Hamiltonian:

$$\begin{aligned} \mathscr{H} &= -\boldsymbol{\mu} \cdot \mathbf{B}_0 \\ &= -\gamma h \mathbf{I} \cdot \mathbf{B}_0/2\pi \\ &= -\gamma h B_0 I_z/2\pi \end{aligned} \tag{2}$$

the direction of the dc magnetic field $\mathbf{B}_0$ being, by convention, chosen to be the z axis of the laboratory coordinate frame. Measurements of the magnetic ("Zeeman") energy in such a situation lead to the values:

$$E = -\gamma h B_0 m_I/2\pi \tag{3}$$

where m_I takes on the $(2I + 1)$ values $\pm I$, $\pm(I - 1), \ldots, \pm 1/2$ or 0, depending on whether I is a half-odd integer or an integer. Successive Zeeman levels are thus displaced in energy by the constant value, $\gamma h B_0/2\pi$. Values of $\gamma B_0/2\pi$ range from 10^0 to 10^3 MHz, depending on the magnetogyric ratio of the isotope in question and the intensity of the external magnetic field $\mathbf{B}_0$. This energy gap, expressed in frequency units, is called the Larmor frequency of the isotope in the field $\mathbf{B}_0$. It in fact is the frequency of precession of the nuclear spins in the magnetic field, originating in the torque exerted by the field on their spin angular moments.

In an ensemble of nuclear spins I, the $(2I + 1)$ allowed energy levels are populated in thermal equilibrium in accordance with the Boltzmann distribution. For $I = 1/2$, for example, the ratio of the number (N_i) of spins per unit volume in the upper energy state ($m_I = -1/2$ if $\gamma > 0$, called the spin-down state, $|\beta\rangle$) to that in the lower state ($m_I = 1/2$ if $\gamma > 0$, called the spin-up state, $|\alpha\rangle$), is given by:

$$N_2/N_1 = \exp[-(E_2 - E_1)/kT] = \exp(-\gamma h B_0/2\pi kT) \tag{4}$$

T being the absolute equilibrium temperature, known as the "lattice" temperature. The spin system in fact requires some time to attain the state of thermal equilibrium after a magnetic field is switched on. The approach to thermal equilibrium is often by a first-order process known as spin–lattice or longitudinal relaxation, characterized by a single time constant, T_1.

In most practical applications, the quantity $(\gamma h B_0/2\pi kT)$ is very much less than 1. With this approximation

$$N_2/N_1 = 1 - \varepsilon \tag{5}$$

where:

$$\varepsilon = \gamma h B_0/2\pi kT \tag{6}$$

This leads at once to:

$$\begin{aligned} N_1 &= N[1 + (\varepsilon/2)]/2 \\ N_2 &= N[1 - (\varepsilon/2)]/2 \end{aligned} \tag{7}$$

correct to terms linear in ε, where:

$$N = \sum_i N_i \tag{8}$$

the total number of spins per unit volume. The difference in population, $(N_1 - N_2)$, is a measure of the "order" induced in the nuclear spin system by the magnetic field: the spins are "polarized" in the field. In this sense there is total disorder when $\mathbf{B}_0$ vanishes, for N_1 is then equal to N_2: the energy levels are populated equally in the absence of the magnetic field. The order induced by the field leads to a bulk magnetization ($\mathbf{M}$) of the nuclear spin ensemble that may be calculated readily. It is given by:

$$\mathbf{M} = \sum_i N_i \boldsymbol{\mu}_i \tag{9}$$

For $I = 1/2$, this turns out to be:

$$\begin{aligned} M &= N\gamma h[[1 + (\varepsilon/2)] - [1 - (\varepsilon/2)]]/8\pi \\ &= N\gamma^2 (h/2\pi)^2 B_0/4kT \end{aligned} \tag{10}$$

In general, the quantity

$$\chi(0) = (N\gamma^2 I(I + 1)/3kT)(h^2/4\pi^2) \tag{11}$$

is termed the Curie susceptibility of the ensemble of nuclear spins. It may be noticed that the magnetization induced in the ensemble of "bare" nuclei at thermal equilibrium is aligned with $\mathbf{B}_0$, and no magnetization exists transverse to $\mathbf{B}_0$. At the level of the individual nuclei in the ensemble, their precession is "incoherent," leading to zero net magnetic moment in the xy plane, while giving rise to $\mathbf{M}$ parallel to $\mathbf{B}_0$ in accordance with Eq. (9). This situation is depicted in Fig. 1.1. Numerically, the dc nuclear paramagnetic susceptibility $\chi(0)$ is so small that the electron diamagnetic susceptibility completely swamps it in closed shell molecules at all but the very lowest attainable temperatures, making it a dauntingly difficult quantity to measure directly. Nuclear mag-

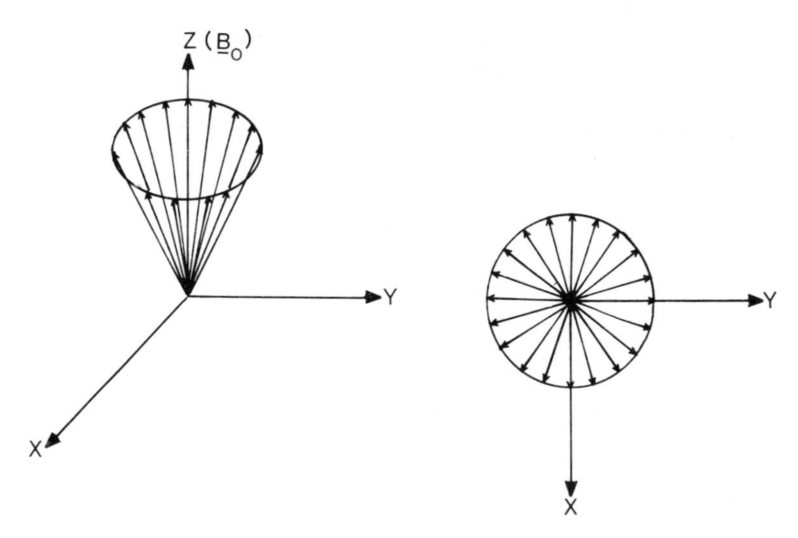

Figure 1.1. Larmor precession of nuclei in a magnetic field $\mathbf{B}_0$.

netic resonance techniques set about this task by measuring a far more amenable quantity, the rf magnetic susceptibility, $\chi(\omega)$, of the nuclear spin ensemble.

To this end, the ensemble of ordered spins is irradiated at the Larmor frequency. The electromagnetic radiation interacts with the spin system through its magnetic dipole component. Induced transitions occur in both directions between neighboring pairs of the Zeeman levels, leading to a net absorption owing to the higher population of the lower levels, provided the radiofrequency power applied is sufficiently small to allow relaxation processes to maintain the equilibrium population differences. This method of exciting an NMR spectrum can be viewed equivalently as a nuclear induction process: the component of the resonant radiofrequency field that has the same sense of rotation as the Larmor precession of the nuclear spins exerts a steady torque on the equilibrium ensemble magnetization. This torque rotates the ensemble magnetization away from its state of alignment with $\mathbf{B}_0$ and into the xy plane, where the nuclear magnetization, precessing at the Larmor frequency, induces an emf; this is the desired NMR signal.

In other words, the resonant rf magnetic field leads to the "coherent" precession of the individual nuclei, giving rise to a nonzero transverse magnetization, a situation corresponding, therefore, to disruption of the thermal equilibrium. In fact, upon removal of the resonant rf the spins lose coherence in the phase of their precession and revert to the equilibrium situation, often by a first-order process known as spin–spin or transverse relaxation, which is characterized by a single time constant, T_2. More frequently, this loss of phase memory is characterized by T_2^*, a time constant that includes the effect of instrument imperfections, such as inhomogeneities in $\mathbf{B}_0$. The power of the resonance method consists in the situation that a weak resonant rf magnetic field $\mathbf{B}_1$—typically weaker than the polarizing dc field $\mathbf{B}_0$ by factors exceeding a million and oriented transverse to it—succeeds in taking control of the orientation of the bulk magnetization of the spins. In other words, the rf magnetic susceptibility of the ensemble of spins is several orders of magnitude higher than its dc magnetic susceptiblity:

$$\text{Im}\,[\chi(\omega)] \sim \chi(0)\omega/\Delta\omega$$

The above relation expresses the imaginary part of the rf magnetic susceptibility, Im $[\chi(\omega)]$ at the resonance frequency ω, in terms of the dc susceptibility $\chi(0)$, ω, and the width of the resonance, $\Delta\omega$. The rf susceptibility has in fact two components, one in phase with the exciting rf and the other in quadrature with it. These are expressed respectively as the real and imaginary parts of $\chi(\omega)$, the imaginary part being responsible for the resonance absorption, whereas the real part contributes a dispersion.

In describing NMR experiments, it is constantly necessary to understand the motion of the ensemble of spins under various influences. These include the rf field $\mathbf{B}_1$, and other magnetic interactions of the nuclear spins in the "real world," including chemical shifts, internuclear scalar coupling, internuclear dipolar coupling, and quadrupolar coupling. These subtleties of motion are all superimposed on their basic precession in $\mathbf{B}_0$. It is very useful therefore to try to follow their motion with reference to their precession in $\mathbf{B}_0$. More conveniently, the transmitter frequency is used as the reference for the motion, rather than the Larmor frequency. This is accomplished experimentally by phase-sensitive detection with respect to the transmitter frequency. In theoretical accounts of this process of referencing, it is known as a transformation from the (static) laboratory frame of reference to the rotating frame of reference, the frequency of rotation being conveniently chosen as the transmitter frequency. It proves possible to describe certain aspects of the motion of the ensemble of spins pictorially in the rotating frame, and these "vector pictures" have played a significant role in understanding the basic phenomena involved. However, certain other subtle features of the motion, which are ever more frequently excited in modern NMR experiments, are very difficult to picture unambiguously in terms of such vector models.

We choose, therefore, to employ a formal method of describing the motion of the spin ensemble under various interactions. This formalism, after being dormant for over two decades, has been developed in the last couple of years into a powerful, yet simple tool capable of providing detailed insight into diverse classes of NMR experiments. The rest of this chapter is devoted to an exposition of the general theory, some powerful results derived from it, and its interpretation in terms of the parameters of the observed spectrum. We shall not omit, however, to point out the correspondences with simple vector pictures, where applicable.

General Theory

State of the Spin System. In a multilevel situation, the spin state $|\Psi)$ is specified in general by:

$$|\Psi) = \sum_{k=1}^{2I+1} c_k |\varphi_k) \tag{12}$$

the $|\varphi_k)$'s being a complete set, e.g., the eigenstates of the Zeeman Hamiltonian; the c_k's are complex coefficients. In words, the pure state $|\Psi)$ is in general a linear combination of the basis states appropriate to the problem. Such a linear combination is called a coherent superposition. The state of an ensemble of such spins is represented by:

$$|\Phi) = \sum_i p^{(i)} |\Psi^{(i)}) \tag{13}$$

In words, the pure states $|\Psi^{(i)})$ occur in an ensemble with the probabilities $p^{(i)}(0 \leqslant p^{(i)} \leqslant 1, \sum_i p^{(i)} = 1)$. The state of the ensemble is termed an incoherent superposition.

The expectation value of an observable O, represented by the operator $\mathbf{O}$, is given for such an ensemble by:

$$\begin{aligned} \langle \mathbf{O} \rangle^{(i)} &= (\Psi^{(i)}|\mathbf{O}|\Psi^{(i)}) = \sum_{m,n} c_m^{*(i)} c_n^{(i)} (\varphi_m|\mathbf{O}|\varphi_n) \\ &= \sum_{m,n} c_m^{*(i)} c_n^{(i)} \mathbf{O}_{mn} \end{aligned} \tag{14}$$

This implies:

$$\begin{aligned} \langle \mathbf{O} \rangle &= \sum_i p^{(i)} \langle \mathbf{O} \rangle^{(i)} = \sum_i \sum_{m,n} p^{(i)} c_m^{*(i)} c_n^{(i)} \mathbf{O}_{mn} \\ &= \sum_{m,n} \sigma'_{nm} \mathbf{O}_{mn} \end{aligned} \tag{15}$$

where:

$$\sigma'_{nm} = \sum_i p^{(i)} c_m^{*(i)} c_n^{(i)} \tag{16}$$

Thus,

$$\langle \mathbf{O} \rangle = \mathrm{Tr}(\sigma' \mathbf{O}) = \mathrm{Tr}(\mathbf{O} \sigma') \tag{17}$$

The state of the ensemble is completely specified by the matrix σ', which has as its entries the pairwise products of the complex coefficients of the chosen eigenstates in terms of which each pure state is represented, weighted by the probability of occurrence of that pure state in the ensemble and summed over all the pure states. σ' is called the density operator; it has, by definition a $(2I + 1) \times (2I + 1)$ dimensional matrix representation and is Hermitian. Computing the ensemble average of an observable O involves taking the trace of the product of $\mathbf{O}$ with σ', in either order.

This formal representation can be applied immediately to a spin-1/2 situation in order to fix the ideas it expresses. In this case, in Boltzmann equilibrium the pure states and probabilities are:

$$|\Psi^{(1)}) = |\alpha); \quad p^{(1)} = [1 + (\varepsilon/2)]/2;$$

$$|\Psi^{(2)}) = |\beta); \quad p^{(2)} = [1 - (\varepsilon/2)]/2$$

$$\sigma'_{\alpha\alpha} = p^{(1)} c_\alpha^{*(1)} c_\alpha^{(1)} + p^{(2)} c_\alpha^{*(2)} c_\alpha^{(2)} = [1 + (\varepsilon/2)]/2; \tag{18}$$

$$\sigma'_{\beta\beta} = p^{(1)} c_\beta^{*(1)} c_\beta^{(1)} + p^{(2)} c_\beta^{*(2)} c_\beta^{(2)} = [1 - (\varepsilon/2)]/2;$$

$$\sigma'_{\alpha\beta} = p^{(1)} c_\beta^{*(1)} c_\alpha^{(1)} + p^{(2)} c_\beta^{*(2)} c_\alpha^{(2)} = 0 = \sigma'_{\beta\alpha}$$

Here, α and β, which represent the "spin up" and "spin down" states of a spin $\frac{1}{2}$ nucleus, have been used as subscripts for the sake of explicitness. Thus, for an ensemble of spin-1/2 nuclei in thermal equilibrium,

$$\sigma' = (1/2)\begin{pmatrix} [1 + (\varepsilon/2)] & 0 \\ 0 & [1 - (\varepsilon/2)] \end{pmatrix} \tag{19}$$

In this state of the system, the ensemble averages of longitudinal and transverse magnetizations can be calculated as:

$$\langle M_z \rangle = \mathrm{Tr}\,(M_z \sigma') = N\gamma h\,\mathrm{Tr}\,(I_z \sigma')/2\pi = (N\gamma\varepsilon/4)(h/2\pi)$$

$$= (N\gamma^2 B_0/4kT)(h/2\pi)^2$$

which is exactly the result given in Eq. (10). Also,

$$\langle M_x \rangle = N\gamma h\,\mathrm{Tr}\,(I_x \sigma')/2\pi = 0; \quad \langle M_y \rangle = N\gamma h\,\mathrm{Tr}\,(I_y \sigma')/2\pi = 0$$

These results confirm that in thermal equilibrium there is no transverse magnetization in the spin system. In fact, σ' would need to have nonzero off-diagonal elements to represent a state with a transverse magnetization component. The calculations outlined above employ the Pauli matrix representation of the spin operators I_x, I_y, and I_z (see Appendix 1).

From Eq. (19) σ' can be rewritten as:

$$\sigma' = (1/2)\mathbf{1} + (\varepsilon/4)\begin{pmatrix} 1 & 0 \\ 0 & -1 \end{pmatrix} = (1/2)\mathbf{1} + \sigma \tag{20}$$

where **1** represents the identity matrix of appropriate dimensions, in this case (2×2).

In other words, in general,

$$\mathrm{Tr}\,(\mathbf{O}\sigma) = (1/2)\,\mathrm{Tr}\,(\mathbf{O}) + \mathrm{Tr}\,(\mathbf{O}\sigma) \tag{21}$$

The first term in σ' is uninteresting because it remains unaffected under the various interactions in an NMR system (*vide infra*) and because we are concerned with traceless operators **O**. It shall henceforth be discarded from the discussion, which will center around σ. From Eq. (20), it is clear that in thermal equilibrium a spin-1/2 ensemble in a magnetic field may be represented by the density operator:

$$\sigma = (\varepsilon/2) I_z \tag{22}$$

We shall find this operator representation of σ extremely convenient and shall build on it.

In fact, we may represent any $(2I + 1) \times (2I + 1)$ matrix in terms of $(2I + 1)^2$ linearly independent matrices. When we recall that each spin operator for a spin I system has a $(2I + 1) \times (2I + 1)$ matrix representation, it is clear that any perfectly general form of the density operator for such a system may be represented as a linear combination of a corresponding set of $(2I + 1)^2$ suitable linearly independent spin operators. For a spin-1/2 system, for instance, $(\mathbf{1}, I_x, I_y, I_z)$ is a suitable set, whereas for spin-1, $(\mathbf{1}, I_x, I_y, I_z, I_x^2, I_z^2, [I_x, I_z]_+, [I_x, I_y]_+, [I_y, I_z]_+)$ is a suitable set of operators, all of them in fact Hermitian and the last three denoting, respectively, the three anticommutators $(I_xI_z + I_zI_x)$, $(I_xI_y + I_yI_x)$, and $(I_yI_z + I_zI_y)$. A two-spin-1/2 ensemble requires 16 operators for the complete representation of its state, whereas a spin-1, spin- 1/2 pair requires 36 operators. (See Appendix 1 for some details for handling such cases, which involve direct product spaces.)

We shall now consider the two-spin-1/2 ensemble in some detail. We may use as eigenbases the simple product functions, with the first entry indicating the state of particle 1, and the second entry that of particle 2. In simplified notation, the eigenbasis is given by $(\alpha\alpha, \alpha\beta, \beta\alpha, \beta\beta)$, which may be labeled as states 1, 2, 3, and 4, respectively. The state of the ensemble will be represented by a (4×4) density matrix. By definition, the elements (σ_{rr}) $(r = 1, \ldots, 4)$ of the density matrix may be interpreted as the (relative) populations of the states r. The occurrence of a nonzero element $\sigma_{rs} (r \neq s)$ indicates that certain pure states have been created in the ensemble, which are a coherent superposition of the eigenstates r and s. If the total magnetic quantum number, $M = \sum_i m_{Ii}$, of states r and s differ by q units, σ_{rs} is said to represent q-quantum coherence. The coherence order q is also a "good" quantum number in high-field NMR, owing to the rotational symmetry of the Hamiltonian. From the foregoing, each transition rs clearly has two coherences, σ_{rs} and σ_{sr}, associated with it, with coherence orders $(M_r - M_s)$ and $(M_s - M_r)$. Thus, $\sigma_{12}, \sigma_{13}, \sigma_{24}, \sigma_{34}$, and their Hermitian conjugates σ_{21}, σ_{31}, σ_{42}, and σ_{43} represent single quantum coherences, whereas σ_{14} and σ_{41} represent double- (or two-) quantum coherence and σ_{23} and σ_{32} represent zero-quantum coherence. The operator representation of such a density matrix will involve the set $(\mathbf{1}, I_{1x}, I_{1y}, I_{1z}, I_{2x}, I_{2y}, I_{2z}, I_{1x}I_{2x}, I_{1x}I_{2y}, I_{1x}I_{2z}, I_{1y}I_{2x}, I_{1y}I_{2y}, I_{1y}I_{2z}, I_{1z}I_{2x}, I_{1z}I_{2y}, I_{1z}I_{2z})$. From the matrix representation of these operators (see Appendix 1), it is clear that I_{1z}, I_{2z} and $I_{1z}I_{2z}$ can be used to represent populations in σ; $I_{1x}, I_{1x}I_{2z}, I_{1y}, I_{1y}I_{2z}, I_{2x}, I_{1z}I_{2x}, I_{2y}, I_{1z}I_{2y}$ represent single-quantum coherences; $I_{1x}I_{2x}, I_{1x}I_{2y}, I_{1y}I_{2x}$, and $I_{1y}I_{2y}$ represent zero- and double-quantum coherences.

Which of these coherences can give rise to observable signals? Applying the prescriptions given above, we seek an answer to this question by looking for nonzero traces of the product of $I_x (= I_{1x} + I_{2x})$ or $I_y (= I_{1y} + I_{2y})$ with each of the 16 basis set operators. It is clear, upon using the matrix representations of the relevant operators given in Appendix 1, that zero-quantum coherences, double-quantum coherences, and longitudinal magnetizations corresponding to the populations cannot lead to an NMR signal. Those parts of the density operator corresponding to an operator representation in terms of I_{1x}, I_{1y}, I_{2x}, and I_{2y} can lead to a signal. The behavior of the $I_{1x}I_{2z}$, $I_{1y}I_{2z}$, $I_{1z}I_{2x}$, and $I_{1z}I_{2y}$ terms is a little more tricky. For instance, $\mathrm{Tr}(I_{1x}I_{1x}I_{2z})$ is zero. This corresponds to zero total intensity of the doublet resonance of spin 1. However, if the two lines of this doublet are resolved, the intensity of the single transitions can be discussed. In this case, the relevant quantities are $\mathrm{Tr}\,[(I_{1x} \pm 2I_{1x}I_{2z})I_{1x}I_{2z}]$ (*vide infra*); i.e., two signals of equal and opposite intensity $[\pm \mathrm{Tr}\,(2I_{1x}I_{2z})]$ are indicated, instead of the normal doublet with two in-phase signals of equal intensity. If the multiplet structure is not resolvable, because of either instrumental considerations or the natural linewidth, the antiphase signals described by $I_{1x}I_{2z}$ collapse to give zero signal.

We conclude that in the two-spin-1/2 system, only single-quantum coherences are observable, involving in-phase transverse magnetizations, and, under favorable circumstances, antiphase transverse magnetizations as well. The correspondence of this result with the magnetic dipole selection rule for NMR transitions, $\Delta M = \pm 1$, may be noted. The above result is valid for multispin systems also, except that in multispin systems with strong coupling, observable single-quantum coherences include "combination lines" as well. For example, in a strongly coupled three-spin system, parts of the density matrix that can be represented by such operators as $I_{1x}(I_{2x}I_{3x} + I_{2y}I_{3y})$ also become observable, where the second factor corresponds to zero-quantum coherence of spins 2 and 3.

Time Evolution of a Spin System. We shall commence a discussion of the motions of spin systems with a brief account of the operator algebra that we shall require for this purpose.

Commutators and Anticommutators. The hallmark of quantum mechanics is its recognition that all the dynamical observables that are specified independently with an arbitrary degree of precision in describing the motion of macroscopic objects together constitute an inadmissibly over-specified set of parameters to describe the state of a microscopic object, such as a molecule, atom, or subatomic particle. In fact, in dealing with such objects, quantum mechanics recognizes limits to the precision with which dynamical variables can be measured or specified and recognizes further the existence of conjugate pairs of variables, only one of which may be specified accurately at a time. Such pairs of variables give widely differing results upon measurement, depending on the order in which they are measured. This property is expressed formally as commutation relations between the quantum mechanical operators corresponding to the observables in question. If two operators commute with each other, measurement of either of the corresponding observables does not constitute a disturbance in the state of the system that will change the outcome of the subsequent measurement of the other observable. On the other hand, pairs of operators that do not commute with each other are conjugate variables, with the property described above. Operators of spin angular momentum obey the following commutation relations:

$$[\mathbf{I}^2, I_i] = (\mathbf{I}^2 I_i - I_i \mathbf{I}^2) = 0$$

$$[I_i, I_j] = (I_i I_j - I_j I_i) = i\varepsilon_{ijk} I_k \qquad (i, j, k = x, y, z)$$

These relations express the fact that the operator of the square of the spin angular momentum commutes with any of the three Cartesian components of the spin. The three components of the spin, however, do not commute with each other: each component, of course commutes with itself; the commutator of each with another component equals the third component multiplied by $\pm i$, depending on whether or not the two components in the commutator are in cyclic order. For spin-1/2 particles anticommutation relations are valid in addition:

$$[I_i, I_j]_+ = (I_i I_j + I_j I_i) = (1/2)\delta_{ij}$$

The squares of spin-1/2 component operators are therefore equal to 1/4. All the above relations may be verified by explicit matrix manipulations, employing the matrix representation of spin operators given in Appendix 1. Some basic properties of commutators are discussed in Appendix 2.

Exponential Operators. Exponential operators are defined as a series expansion:

$$\exp(A) = \sum_{r=0}^{\infty} (A^r/r!) = \mathbf{1} + A + (A^2/2!) + (A^3/3!) + \cdots \tag{23a}$$

This involves the powers of the operator in question, the nth power of an operator representing successive application of the operator n times. From the definition, it follows for example, that for $I = 1/2$,

$$\begin{aligned} \exp(i\theta I_y) &= 1 + i\theta I_y + [(i\theta I_y)^2/2!] + [(i\theta I_y)^3/3!] + [(i\theta I_y)^4/4!] + \cdots \\ &= (1 - [(\theta/2)^2/2!] + [(\theta/2)^4/4!] - \cdots) \\ &\quad + 2i([\theta/2] - [(\theta/2)^3/3!] + \cdots) I_y \\ &= \mathbf{1}\cos(\theta/2) + 2i\sin(\theta/2) I_y \end{aligned} \tag{23b}$$

Also from the definition we find that:

$$\exp(A + B) = \exp(A)\exp(B) = \exp(B)\exp(A)$$

if and only if A and B commute, i.e., $[A, B] = 0$.

A relation that we shall use frequently is the Baker–Campbell–Hausdorff (BCH) formula:

$$\begin{aligned} \exp(-iAt) B \exp(iAt) = B - (it)[A, B] &+ \frac{(it)^2}{2!}[A, [A, B]] \\ &- \frac{(it)^3}{3!}[A, [A, [A, B]]] + \cdots \end{aligned} \tag{23c}$$

From this, it follows, for instance, that:

$$\exp(-i\theta I_y)I_z\exp(i\theta I_y) = I_z - (i\theta)(iI_x) + \frac{(i\theta)^2}{2!}I_z - \frac{(i\theta)^3}{3!}(iI_x) + \cdots$$
$$= I_z(1 - (\theta^2/2!) + \cdots) + I_x(\theta - (\theta^3/3!) + \cdots) \quad (23d)$$
$$= I_z\cos\theta + I_x\sin\theta$$

regardless of the value of I. The BCH formula is explicitly evaluated for various forms of A and B in Appendix 2. In particular,

$$\exp(-iAt)B\exp(iAt) = B,$$

if A and B commute.

Time Evolution. The basic description of time evolution of a system is given by the Schrödinger equation:

$$i\frac{\delta|\Psi)}{\delta t} = +(2\pi/h)\mathscr{H}|\Psi) \quad (24a)$$

For a spin in a Zeeman field, this takes the form:

$$i\frac{\delta|\Psi)}{\delta t} = \omega_0 I_z|\Psi) \quad (24b)$$

where $\boldsymbol{\omega}_0 = -\gamma\mathbf{B}_0$.

Because the Hamiltonian is time independent in this case, Eqs. (24a) and (24b) may be integrated between time 0 and t to give:

$$|\Psi(t)) = \exp[-2\pi i\mathscr{H}t/h]|\Psi(0))$$
$$= \exp[-i\omega_0 I_z t]|\Psi(0)) \quad (25)$$

This equation gives formal expression to the time evolution of the spin system originating in its Larmor precession. If one were to reference all the motional modes of the spins to this precession, one would be interested in the time evolution of $|\Psi_{\text{rot}}(t)) = \exp(i\omega_0 I_z t)|\Psi(t))$, i.e., the wavefunction in the rotating frame where the explicit time dependence from the Larmor precession has been removed. More conveniently, the frequency ω of the rotating frame is chosen to equal the transmitter frequency, rather than the Larmor frequency ω_0. Because:

$$\frac{\delta|\Psi_{\text{rot}})}{\delta t} = \exp(i\omega I_z t)\frac{\delta|\Psi)}{\delta t} + (i\omega I_z)|\Psi_{\text{rot}}) \quad (26)$$

we have:

$$i\frac{\delta|\Psi_{\text{rot}})}{\delta t} = [(2\pi\mathscr{H}'/h) - \omega I_z]|\Psi_{\text{rot}}) \quad (27)$$

where:

$$\mathscr{H}' = \exp(+i\omega I_z t)\mathscr{H}\exp(-i\omega I_z t) \quad (28)$$

In the case that the spin system is subject to a Zeeman field $\mathbf{B}_0$ oriented along the z axis and a linearly polarized rf field $2B_1$ of frequency ω, oriented along the x axis,

$$2\pi\mathscr{H}/h = \omega_0 I_z + 2\omega_1 I_x \cos(\omega t) \tag{29}$$

where:

$$\omega_1 = -\gamma B_1 \tag{30}$$

Equation (28) expresses a rotation of $\mathscr{H}$ about the z axis, with a frequency ω. Such a rotation obviously leaves invariant all terms in $\mathscr{H}$ that are along the rotation axis and carries the other off-axis terms into one another.

Employing the expressions for exponential operators derived in Appendix 2, we have:

$$2\pi\mathscr{H}'/h = (\omega_0 I_z + \omega_1 I_x) \tag{31}$$

which discards the term $\omega_1(I_x \cos 2\omega t - I_y \sin 2\omega t)$, since this "counter-rotating" component of the rf field does not, to first order, exert any influence on the evolution of the spins.

In the rotating frame, therefore, we have:

$$\begin{aligned} i\frac{\delta|\Psi_{\text{rot}})}{\delta t} &= [(\omega_0 - \omega)I_z + \omega_1 I_x]|\Psi_{\text{rot}}) \\ &= (2\pi/h)\mathscr{H}_{\text{rot}}|\Psi_{\text{rot}}) \end{aligned} \tag{32}$$

By analogy, when two spin species I and S with different Larmor frequencies are involved, the transformation is to a "doubly rotating" frame:

$$|\Psi_{\text{rot}}(t)) = \exp[i(\omega_I I_z + \omega_S S_z)t]|\Psi(t)) \tag{33}$$

Now, discarding nonsecular terms once again,

$$i\frac{\delta|\Psi_{\text{rot}})}{\delta t} = (2\pi/h)\mathscr{H}_{\text{rot}}|\Psi_{\text{rot}}) \tag{34}$$

where

$$\mathscr{H}_{\text{rot}} = (h/2\pi)[(\omega_{0I} - \omega_I)I_z + (\omega_{0S} - \omega_S)S_z + \omega_{1I}I_x + \omega_{1S}S_x] \tag{35}$$

Homonuclear interactions of the form $\mathbf{I}_1 \cdot \mathbf{I}_2$ are invariant to the transformation to the rotating frame, whereas heteronuclear interactions of the form $\mathbf{I} \cdot \mathbf{S}$ get truncated to $I_z S_z$ in the doubly rotating frame, because all other terms are nonsecular with frequency $(\omega_I - \omega_S)$. If $\omega_{1I} = \omega_{1S} = \omega \gg 0$, on the other hand, the interaction takes the interesting form $1/2(I_z S_z + I_y S_y)$ in a synchronized doubly rotating frame, rotating about the x axis with frequency ω, where we discard terms with frequency 2ω. We may refer to this operator form of the coupling as "anisotropic strong coupling" because it corresponds neither to the weakly coupled, axially symmetrical situation characterized by $I_z S_z$ nor to the strongly coupled isotropic situation given by $\mathbf{I} \cdot \mathbf{S} = I_x S_x + I_y S_y + I_z S_z$.

We employ this nomenclature without any reference to the nature of the coupling parameter, which may be isotropic or anisotropic indirect nuclear spin–spin coupling, or anisotropic dipolar coupling. All discussions of the evolution of spin systems shall henceforth refer to the rotating frame and we shall therefore drop the suffix "rot," for convenience of notation.

Equation (32) indicates that in the rotating frame of reference, the spins feel a "reduced" field along the z axis, corresponding to the precessional frequency $(\omega_0 - \omega)$.

At resonance, i.e., when $\omega_0 = \omega$, Eq. (32) takes on the interesting form:

$$i\frac{\delta|\Psi)}{\delta t} = \omega_1 I_x|\Psi) \tag{36}$$

In words, the spins precess around the effective field, which at resonance is along the x axis of the rotating frame, with frequency $|\omega_{\text{eff}}| = [(\omega_0 - \omega)^2 + \omega_1^2]^{1/2}$, which reduces at resonance to $|\omega_1| = \gamma H_1$.

The corresponding equations of motion for an ensemble of spins involve the density operator:

$$i\frac{\delta\sigma'}{\delta t} = (2\pi/h)[\mathscr{H}, \sigma'] \tag{37}$$

which for $\mathscr{H}$ constant during the time interval 0 to t integrates to:

$$\sigma'(t) = \exp(-2\pi i\mathscr{H}t/h)\sigma'(0)\exp(2\pi i\mathscr{H}t/h) \tag{38}$$

From Eq. (38), it is clear that the first term of σ' [Eq. (20)] does not evolve in time, no matter what the precise form of the operative Hamiltonian may be.

If the Hamiltonian is composed of commuting parts $\mathscr{H}_i$, it is clear from the definition of exponential operators (*vide supra*) that Eq. (38) may be viewed as a cascade of evolutions under the commuting $\mathscr{H}_i$'s, taken in any order. Thus, the effects of chemical shifts and weak couplings may be computed separately and in any order; however, chemical shifts and strong couplings may *not* be separated because the two Hamiltonians in question fail to commute (see Appendix 2). Similarly, the effects of weak and isotropic strong coupling may be cascaded because the corresponding Hamiltonians commute with each other if $I = 1/2$.

We shall now work out from first principles the effects of some important interactions on the motion of spins in the rotating frame. It is important to note, as is demonstrated in what follows, that only rf pulses effect changes in coherence orders; all high-field Hamiltonians operative during free evolution of the spins conserve the coherence order. It may be noted also that if $\sigma(0)$ represents transverse magnetization, and $\mathscr{H}$ the complete NMR Hamiltonian, Eq. (38) describes the free evolution (or free precession) of the coherent spins, which generates all the frequencies and multiplet patterns observed in the NMR spectrum. Equation (38) therefore represents the direct method of calculating the spectrum, in contrast to the usual indirect method of solving for the energy eigenvalues and eigenfunctions and applying the magnetic dipole selection rule to derive thence frequencies and intensities of allowed transitions.

Effect of Radiofrequency Pulses. At resonance in the rotating frame, the effective Hamiltonian during a "hard" rf pulse of duration t, applied along the rotating frame y axis, is given by:

$$\mathscr{H} = -\gamma h B_1 I_y/2\pi = h\omega_1 I_y/2\pi \tag{39}$$

The effect of this pulse on the spins is readily calculated employing Eq. (36), which integrates to:

$$|\Psi(t)) = \exp(-i\omega_1 I_y t)|\Psi(0)) = \exp(-i\theta I_y)|\Psi(0)) \tag{40}$$

since $\mathscr{H}$ is constant for the duration of the pulse, 0 to t. For the two-level system of spin-1/2 particles,

$$\exp(\pm i\theta I_y) = \mathbf{1}\cos(\theta/2) \pm 2i\sin(\theta/2)I_y \tag{41}$$

employing the definition of exponential operators and the recursion $(I_y)^{2n} = (I_y)^{2(n-1)}/4$, n being a positive, nonzero integer. When $\theta = \pi/2$, i.e., for a 90° pulse,

$$\exp(\pm i\pi I_y/2) = \mathbf{1}(1/\sqrt{2}) \pm \sqrt{2}iI_y \tag{42}$$

Thus,

$$\exp(-i\pi I_y/2)|\alpha) = (1/\sqrt{2})(|\alpha) + |\beta))$$

and

$$\exp(-i\pi I_y/2)|\beta) = (1/\sqrt{2})(|\beta) - |\alpha)) \tag{43}$$

The pure states $|\Psi^{(1)}) = |\alpha)$ and $|\Psi^{(2)}) = |\beta)$ have therefore been converted by the pulse into the pure states $|\Psi^{(1)}(t))$ and $|\Psi^{(2)}(t))$, which are coherent superpositions of $|\alpha)$ and $|\beta)$. The probabilities $p^{(i)}$ of $|\Psi^{(1)})$ and $|\Psi^{(2)})$, however, are unaffected during the short pulse duration t (which is typically of the order of 10 μs), since the relaxation processes responsible for establishing $p^{(i)}$'s in accordance with the Boltzmann distribution operate typically on far more extended time scales. The ensemble is now characterized thus by the density operator:

$$\begin{aligned} \sigma'_{\alpha\alpha}(t)\,&[(1 + (\varepsilon/2)) + (1 - (\varepsilon/2))]/4 = 1/2; \\ \sigma'_{\beta\beta}(t)\,&[(1 + (\varepsilon/2)) + (1 - (\varepsilon/2))]/4 = 1/2; \\ \sigma'_{\alpha\beta}(t)\,&[(1 + (\varepsilon/2)) - (1 - (\varepsilon/2))]/4 = \varepsilon/4 = \sigma'_{\beta\alpha}(t) \end{aligned} \tag{44}$$

Thus,

$$\sigma'(t) = (1/2)\begin{pmatrix} 1 & \varepsilon/2 \\ \varepsilon/2 & 1 \end{pmatrix} = (1/2)\mathbf{1} + (\varepsilon/2)I_x \tag{45}$$

The same result may be derived by employing Eqs. (38), (20), and (22):

$$\begin{aligned}\sigma'(t) &= \exp(-i\theta I_y)\sigma'(0)\exp(i\theta I_y)\\ &= (1/2)\mathbf{1} + (\varepsilon/2)\exp(-i\theta I_y)I_z\exp(i\theta I_y)\\ &= (1/2)\mathbf{1} + (\varepsilon/2)(I_x\sin\theta + I_z\cos\theta)\end{aligned} \tag{46}$$

the last equality following from the Hausdorff formula regardless of the magnitude of $\mathbf{I}$, as discussed earlier. Working with the interesting part of σ' expressed in terms of operators we can show similarly that on resonance pulses of "flip angle" θ applied along the rotating frame x or y axis have the effect:

$$\begin{aligned}&I_z \xrightarrow{\theta I_x} I_z\cos\theta - I_y\sin\theta\\ &I_y \xrightarrow{\theta I_x} I_y\cos\theta + I_z\sin\theta\\ &I_z \xrightarrow{\theta I_y} I_z\cos\theta + I_x\sin\theta\\ &I_x \xrightarrow{\theta I_y} I_x\cos\theta - I_z\sin\theta\\ &I_x \xrightarrow{\theta I_x} I_x\\ &I_y \xrightarrow{\theta I_y} I_y\end{aligned} \tag{47}$$

with the shorthand notation $I_i \xrightarrow{\theta I_j}$ employed for $\exp(-iI_j\theta)I_i\exp(iI_j\theta) = \ldots$. More complicated expressions can be readily worked out to describe the effect of off-resonance rf pulses of arbitrary phase.

Effect of Resonance Offsets, Chemical Shifts, and Magnetic Field Inhomogeneities. The Hamiltonian in this case can be written quite generally as:

$$(2\pi/h)\mathscr{H} = (-\gamma[B_0 + \delta B_0(z)][1-\sigma] - \omega)I_z$$

i.e.,

$$\begin{aligned}\mathscr{H} &= (h/2\pi)([\omega_0 + \delta\omega_0][1-\sigma] - \omega)I_z\\ &= h\Delta I_z/2\pi\end{aligned} \tag{48}$$

The motion of the spins under this interaction is given by the evolution of the density operator:

$$\begin{aligned}\sigma'(t) &= \exp(-i\Delta I_z t)\sigma'(0)\exp(i\Delta I_z t)\\ &= \exp(-i\Delta I_z t)((1/2)\mathbf{1} + \sigma(0))\exp(i\Delta I_z t)\end{aligned} \tag{49}$$

Employing the Hausdorff formula, the following equations of motion can be derived (see Appendix 2), which are again independent of the magnitude of $\mathbf{I}$:

$$\begin{aligned}&I_x \xrightarrow{\Delta I_z t} I_x\cos\Delta t + I_y\sin\Delta t\\ &I_y \xrightarrow{\Delta I_z t} I_y\cos\Delta t - I_x\sin\Delta t\\ &I_z \xrightarrow{\Delta I_z t} I_z\end{aligned} \tag{50}$$

This corresponds to the well-known vector picture description (see Fig. 1.2).

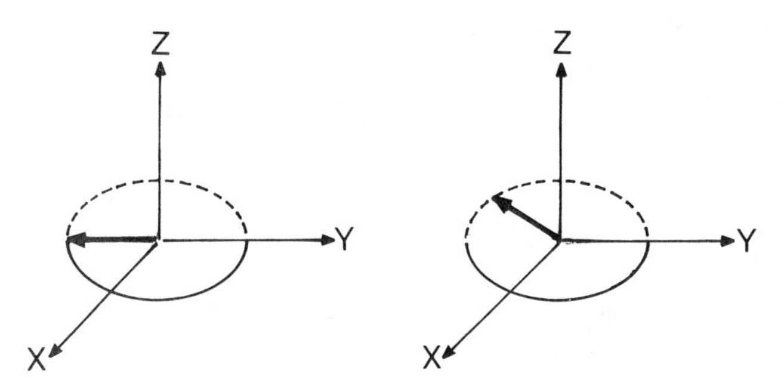

Figure 1.2. Precession of transverse magnetization in the rotating frame under offsets and chemical shifts.

Equations (50) also describe the effect of phase shifts on transverse and (longitudinal) z magnetizations, with the phase shift φ replacing Δt.

It is interesting to observe at this point the effect of offsets, $\mathbf{B}_0$ inhomogeneities, and phase shifts on multiple-quantum coherences. In a homonuclear two-spin system, zero-quantum coherence may be represented in the density operator by $(I_{1x}I_{2x} + I_{1y}I_{2y})$ and $(-I_{1x}I_{2y} + I_{1y}I_{2x})$. If Δ_1 and Δ_2 represent the offsets of the two spins, it can be shown upon applying Eqs. (50) that:

$$(I_{1x}I_{2x} + I_{1y}I_{2y}) \xrightarrow{(\Delta_1 I_{1z} + \Delta_2 I_{2z})t} (I_{1x}I_{2x} + I_{1y}I_{2y})\cos(\Delta_1 - \Delta_2)t + (I_{1y}I_{2x} - I_{1x}I_{2y})\sin(\Delta_1 - \Delta_2)t \tag{51}$$

and

$$(-I_{1x}I_{2y} + I_{1y}I_{2x}) \xrightarrow{(\Delta_1 I_{1z} + \Delta_2 I_{2z})t} (I_{1y}I_{2x} - I_{1x}I_{2y})\cos(\Delta_1 - \Delta_2)t - (I_{1x}I_{2x} + I_{1y}I_{2y})\sin(\Delta_1 - \Delta_2)t \tag{52}$$

Equations (51) and (52) describe the evolution of the x and y components, respectively, of zero-quantum coherence and show that they evolve at a frequency involving the *difference* of offsets/chemical shifts. With the substitution $\Delta_1 t = \Delta_2 t = \varphi$, it is also evident from Eq. (51) and (52) that zero-quantum coherences (ZQC) are completely invariant to phase shifts and are totally unaffected by inhomogeneities in $\mathbf{B}_0$, properties they share with z magnetization. Applying Eqs. (50) to the evolution of double-quantum coherence, on the other hand, it follows that:

$$(I_{1x}I_{2x} - I_{1y}I_{2y}) \xrightarrow{(\Delta_1 I_{1z} + \Delta_2 I_{2z})t} (I_{1x}I_{2x} - I_{1y}I_{2y})\cos(\Delta_1 + \Delta_2)t + (I_{1x}I_{2y} + I_{1y}I_{2x})\sin(\Delta_1 + \Delta_2)t \tag{53}$$

and:

$$(I_{1x}I_{2y} + I_{1y}I_{2x}) \xrightarrow{(\Delta_1 I_{1z} + \Delta_2 I_{2z})t} (I_{1x}I_{2y} + I_{1y}I_{2x})\cos(\Delta_1 + \Delta_2)t - (I_{1x}I_{2x} - I_{1y}I_{2y})\sin(\Delta_1 + \Delta_2)t \tag{54}$$

Equations (53) and (54) describe the evolution of the x and y components, respectively, of double-quantum coherence and show that they evolve at a frequency involving the *sum* of offsets/chemical shifts. Substituting $\Delta_1 t = \Delta_2 t = \varphi$, it is clear from Eqs. (53) and (54) additionally that the phase of double-quantum coherence (DQC) is shifted by 2φ radians when a phase shift of φ radians is applied to the spin system. So also, the effect of $\mathbf{B}_0$ inhomogeneities is doubled. In summary, for the two-spin system,

$$\begin{aligned}
(\mathrm{ZQC})_x &\xrightarrow{(\Delta_1 I_{1z}+\Delta_2 I_{2z})t} (\mathrm{ZQC})_x \cos(\Delta_1 - \Delta_2)t + (\mathrm{ZQC})_y \sin(\Delta_1 - \Delta_2)t \\
(\mathrm{ZQC})_y &\xrightarrow{(\Delta_1 I_{1z}+\Delta_2 I_{2z})t} (\mathrm{ZQC})_y \cos(\Delta_1 - \Delta_2)t - (\mathrm{ZQC})_x \sin(\Delta_1 - \Delta_2)t \\
(\mathrm{DQC})_x &\xrightarrow{(\Delta_1 I_{1z}+\Delta_2 I_{2z})t} (\mathrm{DQC})_x \cos(\Delta_1 + \Delta_2)t + (\mathrm{DQC})_y \sin(\Delta_1 + \Delta_2)t \\
(\mathrm{DQC})_y &\xrightarrow{(\Delta_1 I_{1z}+\Delta_2 I_{2z})t} (\mathrm{DQC})_y \cos(\Delta_1 + \Delta_2)t - (\mathrm{DQC})_x \sin(\Delta_1 + \Delta_2)t
\end{aligned} \tag{55}$$

Equations (55) represent the evolution of zero- and double-quantum coherence in their respective (zero and double quantum) frames of reference.

In general, expressions for multiple-quantum coherences are given more conveniently in terms of the raising and lowering operators:

$$I^+ = I_x + iI_y; \qquad I^- = I_x - iI_y \tag{56}$$

rather than the Cartesian components of $\mathbf{I}$.

In terms of these operators, Eqs. (50) may be rewritten:

$$\begin{aligned}
I^+ &\xrightarrow{\Delta I_z t} I^+ \exp(-i\Delta t) \\
I^- &\xrightarrow{\Delta I_z t} I^- \exp(i\Delta t)
\end{aligned} \tag{57}$$

One can then discuss readily the q quantum coherence arising from a set of n spins; it can be shown that:

$$\begin{aligned}
(n\text{-spin } q\mathrm{QC})_x &\xrightarrow{\sum \Delta_k I_{kz} t} (n\text{-spin } q\mathrm{QC})_x \cos \Delta_{\mathrm{eff}} t \\
&\quad + (n\text{-spin } q\mathrm{QC})_y \sin \Delta_{\mathrm{eff}} t
\end{aligned} \tag{58}$$

where

$$n = \sum_k |\Delta m_k|, \quad q = \Delta M = \sum_k \Delta m_k, \quad \Delta_{\mathrm{eff}} = \sum_k \Delta m_k \Delta_k, \quad \Delta m_k = \pm 1$$

Effect of Spin–Spin Couplings. In this situation, the interaction may be represented by the Hamiltonian:

$$\mathscr{H}_{\mathrm{sc}} = h \sum_{i<j}\sum J_{ij} \mathbf{I}_i \cdot \mathbf{I}_j \tag{59}$$

for scalar coupling, and

$$\mathscr{H}_{\mathrm{d}} = \sum_{i<j}\sum [\gamma_i \gamma_j h^2/4\pi^2 r_{ij}^3][\mathbf{I}_i \cdot \mathbf{I}_j - (3/r_{ij}^2)(\mathbf{I}_i \cdot \mathbf{r}_{ij})(\mathbf{I}_j \cdot \mathbf{r}_{ij})] \tag{60}$$

for dipolar coupling. Several cases can be distinguished, as below.

Couplings Among Equivalent Spins. Consider an $I_N S_M$ spin system including the strong scalar couplings J_{II}, J_{SS}, the weak couplings J_{IS}, the resonance offset interactions $\Delta_I \sum_i I_{iz}$ and $\Delta_S \sum_j S_{jz}$, and the rf interactions during the pulses, $\omega_{1I} \sum_i I_{ix}$ and $\omega_{1S} \sum_j S_{jx}$. During the pulses, only the rf interaction is active in the strong pulse situation. The strong couplings among the equivalent I spins, $2\pi J_{II} \sum\sum_{i<j} \mathbf{I}_i \cdot \mathbf{I}_j$ and among the equivalent S spins, $2\pi J_{SS} \sum\sum_{k<l} \mathbf{S}_k \cdot \mathbf{S}_l$, commute with the rf Hamiltonian, the offset Hamiltonian, as well as with the weak coupling Hamiltonian between the I and S spins.

The offset and weak coupling Hamiltonians also commute with each other. The effect of the couplings among the equivalent spins on the evolution of the density operator may therefore be predicted solely on the basis of their effect on the initial density operator. The initial density operator, which is its thermal equilibrium form, is proportional to $(\beta \sum_i I_{iz} + \sum_j S_{jz})$, which commutes with the two strong coupling Hamiltonians and therefore does *not* evolve under their influence. We obtain, therefore, the well-known result that mutual couplings among equivalent spins are not observable in the spectrum of the spin system.

Weak Scalar Coupling Between a Pair of Nuclei. The interaction:

$$\mathscr{H} = hJI_{1z}I_{2z} \tag{61}$$

leads to an evolution that depends on the spin quantum numbers of the nuclei I_1 and I_2. Employing the Hausdorff formula, it can be shown that (see Appendix 2):

$$\begin{aligned}
I_{1x} &\xrightarrow{2\pi J I_{1z} I_{2z} t} I_{1x} \cos \pi Jt + 2I_{1y}I_{2z} \sin \pi Jt \\
I_{1x}I_{2z} &\xrightarrow{2\pi J I_{1z} I_{2z} t} I_{1x}I_{2z} \cos \pi Jt + (1/2)I_{1y} \sin \pi Jt \\
I_{1y} &\xrightarrow{2\pi J I_{1z} I_{2z} t} I_{1y} \cos \pi Jt - 2I_{1x}I_{2z} \sin \pi Jt \\
I_{1y}I_{2z} &\xrightarrow{2\pi J I_{1z} I_{2z} t} I_{1y}I_{2z} \cos \pi Jt - (1/2)I_{1x} \sin \pi Jt
\end{aligned} \tag{62}$$

provided nucleus 2 is a spin-1/2 particle, no matter what the spin quantum number of particle 1 may be. If nucleus 2 is a spin-1 particle, on the other hand, the following equations of motion may be derived regardless of the spin of nucleus 1:

$$\begin{aligned}
S_x &\xrightarrow{2\pi J t I_z S_z} \sin(2\pi Jt)S_yI_z + (1 + I_z^2(\cos 2\pi Jt - 1))S_x \\
S_y &\xrightarrow{2\pi J t I_z S_z} -\sin(2\pi Jt)S_xI_z + (1 + I_z^2(\cos 2\pi Jt - 1))S_y \\
S_yI_z &\xrightarrow{2\pi J t I_z S_z} \cos(2\pi Jt)S_yI_z - \sin(2\pi Jt)S_xI_z^2 \\
S_xI_z &\xrightarrow{2\pi J t I_z S_z} \cos(2\pi Jt)S_xI_z + \sin(2\pi Jt)S_yI_z^2 \\
S_xI_z^2 &\xrightarrow{2\pi J t I_z S_z} \cos(2\pi Jt)S_xI_z^2 + \sin(2\pi Jt)S_yI_z \\
S_yI_z^2 &\xrightarrow{2\pi J t I_z S_z} \cos(2\pi Jt)S_yI_z^2 - \sin(2\pi Jt)S_xI_z
\end{aligned} \tag{63}$$

Equations (62) and (63) describe the motion of transverse magnetization under

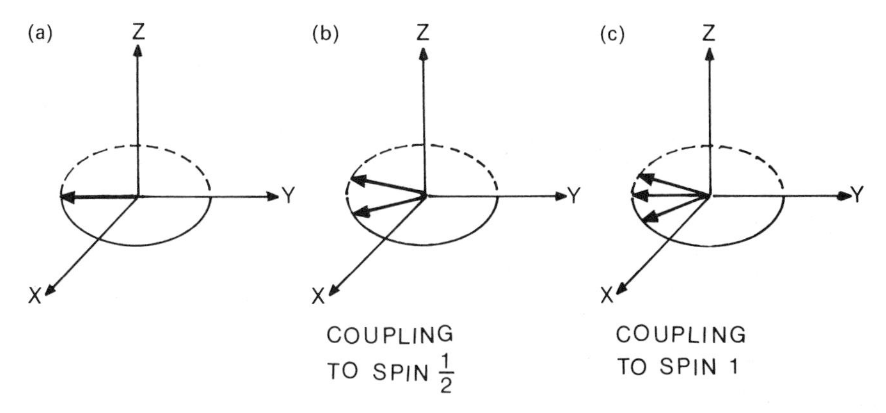

Figure 1.3. Precession of transverse magnetization in the rotating frame under coupling to a spin-$\frac{1}{2}$ and a spin-1 nucleus.

coupling to a spin-1/2 and a spin-1 nucleus, respectively. These give formal expression to the evolution under J as given by the well-known vector pictures (Fig. 1.3). In particular, the first two equations of motion in Eqs. (63) indicate the presence of a component that does not evolve, corresponding to the spin state $m_I = 0$ of the coupled nucleus I. Employing Eqs. (62), it can be shown that:

$$I_{1i}I_{2j} \xrightarrow{2\pi J I_{1z} I_{2z} t} I_{1i}I_{2j}(i,j = x, y) \tag{64}$$

provided both 1 and 2 are spin-1/2 particles. In other words, zero- and double-quantum coherences of two spin-1/2 particles are invariant to weak scalar coupling between them. If nucleus I has spin 1 and nucleus S has spin 1/2, on the other hand, it turns out that:

$$\begin{aligned} I_i S_x &\xrightarrow{2\pi J I_{1z} I_{2z} t} \cos(\pi J t) I_i S_x + \sin(\pi J t) S_y [I_i, I_z]_+ \\ I_i S_y &\xrightarrow{2\pi J I_{1z} I_{2z} t} \cos(\pi J t) I_i S_y - \sin(\pi J t) S_x [I_i, I_z]_+ \end{aligned} \tag{65}$$

Equations (64) and (65) merely reflect the dependence (or otherwise) on J of the energy difference between the pairs of levels between which coherence is established, as represented by the operators $I_i S_j$, as a function of the spin quantum numbers of the two particles (see Fig. 1.4). It may also be noted that in a spin-1/2 system of the type $I_N S$, zero- and double-quantum coherences of the type $I_{1i} S_j$ do evolve under the coupling between particles S and $I_2, \ldots, I_n$. For an n-spin q-quantum coherence,

$$\begin{aligned} (n\text{-spin } q\text{QC})_x \xrightarrow{\sum_k 2\pi J_{km} I_{kz} I_{mz} t} & (n\text{-spin } q\text{QC})_x \cos \pi J_{\text{eff}} t \\ & + 2I_{mz}(n\text{-spin } q\text{QC})_y \sin \pi J_{\text{eff}} t \end{aligned} \tag{66}$$

with

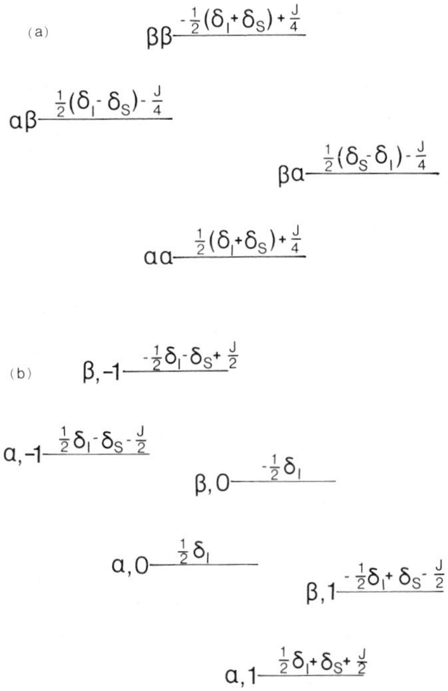

(a)

ββ $-\frac{1}{2}(\delta_I+\delta_S)+\frac{J}{4}$

αβ $\frac{1}{2}(\delta_I-\delta_S)-\frac{J}{4}$

βα $\frac{1}{2}(\delta_S-\delta_I)-\frac{J}{4}$

αα $\frac{1}{2}(\delta_I+\delta_S)+\frac{J}{4}$

(b)

β,−1 $-\frac{1}{2}\delta_I-\delta_S+\frac{J}{2}$

α,−1 $\frac{1}{2}\delta_I-\delta_S-\frac{J}{2}$

β,0 $-\frac{1}{2}\delta_I$

α,0 $\frac{1}{2}\delta_I$

β,1 $-\frac{1}{2}\delta_I+\delta_S-\frac{J}{2}$

α,1 $\frac{1}{2}\delta_I+\delta_S+\frac{J}{2}$

Figure 1.4. Energy levels in the rotating frame: (a) for a pair of coupled spin-$\frac{1}{2}$ nuclei I and S; (b) for a spin-$\frac{1}{2}$ nucleus I coupled to a spin-1 nucleus S.

$$J_{\text{eff}} = \sum_k \Delta m_k J_{km} \qquad \text{(nucleus } m \text{ passive)} \tag{67}$$

Anisotropic Strong Coupling Between a Pair of Nuclei. As described in the section on rotating-frame transformations, a heteronuclear spin system with scalar coupling J takes on strong coupling characteristics when subjected to a matched pair of resonant rf fields. This matching, which equalizes the Larmor frequency of the two nuclei in the rotating frame (see Eq. (36)) such that $\gamma_I H_{1I} = \gamma_S H_{1S}$, is known as the Hartmann–Hahn condition. The effective Hamiltonian under which the spins evolve in now given by:

$$\mathscr{H} = (hJ/2)(I_z S_z + I_y S_y) \tag{68}$$

assuming both spins I and S are on resonance. The relevant equation of motion under such anisotropic strong coupling can again be derived from the Hausdorff formula and depends on the spin quantum numbers I and S (see Appendix 2). For $I = S = 1/2$,

$$\begin{aligned} I_x \xrightarrow{\pi J(I_z S_z + I_y S_y)t} & (1/2)I_x(1 + \cos \pi Jt) + (1/2)S_x(1 - \cos \pi Jt) \\ & + (I_y S_z - I_z S_y) \sin \pi Jt \end{aligned} \tag{69}$$

For $I = 1$ and $S = 1/2$,

$$
\begin{aligned}
I_x \xrightarrow{\pi J(I_zS_z + I_yS_y)t} & (1/4)I_x(3 + \cos\sqrt{2}\pi Jt) \\
& + (1/2)(\mathbf{I}^2 - I_x^2)S_x(1 - \cos\sqrt{2}\pi Jt) \\
& + (1/\sqrt{2})(I_yS_z - I_zS_y)\sin\sqrt{2}\pi Jt
\end{aligned} \tag{70}
$$

$$
\begin{aligned}
S_x \xrightarrow{\pi J(I_zS_z + I_yS_y)t} & (1/4)I_x(1 - \cos\sqrt{2}\pi Jt) + S_x\cos\sqrt{2}\pi Jt \\
& + (1/2)I_x^2 S_x(1 - \cos\sqrt{2}\pi Jt) \\
& + (1/\sqrt{2})(I_zS_y - I_yS_z)\sin\sqrt{2}\pi Jt
\end{aligned} \tag{71}
$$

Regardless of the spin quantum numbers, however,

$$I_x + S_x \rightarrow I_x + S_x \tag{72}$$

Equations (70) and (71) are valid also for the I_2S system with $I = S = 1/2$.

It is instructive to compare Eqs. (69), (70) and (71) with the corresponding members from Eqs. (62) and (63) that are valid for weak coupling. Whereas the evolution frequencies are identical for the situations expressed in Eqs. (62) and (69), evolution under anisotropic strong coupling [Eq. (69)] generates S_x from I_x, which is not the case for the weak coupling situation. Comparison of the relevant members of Eqs. (62) and (63) with Eqs. (70) and (71) points to an additional interesting feature. Under weak coupling in an I_2S system, the evolution frequencies are different for the I and S spins, being, respectively, $\pm J/2$ and 0, $\pm J$ Hz, whereas under anisotropic strong coupling the evolution frequencies are identical for both I and S spins, being, in fact, 0.707 J Hz. Evolution under anisotropic strong coupling is therefore a process exhibiting collective modes of the spin system, whereas evolution under weak coupling exhibits single-spin modes. It may be noted that the same anisotropic operator form also results when Hartmann–Hahn matching is performed on a dipolar-coupled heteronuclear spin pair.

Isotropic Strong Coupling Between a Pair of Nuclei. When the situation described in the previous subsection is applied to a homonuclear spin system, we have:

$$\mathscr{H} = hJ\mathbf{I}\cdot\mathbf{S} \tag{73}$$

The coupling between the spins is now strong and isotropic. The relevant equation of motion under isotropic strong coupling can again be derived employing the Hausdorff formula, and depends on the spin quantum numbers I and S. We have regardless of the spin quantum numbers I and S,

$$I_i + S_i \xrightarrow{2\pi J\mathbf{I}\cdot\mathbf{S}t} I_i + S_i \tag{74}$$

For $I = S = 1/2$, we find:

$$I_i \xrightarrow{2\pi J\mathbf{I}\cdot\mathbf{S}t} (1/2)I_i(1+\cos 2\pi Jt) + (1/2)S_i(1-\cos 2\pi Jt) + (I_jS_k - I_kS_j)\sin 2\pi Jt \tag{75a}$$

and

$$S_i \xrightarrow{2\pi J\mathbf{I}\cdot\mathbf{S}t} (1/2)I_i(1-\cos 2\pi Jt) + (1/2)S_i(1+\cos 2\pi Jt) + (S_jI_k - S_kI_j)\sin 2\pi Jt \tag{75b}$$

For $I = 1$ and $S = 1/2$, on the other hand, we find:

$$\begin{aligned} I_i \xrightarrow{2\pi J\mathbf{I}\cdot\mathbf{S}t} & (1/9)I_i(7 + 2\cos 3\pi Jt) + (4/9)(\mathbf{I}^2 - I_i^2)S_i(1-\cos 3\pi Jt) \\ & - (2/9)(S_j[I_i, I_j]_+ + S_k[I_i, I_k]_+)(1-\cos 3\pi Jt) \\ & + (2/3)(I_jS_k - I_kS_j)\sin 3\pi Jt \end{aligned} \tag{76a}$$

and

$$\begin{aligned} S_i \xrightarrow{2\pi J\mathbf{I}\cdot\mathbf{S}t} & (2/9)I_i(1-\cos 3\pi Jt) + (1/9)S_i(1+8\cos 3\pi Jt) \\ & + (4/9)I_i^2S_i(1-\cos 3\pi Jt) + (2/3)(S_jI_k - S_kI_j)\sin 3\pi Jt \\ & + (2/9)(S_j[I_i, I_j]_+ + S_k[I_i, I_k]_+)(1-\cos 3\pi Jt) \end{aligned} \tag{76b}$$

Equations (75) and (76) are valid also with cyclic permutations of the indices i, j, and k which equal x, y, and z.

Isotropic strong coupling between spins also leads to collective modes of evolution of the system in the sense discussed in the previous subsection. However the frequencies active in the evolution are in general different from the anisotropic strong coupling situation discussed there. Suitable simultaneous pulsing on both channels of a heteronuclear spin system leads to such isotropic strong coupling on an average. Here the frequencies of evolution are $J/3$ and $J/2$, respectively, for IS and I_2S systems ($I = S = 1/2$), which should be compared with the corresponding frequencies for evolution under anisotropic strong coupling ($J/2$ and $0.707\ J$, respectively, as given in the previous subsection).

In the homonuclear situation, isotropic strong coupling leads to frequencies of evolution of J and $3J/2$ Hz for the spin-1/2 systems IS and I_2S respectively, in contrast to the weak coupling situation of $J/2$ Hz for IS and $J/2$ and J Hz for I_2S.

It may be noted that single spin modes conserve the coherence order of the individual spins in a coupled network; isotropic mixing conserves the coherence order of the entire coupled network, but not of the individual spins involved. On the other hand, cross polarization, which leads to anisotropic strong coupling, does not conserve even the overall coherence order.

Dipolar Couplings. For a heteronuclear spin pair with a dipolar coupling Eq. (60) takes on the simple form:

$$\mathscr{H}_{\mathrm{d}} = \frac{\gamma_I\gamma_S h^2}{4\pi^2 r_{12}^3} I_zS_z(1 - 3\cos^2\theta) \tag{77}$$

θ being the angle between $\mathbf{B}_0$ and the internuclear vector $\mathbf{r}_{12}$; all other terms are nonsecular in the doubly rotating frame. The form of this Hamiltonian is identical in its operator content to the weak coupling Hamiltonian, Eq. (61). The evolution of spins under Eq. (77) is therefore identical to that described earlier, with J replaced by $D = [\gamma_I \gamma_S h(1 - 3\cos^2\theta)/4\pi^2 r^3]$. Under conditions of Hartmann–Hahn matched irradiation, $\mathscr{H}_\mathrm{d}$ takes on the form of the anisotropic strong coupling Hamiltonian discussed earlier in the chapter.

For a homonuclear spin pair, Eq. (60) takes on the form:

$$\begin{aligned} \mathscr{H}_\mathrm{d} &= \frac{\gamma^2 h^2}{4\pi^2 r_{12}^3}(I_z S_z - (1/2)(I_x S_x + I_y S_y))(1 - 3\cos^2\theta) \\ &= \frac{\gamma^2 h^2}{8\pi^2 r_{12}^3}(1 - 3\cos^2\theta)(3I_z S_z - \mathbf{I}\cdot\mathbf{S}) \end{aligned} \tag{78}$$

This Hamiltonian is composed of a weak coupling term and an isotropic strong coupling term, evolutions under both of which have been considered separately in previous subsections. The effects of the two terms may be cascaded during evolution under $\mathscr{H}_\mathrm{d}$ provided spin-1/2 nuclei are being considered. This leads, for example, to:

$$\begin{aligned} I_x \xrightarrow{\pi D(3I_z S_z - \mathbf{I}\cdot\mathbf{S})t} &(1/2)I_x(\cos(3\pi Dt/2) + \cos(\pi Dt/2)) \\ &+ (1/2)S_x(\cos(3\pi Dt/2) - \cos(\pi Dt/2)) \\ &+ (I_y S_z)(\sin(3\pi Dt/2) + \sin(\pi Dt/2)) \\ &+ (I_z S_y)(\sin(3\pi Dt/2) - \sin(\pi Dt/2)) \\ (I_z + S_z) \rightarrow &(I_z + S_z) \\ (I_x + S_x) \rightarrow &(I_x + S_x)\cos(3\pi Dt/2) + 2(I_y S_z + I_z S_y)\sin(3\pi Dt/2) \\ (I_z - S_z) \rightarrow &(I_z - S_z)\cos(\pi Dt) - 2(I_x S_y - I_y S_x)\sin(\pi Dt) \\ (I_x - S_x) \rightarrow &(I_x - S_x)\cos(\pi Dt/2) + 2(I_y S_z - I_z S_y)\sin(\pi Dt/2) \end{aligned} \tag{79}$$

These equations predict also the zero field evolution, provided all frequencies are scaled up by a factor of two (see Chapter 6).

It is also to be noted that dipolar couplings do lead to an evolution of the spin system even when a pair of equivalent spins is considered, unlike the scalar coupling situation. This behaviour stems from the presence of the zz term in the dipolar Hamiltonian, which does not commute with the rf Hamiltonian.

Effect of Quadrupolar Couplings. The quadrupolar interaction comes into play for nuclei with spin quantum numbers greater than 1/2. For spin-1 nuclei in an axially symmetric situation ($\eta = 0$), the interaction takes the form:

$$\mathscr{H}_\mathrm{Q} = (e^2 qQ/4)(3I_z^2 - \mathbf{I}^2) = (h/2\pi)\omega_\mathrm{Q}[I_z^2 - (1/3)\mathbf{I}^2] \tag{80}$$

ω_Q being one half the quadrupolar splitting. $\mathscr{H}_Q$ commutes with both chemical shift and weak coupling Hamiltonians. Evolution under the influence of $\mathscr{H}_Q$ leads to:

$$\begin{aligned} I_x &\xrightarrow{\omega_Q I_z^2 t} I_x \cos \omega_Q t + [I_y, I_z]_+ \sin \omega_Q t \\ I_y &\xrightarrow{\omega_Q I_z^2 t} I_y \cos \omega_Q t - [I_z, I_x]_+ \sin \omega_Q t \end{aligned} \tag{81}$$

These relations can be obtained readily employing the Hausdorff formula and the equalities:

$$I_z^3 = I_z; \qquad I_z I_y I_z = I_z I_x I_z = 0$$

which are valid for $I = 1$ (see Appendix 1). The matrix representations of the anti-commutators $[I_x, I_z]_+$ and $[I_y, I_z]_+$ are also given in Appendix 1 and correspond to anti-phase single-quantum magnetization of the quadrupolar doublet along the x axis and y axis, respectively.

It is especially interesting to note that the double-quantum coherence in spin-1 systems does *not* evolve under the quadrupole coupling, since

$$[I_z^2, [I_x, I_y]_+] = [I_z^2, (I_x^2 - I_y^2)] = 0$$

Interpretation of the Theory in Terms of the Observed NMR Spectrum

We shall give a one-to-one correspondence in this section between the general theory developed in the previous subsections and the observed spectral features, including frequency, phase, and multiplet structure. Some basic properties of the Fourier transform, which we shall apply here, are summarized in Appendix 3.

Absorption and Dispersion. We commence the discussion with an ensemble of spin-1/2 particles that are Δ radians per second off resonance. In other words,

$$\mathscr{H} = h(\omega_0 - \omega)I_z/2\pi = h\Delta I_z/2\pi$$

The initial state of the system is characterized by: $\sigma^- = \varepsilon I_z/2$. When this system is excited with a resonant 90° pulse along the y axis of the rotating frame, $\sigma^+ = \varepsilon I_x/2$. Following the pulse, the spins undergo free precession, giving rise to the free induction decay (FID) (Fig. 1.5):

$$\begin{aligned} \sigma(t) &= \exp(-2\pi i \mathscr{H} t/h)\sigma^+ \exp(2\pi i \mathscr{H} t/h) \\ &= (\varepsilon/2)\exp(-i\Delta I_z t) I_x \exp(i\Delta I_z t) \\ &= (\varepsilon/2)(I_x \cos \Delta t + I_y \sin \Delta t) \end{aligned} \tag{82}$$

The spectrometer reference phase may be set to detect either the x or the y component of the magnetization in the rotating frame, the detected signals being proportional, respectively, to $\mathrm{Tr}\,(\sigma(t)I_x)$ or $\mathrm{Tr}\,(\sigma(t)I_y)$, depending on the choice of the receiver reference phase. From Eq. (82),

$$\mathrm{Tr}\,(\sigma(t)I_x) \sim \mathrm{Tr}\,(I_x^2)\cos \Delta t + \mathrm{Tr}\,(I_y I_x)\sin \Delta t \tag{83a}$$

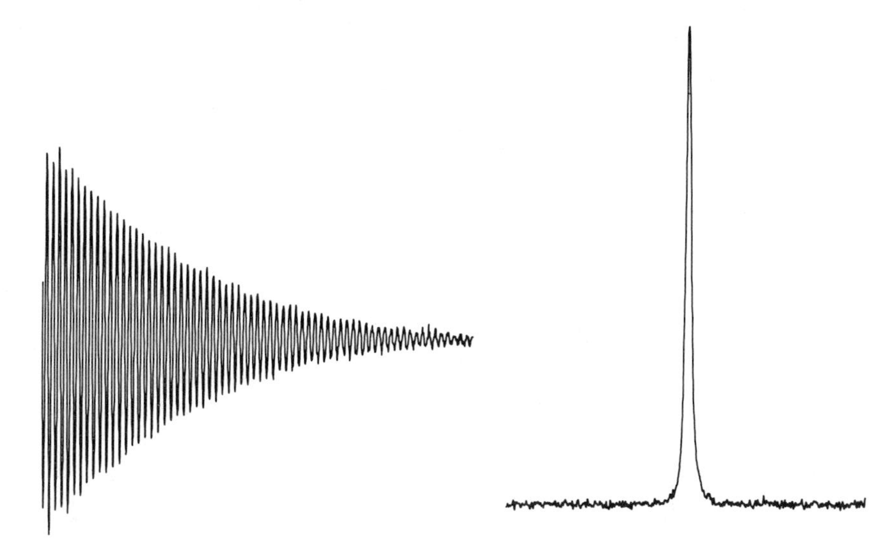

Figure 1.5. Free induction decay: proton-decoupled ^{13}C magnetization of benzene at 22.64 MHz, measured off resonance (T_2^* ca. 1.6 s).

and

$$\mathrm{Tr}\,(\sigma(t)I_y) \sim \mathrm{Tr}\,(I_xI_y)\cos\Delta t + \mathrm{Tr}\,(I_y^2)\sin\Delta t \tag{83b}$$

According to Appendix 1, only the square operators have a nonzero trace and therefore give rise to an observable signal. Because the excitation was along the y axis, detection of I_x corresponds to sensing the response in quadrature with the excitation, which gives rise to the absorption at frequency Δ upon Fourier transformation of the FID, which is actually of the form:

$$S(t) \sim \mathrm{Tr}\,(I_x^2)u(t)\exp(-t/T_2^*)\cos\Delta t \tag{84}$$

when due account is taken of the loss of phase coherence during acquisition of the FID, $u(t)$ being the unit step function insuring the expression of the fact that the FID has not preceded the pulse at $t = 0$. Note the occurrence of the cosine term for absorption mode. It is worthwhile to note further from Eqs. (83) that if $\Delta \neq 0$, there is also a dispersion signal, $\mathrm{Tr}\,(\sigma(t)I_y)$, where the sine term occurs. The time domain dispersion vanishes for a one-spin ensemble on resonance. We next consider a homonuclear two-spin liquid system with no coupling of the spins. The most general Hamiltonian is in this case:

$$\mathscr{H} = (h/2\pi)[(\Delta + (\delta/2))I_{1z} + (\Delta - (\delta/2))I_{2z}] \tag{85}$$

where the spins are off resonance by Δ radians per second and have a relative chemical shift of δ radians per second. When a nonselective $(\pi/2)_y$ pulse is applied,

$$\sigma^- \sim (I_{1z} + I_{2z}); \qquad \sigma^+ \sim (I_{1x} + I_{2x}) \tag{86}$$

Subsequent free evolution generates:

$$\sigma(t) = I_{1x}\cos(\Delta + (\delta/2))t + I_{1y}\sin(\Delta + (\delta/2))t + I_{2x}\cos(\Delta - (\delta/2))t + I_{2y}\sin(\Delta - (\delta/2))t \quad (87)$$

We now have:

$$\mathrm{Tr}(\sigma(t)I_x) = \mathrm{Tr}(\sigma(t)(I_{1x} + I_{2x})) = \mathrm{Tr}(I_{1x}^2)\cos(\Delta + (\delta/2))t + \mathrm{Tr}(I_{2x}^2)\cos(\Delta - (\delta/2))t \quad (88)$$

This FID, $S(t) = \mathrm{Tr}(\sigma(t)I_x)u(t)\exp(-t/T_2^*)$ (assuming a common T_2^*) leads upon Fourier transformation to absorption mode signals of equal intensity at frequencies $(\Delta + (\delta/2))$ and $(\Delta - (\delta/2))$, since cosine terms occur. Dispersion is again given by $\mathrm{Tr}(\sigma(t)I_y)$, where sine terms occur; there is no way of nulling it in this case of two chemically shifted spins.

Multiplet Patterns. We next consider a heteronuclear system of spin-1/2 particles, IS_2, with a nonzero scalar coupling J_{IS}. In this case,

$$\sigma^- \sim (\beta I_z + S_{1z} + S_{2z}) \quad (89)$$

which, when a $(\pi/2)_y$ pulse is applied to the I spins, becomes:

$$\sigma^+ \sim (\beta I_x + S_{1z} + S_{2z}) \quad (90)$$

The I signal originates from:

$$\sigma(t) \sim \exp(-i\Delta I_z t)\exp(-2\pi i J I_z(S_{1z} + S_{2z})t) \times I_x \exp(2\pi i J I_z(S_{1z} + S_{2z})t)\exp(i\Delta I_z t) \quad (91)$$

From Eqs. (62), because $[I_z S_{1z}, I_z S_{2z}] = 0$, we have,

$$\sigma(t) \sim \exp(-i\Delta I_z t)[(I_x/2)(1 + \cos 2\pi Jt) + 2I_x S_{1z} S_{2z}(\cos 2\pi Jt - 1) + I_y(S_{1z} + S_{2z})\sin 2\pi Jt]\exp(i\Delta I_z t) \quad (92)$$

With the offset Δ operative,

$$\sigma(t) \sim [\tfrac{1}{2}(1 + \cos 2\pi Jt) + 2S_{1z}S_{2z}(\cos 2\pi Jt - 1)](I_x\cos\Delta t + I_y\sin\Delta t) + (S_{1z} + S_{2z})\sin 2\pi Jt(I_y\cos\Delta t - I_x\sin\Delta t) \quad (93)$$

Hence,

$$\mathrm{Tr}(\sigma(t)I_x) \sim \tfrac{1}{2}(\cos\Delta t + \tfrac{1}{2}\cos(\Delta + 2\pi J)t + \tfrac{1}{2}\cos(\Delta - 2\pi J)t) + 2S_{1z}S_{2z}(\tfrac{1}{2}\cos(\Delta + 2\pi J)t + \tfrac{1}{2}\cos(\Delta - 2\pi J)t - \cos\Delta t) - \tfrac{1}{2}(S_{1z} + S_{2z})(\cos(\Delta - 2\pi J)t - \cos(\Delta + 2\pi J)t) \quad (94)$$

From Eq. (94), it is clear that three absorption mode signals result, at frequencies Δ, $(\Delta + 2\pi J)$, and $(\Delta - 2\pi J)$ radians per second. The intensities of the three signals may be calculated by associating with each frequency the microstates of the S spins, corresponding to S_{1z} and S_{2z}. Thus, two microstates contribute to the signal at Δ, for which $S_{1z} + S_{2z} = 0$ and $S_{1z}S_{2z} = -1/4$ per microstate, whereas for the signal at $(\Delta + 2\pi J)$, one microstate contributes, with $S_{1z} + S_{2z} = 1$ and $S_{1z}S_{2z} = 1/4$. For the signal at $(\Delta - 2\pi J)$, one microstate contributes, with $S_{1z} + S_{2z} = -1$ and $S_{1z}S_{2z} = 1/4$. These considerations based on Eq. (94) lead to a triplet of signals at $(\Delta - 2\pi J, \Delta, \Delta + 2\pi J)$, with intensities $(1, 2, 1)$. They are also readily obtained upon setting up the relevant direct product matrices.

Phase Anomalies. We shall now investigate the effect of a density operator of the form:

$$\sigma^+ \sim I_x + \alpha I_y S_z \tag{95}$$

In this case, collecting the I-spin signal with offset Δ and coupling J to the S spin, we have:

$$\begin{aligned}\sigma(t) \sim \tfrac{1}{2}I_x(&\cos(\Delta + \pi J)t + \cos(\Delta - \pi J)t)\\ &- I_x S_z(\cos(\Delta - \pi J)t - \cos(\Delta + \pi J)t)\\ &+ \tfrac{1}{2}I_y(\sin(\Delta + \pi J)t + \sin(\Delta - \pi J)t)\\ &- I_y S_z(\sin(\Delta - \pi J)t - \sin(\Delta + \pi J)t)\end{aligned} \tag{96}$$

from the first term of Eq. (95), so that:

$$\mathrm{Tr}(\sigma(t)I_x) \sim (\cos(\Delta + \pi J)t + \cos(\Delta - \pi J)t) \tag{97}$$

from the first two terms of Eq. (96), indicating the expected $1:1$ doublet at $(\Delta - \pi J, \Delta + \pi J)$.

The second term of Eq. (95) leads to:

$$\begin{aligned}\alpha^{-1}\sigma(t) \sim \tfrac{1}{4}I_x(&\sin(\Delta - \pi J)t - \sin(\Delta + \pi J)t)\\ &- \tfrac{1}{2}I_x S_z(\sin(\Delta + \pi J)t + \sin(\Delta - \pi J)t)\\ &+ \tfrac{1}{4}I_y(\cos(\Delta + \pi J)t - \cos(\Delta - \pi J)t)\\ &+ \tfrac{1}{2}I_y S_z(\cos(\Delta + \pi J)t + \cos(\Delta - \pi J)t)\end{aligned} \tag{98}$$

Hence,

$$\mathrm{Tr}(\sigma(t)I_x) \sim (\alpha/2)(\sin(\Delta - \pi J)t - \sin(\Delta + \pi J)t) \tag{99}$$

Notice the occurrence of the sine terms. The second term of σ in Eq. (95) leads therefore to a dispersive component in the detected x magnetization, which has the opposite sign for the two lines of the doublet. For a small value of the admixture parameter α, the spectrum appears as in Fig. 1.6. It should be borne in mind that this admixture is detected in spite of the receiver phase-sensitive detector having been set for absorption mode. Such phase anomalies cannot be corrected by the normal phase correction routines of an FT-NMR spectrometer because they do not exhibit a linear dependence on frequency, as is clear from Fig. 1.6. This is to be contrasted with the situation discussed earlier with $\sigma^+ \sim I_x$. In this case magnetization is oriented entirely along one axis just prior to signal acquisition; free evolution during acquisition does generate the

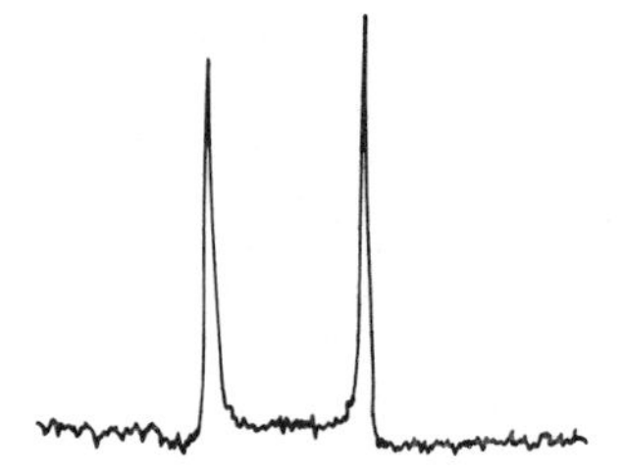

Figure 1.6. Phase anomaly in multiplets. Notice that the two lines of the doublet have a small admixture of dispersion, which has changed sign across the doublet.

I_y component also, but this is not detected if the reference is set for detection of I_x, so that signals result without any phase anomalies.

Decoupling. It is instructive to consider a weakly coupled two-spin-1/2 system, where the signal is acquired while one of the spins is irradiated on resonance following a pulse that creates transverse magnetization. In this case, the Hamiltonian under which the spins evolve is given in the doubly rotating frame by:

$$(2\pi/h)\mathscr{H} = \Delta_I I_z + 2\pi J I_z S_z + \omega_2 S_y \tag{100}$$

with the observed I spins Δ_I off resonance, whereas the S spins are irradiated on resonance. Note that the first term of $\mathscr{H}$ commutes with the other two, but the second and third terms fail to commute with each other. We may therefore write

$$(2\pi/h)\mathscr{H} = \Delta_I I_z + (2\pi J I_z S_z + \omega_2 S_y) = 2\pi(\mathscr{H}_1 + \mathscr{H}_2)/h \tag{101}$$

and cascade the evolutions due to $\mathscr{H}_1$ and $\mathscr{H}_2$ in either order. The initial density operator that is to evolve under this Hamiltonian is given by:

$$\sigma^+ = I_x + \alpha S_i \tag{102}$$

where the spin system has been prepared with a 90° pulse on the I spins, along the y axis. Here α is the ratio (γ_S/γ_I); in the homonuclear situation, $\alpha = 1$ and $i = x$, following a "nonselective" preparation pulse; in the heteronuclear situation, $\alpha \neq 1$ and $i = z$ are the conditions encountered often in practice. Because ω_2 is constant throughout acquisition, Eq. (38) may be employed to calculate the FID. The following equations of motion under $\mathscr{H}_2$ can be derived applying the Hausdorff formula:

$$\begin{aligned}
I_x &\rightarrow I_x[1 - (\pi^2 J^2/\beta)] + (\pi^2 J^2/\beta) I_x \cos(\sqrt{\beta}\,t) \\
&\quad - 2\pi J \omega_2 I_y S_x(\cos(\sqrt{\beta}\,t) - 1)/\beta + (2\pi J/\sqrt{\beta}) I_y S_z \sin(\sqrt{\beta}\,t) \\
S_x &\rightarrow S_x \cos(\sqrt{\beta}\,t) - (\omega_2/\sqrt{\beta}) S_z \sin(\sqrt{\beta}\,t) + (2\pi J/\sqrt{\beta}) I_z S_y \sin(\sqrt{\beta}\,t) \\
S_z &\rightarrow S_z[1 - (\omega_2^2/\beta)] + (\omega_2^2/\beta) S_z \cos(\sqrt{\beta}\,t) + (\omega_2/\sqrt{\beta}) S_x \sin(\sqrt{\beta}\,t) \\
&\quad - (2\pi J \omega_2/\beta) I_z S_y(\cos(\sqrt{\beta}\,t) - 1)
\end{aligned} \tag{103}$$

Here, $\beta = (\pi^2 J^2 + \omega_2^2)$, the square of the effective field on the S spins in the rotating frame. It is to be noted that the coherence order is not conserved under the action of $\mathscr{H}_2$, because this Hamiltonian includes the effect of the irradiating rf field. In particular, the creation of zero- and double-quantum coherences, given by $I_y S_x$, is noteworthy.

In case the irradiating field is applied along the x axis instead of the y axis, the above equations are modified:

$$\begin{aligned} I_x &\to I_x[1 - (\pi^2 J^2/\beta)] + (\pi^2 J^2/\beta) I_x \cos(\sqrt{\beta}\, t) \\ &\quad + (2\pi J \omega_2/\beta) I_y S_y (\cos(\sqrt{\beta}\, t) - 1) + (2\pi J/\sqrt{\beta}) I_y S_z \sin(\sqrt{\beta}\, t) \\ S_x &\to S_x[1 - (\pi^2 J^2/\beta)] + (\pi^2 J^2/\beta) S_x \cos(\sqrt{\beta}\, t) \\ &\quad + (2\pi J/\sqrt{\beta}) I_z S_y \sin(\sqrt{\beta}\, t) - (2\pi J \omega_2/\beta) I_z S_z (\cos(\sqrt{\beta}\, t) - 1) \\ S_z &\to S_z[1 - (\omega_2^2/\beta)] + (\omega_2^2/\beta) S_z \cos(\sqrt{\beta}\, t) - (\omega_2/\sqrt{\beta}) S_y \sin(\sqrt{\beta}\, t) \\ &\quad - (2\pi J \omega_2/\beta) I_z S_x (\cos(\sqrt{\beta}\, t) - 1) \end{aligned} \tag{104}$$

In either case, $\mathscr{H}_1$ has no effect on the overall evolution of S_x and S_z, as is evident from the above. The equations for the evolution of I_x, however, are modified:

$$\begin{aligned} I_x &\to [1 - (\pi^2 J^2/\beta)](I_x \cos \Delta_I t + I_y \sin \Delta_I t) \\ &\quad + (\pi^2 J^2/\beta)(I_x \cos \Delta_I t + I_y \sin \Delta_I t) \cos(\sqrt{\beta}\, t) \\ &\quad + (2\pi J/\sqrt{\beta}) S_z (I_y \cos \Delta_I t - I_x \sin \Delta_I t) \sin(\sqrt{\beta}\, t) \end{aligned} \tag{105}$$

for irradiation along either axis, omitting the evolution under $\mathscr{H}_1$ of the zero and double quantum coherences, which remain unobservable because $\mathscr{H}_1$ conserves the coherence order.

With the receiver phase detector set for detection of I_x, we have for the signal:

$$\begin{aligned} S(t) &\sim (1 - [\pi^2 J^2/\beta]) \cos \Delta_I t + (\pi^2 J^2/\beta) \cos(\Delta_I t) \cos(\sqrt{\beta}\, t) \\ &\quad - (2\pi J/\sqrt{\beta}) S_z \sin(\Delta_I t) \sin(\sqrt{\beta}\, t) \end{aligned} \tag{106}$$

There is thus an absorption mode signal of intensity $2[1 - (\pi^2 J^2/\beta)]$ at frequency Δ_I, and an absorption pair of equal intensity, $(\pi J/2\sqrt{\beta})(1 + [\pi J/\sqrt{\beta}])$, at $(\Delta_I \pm \sqrt{\beta})$.

It is interesting to consider the following limiting cases.

CASE 1, $\omega_2 = 0$. The parameter β now equals $\pi^2 J^2$. The signal at Δ_I therefore vanishes, and a (1, 1) doublet results at the frequencies $(\Delta_I \pm \pi J)$. This corresponds to the normal spectrum with weak coupling. In the homonuclear case, the S spectrum is also seen to be a doublet at $(\pm \pi J)$.

CASE 2, $\pi J/\omega_2 \to 0$. The parameter β now tends to ω_2^2. A signal of two units intensity results at Δ_I, and the signals at $(\Delta_I \pm \omega_2)$ have vanishing intensities. The I spin has thus been "decoupled" from the S spin. Irradiation with a strong

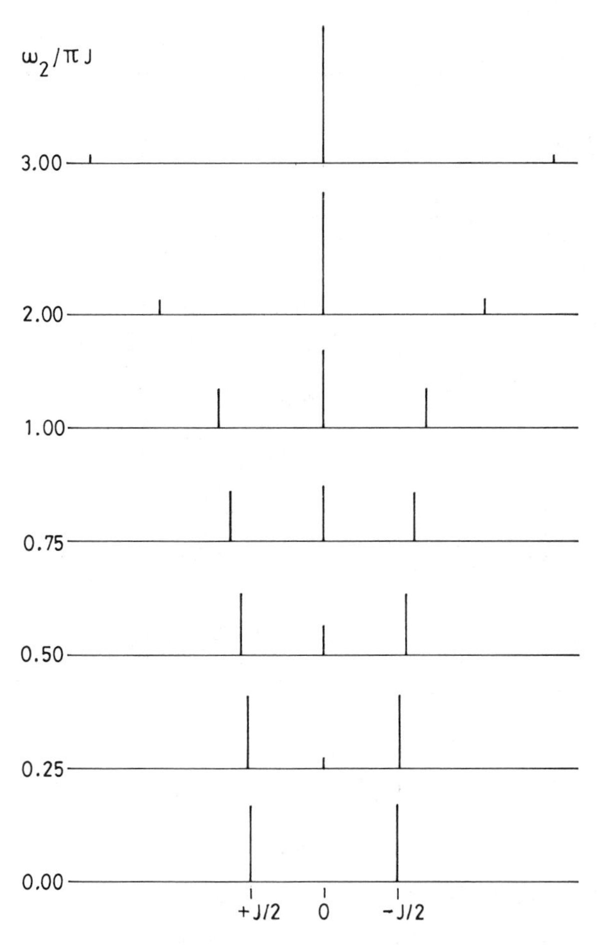

Figure 1.7. Stick plot of the A region of the NMR spectrum of a spin-$\frac{1}{2}$ AX system, with the X nuclei irradiated on resonance; the plot has been made as a function of the irradiation amplitude B_2 expressed in units of the coupling J.

rf field is very frequently used to perform such decoupling experiments. Figure 1.7 gives the I spectrum as a function of the relative amplitude $(\omega_2/\pi J)$ of the on-resonance rf irradiation of the S spin. In the heteronuclear case, the S spins clearly precess in the plane perpendicular to the irradiation axis, with frequency ω_2. In the homonuclear case, there is an S absorption at $(\pm\omega_2)$ if irradiation is along the y axis; if irradiation is along the x axis, on the other hand, the S absorption occurs at zero frequency, corresponding to a selective spin lock on the S spins. This description of the I and S spectra ignores, of course, the saturating effect of the strong irradiation, ω_2, and cross-relaxation between I and S spins, which is discussed in Chapter 2.

Antiphase Multiplets. We shall consider finally the effect of a density operator of the form:

$$\sigma^+ \sim I_x S_z \tag{107}$$

In this case, during acquisition of the I signal,

$$\begin{aligned}\sigma(t) &\sim \exp(-i\Delta I_z t)(I_x S_z \cos \pi J t + (1/2) I_y \sin \pi J t) \exp(i\Delta I_z t)\\ &\sim \tfrac{1}{2} I_x S_z (\cos(\Delta + \pi J)t + \cos(\Delta - \pi J)t)\\ &\quad - \tfrac{1}{4} I_x (\cos(\Delta - \pi J)t - \cos(\Delta + \pi J)t)\\ &\quad + \tfrac{1}{2} I_y S_z (\sin(\Delta + \pi J)t + \sin(\Delta - \pi J)t)\\ &\quad - \tfrac{1}{4} I_y (\sin(\Delta - \pi J)t - \sin(\Delta + \pi J)t)\end{aligned} \tag{108}$$

Hence,

$$\mathrm{Tr}(\sigma(t) I_x) \sim \tfrac{1}{2}[\cos(\Delta + \pi J)t - \cos(\Delta - \pi J)t] \tag{109}$$

Upon Fourier transforming, a doublet results with $(-\frac{1}{2}, \frac{1}{2})$ intensity at frequencies $(\Delta - \pi J, \Delta + \pi J)$. This is the state of anti-phase magnetization considered earlier.

In Tables 1.1 A and B we summarize the multiplet patterns for some common spin systems, generated by various forms of the density operator just prior to acquisition. They may all be derived by the explicit arguments presented above. In general, for operators of the type $I_x \prod_i S_{i\alpha}$, detectability of I_x demands that the product of operators $\prod_i S_{i\alpha}$ ($\alpha = x, y, z$) have diagonal elements in its matrix representation in the direct product space of the composite spin system IS_N. When the spins S_i are all spin-1/2 particles, this requires that each of the operators in the product be the z component, S_{iz}. In a weakly coupled system, the occurrence of even a single transverse operator, $S_{ix,y}$, renders the operator $I_x \prod_i S_{i\alpha}$ unobservable. When the spins S are spin-1 particles, on the other hand, the squares of the transverse operators, $S_{ix,y}^2$, do have diagonal elements, although they are not proportional to the identity as in the spin-1/2 case.

Table 1.1A. Multiplet Patterns for N Coupled Spin-1/2 Nuclei S

System	Operator	Intensity Pattern
IS	I_x	(1/2)(1, 1)
	$I_x S_z$	(1/2)(−1/2, 1/2)
IS_2	I_x	(1/4)(1, 2, 1)
	$I_x S_{1z} S_{2z}$	(1/4)(1/4, −1/2, 1/4)
	$I_x(S_{1z} + S_{2z})$	(1/4)(−1, 0, 1)
IS_3	I_x	(1/8)(1, 3, 3, 1)
	$I_x S_{1z} S_{2z} S_{3z}$	(1/8)(−1/8, 3/8, −3/8, 1/8)
	$I_x(S_{1z} + S_{2z} + S_{3z})$	(1/8)(−3/2, −3/2, 3/2, 3/2)
	$I_x(S_{1z}S_{2z} + S_{1z}S_{3z} + S_{2z}S_{3z})$	(1/8)(3/4, −3/4, −3/4, 3/4)

Table 1.1B. Multiplet Patterns for N Coupled Spin-1 Nuclei S

System	Operator	Multiplet
IS	I_x	(1/3)(1, 1, 1)
	$I_xS_z^2$	(1/3)(1, 0, 1)
	$I_xS_x^2$	(1/6)(1, 2, 1)
	I_xS_z	(1/3)(−1, 0, 1)
IS_2	I_x	(1/9)(1, 2, 3, 2, 1)
	$I_x(S_{1z} + S_{2z})$	(1/9)(−2, −2, 0, 2, 2)
	$I_xS_{1z}S_{2z}(S_{1z} + S_{2z})$	(1/9)(−2, 0, 0, 0, 2)
	$I_xS_{1z}S_{2z}$	(1/9)(1, 0, −2, 0, 1)
	$I_x(S_{1x}^2S_{2z} + S_{1z}S_{2x}^2)$	(1/9)(−1, −2, 0, 2, 1)
	$I_x(S_{1z}^2 + S_{2z}^2)$	(1/9)(2, 2, 4, 2, 2)
	$I_x(S_{1x}^2 + S_{2x}^2)$	(1/9)(1, 3, 4, 3, 1)
	$I_xS_{1z}^2S_{2z}^2$	(1/9)(1, 0, 2, 0, 1)
	$I_x(S_{1x}^2S_{2z}^2 + S_{1z}^2S_{2x}^2)$	(1/9)(1, 2, 2, 2, 1)
	$I_xS_{1x}^2S_{2x}^2$	(1/36)(1, 4, 6, 4, 1)
IS_3^a	I_x	(1/27)(1, 3, 6, 7, 6, 3, 1)
	$I_x\mathrm{Sym}\,(S_{iz})$	(1/27)(−3, −6, −6, 0, 6, 6, 3)
	$I_xS_{1z}S_{2z}S_{3z}$	(1/27)(−1, 0, 3, 0, −3, 0, 1)
	$I_x\mathrm{Sym}\,(S_{iz}^2(S_{jz} + S_{kz}))$	(1/27)(−6, −6, −6, 0, 6, 6, 6)
	$I_x\mathrm{Sym}\,(S_{ix}^2(S_{jz} + S_{kz}))$	(1/27)(−3, −9, −9, 0, 9, 9, 3)
	$I_x\mathrm{Sym}\,(S_{iz}^2S_{jz}^2S_{kz})$	(1/27)(−3, 0, −3, 0, 3, 0, 3)
	$I_x\mathrm{Sym}\,(S_{kz}(\mathrm{Sym}\,(S_{iz}^2S_{jx}^2)))$	(1/27)(−3, −6, −3, 0, 3, 6, 3)
	$I_x\mathrm{Sym}\,(S_{ix}^2S_{jx}^2S_{kz})$	(1/108)(−3, −12, −15, 0, 15, 12, 3)
	$I_x\mathrm{Sym}\,(S_{iz}^2)$	(1/27)(3, 6, 12, 12, 12, 6, 3)
	$I_x\mathrm{Sym}\,(S_{ix}^2)$	(3/54)(1, 4, 8, 10, 8, 4, 1)
	$I_x\mathrm{Sym}\,(S_{iz}^2S_{jz}^2)$	(1/27)(3, 3, 9, 6, 9, 3, 3)
	$I_x\mathrm{Sym}\,(S_{ix}^2S_{jx}^2)$	(1/108)(3, 15, 33, 42, 33, 15, 3)
	$I_x\mathrm{Sym}\,(S_{iz}^2S_{jx}^2)$	(1/27)(3, 9, 15, 18, 15, 9, 3)
	$I_xS_{1z}^2S_{2z}^2S_{3z}^2$	(1/27)(1, 0, 3, 0, 3, 0, 1)
	$I_xS_{1x}^2S_{2x}^2S_{3x}^2$	(1/216)(1, 6, 15, 20, 15, 6, 1)
	$I_x\mathrm{Sym}\,(S_{iz}^2S_{jz}^2S_{kx}^2)$	(1/54)(3, 6, 9, 12, 9, 6, 3)
	$I_x\mathrm{Sym}\,(S_{iz}^2S_{jx}^2S_{kx}^2)$	(1/108)(3, 12, 21, 24, 21, 12, 3)
	$I_x\mathrm{Sym}\,(S_{iz}S_{jz})$	(1/27)(3, 3, −3, −6, −3, 3, 3)
	$I_x\mathrm{Sym}\,(S_{iz}S_{jz}S_{kz}^2)$	(1/27)(3, 0, −3, 0, −3, 0, 3)
	$I_x\mathrm{Sym}\,(S_{iz}S_{jz}S_{kx}^2)$	(1/54)(3, 6, −3, −12, −3, 6, 3)

[a] For the IS_3 system, a shorthand notation, Sym (.....), has been employed, indicating a symmetrized sum over the indices i, j, and k, where $i, j, k = 1, 2, 3$ and $i \neq j \neq k$, each term occurring exactly once in the sum. For example:
(1) $\mathrm{Sym}\,(S_{kz}(\mathrm{Sym}\,(S_{iz}^2S_{jx}^2))) = (S_{1z}^2S_{2x}^2 + S_{1x}^2S_{2z}^2)S_{3z} + (S_{1z}^2S_{3x}^2 + S_{1x}^2S_{3z}^2)S_{2z} + (S_{2z}^2S_{3x}^2 + S_{2x}^2S_{3z}^2)S_{1z}$
(2) $\mathrm{Sym}\,(S_{iz}^2S_{jx}^2) = S_{1z}^2S_{2x}^2 + S_{1x}^2S_{2z}^2 + S_{1z}^2S_{3x}^2 + S_{1x}^2S_{3z}^2 + S_{2z}^2S_{3x}^2 + S_{2x}^2S_{3z}^2$.

CHAPTER 2

One-Dimensional Experiments in Liquids

Introduction

In this chapter we discuss the various one-dimensional pulse NMR experiments in liquids. By one-dimensional experiments we mean those experiments involving the manipulation of spin systems wherein the final spectra are presented in the conventional way, in terms of intensity versus frequency. By applying suitable rf pulses to a system of homo- or heteronuclear coupled spins we can manipulate it in a number of ways toward specific ends. We restrict ourselves in this chapter to spectra that are obtained by a single Fourier transform (FT) of the acquired free induction decay. Many of these one-dimensional experiments have been evolved to extract certain specific or additional information that are otherwise not possible in a simple minded single-pulse excitation followed by Fourier transform of the free induction decay that is well known to be equivalent to the slow passage continuous wave (CW) spectrum. In particular, we shall be concerned with such techniques as removal of $\mathbf{B}_0$ inhomogeneity using π refocusing pulses (J spectroscopy, spin echo, echo train), enhancement of ^{13}C spectra using broad-band decoupling (nuclear Overhauser effect, NOE), elucidating J connectivities (off-resonance decoupling), introducing a certain amount of selectivity in excitations (selective excitation, delays alternating with nutation for tailored excitation, DANTE), editing subspectra using J discrimination (spin echo Fourier transform, SEFT), suppressing solvent peaks, as well as simplifying ^{1}H-coupled ^{13}C spectra by uniform J scaling.

Wherever possible, classical vector pictures of the precession of magnetization in the rotating frame are given, although, in our opinion these vector pictures should not be extended indiscriminately to all situations. We are as yet not convinced about taking the vector pictures beyond representing single-quantum coherences. On the other hand, the representation of the state of the system by the density operator $\sigma(t)$ in terms of product spin operators can be followed systematically throughout the various evolution pathways and gives a clear-cut physical meaning. Throughout this book this product operator formalism, as laid out in Chapter 1, is followed. While it is not possible—nor is it the intention of the present authors—to cover all these techniques exhaustively, an attempt is made to outline the important techniques and their implications.

To summarize, this chapter is concerned with those aspects of high-resolution NMR spectroscopy in the pulsed FT mode wherein we deal with the creation, manipulation, and detection of single-quantum coherences, the manipulations leading to, in suitable systems, resolution and sensitivity enhancement, selective excitation, and signal assignments.

Transverse Magnetization

Upon excitation of the spin system by an rf pulse along the x axis such that

$$\gamma B_1 t_p = \alpha \tag{1}$$

where $2B_1$ is the amplitude of the linear rf magnetic field and t_p is the pulsewidth, the magnetization is tilted by an angle α about this axis. This single pulse generates transverse magnetization that is detectable single quantum coherence and evolves freely under the influence of chemical shift and spin–spin coupling. In the absence of any other phenomenon it would continue to do so indefinitely. However, in the presence of spin–spin interactions the transverse component decays with a characteristic time constant T_2, the spin–spin relaxation time obeying first-order kinetics:

$$M_{xy}(t) = M_{xy}(0)\exp(-t/T_2) \tag{2}$$

For a 90° pulse along the x axis the on-resonance y magnetization decays as:

$$\sigma(t) = -I_y \exp(-t/T_2) \tag{3}$$

and FT of this single exponential gives a Lorentzian absorption with width at half-height of $(\pi T_2)^{-1}$ Hz.

This ideal picture is seldom realized in practical situations, wherein, because of magnetic field inhomogeneity, the magnetization vectors (isochromats) in different regions are subject to different Larmor precessional frequencies and the initial phase coherence at the end of the 90° pulse is soon lost by rapid dephasing (Fig. 2.1). This leads to loss of magnetization governed by a relaxation time T_2^* so that,

$$\sigma(t) = -I_y \exp(-t/T_2^*) \tag{4}$$

where:

$$1/T_2^* = (1/T_2)_{\text{natural}} + (1/T_2)_{\text{external}} \tag{5}$$

where the contribution to linewidth from external factors includes magnetic field inhomogeneity as well as the instability of the field/frequency lock.

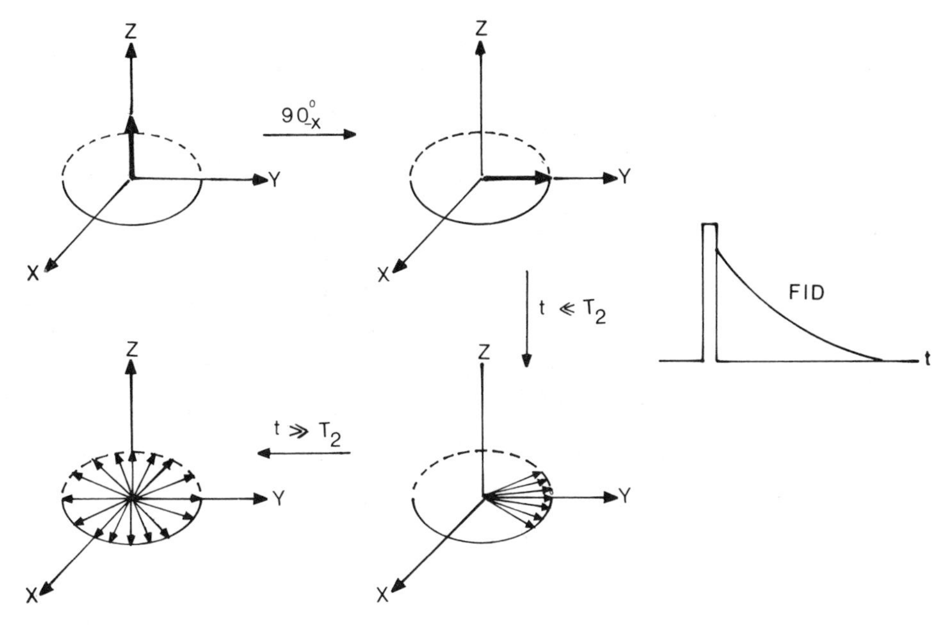

Figure 2.1. Dephasing of magnetization vectors in the transverse plane by spin–spin relaxation and the corresponding free induction decay on resonance.

The measured linewidths are therefore to be corrected for such external contributions.

In addition, immediately following the pulse, the spectrometer electronics requires a certain recovery time before the receiver can acquire the free induction decay (FID). During this recovery time, depending upon the different precessional frequencies present in the system, the transverse magnetizations of different chemically shifted isochromats make differing phase angles so that FT of the FID leads to frequency-dependent phase errors in the spectrum. Assuming that during the receiver dead time three spins, A, B, and C, excurse through 0°, 45°, and 90° with respect to the receiver axis, FT gives rise to lineshapes that are, respectively, pure absorption, a 50–50 mixture of absorption and dispersion, and pure dispersion (see Fig. 2.2). This happens all the time in practice and a phase correction is applied to a spectrum after FT that is a function of the resonance offset, as long as no aliasing has occurred (see Appendix 3).

Homo- and Heteronuclear Spin–Spin Coupling and the Effect of a π Pulse on the Transverse Magnetization. While the loss of transverse coherence brought about by spin–spin relaxation is irreversible, that due to the inhomogeneity of the dc field is reversible. A second $\pi/2$ pulse at a time τ after the initial pulse partially refocuses the dephasing isochromats at a time τ after the second pulse, corresponding to an enhanced transverse magnetization compared to the normal single-pulse FID. The formation of a spin echo, as this phenomenon is called, can be better understood if the second pulse is a π pulse. The

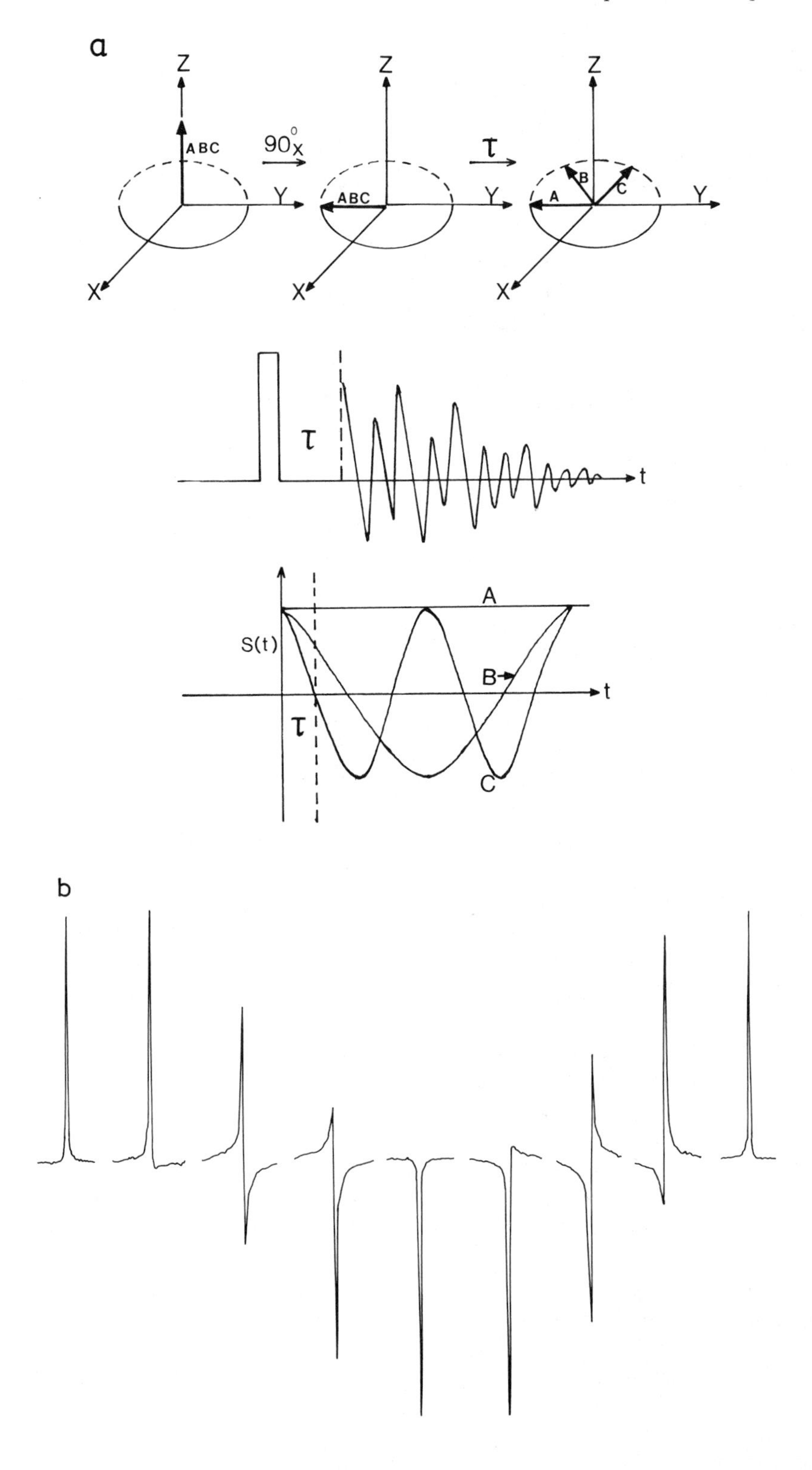
a
Z
ABC
90^0_x
Y
X
Z
ABC
Y
X
τ
Z
A
B
C
Y
X
τ
t
A
S(t)
B
t
τ
C
b

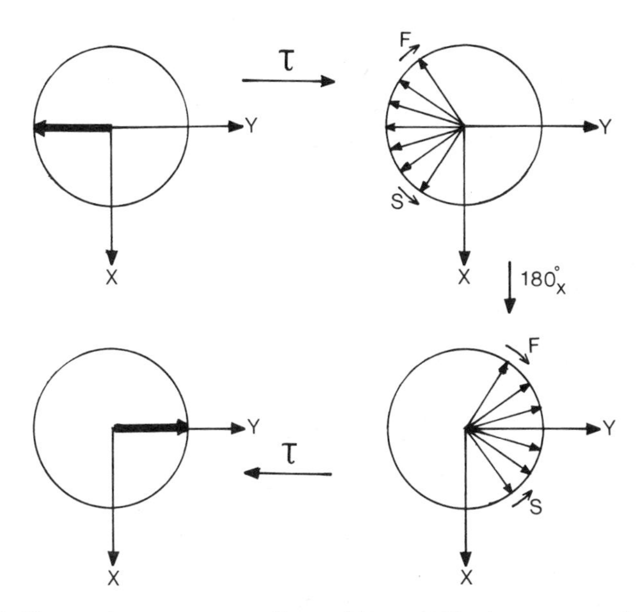

Figure 2.3. Vector picture corresponding to 90°_x–τ–180°_x echo sequence showing that the echo at 2τ occurs with the negative phase.

90°_x–τ–180°_x pulse sequence leads to an echo with a negative amplitude and can be described in terms of the classical vector picture (Fig. 2.3).

In systems where only chemical shifts are considered, phase anomalies introduced by the receiver dead time can be eliminated by using a 90°–τ–180° echo sequence and acquiring the FID from the top of the echo starting at 2τ. This elminates offset-dependent phase errors. Because evolution under chemical shift during the first τ period is refocused at the end of the 2τ period by the π pulse, there is no "net" evolution under the shift Hamiltonian. Considering Fig. 2.4 and for an AX (A = I_1, X = I_2) system:

$$\begin{aligned}
\sigma_0 &= I_{1z} + I_{2z} \\
\sigma_1 &= -I_{1y} - I_{2y} \\
\sigma_2 &= -c_{2\delta_1}(\tau)I_{1y} + s_{2\delta_1}(\tau)I_{1x} - c_{2\delta_2}(\tau)I_{2y} + s_{2\delta_2}(\tau)I_{2x} \\
\sigma_3 &= -c_{2\delta_1}(\tau)I_{1y} - s_{2\delta_1}(\tau)I_{1x} - c_{2\delta_2}(\tau)I_{2y} - s_{2\delta_2}(\tau)I_{2x} \qquad (6) \\
\sigma_4 &= -c^2_{2\delta_1}(\tau)I_{1y} + c_{2\delta_1}(\tau)s_{2\delta_1}(\tau)I_{1x} - s_{2\delta_1}(\tau)c_{2\delta_1}(\tau)I_{1x} - s^2_{2\delta_1}(\tau)I_{1x} \\
&\quad - c^2_{2\delta_2}(\tau)I_{2y} + c_{2\delta_2}(\tau)s_{2\delta_2}(\tau)I_{2x} - s_{2\delta_2}(\tau)c_{2\delta_2}(\tau)I_{2x} - s^2_{2\delta_2}(\tau)I_{2x} \\
&= -I_{1y} - I_{2y} = \sigma_1
\end{aligned}$$

with $c_{2\delta}(\tau) = \cos 2\pi\delta\tau$ and $s_{2\delta}(\tau) = \sin 2\pi\delta\tau$. (7)

◀ Figure 2.2. (a) Effect of delayed acquisition following a pulse. Classical vector picture, the corresponding schematic FID and expanded $S(t)$ for three isochromats, A, B, C, that differ in phase with respect to the receiver by 0°, 45°, and 90°. (b) The effect of progressively delayed acquisition on the lineshape of NMR.

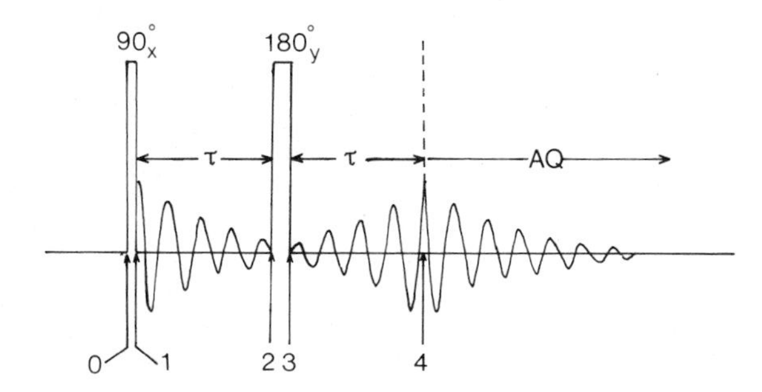

Figure 2.4. Focusing of chemical shift/offset at the echo maximum for a 90°–τ–180° echo sequence.

The situation is quite different when homo- or heteronuclear spin–spin coupling is present in addition, and it is very important to note this difference because it is the principal vehicle for polarization transfer between spins, the subject of the next chapter.

In describing pulse interrupted free precession, in general, an alternative equivalent method treats the evolution in terms of propagators of the whole sequence, operating on the initial density matrix. Thus, the effect of a τ–$\pi(I, y)$–τ sequence on a two spin IS system may be computed as:

$$\begin{aligned}\sigma(2\tau) &= \exp(-i\mathscr{H}\tau)\exp(-i\pi I_y)\exp(-i\mathscr{H}\tau)\sigma(0) \\ &\quad\times \exp(i\mathscr{H}\tau)\exp(i\pi I_y)\exp(i\mathscr{H}\tau) \\ &= \exp(-i\mathscr{H}\tau)\exp(-i\pi I_y)\exp(-i\mathscr{H}\tau)\exp(i\pi I_y) \\ &\quad\times \exp(-i\pi I_y)\sigma(0)\exp(i\pi I_y)\exp(-i\pi I_y) \\ &\quad\times \exp(i\mathscr{H}\tau)\exp(i\pi I_y)\exp(i\mathscr{H}\tau)\end{aligned} \tag{8}$$

For a weakly coupled system,

$$\frac{2\pi}{h}\mathscr{H} = \Delta_I I_z + \Delta_S S_z + 2\pi J I_z S_z \tag{9}$$

leading to

$$\begin{aligned}\sigma(2\tau) &= U\sigma(0)U^\dagger \\ &= \exp(-2\mathrm{i}\Delta_S S_z\tau)\exp(-\mathrm{i}\pi I_y)\sigma(0)\exp(\mathrm{i}\pi I_y)\exp(2\mathrm{i}\Delta_S S_z\tau)\end{aligned} \tag{10}$$

since the three terms of $\mathscr{H}$ commute, and the propagator is given by:

$$U(2\tau) = \exp(-2\mathrm{i}\Delta_S S_z\tau)\exp(-\mathrm{i}\pi I_y) \tag{11}$$

We shall occasionally employ this mode of description of spin evolutions, as well.

In homonuclear situation, say in a two-spin AX system, the second π pulse is not selective. Whereas the π pulse can reverse the evolution under chemical

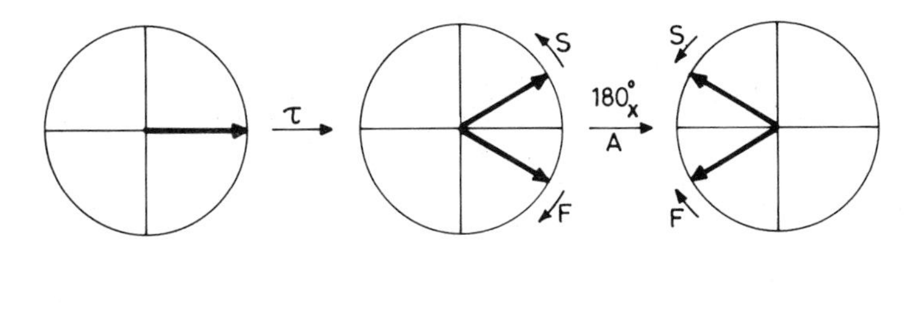

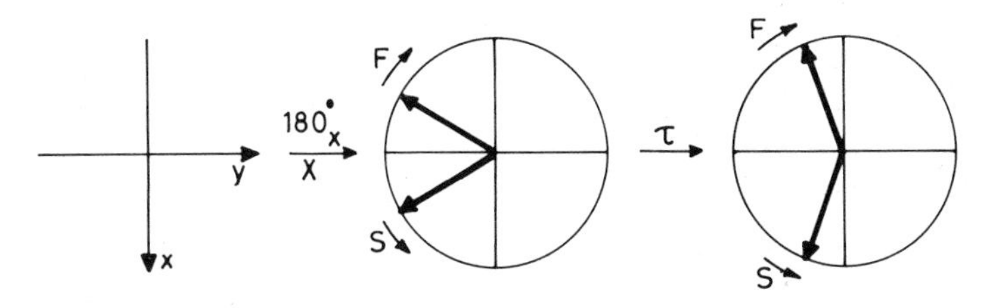

Figure 2.5. Vector picture showing that in homonuclear coupled systems the 90°–τ–180° sequence does not refocus the evolution under coupling and leads to modulation of the echo. In this picture F = fast and S = slow relative to the rotating reference frame.

shift, the multiplet components evolving under scalar coupling, $JI_{1z}I_{2z}$, are unaffected, leading to a modulation of the echo (see Fig. 2.5). In terms of product operators,

$$\begin{aligned}\sigma_0 &= I_{1z} + I_{2z} \\ \sigma_1 &= -I_{1y} - I_{2y} \\ \sigma_4 &= c_J(I_{1y} + I_{2y}) - 2s_J(I_{1x}I_{2z} + I_{2x}I_{1z})\end{aligned} \tag{12}$$

with $c_J = \cos \pi Jt$ and $s_J = \sin \pi Jt$.

Thus the transverse magnetization at $t = 2\tau$ is not a simple echo but is modulated as a function of τ so that a FT leads to a frequency-domain signal that is a mixture of absorption and anti-phase dispersion. This phenomenon, known as echo modulation, is very important in homonuclear spin systems.

Repeated π pulses can be applied at τ, 3τ, 5τ, ..., giving rise to the Carr–Purcell sequence. This latter gives echoes at 2τ, 4τ, 6τ, ..., with echo amplitudes alternating in sign with amplitude governed by true spin–spin relaxation, except for self-diffusion and chemical exchange during the 2τ period. In a Carr-Purcell (CP) sequence any error in setting up the π pulse will affect echo amplitude cumulatively depending on the length of the pulse train.

In order to avoid the alternation of the phase of the successive echoes the 180° pulses are applied in quadrature to the initial 90° pulse in the so-called Meiboom–Gill modification (Fig. 2.6). The decay of the echo amplitudes is governed by,

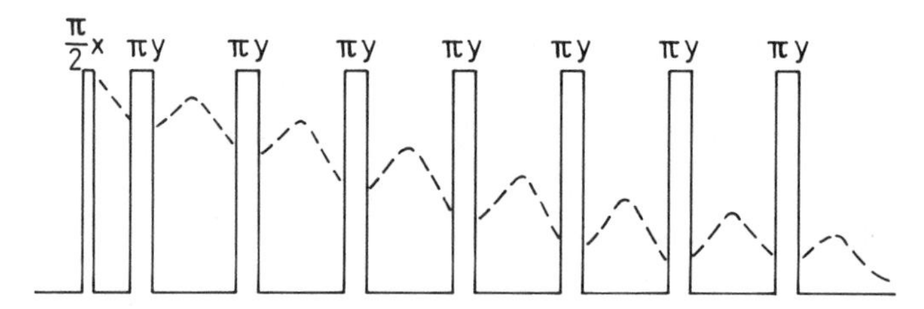

Figure 2.6. Meiboom–Gill modification of the Carr–Purcell sequence to produce successive echo maxima in phase.

$$M(t) = M_0 \exp(-2\tau/T_2 - \tfrac{2}{3}\gamma^2 G^2 D C^3) \qquad (13)$$

where **G** is the field gradient in the z direction, $d\mathbf{B}_0/dz$, and D is the diffusion coefficient. In fact, in trying to measure T_2 in systems with rapid self-diffusion, the apparent T_2 is measured as a function of the delay τ between the $\pi/2$ and π pulses and extrapolated to $\tau = 0$, where the diffusion effects are removed.

The above modification of the Carr–Purcell sequence, known as the CPMG, not only eliminates the effects of static field inhomogeneities but compensates for pulse imperfections also. This is true only when there is no spin–spin coupling. If spin–spin coupling is present then composite π pulses (see Chapter 7) are to be employed to compensate for pulse errors.

In an AX system (both with $I = 1/2$) the CPMG sequence leads to two echo components with a phase difference of $4\pi J\tau$ after a single focusing pulse or $4n\pi J\tau$ at the end of an n-pulse train. When an even number N of X nuclei are equally coupled to A, the central component of A spin magnetization shows no modulation, whereas the progressively flanking pairs show phase modulations at frequencies $\pm J$, $\pm 2J$, etc. The corresponding frequencies for N odd are $\pm J/2$, $\pm 3J/2$, $\pm 5J/2$, etc.

Fourier transformation of the last half of the end echo in a CPMG train for various total lengths of the train should give high-resolution (HR) spectra where the intensities of individual groups of equivalent spins are governed by the true spin–spin relaxation, in the absence of resolved coupling. In the presence of coupling, however, J modulation renders the measurement of intensities rather difficult, especially when the number of the coupled spin-1/2 nuclei is odd (if it is even one can follow the unmodulated central component). One way to circumvent this problem is by resorting to the absolute value mode $(u^2 + v^2)^{1/2}$. However, the presence of dispersion mixing leads to long tails and the measured intensities are likely to be in error.

One way to avoid this problem is to increase the pulsing rate. If the pulse repetition rate is large compared to chemical shift difference, the amplitude of the nth echo is given by:

$$I_n = \cos n\pi J\tau \left(1 - \frac{\sin 2\pi\delta\tau}{2\pi\delta\tau}\right) \exp(-2n\tau/T_2) \qquad (14)$$

where the phase modulation is reflected as an amplitude modulation with a $(\sin x)/x$ dependence of modulation frequency on the pulse interval. It is also to be noted that when the pulse repetition rate is fast compared to any chemical exchange present, the effect of this on the transverse relaxation can be eliminated; concomitantly exchange information also is lost. The rapid pulsing situation leads to spin locking of the transverse magenetization, leading to relaxation in the rotating frame ($T_{1\rho}$) and, if a sufficiently long pulse train is used, very little phase modulation results for the end FID ($T_{1\rho}$ approaches T_2 when the rf power level is very low). It should be pointed out, however, that if the pulse train is very short, *J* modulations at enhanced frequencies result via isotropic mixing, cause homonuclear coherence transfer, and, depending on the coupling network, introduce intensity and phase anomalies (see Chapter 3).

J Spectroscopy

In the high-resolution mode it is possible and often advantageous to measure the spin–spin relaxation times of nuclei in different parts of the molecule. One way of doing this has been mentioned previously, where spectra are measured as a function of the Carr–Purcell pulse train length after FT of the last half-echo. It is also possible to use analog or digital filtration techniques to separate the different frequency components, each associated with an individual decay constant. However, this can be better achieved indirectly by double Fourier transformation (see Chapter 4). In an one-dimensional sense one can also resort to the following technique, known as spin-echo FT. In this technique the peak of each echo is followed as far along the echo train as possible and Fourier transformed as a function of the pulse train duration (Fig. 2.7). This spin-echo spectrum, of course, covers only a narrow range, as only *J* modulation frequencies are involved. Also, the linewidths are governed only by true T_2 and self-diffusion/chemical exchange, if any. The same technique applied to liquid crystalline medium gives spectra at frequences corresponding to anisotropically averaged dipolar coupling constants. To avoid aliasing, the pulse repetition rate should be at least twice that of the highest *J* modulation present in the system. The spectral linewidths should directly relate to T_2

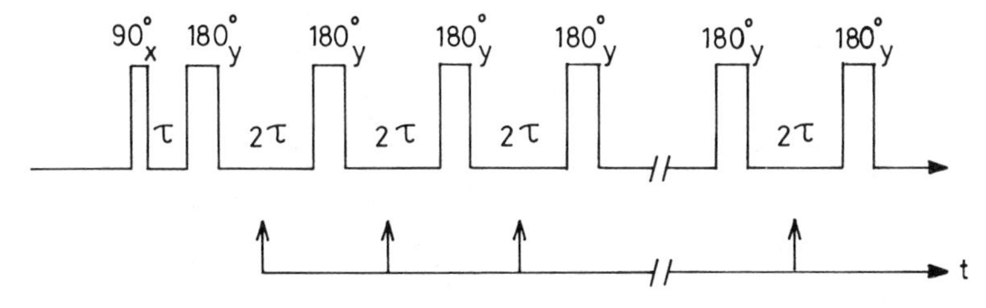

Figure 2.7. Pulse sequence for producing the *J* spectrum. The FID's are sampled midway between the π pulses and Fourier transformed as a function of the pulse train length.

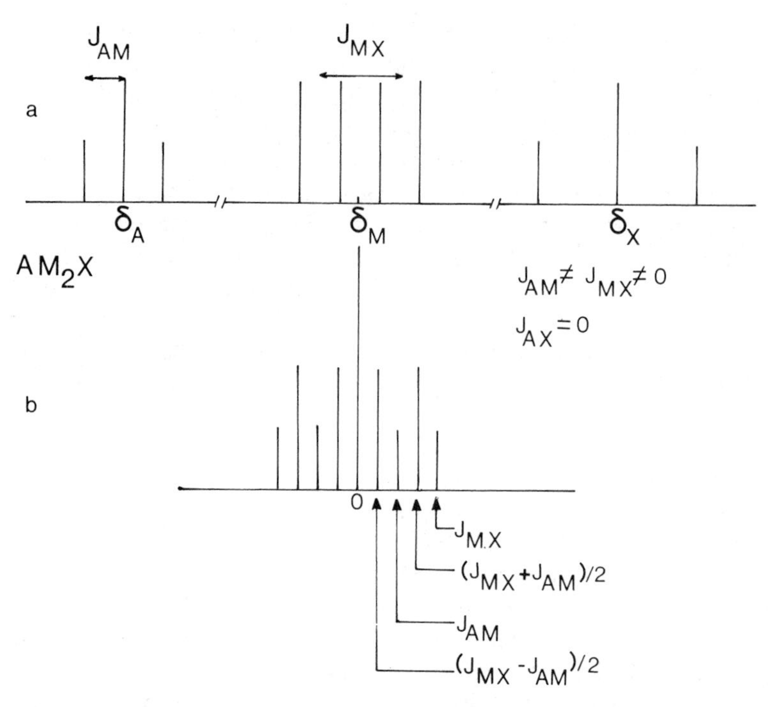

Figure 2.8. (a) Ordinary and (b) J spectrum of an AM_2X system. The J spectrum in (b) can be obtained by superimposing the multiplets on the top of each other.

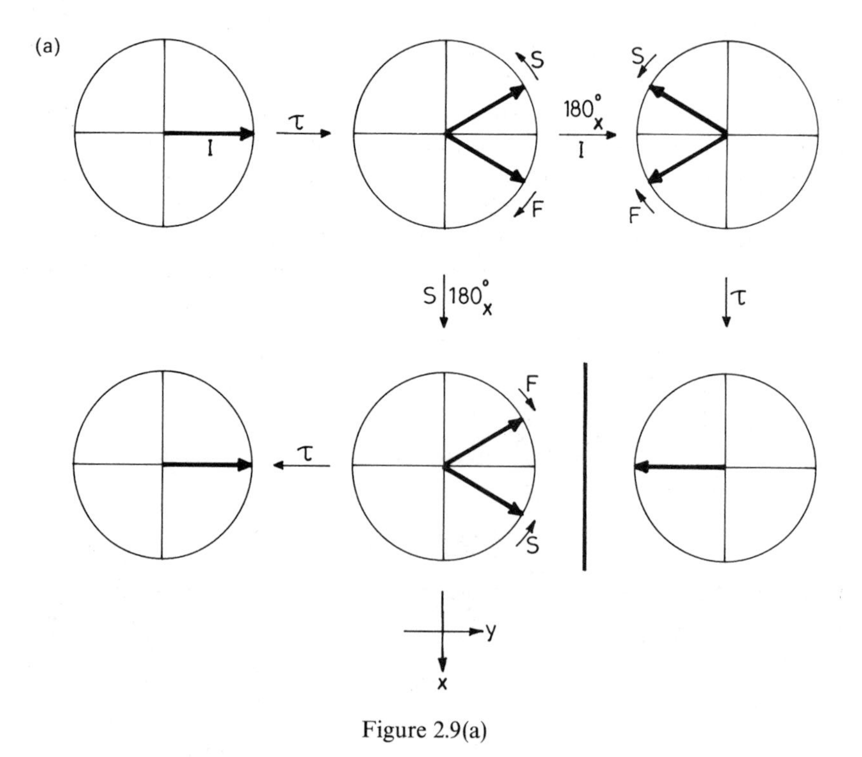

Figure 2.9(a)

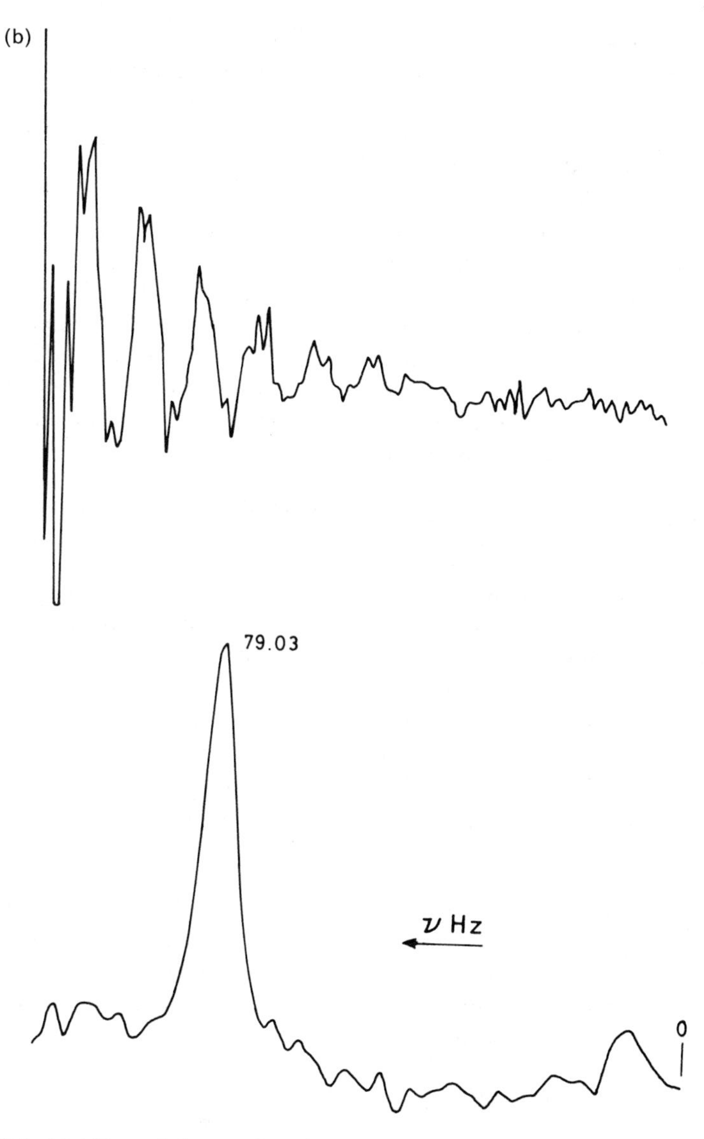

Figure 2.9. (a) Effect of the π refocusing pulse in a heteronuclear IS system. Both chemical shifts and spin coupling are refocused. The echo amplitude is reversed depending on whether the π pulse is on the I or the S channel. Note that if the π pulse is applied synchronously on both channels, J modulation persists. (b) J spectrum of C_6H_6.

and the intensity pattern is also governed by a binomial distribution, all of which is centrosymmetric with respect to zero frequency (incidentally, this zero frequency is the absolute zero). In fact, the J spectrum can be derived from the normal NMR spectrum by superposing all multiplets of different chemically shifted groups on the top of each as if they all had zero chemical shift (see Fig. 2.8).

The situation is quite different in the case of heteronuclear coupled systems. Here, the single π pulse applied after an initial $\pi/2$ pulse is selective to only the resonant spin and as such both chemical shifts and heteronuclear spin couplings are refocused (Fig. 2.9). If the refocusing π pulse is applied to the coupled spin, only evolution under hetero J coupling is refocused, with no effect on the observed spin chemical shift evolution. For on-resonance observation of a heteronuclear IS system the echo appears 180° out of phase depending upon the refocusing π pulse channel.

Spin-Echo Fourier Transform (SEFT) and the "Editing" of ^{13}C Subspectra

Heteronuclear J modulation of spin echo can be used in an ingenious way, for example, in ^{13}C NMR spectroscopy to distinguish between different coupling networks, such as CH_3, CH_2, CH, and quaternary carbons in a complex spectrum. One way of doing this is off-resonance decoupling (*vide infra*) where one uses the scaled down J multiplets to assign the different types of C—H networks. The current method is superior to off-resonance decoupling when too many lines overlap. The method also uses the gated ^{1}H-decoupling mode to retain Overhauser enhancements (*vide infra*). The pulse sequence is given in Fig. 2.10.

The ^{13}C spins are subject to a $90°-\frac{1}{2}\tau-180°-\frac{1}{2}\tau$–echo sequence with the ^{1}H broadband decoupler off only during the second $\frac{1}{2}\tau$ period. Whereas ^{13}C chemical shift evolutions are refocused irrespective of the different types of ^{13}C nuclei, J modulation leads to different phase relationships with respect to the detector reference for different types of coupling networks. Assuming the ideal situation, in which all CH coupling constants J_{CH} (CH, CH_2, CH_3)

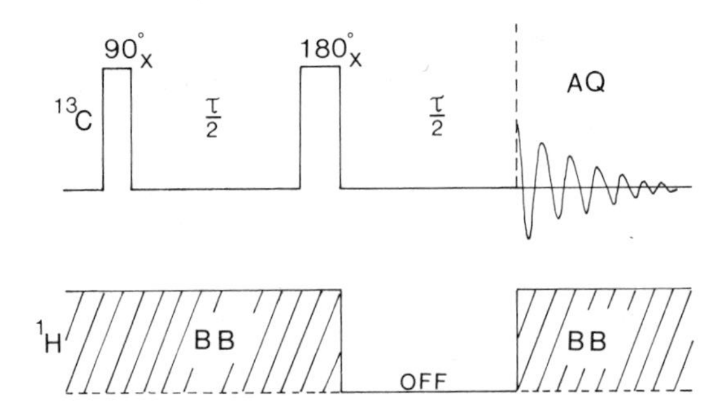

Figure 2.10. Spin-echo Fourier transform (SEFT) sequence.

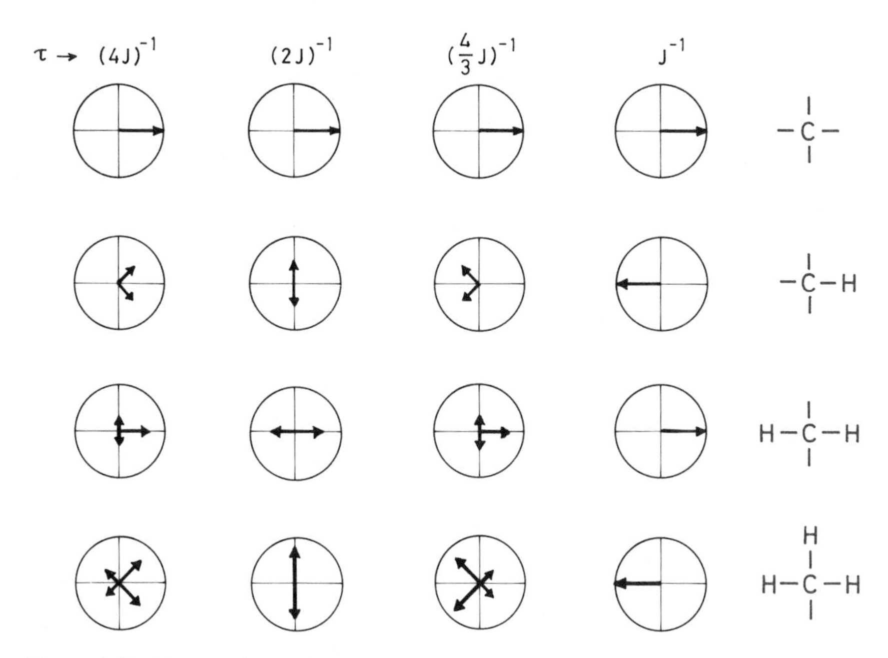

Figure 2.11. Vector picture for transverse magnetizations of spin multiplets for quaternary, methine, methylene, and methyl groups and their dependence on the evolution period τ (see Fig. 2.10). The diagram depicts the situation at the start of the acquisition.

are the same, and neglecting relaxation effects, it is possible to depict the various transverse components for different groups as a function of the period τ (Fig. 2.11).

For —C—, there is no evolution under J at $t = \tau$. Hence the transverse magnetization from quaternary carbons remains along the y axis.

For C—H, evolution under J_{CH} is such that the two transverse components corresponding to $I_z(^1H) = \pm 1/2$ are modulated at frequencies $\pm J/2$ and make angles $\pm \pi J \tau$ with the y axis.

For CH_2, the transverse magnetization is split into three components, corresponding to m_I ($I = \sum_i I_i$) $= \pm 1, 0$. Hence the central component is not modulated and behaves like a quaternary carbon resonance, while the other two components undergo modulation at $\pm J$ Hertz and hence make phase angles $\pm 2\pi J \tau$ with the detector axis, y.

For CH_3, on similar grounds, two of the transverse components behave similarly to those of CH, whereas the remaining two have modulation at $\pm 3J/2$ and hence make phase angles $\pm 3\pi J \tau$.

By evolving the I spin magnetization in IS_n system under spin coupling alone for a period τ, it can be shown that the density operator at the end of evolution corresponding to I_y operator is given by:

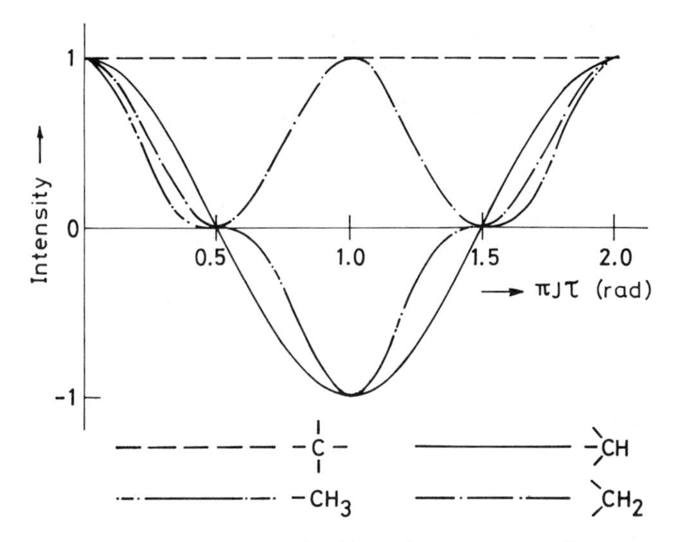

Figure 2.12. Time dependence of the ^1J ($^{13}C - {}^1H$) modulated ^{13}C magnetizations under subsequent decoupled acquisition.

$$IS_0 \quad : \quad \sigma_f = I_y$$
$$IS_1 \quad : \quad \sigma_f = \cos(\pi J\tau)I_y$$
$$IS_2 \quad : \quad \sigma_f = \tfrac{1}{2}(1 + \cos(2\pi J\tau))I_y$$

and

$$IS_3 \quad : \quad \sigma_f = \tfrac{1}{4}(3\cos(\pi J\tau) + \cos(3\pi J\tau))I_y \tag{15}$$

This is indicated graphically in Fig. 2.12.

Looking at the intensities of various types of ^{13}C, it is at once evident that when $\tau = J^{-1}$ (last column of Fig. 2.11) the signals from quaternary carbons and methylene carbons, on the one hand, can be distinguished from methyl and methine carbons because they appear as pure absorptions 180° out of phase. When $\tau = (2J)^{-1}$, all the carbons except the quaternary ones have only anti-phase transverse magenetizations and the decoupled spectrum should give only signals from the quaternary carbons. With this we could have distinguished quaternary carbons and the methylenes. It is rather difficult to distinguish between the methyl and methine carbons because there is no unique recipe, as is obvious from the close similarity in their intensity dependence on the τ value. For $\pi J\tau$ corresponding to the magic angle (54°44′) the intensity difference between ^{13}CH and $^{13}CH_3$ is maximized. It can be seen that τ values on either side of $(4J)^{-1} \pm 20\%$ should lead to change of sign of CH protons while little affecting the methyl ^{13}C resonances.

To achieve reasonable signal to noise ratio in ^{13}C spectra one must accumulate several hundred FIDs and one must wait at least $5T_1$ between the pulses. Besides, in most high-resolution methods one seldom uses 90° pulses to "prepare" the system and generally small flip angles are used. This results in a large $-z$ magnetization at the end of the focusing period so that introduction of an additional t–180°–t ($t \simeq 1$ ms) returns this to the $+z$ direction, indirectly aiding the T_1 process.

It is also possible to use the SEFT sequence to eliminate signals from deuterated solvents. Irrespective of the type of the deuterated group, i.e., CD, CD_2, or CD_3, both vector picture and product operator algebra show that when the J evolution progresses through $\tau = (3J)^{-1}$, the net y magnetization is zero, so that acquisition of ^{13}C with ^2H decoupling completely obliterates to solvent ^{13}C signals. This, of course, necessitates that the instrument be capable of providing simultaneous decoupling of both ^2H and ^1H during the acquisition. It is also possible to extend this technique to the analysis of deuterated and partially deuterated ^{13}C NMR spectra, and it is in principle possible to distinguish the different groups, such as C, CH, CH_2, CH_3, CHD, CH_2D, CHD_2, and CD_3. Another variation is to utilize the fact that singlets (—C—, with vertical bonds above and below C) and, the central triplet of (CH_2) groups are not subject to J modulation and hence are phase invariant to the refocusing period τ. Thus, coadding a number of FIDs with differing τ values and Fourier transforming the resulting time signal causes the elimination of all lines except the quaternary carbons and the central line of the methylene triplets. This is a simplification that perhaps may help in the analysis of ^{13}C spectra. There are more elegant one- and two-dimensional techniques used in subspectral editing of ^{13}C spectra with the additional bonus of polarization transfer from abundant ^1H to the rare ^{13}C; these are highlighted in the next two chapters.

Spin Decoupling

Whereas spin–spin coupling gives valuable additional fine structural data, and hence information on the spin system, it can be at times rather embarrassingly complicated and leads to overlap of resonances from nuclei in closely chemically related environments in systems with hetero- or homonuclear couplings. Also, when chemical shift differences are of the order of coupling constants, I_z's are no longer good quantum numbers and hence interesting intensity and multiplet anomalies can occur. Although it is possible in suitable cases to substitute the coupled nuclei chemically with its nonmagnetic isotope or an isotope of much lower magnetic moment, these techniques are rather cumbersome, time consuming, and in certain cases not possible. Of course, selective substitution by isotopes is a very powerful direct tool in the analysis of complex spectra. However, it is possible to decouple the otherwise coupled nucleus from the "observe" spins when the system is simultaneously subject to a second irradiation at or near the Larmor frequency of the nuclei whose coupling one wants to eliminate or attenuate. From the experimental standpoint, decoupling of heteronuclear spin systems is quite distinctly different from decoupling in homonuclear systems. In the former case the decoupling rf is so far removed in frequency that it does not interfere with the observe frequency. The mechanism of homo- and heteronuclear decoupling using the density operator formalism has been dealt with in Chapter 1. It is perhaps not too difficult to understand the mechanism of decoupling phenomenologically. In an AX system with $J_{AX} \neq 0$, irradiation of X spin at its chemical shift frequency induces transitions among all its m_I levels, thus scrambling the populations of the various m_I levels of X. If this scrambling takes place at a frequency much faster than the spin–spin coupling constant, then the effective magnetization due to X spin vanishes, because

$$\sum_i I_{zi} = 0$$

In other words, the spin system whose coupling we want to nullify is subject to a time-dependent perturbation such that the "average" coupling Hamiltonian vanishes.

The decoupling frequency may be selective, coherent, broadband, or on or off resonance. A density matrix treatment leads to a conception that decoupling leads to randomization of the orientation of magnetization vectors of irradiated spins associated with the observe multiplets. This randomization can be in a plane (noise decoupling) or over a sphere (decoupler being switched randomly along $+x$, $-x$, $+y$, $-y$ axes of the rotating frame). In modern spectrometers broadband decoupling is achieved by phase modulation of the rf, which is on continuously. The modulation can be randomly switched around the transverse plane to achieve spherical randomization. One important bonus during decoupling is transfer of polarization via the dipolar link, and we shall now briefly examine the aspects of relaxation and the nuclear Overhauser effect.

Dipolar Relaxation. Any disturbance of a spin system in a magnetic field by an applied rf to create a nonequilibrium state is followed by the T_1 and T_2 relaxation processes to restore equilibrium conditions. The transfer of energy between the spins and the surroundings can take place only when the surrounding medium also executes fluctuating motion at the appropriate Larmor frequencies. The "power spectrum" at the frequencies of interest dictates the efficiency of relaxation. Because we are talking about a Larmor frequency in the radiofrequency region ($\sim$megahertz), electronic motions or molecular vibrations, which are fast motions, do not really contribute to nuclear relaxation. However, Brownian motion and torsional motions (rotational, librational) of segments of molecules provide the necessary fluctuating local field $B_{\text{loc}}(t)$, as shown in the Appendix 4. The power spectrum $J(\omega)$ of this field is given by:

$$J(\omega) = \langle B_{\text{loc}}(0) \rangle^2 \int_{-\infty}^{+\infty} e^{-|\tau|/\tau_c + i\omega\tau} \, d\tau \tag{16}$$

which is the Fourier transform of the autocorrelation function $G(\tau)$ of a rotational motion with correlation time τ_c and can be shown to be

$$J(\omega) = \langle B_{\text{loc}}(0) \rangle^2 \frac{2\tau_c}{1 + \omega^2 \tau_c^2} \tag{17}$$

A plot of $J(\omega)$ versus the frequency gives the spectral density profiles and this is shown in Appendix 4 for three regimes, namely, long τ_c (viscous liquids, solids), intermediate correlation time, and short correlation time (fast tumbling in isotropic media).

When the spectral density is appreciable at the Larmor frequency, then the relaxation process is efficient so that relaxation times whose origin is governed by rotational Brownian motion are expected to go through a minimum as a function of temperature.

Dipolar Relaxation in the "White" Spectral Limit. The principal source of relaxation for spin-1/2 particles is via the dipole–dipole interactions, which are randomly modulated by the molecular rotation and translational diffusion in liquids or by lattice vibrations in solids. In the case of organic compounds the ^{13}C spin–lattice relaxation is supposed to originate mainly from the fluctuating dipolar coupling to the protons attached to it and those other intramolecular protons that are close to it, as dictated by the molecular conformation. For illustrating the dipolar mechanism of relaxation we shall consider a two-spin system, say, a $\overset{|}{\underset{|}{-\text{C}-}}\text{H}$ group, and the energy level diagram given in Appendix 4.

Let $W_{1\text{C}}$ and $W_{1\text{H}}$ be the single-quantum transition rates of carbon and hydrogen, while W_0 and W_2 are probabilities of zero-quantum and double-quantum transitions. As shown in Appendix 4, these probabilities can be calculated in terms of the square of the matrix elements P_{ij} of the dipole–dipole interaction operator such that:

$$W_{ij} = |P_{ij}|^2 J_m(\omega_{ij}) \tag{18}$$

The various transition probabilities are given in Appendix 4 for ^{13}C resonances under proton decoupling. The longitudinal magnetization is governed by

$$T_{1\text{C}}^{\text{dec}} = (W_0 + 2W_{1\text{C}} + W_2)^{-1}$$

whereas in the absence of decoupling the magnetization is governed by:

$$T_{1\text{C}}^{\text{cou}} = T_{1\text{C}}^{\text{dec}} + (W_2 - W_0)^{-1} \tag{19}$$

and hence the rate of dipolar relaxation $T_{1\text{DD}}^{-1}$ is given in terms of spectral densities at the appropriate frequencies

$$\frac{1}{T_{1\text{DD}}} = \frac{h^2 \gamma_{\text{H}}^2 \gamma_{\text{C}}^2}{80\pi^2 r_{\text{C-H}}^6} \left\{ J_0(\omega_{\text{H}} - \omega_{\text{C}}) + 3J_1(\omega_{\text{C}}) + 6J_2(\omega_{\text{H}} + \omega_{\text{C}}) \right\} \tag{20}$$

where $r_{\text{C-H}}$ is the distance between the dipolar-coupled partners. In the so-called "extreme narrowing limit" or "white spectral limit," where we assume that the $J_i(\omega)$ is independent of frequency this expression, simplifies to:

$$\frac{1}{T_{1DD}} = \frac{h^2\gamma_C^2\gamma_H^2}{4\pi^2} \cdot r_{C-H}^{-6} \cdot \tau_c \tag{21}$$

and when there are more than one coupled nuclei,

$$\frac{1}{T_{1DD}} = \frac{h^2\gamma_C^2\gamma_H^2}{4\pi^2} \cdot \sum_i r_{C-H}^{-6}\tau_c$$

Therefore, proton-bearing carbons would relax at a faster rate than quaternary carbons. For the sake of completeness, it can be shown that the spin–spin relaxation requiring the spectral densities at $J_1(\omega_H)$ and $J(0)$ is given by:

$$\frac{1}{T_{2DD}} = \frac{h^2\gamma_C^2\gamma_H^2}{80\pi^2 r_{C-H}^6}\left\{2J(0) + \frac{1}{2}J_0(\omega_H - \omega_C) + \frac{3}{2}J_1(\omega_C) + 3J_1(\omega_H) + 3J_2(\omega_H + \omega_C)\right\} \tag{22}$$

and, in the white spectral limit, by:

$$\frac{1}{T_{2DD}} = \frac{1}{T_{1DD}} = \frac{h^2\gamma_C^2\gamma_H^2}{4\pi^2}\sum_i r_{C-H}^{-6}\tau_c \qquad T_1 = T_2 \tag{23}$$

***Nuclear Overhauser Effect* (*NOE*).** The transfer of polarization from a given spin to another within the same molecule or in a nearby molecule via the dipolar Hamiltonian as the link is called the nuclear Overhauser effect (NOE). This can be either homo- or heteronuclear, inter- or intramolecular. The resulting non-Boltzmann distribution of the observe spin population, while saturating the other spins, can lead to enhancement, decrease, or change in sign of the signals depending on the relative signs and magnitude of the γ's and the nature of the power spectrum corresponding to the fluctuating dipolar interaction. When dipolar interaction is the main thermal link between spins then the r^{-6} dependence of the relaxation rates also leads to NOE buildup rates with the same r^{-6} dependence. The NOE is expressed by the symbol η and is defined for an AX coupled system as (see Appendix 4),

$$\eta = \frac{S_{dec}}{S_{normal}} = 1 + \frac{W_2 - W_0}{(W_0 + 2W_1 + W_2)} \cdot \frac{\gamma_X}{\gamma_A} \tag{24}$$

where W_1 is the A spin single-quantum transition probability and γ_A and γ_X are the magnetogyric ratios of observe and decoupled nuclei, respectively. For $^{13}C-^1H$ interaction, in the extreme narrowing limit,

$$W_0 : W_1 : W_2 = 1/6 : 1/4 : 1$$

and

$$\eta = 1 + 0.5\frac{\gamma_H}{\gamma_C} \approx 3 \tag{25}$$

With ^{15}N and ^{29}Si, proton decoupling leads to negative η because their magnetogyric ratios are negative.

In order to simplify ^{13}C spectra and to make use of the NOE one normally uses noise decoupling of ^1H in a broadbanded sense such that all protons irrespective of their chemical shifts are decoupled. However, it is possible to apply the decoupling frequency in a coherent mode slightly off resonance, which produces the J multiplets on ^{13}C, but with reduced coupling constants, still retaining most of the NOE. In a heteronuclear situation one can use perturbation theory to understand the mechanism of on-resonance/off-resonance low-power/high-power decoupling. Considering an AX system let us try to decouple X and observe the A resonances. By transforming the spin Hamiltonian to a doubly rotating frame rotating with a frequency ν_2 of the decoupler, the time-independent part of the Hamiltonian is given by:

$$\begin{aligned}\mathscr{H}_0 = & -\frac{h}{2\pi}(\nu_A - \nu_2)I_z(A) + \frac{h}{2\pi}JI_z(A)I_z(X) \\ & -\frac{h}{2\pi}\gamma(X)B_2I_x(X) - \frac{h}{2\pi}(\nu_X - \nu_2)I_z(X)\end{aligned} \tag{26}$$

Considering the secular part of the coupling term and the fact that I_x connects only transitions involving levels of X spin differing by $_{\Delta m_I(X)} = \pm 1$, we can rewrite the above equation as:

$$\begin{aligned}\mathscr{H}_0 = & -\frac{h}{2\pi}(\nu_A - \nu_2)I_z(A) - \frac{h}{2\pi}\left[(\nu_X - \nu_2) - JI_z(A)\right]I_z(X) \\ & -\frac{h}{2\pi}\gamma(X)B_2I_x(X)\end{aligned} \tag{27}$$

It follows, therefore, that:

$$\begin{aligned}\gamma_X B_{\text{eff}}^{\pm} &= [(\Delta\nu \pm J/2)^2 + (\gamma_X B_2/2\pi)^2]^{1/2} \\ \text{E}(I_z(A), I_z(X)) &= I_z(A)\nu_A + (\gamma_X B_{\text{eff}}^{\pm}/2\pi)I_z(X)\end{aligned} \tag{28}$$

Transitions involving change in $I_z(X)$ are forbidden under strong decoupling fields. When the resonance offset $\Delta\nu$ is much smaller than rf power expressed in frequency units, a doublet results with a reduced splitting J^r given by:

$$J^r = J\Delta\nu/(\gamma_X B_2/2\pi) \tag{29}$$

A vector picture for off-resonance decoupling is shown in Fig. 2.13. Also included in the figure is the effect of off-resonance decoupling as a function of the offset at constant decoupler power.

Gated and Inverse Gated Decoupling. The buildup and decay of the NOE is proportional to $(T_{1DD})^{-1}$, the dipolar spin–lattice relaxation rate. Because

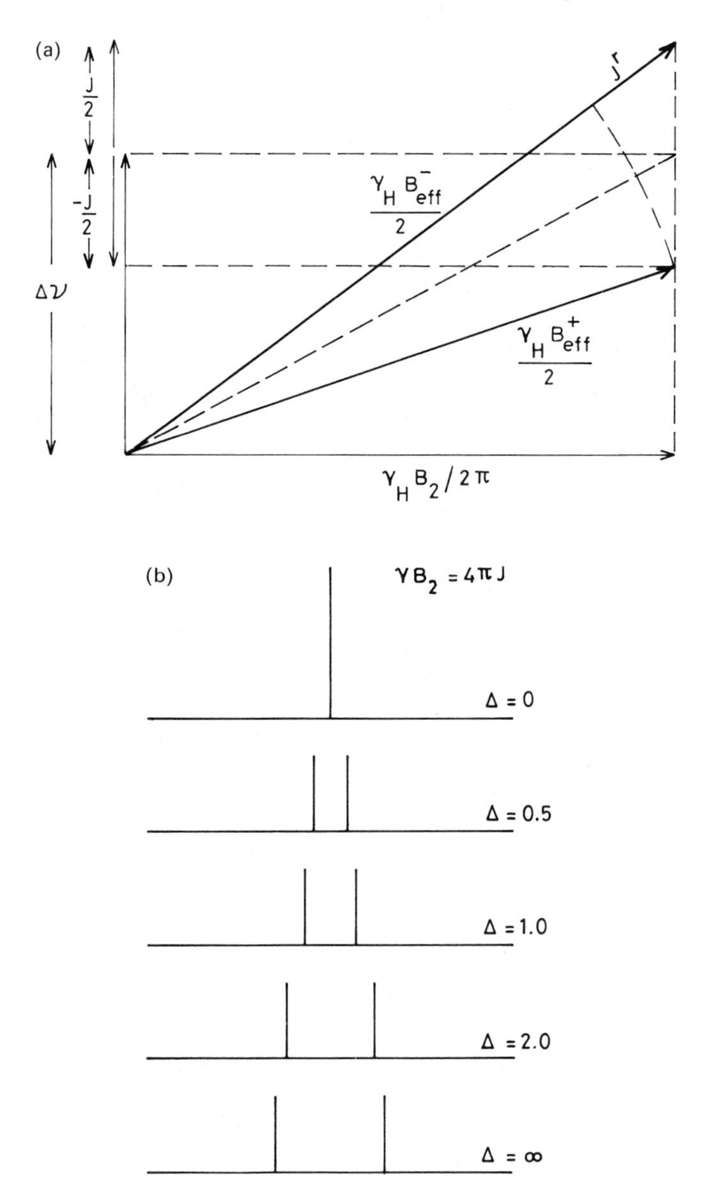

Figure 2.13. (a) Vector representation of off-resonance decoupling. $\Delta\nu$ is the offset and B_2 is the amplitude of the rf field. J^r is the reduced spin–spin coupling. (b) Schematic representation of the effect of decoupler offset at constant power on the multiplet intensity and separation.

Overhauser enhancement and decoupling take place simultaneously, it is possible, in favorable cases, to assess the buildup of the NOE quantitatively in specific resonances of the spectra. Thus, useful information can be obtained on the distance parameters and hence some conformational information.

In this connection, it should be remembered that as soon as the decoupler is on the spin–spin coupling information is lost at the rate of inverse Larmor frequencies (in micro- to nanoseconds), whereas the NOE builds up typically in a few seconds ($T_1 = T_2 \sim$ seconds in liquids for ^{13}C). When we want to integrate the ^{13}C spectra to get the relative ^{13}C spin counts in a molecule, we cannot use a decoupled spectrum with NOE because, depending upon the environment of the ^{13}C with respect to the dipolar network, the NOE will be different and intensities are not governed by specific concentration.

In such cases we use inverse gated decoupling, wherein the ^1H decoupler is on only during the acquisition and is switched off between pulse delays. Hence, at the time of ^{13}C excitation only the natural Boltzmann distribution governs the signal intensities, and during the FID, because the decoupling field is present (typically for 1 s), there is no appreciable buildup of NOE, provided $T_1 >$ acquisition time. However, one must cope with a poor signal to noise ratio.

If, on the other hand, one wants to take full advantage of the NOE but at the same time desires to retain the natural spin multiplets from C–H coupling, one can use the gated decoupling strategy. In this the ^1H noise decoupler is on throughout except during acquisition. During the pulse delays full NOE builds up and continues to persist during the short acquisition time, even when the decoupler is switched off (see Fig. 2.14).

In order to optimize the pulse delay and acquisition time to get maximum or minimum NOE it is better first to estimate T_1 of the shortest relaxing carbon of interest. Then one can calculate the buildup or decay of the NOE by adjusting both the pulse delay and the acquisition time, keeping in mind that for repeated acquisitions in high-resolution NMR it is better to use the Ernst optimal flip angle, given by:

$$\cos\alpha = \exp(-\tau/T_1) \tag{30}$$

where τ is the recycle delay. Obviously, such considerations do not work if T_1 is on the order of the acquisition time, and so it is difficult to achieve the separation of NOE and decoupling effects in polymer solutions.

Homonuclear Decoupling. The situation in homonuclear systems for selective decoupling is rather different. Because the decoupling and observe frequencies are essentially the same, there is a possibility that decoupling irradiation will interact with the receiver, causing considerable overload and distortion of the signal. This is especially severe in proton NMR because both the chemical shifts and the spin–spin couplings span a narrow range. Homonuclear decoupling in the pulsed mode, therefore, is more tricky than in the continuous wave (CW) mode. Although we are concerned with FT methods it is perhaps

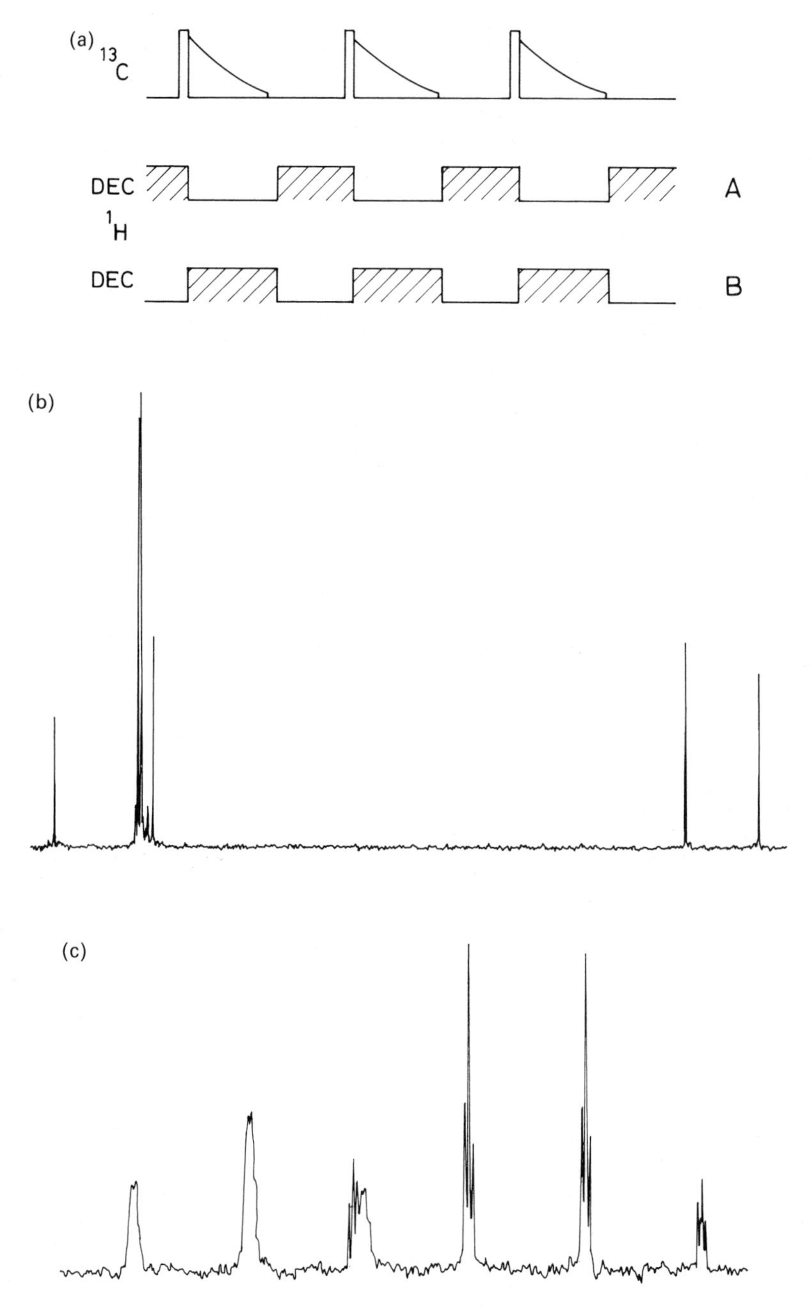

Figure 2.14. Gated (A) and inverse gated (B) decoupling schemes. In (A) NOE is retained while the spectra show spin–spin multiplets and in (B) decoupled spectra with no NOE are obtained. (b) Carbon-13 spectrum of 80% ethyl benzene at 22.64 MHz using the inverse gated decoupling scheme. Spectral width ≈ 3135 Hz. (c) Same as in (b) using the gated decoupling scheme, showing only the aliphatic region.

not out of place to summarize some of the techniques used in the CW mode and appreciate why such techniques cannot be transcribed into FT methods.

In the CW mode one can resort to "spin tickling," which corresponds to irradiating coherently at the resonance frequency ν_2, say, of a given proton and simultaneously sweeping the observe frequency, or subjecting the system to two frequencies ν_1 and ν_2 and varying the magnetic field in the so-called field-swept mode. To understand the effect of tickling it is pertinent here to define what are known as progressively and regressively connected transitions. If there are two transitions that share a common level, then the transitions are said to be progressively connected if the net change in z component of total spin angular momentum is ± 2 and regressively connected if 0 (Fig. 2.15).

The effect of double irradiation of a spin-1/2 AX system as a function of the decoupling power has been described in detail in Chapter 1, employing the density operator method. Here, we look at the detail of decoupling in an AX_n system. To understand the effect of irradiating the system at the second frequency ν_2 one can transform the Hamiltonian to a frame rotating with ν_2. Considering the spin system AX_n (A spin angular momentum being represented by S and X by I) when we irradiate A spins and observe X spins, the transformation into the rotating frame is given by:

$$\Psi^{R} = U\Psi \tag{31}$$

$$U = \exp(+i\omega_2 t(S_z + \sum_n I_{zn})) \tag{32}$$

where $\omega_2 = 2\pi\nu_2$.

Considering the total Hamiltonian,

$$\mathscr{H} = \mathscr{H}^0 + \mathscr{H}'(t) + \mathscr{H}''(t) \tag{33}$$

where $\mathscr{H}^0$ is the chemical shift and spin–spin coupling Hamiltonians, $\mathscr{H}''(t)$ corresponds to weak observe radiation, and $\mathscr{H}'(t)$ is the irradiating field. In the rotating frame the time-independent part is:

$$\mathscr{H}^{R} = \mathscr{H}^0_{R} + \mathscr{H}'_{R} \tag{34}$$

where

$$\mathscr{H}^0_{R} = \sum_n -h(\nu_I - \nu_2)I_z + h\sum_n J_{IS}\mathbf{I}_n\cdot\mathbf{S} \tag{35}$$

and

$$\mathscr{H}'_{R} = -\frac{h}{2\pi}\gamma_S B_2 S_x \tag{36}$$

where B_2 is the amplitude of the irradiating field. The time-dependent weak perturbation $\mathscr{H}''_{R}$ is:

$$\mathscr{H}''_{R} = \sum_n -\frac{h}{2\pi}\gamma_I I_{xn} B_1 \cos(\omega_1 - \omega_2)t - \frac{h}{2\pi}\gamma_I I_{yn} B_1 \sin(\omega_1 - \omega_2)t \tag{37}$$

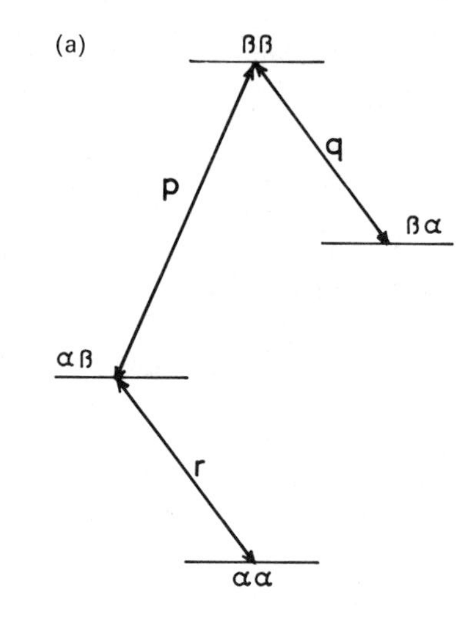

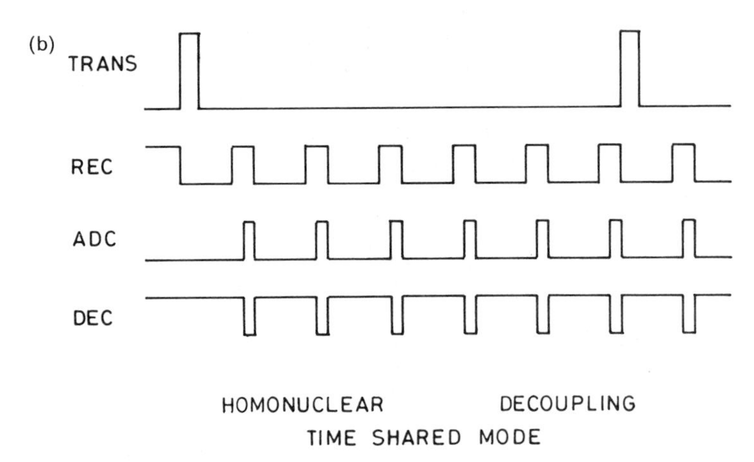

Figure 2.15. Schematic energy level diagram showing progressively connected (p and r) and regressively connected (p and q) transitions. (b) Homonuclear decoupling, time-shared mode.

Thus the S spins are subject to a magnetic field B_2 along the rotating x axis and $\frac{2\pi}{\gamma_S}(\nu_S - \nu_2) - \sum_n J_{IS} I_{zn}$ along the z axis. A rotation about the rotating y axis by θ will bring the z axis into alignment with the resultant of these two fields, θ depending upon the z component of the observe spin. This rotation:

$$T = \exp(i\theta(m_{I_n})S_y) \tag{38}$$

with θ depending on the strength of B_2 and the offset ν_2, and the z components of all the I spins coupled to S. This second rotation, which is necessary to make the Hamiltonian stationary, affects only the S spins. In the tilted doubly rotating frame,

$$\mathscr{H}^{\mathrm{TR}} = T\mathscr{H}^{\mathrm{R}}T^{-1} \tag{39}$$

The rf field B_1 applied to the I spins is weak and we need consider only I_x and I_y operators in $\mathscr{H}''_{\mathrm{R}}$; also, because I_x and I_y commute with S_y, the time-dependent part $\mathscr{H}''_{\mathrm{TR}}$ in the tilted doubly rotating frame is the same, i.e.,

$$\mathscr{H}''_{\mathrm{TR}} = T\mathscr{H}''_{\mathrm{R}}T^{-1} = \mathscr{H}''_{\mathrm{R}} \tag{40}$$

Also, the time-independent Hamiltonian $\mathscr{H}^{\mathrm{TR}}$ is diagonal in all magnetic quantum numbers:

$$\mathscr{H}^{\mathrm{TR}} = -\sum_n (\nu_I - \nu_2)I_{zn} - \mathrm{A}S_z$$

where:

$$\mathrm{A} = \left[(\nu_S - \nu_2 - \sum_n J_{IS}I_{zn})^2 + \left(\frac{\gamma_S B_2}{2\pi}\right)^2\right]^{1/2} \tag{41}$$

and now,

$$\mathscr{H}^{\mathrm{TR}}\Psi^{\mathrm{TR}} = E_{\mathrm{TR}}\Psi^{\mathrm{TR}} \tag{42}$$

with the eigenfunctions given by the product of spin functions

$$\Psi^{\mathrm{TR}} = \prod_n U(I_n, I_{zn})(S, S_z)$$

and

$$E_{\mathrm{TR}} = -\sum_n (\nu_I - \nu_2)m_{I_n} - \mathrm{A}M_S \tag{43}$$

in frequency units.

In calculating the transition frequencies one should bear in mind that the axis of quantization of S spins depends on the resultant quantum number I_z of the I spins; i.e., the set of functions $U(S, M_s)$ are not orthogonal unless they refer to the same z component of I spins. Thus, transition induced by weak rf field B_1 at the I resonance frequency changes the magnetic quantum number $\Delta M_I = \pm 1$, whereas the double rotation of the S spins about the x and y axes corresponds to $\Delta M_S = 0, \pm 2, \pm 4$, up to $\pm 2S$. When the rf power at ν_2 is high, the effective field is perpendicular to the Zeeman field. Considering the transition moment,

$$\begin{aligned} P_{m_I}, m_I + 1 &= |\langle U(I, m_I)| I_x | U(I, m_I + 1)\rangle|^2 \\ &\quad \times |\langle U(S, M_S) \exp\{(i\theta(m_I) - i\theta(m_I + 1)S_y\} U(S, M'_S)\rangle|^2 \end{aligned} \tag{44}$$

The transitions take place with the normal selection rule $\Delta m_I = \pm 1$, $\Delta M_S = 0$, but the frequency of transitions $m_I \rightarrow m_I - 1$ is at:

$$\nu_I = \nu_{m_I} - \frac{2\pi J_{IS}}{\gamma_S B_2}(\nu_S - \nu_2 - J_{IS}(m_I - 1/2))M_S \tag{45}$$

and the multiplet structure collapses for $m_I = \pm 1/2$. For $I > 1/2$ the spin couplings are reduced by a factor $2\pi J_{IS}/\gamma_S B_2$ and strong irradiation fields are required for complete decoupling.

When there is weak irradiation at ν_2 because of incomplete decoupling we can show that the progressively connected transitions are shifted to high field and split, whereas those that are regressively connected are shifted to low field and split. The splittings are sharper in the latter. The magnitude of the splitting is $(\gamma_S B_2/2\pi)$ and the multiplicity in the splitting is $(2S + 1)$. It may be noted that this is an accurate method of calibrating the decoupler rf power regardless of the irradiation offset and the spin–spin coupling. Thus, tickling experiments correspond to just weakly perturbing a given transition, only mixing basis functions corresponding to the irradiated and its connected transitions. The heteronuclear analog of spin tickling can be used to monitor hidden and weak resonances. For tickling, the strength of the irradiating power is of the order of the linewidths.

Another technique to get an idea of the homonuclear J connectivities and the relative signs of J is to sweep the spectrum with a strong irradiation field while continuously monitoring the intensity of a given transition, a technique called INDOR. INDOR can be done at a strong rf power to produce tickling effects and at a lower power corresponding to a situation leading to redistribution of populations. Using a generalized Overhauser effect one can show that the intensity is positive for progressively connected transitions and negative for regressively connected transitions. There is no strict parallel for INDOR in the FT regime, although similar results can be achieved in the so-called selective population inversion technique, SPI (Chapter 3). Both tickling and INDOR were used mainly to discern connectivities in the energy level diagrams. As discussed at length in Chapter 4, however, these methods have been superseded by two-dimensional correlation techniques.

As far as general decoupling techniques for homonuclear systems in the pulsed mode are concerned, one uses the so-called time-shared mode to avoid receiver overloading. The sharing takes place during the acquisition of the FID. The decoupler is off during the actual conversion period of the analog to digital converter (ADC) and on during the rest of the dwell time. At large sweep widths, which sets the dwell time, it is clear that decoupler power is inadequate. The decoupler on time, which is the difference between the dwell time and the ADC, is in the range 10^{-4} to 10^{-5} s. Although the switching of the decoupler generates sidebands these do not fall within the spectral range of interest; however, power gets dissipated into the sidebands and the available decoupler power is $(B_1 \tau)^2$ where $\tau \leqslant [2SW]^{-1}$. The efficiency of decoupling is poorer for large sweep widths. (See Fig. 2.15.(b).)

Scaling of J

The idea of scaling down isotropic chemical shifts in liquids to simulate "strong coupling" conditions is achieved by a multiple-pulse technique known as the "concertina effect." The system is subjected to a series of pairs of pulses whose phases are alternated. Irrespective of the flip angle it can be shown using the average Hamiltonian (see Appendix 5) theory that the chemical shifts are scaled down. In this section we shall restrict ourselves to scaling of heteronuclear spin–spin coupling. It has been mentioned previously that off-resonance decoupling can lead to scaling of coupling constants. However, this technique has the difficulty that for different chemically shifted decoupled nuclei the offsets are different, and hence the scaling down of the multiplets is nonuniform. If one can use a method that provides a fairly uniform J scaling for a reasonable spectral width, then the overlapping multiplets in the whole of the spectra "separate out" uniformly for ease of interpretation.

The method consists in subjecting the spin system to a repetitive pulse sequence that fulfils the cyclic property (Appendix 5); i.e., the pulse sequence takes the Hamiltonian back to its initial condition in the spin operator space representation. The magnetization of the system is observed synchronously with the pulsing process and appears to evolve under the average Hamiltonian, which can be easily computed using the Magnus expansion correct to any order of accuracy. Two pulse sequences have been suggested to achieve uniform J scaling. These are (a) the repetitive phase-alternated separated pulse pairs or (b) a sequence of pairs of joined phase-alternated pulses (Fig. 2.16).

The Hamiltonian for a heteronuclear system for n spins (e.g., ^1H) coupled to a single I spin (e.g., ^{13}C) is:

$$\mathscr{H}_0 = \frac{h}{2\pi}\Delta_I I_z + \sum_n \left\{ \frac{h}{2\pi}\Delta_{S_n} + hJ_{IS_n} I_z \right\} S_{nz} + \sum_{i<j} hJ_{ij}\mathbf{S}_i \cdot \mathbf{S}_j \tag{46}$$

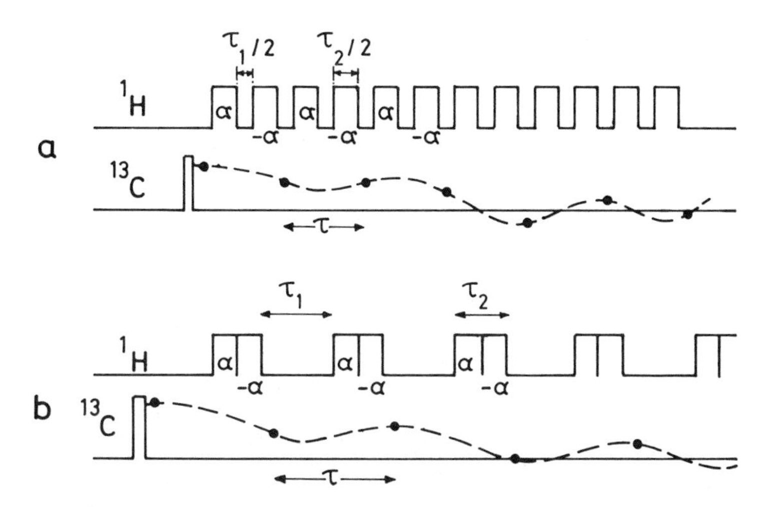

Figure 2.16. Pulse scheme for uniform J scaling. (a) Separated phase-alternant pulse train with flip angle α and $-\alpha$. (b) Same as above with joined pairs of phase-alternant pulses. Cycle time $\tau = \tau_1 + \tau_2$ in both schemes. During the FID of ^{13}C, the pulse sequence is applied synchronously with the sampling.

In the presence of the periodic pulse sequence applied to the S spins this leads to the average Hamiltonian:

$$\bar{\mathscr{H}}_0 = \Delta_I I_z + \sum_n (\Delta_{S_n} + 2\pi J_{IS_n} I_z)(aS_{nx} + bS_{ny} + cS_{nz}) + \sum_{i<j} 2\pi J_{ij} \mathbf{S}_i \cdot \mathbf{S}_j \quad (47)$$

The average $\bar{\mathscr{H}}_0$ is identical in form to the original Hamiltonian if we rewrite the above expression as

$$\bar{\mathscr{H}}_0 = \Delta_I I_z + (\mathrm{SF}) \sum_n \left\{ \Delta_{S_n} + 2\pi J_{IS_n} I_z \right\} S_{nz} + 2\pi \sum_{i<j} J_{ij} \mathbf{S}_i \cdot \mathbf{S}_j \quad (48)$$

where $\mathrm{SF} = (a^2 + b^2 + c^2)^{1/2}$ is the uniform scale factor for heteronuclear spin–spin coupling. For the two sequences given,

$$\mathrm{SF}(a) = \frac{1}{\tau_1 + \tau_2} \left[\frac{1}{2} \tau_1^2 (1 + \cos\alpha) + \frac{2\tau_2^2}{\alpha^2} (1 + \cos\alpha) + \frac{2}{\alpha} \cdot \tau_1 \tau_2 \sin\alpha \right]^{1/2} \quad (49)$$

and

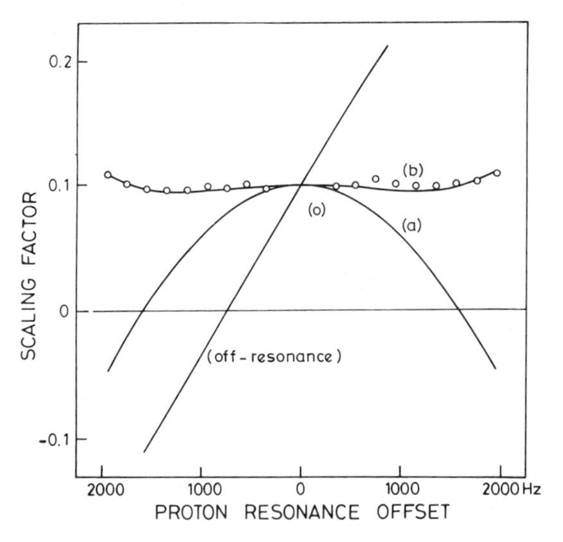

Figure 2.17. Scale factors for an AX spin-$\frac{1}{2}$ system for the pulse schemes (a) and (b) of Fig. 2.16; also included is the scale factor for off-resonance decoupling, with the offset so adjusted to give the same scale factor at the center of the diagram. For sequence (b) the optimum combination of pulse length and amplitude for getting minimum offset dependence of scaling was $\tau_2/2 = 112\ \mu s$ and $\gamma B_2/2\pi = 7500$ Hz for a cycle time of 278 μs. [Reproduced by permission. W.P. Aue and R.R. Ernst, *J. Magn. Reson.*, **31**, 533 (1978), copyright 1978, Academic Press, New York]

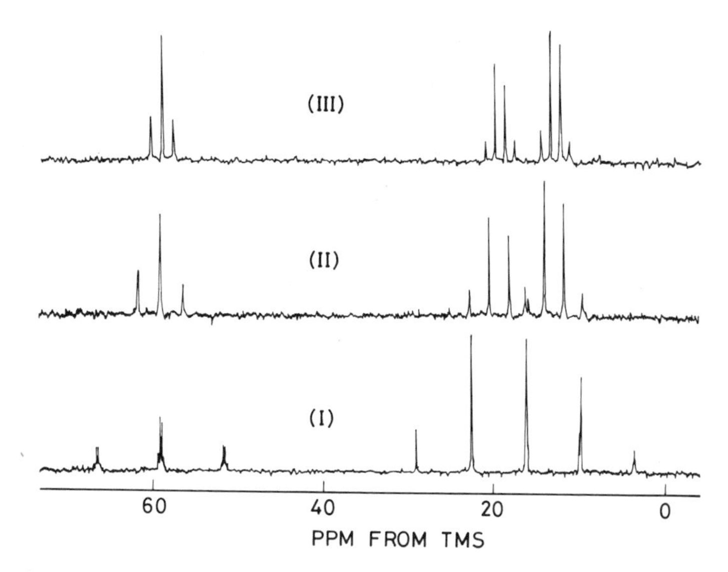

Figure 2.18. Uniform J scaling in ethyl acetate using sequence (b) of Fig. 2.16. (I) No decoupling, (II) 35.6% scaling, and (III) 17.8% scaling. The decoupler was positioned 350 Hz downfield from the CH_2 resonance. [Reproduced by permission. W.P. Aue and R.R. Ernst, *J. Magn. Reson.*, **31**, 533 (1978), copyright 1978, Academic Press, New York]

$$\mathrm{SF}(b) = \frac{1}{\tau_1 + \tau_2}\left[\tau_1^2 + 2\frac{\tau_2^2}{\alpha^2}(1 + \cos\alpha) + \frac{2}{\alpha}\tau_1\tau_2\sin\alpha\right]^{1/2} \tag{50}$$

It is also to be noted that proton chemical shifts are also scaled down by the same factor. By proper selection of the flip angle α and the periods τ_1 and τ_2, both sequences can be used to provide any arbitrary scale factor. However, the "joined" pulse pair in sequence (*b*) is superior with respect to insensitivity of scale factor to rf inhomogeneity. In fact, for strong rf fields (and hence large α) sequence (*b*) leads to scale factors independent of α and rf inhomogeneity. Also, sequence (*b*) leads to scale factors less sensitive (1.5%) to offset, whereas sequence (*a*) leads to nearly 40% changes for an offset of about 2000 Hz. These are summarized in Fig. 2.17 and illustrated by the spectra given in Fig. 2.18.

Suppression of Strong Solvent Peaks

In Fourier transform spectroscopy one is often faced with a problem of eliminating strong solvent peaks. This is especially so in trying to measure ^1H NMR of very dilute solutions of biomolecules in D_2O, where one is faced with a very strong line from residual HOD. This poses the problem of dynamic range as well as the word length of computer memory (see Appendix 3). We are concerned here with techniques that can "scale down" the large signals appearing at the digitization stage. The ideal situation is to use solvents that do not give resonances in the area of interest. However, this is not always possible and many experiments often have to be carried out in D_2O at optimal pH ranges to simulate certain biological situations. In this circumstance, there are a few techniques, some of which are outlined below, to suppress strong solvent peaks.

One method is to use the inversion–recovery sequence ($180°$–τ–$90°$–Acquire–$T_d)_n$. The idea is that the delay τ between the two pulses is such that when transverse magnetization is created, the solvent just "zero crosses", so that the solvent signal is eliminated from the time domain response. The solute molecules, however, by this time have reached the equilibrium Boltzmann

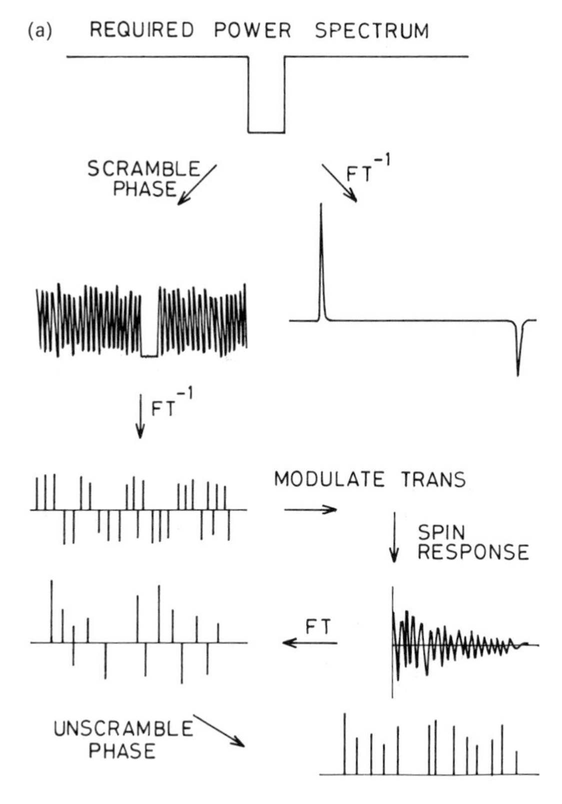

Figure 2.19. (a) Tailored excitation scheme for solvent suppression (see text for an explanation of the figure). (b) Example of suppression of the intense HOD resonances in a D_2O solution of tyrosine. a The intense HOD line in the middle with a very high S/N ratio compared to signals from the solute under normal stochastic excitation. b The results using tailored excitation where a notch was cut in the frequency power spectrum at the HOD resonance frequency. A dramatic reduction in the solvent line intensity is seen. [Reproduced by permission. L. Tomlinson and H.D.W. Hill, *J. Chem. Phys.*, **59**, 1775 (1973), American Institute of Physics, New York]

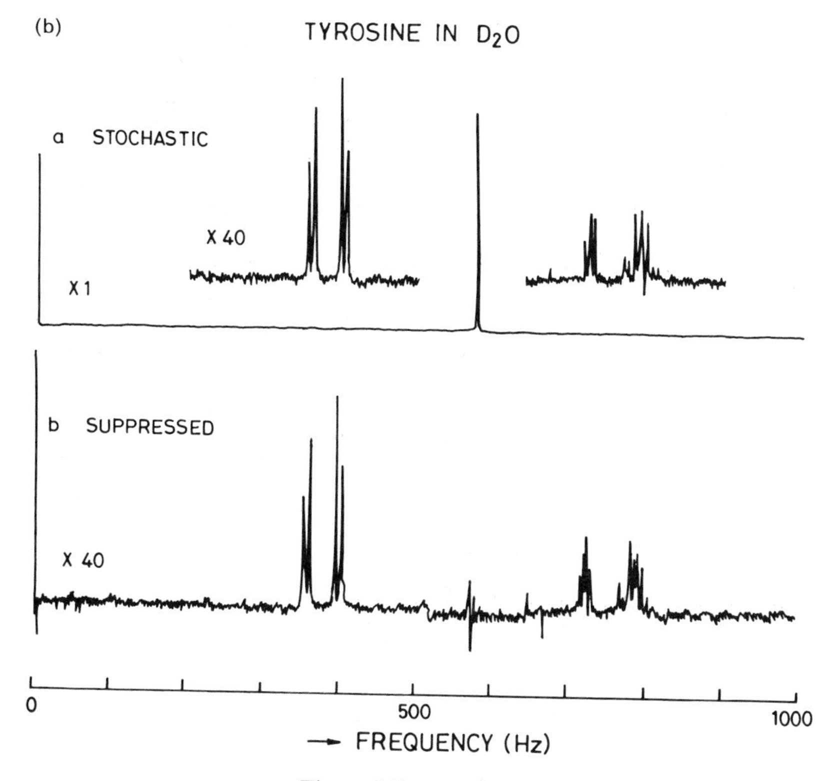

Figure 2.19. (Continued)

distribution because of their shorter spin–lattice relaxation time. An additional homogeneity–spoil pulse is applied immediately after the 180° pulse to eliminate any transverse magnetization that may be created by an imperfect 180° pulse. The sequence is repeated after five times the solvent relaxation (T_d). Unfortunately the T_1 of HOD is on the order of 20 s, it is not efficient timewise to produce the desired signal to noise ratio. To increase the efficiency one can use the so-called steady-state condition by adjusting τ and T_D (which will be the sum of T_d + Acquisition in the above scheme) so that the magnetization from solvent protons is zero immediately prior to the 90° observe pulse. The steady-state condition is:

$$\exp(-T_{1s}\ln 2 + T_D) + \exp(-(T_{1s}\ln 2 - \tau)/T_{1s}) = 1 \tag{51}$$

One should know or determine the T_{1s} of the solvent. The success of the method again relies on the longer spin–lattice relaxation of the solvent compared to that of the solute. Hence, one should rigorously avoid paramagnetic impurities or dissolved oxygen.

Alternatively, a series of closely spaced 90° pulses is applied and all spin systems are subject to saturation. The magnetization recovers with the spin–

lattice relaxation rate and, as in the $180°-\tau-90°$ technique, the method relies on the longer solvent T_1 for its success.

A third way of eliminating solvent peaks is to tailor the power spectrum of the pulse excitation such that there is a notch in the region of solvent resonance. The method consists of first defining the required power spectrum as a function of frequency (say in this case with a notch at the solvent resonance frequency).

This spectral function $f(\omega)$ is then Fourier transformed to get a time-domain function that leads to N discrete amplitudes as a function of time. This time-domain function is used to modulate a regular sequence of pulses τ_p apart so as to produce a series of N pulses $N\tau_p$ long where $N\tau_p$ is the time required to attain the desired resolution of notch. The excitation function is thus a modulated series of pulses with the height corresponding to the modulation amplitude and the phase shift $\pm 90°$ representing the sign. Once this sequence is ready this can be applied in a repeated fashion for time averaging as in normal pulse FT mode. It is possible to do tailored excitation in the stochastic mode. After defining the desired spectrum in the frequency domain the phases are scrambled at each point by a random phase generator before doing the FT into the excitation function (the use of phase scrambling reduces the dynamic range required). The phase angles introduced by random scrambling are transferred to the excitation function. If one now deconvolutes the spectra after FT by the function (corresponding to random phase angle added to make the excitation stochastic), the resulting spectrum will be normal (see Fig. 2.19). Thus, one can use specially tailored pulse sequences to produce a frequency spectrum that excites only the solute resonances and does not create transverse magnetization from solvent spins. By the same token one can also define a frequency–power spectrum with a large spike at the solvent resonance frequency, in which case the solvent resonances alone get saturated and the rest of the resonances come through in the FT spectrum.

Selective Excitation of Resonances Using Delays Alternating with Nutation for Tailored Excitation (DANTE)

In the previous section on solvent signal suppression, we described a pulsed technique for selectively exciting a few individual resonances. This method is especially useful in the analysis of ^{1}H-coupled ^{13}C spectra where, because of

Figure 2.20. (a) Pulse sequence used for the DANTE scheme. (b) Selective excitation of successive multiplets in the coupled ^{13}C spectrum of menthone showing each ^{13}C submultiplet singly at a time. The spectrum at the bottom is the fully coupled ^{13}C spectrum; on the top, the proton broadband decoupled spectrum is shown. [Reproduced with permission. G.A. Morris and R. Freeman, *J. Magn. Reson.*, **29**, 433 (1978), copyright 1978, Academic Press, New York] ▶

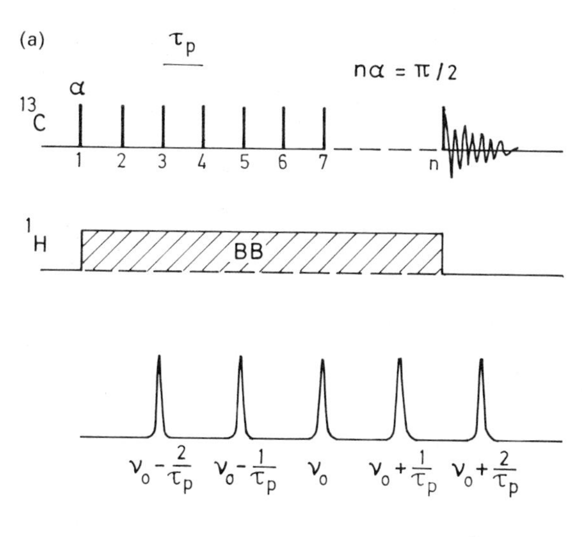
(a)
τ_p
$n\alpha = \pi/2$
α
^{13}C
1 2 3 4 5 6 7
n
^1H
BB
$\nu_0 - \frac{2}{\tau_p}$ $\nu_0 - \frac{1}{\tau_p}$ ν_0 $\nu_0 + \frac{1}{\tau_p}$ $\nu_0 + \frac{2}{\tau_p}$
$\tau_p \approx 2$ ms ; n = 150−200 ; α = 0.6 − 0.4°

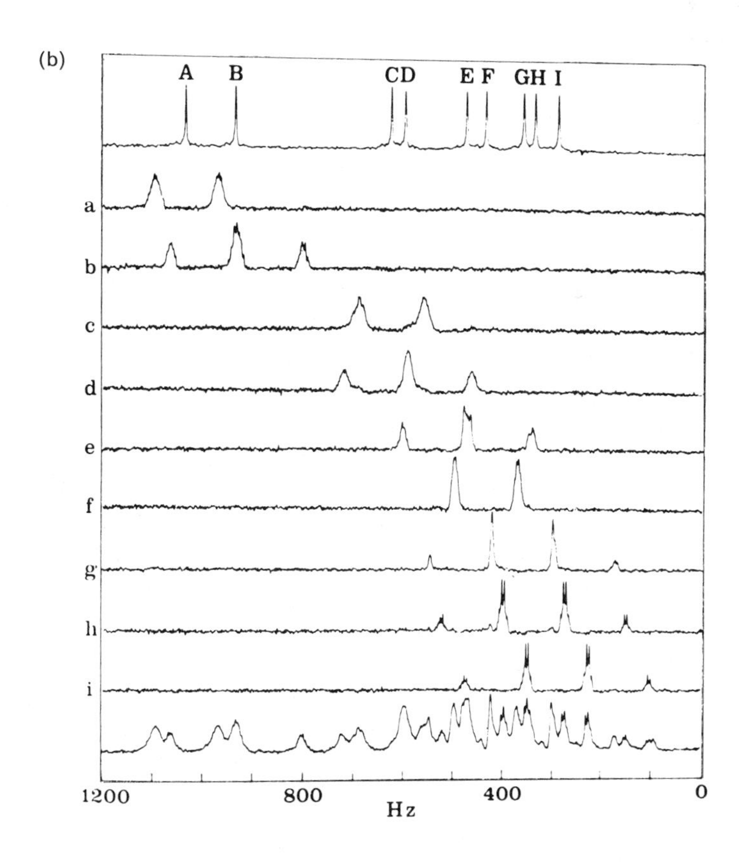
(b)
A B CD E F GH I
a
b
c
d
e
f
g
h
i
1200 800 400 0
Hz

the comparative magnitudes of ^{13}C chemical shift differences and ^{1}H–^{13}C coupling constants, spectra can at times be very complicated. In DANTE, the method consists of applying a series of very small flip angle pulses at a carrier frequency with time delays τ_p. Such an rf pulse modulation corresponds to a frequency spectrum that has sidebands around the carrier frequency ν_0, at $\nu_0 \pm n/\tau_p$, $n = 1, 2, \ldots$. Thus with a single transmitter we simulate the effect of several selective pulse transmitters. If the cumulative pulse angle for the pulse train is 90° then only nuclei that resonate at $\nu_0 \pm n/\tau_p$ are tilted into the xy plane, and the trajectories of the magnetization vectors of other nuclei not falling at the above discrete frequencies lead to no transverse components (see Fig. 2.20).

CHAPTER 3

Coherence Transfer

Introduction

The poor sensitivity of NMR spectroscopy has always been a matter of much concern to users of NMR. Sensitivity improvement therefore has been a constant preoccupation of both the NMR engineer and the NMR spectroscopist. We shall be concerned in this chapter with the spectroscopist's approach in alleviating the problem.

In the more general context, the spectroscopist has the dual concerns of improving the spectral signal to noise ratio as well as of spectral simplification. A powerful class of methods developed for both these purposes is based on coherence transfer, and this is the subject of our study here.

Signal to Noise Ratios

Following a 90° pulse along the rotating frame y axis, the detected NMR signal is proportional to:

$$\begin{aligned}\frac{dM_x}{dt} &= \frac{d}{dt}[\mathrm{Tr}(M_x \exp(-i\omega_0 I_z t)\sigma \exp(i\omega_0 I_z t))] \\ &= \frac{d}{dt}[\mathrm{Tr}(\exp(i\omega_0 I_z t) M_x \exp(-i\omega_0 I_z t)\sigma)] \\ &\sim N\gamma^3 B_0^2\end{aligned} \tag{1}$$

upon employing the necessary identities from Chapter 1 (Eqs. 1.6, 1.45, and 1.50). Here M_x is the transverse component of magnetization, N is the total number of spins, γ the magnetogyric ratio and B_0 is the strength of the dc field. The noise ($\mathcal{N}$) detected by the receiver, on the other hand, is proportional to:

$$\mathcal{N} \approx \gamma^{1/2} B_0^{1/2} \tag{2}$$

This gives for the signal to noise ratio ($\mathcal{S}/\mathcal{N}$):

$$(\mathcal{S}/\mathcal{N}) \sim N\gamma^{5/2} B_0^{3/2} \tag{3}$$

When the signals from N spin-I nuclei per unit volume at the field B_0 are compared, a $\gamma^{5/2}$ dependence clearly emerges; when the NMR signal from an ensemble of spins is computed as a function of field strength, on the other hand, a $B_0^{3/2}$ dependence is evident.

Clearly then, the key to improved sensitivity of NMR spectra is the measurement of high-γ nuclei at high fields, with large concentrations of spins per unit volume. Protons are in this sense the most favorable NMR nuclei.

Even in measuring proton NMR, however, signal averaging is frequently necessary as a primary method of improving signal to noise ratios. It is based on the fact that the NMR signal is a coherent phenomenon that repeats itself identically on successive excitation of the spin system, subject to the establishment of equilibrium or at least a steady state. The noise detected by the receiver, on the other hand, is a random quantity that changes on successive "scans" of the spectrum following repeated excitations of the spins. Upon coadding the result of N scans, therefore, the total signal equals $N\mathcal{S}$, while the noise $\mathcal{N}$ equals $\sqrt{N}\mathcal{N}$.

We therefore have for the signal to noise ratio achieved after N scans:

$$(\mathcal{S}/\mathcal{N})_N \approx \sqrt{N}(\mathcal{S}/\mathcal{N}) \tag{4}$$

For basically the same reasons, we also have for equal total times of measurement by the pulse and continuous wave (CW) methods:

$$\frac{(\mathcal{S}/\mathcal{N})_{\mathrm{Pulse}}}{(\mathcal{S}/\mathcal{N})_{\mathrm{CW}}} \leqslant \sqrt{\frac{(SW)}{(\Delta\nu)_{1/2}}}. \tag{5}$$

SW being the spectral width and $(\Delta\nu)_{1/2}$ the typical linewidth in the spectrum. The maximal improvement in signal to noise ratio for the pulsed method over the CW method is seldom realized in practice owing to the time required between scans in pulsed NMR to allow the spin system to relax back to equilibrium.

Signal averaging in the pulsed NMR mode is basic to modern NMR mearurements; this approach has made possible the study of nuclei that are far less NMR receptive than protons, but that frequently exhibit a higher figure of merit than ^1H in terms of the measurability of changes in their NMR spectra as a function of subtle changes in the molecular environment.

Polarization Transfer

It is common to have to measure NMR spectra of such low-γ nuclei as ^{13}C, ^{29}Si, and ^{15}N, many of which have low natural abundance as well. For measurements at any given value of B_0, it would be an advantage, clearly, if the polarization of a high-γ/high-abundance nucleus, such as ^1H, could be

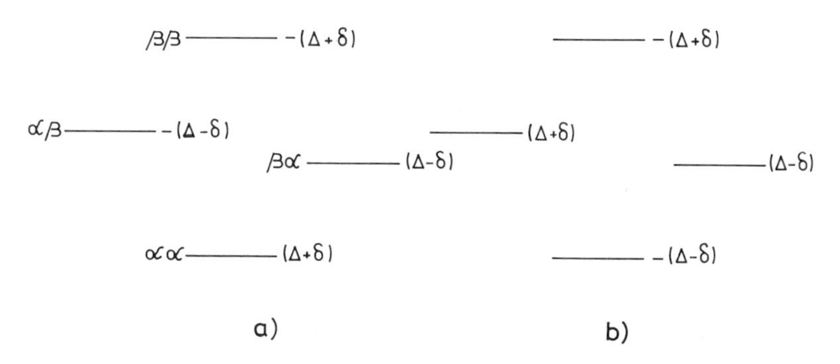

Figure 3.1. Relative populations in a two-spin-$\frac{1}{2}$ system, Δ and δ being the respective Boltzmann factors, (a) in thermal equilibrium, (b) following inversion of the $\alpha\alpha \leftrightarrow \alpha\beta$ transition.

transferred to the less receptive nucleus under observation. An important strategy to effect such polarization transfer involves selective population inversion in coupled spin systems.

Consider, for example, a weakly coupled AX spin system composed of spin-1/2 particles. The A and X resonances are both doublets owing to the spin–spin coupling, J_{AX}. Provided the coupling is resolvable, one may selectively invert one of the two X transitions, for example. Referring to Fig. 3.1, which sets out the relative populations of the four levels, it is clear that after such a selective population inversion across a transition, the population differences across the transitions of the coupled A spin are not given by their equilibrium values. In fact, the new population differences across the A transitions are the original values plus (or minus) those across the X transitions. In a ^{13}C–^{1}H spin system, for instance, the ratio $\gamma(^{1}H)/\gamma(^{13}C)$ is about 4; the equilibrium population differences across the ^{13}C transitions are (1, 1), whereas they become $(-3, 5)$ after one of the coupled proton transitions is selectively inverted. In a ^{15}N–^{1}H system, in fact, the changes are even more dramatic: the $\gamma(^{1}H)/\gamma(^{15}N)$ ratio is about 10, and a (1, 1) ^{15}N doublet becomes a $(-9, 11)$ doublet.

The selective population inversion (SPI) experiment therefore leads to anti-phase multiplets with an asymmetry in the absolute intensities as well, but to a signal enhancement by a factor $\gamma_{inv}/\gamma_{obs}$, i.e., the ratio of the magnetogyric ratios of the inverted and the observed spins. Experimentally one is required to irradiate by a long, weak pulse one of the lines in the proton satellite spectrum to achieve its inversion, corresponding to the application of a selective 180° pulse; this is then followed by giving a "read" pulse, e.g., a 90° pulse, to the observed spins and acquiring their signal without decoupling. Improvement of sensitivity across an entire spectrum would clearly be a laborious task by this SPI process, and this circumstance has led to the development of a method of achieving SPI that is nonselective to the chemical shift spectrum.

INEPT Class of Sequences

Insensitive nuclei enhanced by polarization transfer, facetiously given the acronym INEPT, is a five-pulse experiment that achieves the required non-selectivity; Fig. 3.2 gives the pulse sequence.

The effect of the "simultaneous" pair of π pulses on the two spins is to refocus the chemical shifts of the polarization source spins (usually protons), while retaining their coupling to the observed spins, as has been discussed in detail in Chapter 2. It is by the use of hard pulses and this chemical shift refocusing strategy that INEPT works as a chemical shift-insensitive SPI.

It is instructive to trace the evolution of the spin system when subjected to such a pulse sequence. We shall consider a heteronuclear two-spin-1/2 system and calculate the density operator at the various instants of time marked in Fig. 3.2.

$$\sigma_0 = I_z + \alpha S_z \tag{6}$$

in the initial state of thermal equilibrium;

$$\sigma_1 = -I_y + \alpha S_z \tag{7}$$

following the $(\pi/2)_x$ pulse on the I spins;

$$\sigma_2 = -c_1 I_y + 2s_1 I_x S_z + \alpha S_z \tag{8}$$

following free evolution for τ_1 seconds under weak coupling. The second term of σ_2 represents a J-ordered state of the I spins.

$$\sigma_3 = -c_1 I_y + 2s_1 I_z S_y - \alpha S_y \tag{9}$$

following the final 90° pulse pair. The corresponding vector diagrams are given in Fig. 3.3. In the above expressions, c_1 denotes $\cos \pi J \tau_1$ and s_1 denotes $\sin \pi J \tau_1$; α is the ratio $\gamma(^{13}C)/\gamma(^{1}H)$. If τ_1 is selected to be $(2J)^{-1}$,

$$\sigma_3 = (2I_z - \alpha)S_y \tag{10}$$

The first term in this expression gives rise to an anti-phase doublet $1/2(-1, 1)$ in accordance with the interpretation discussed in Chapter 1. The second term, which originates in the natural polarization of the X nuclei, leads to the asymmetry in the anti-phase doublet discussed above. It can be suppressed by subtracting a second FID acquired under identical conditions except that the phase of either one of the 90° pulses on the I spins is shifted 180°, changing the sign of the first term in σ_3 but leaving the second invariant. Subtraction now reinforces the first term, i.e., the anti-phase doublet, and cancels the second term, i.e., the contribution of the natural polarization of the X spins. This is an example of a recurring feature in modern NMR experiments: undesired signals are suppressed and others reinforced by combining FID's resulting from pulse sequences that are identical but for the change in phase of one or a group of pulses. Such "phase cycling" selects a specific coherence transfer pathway while discriminating against other pathways that are simultaneously operative in generating each FID signal.

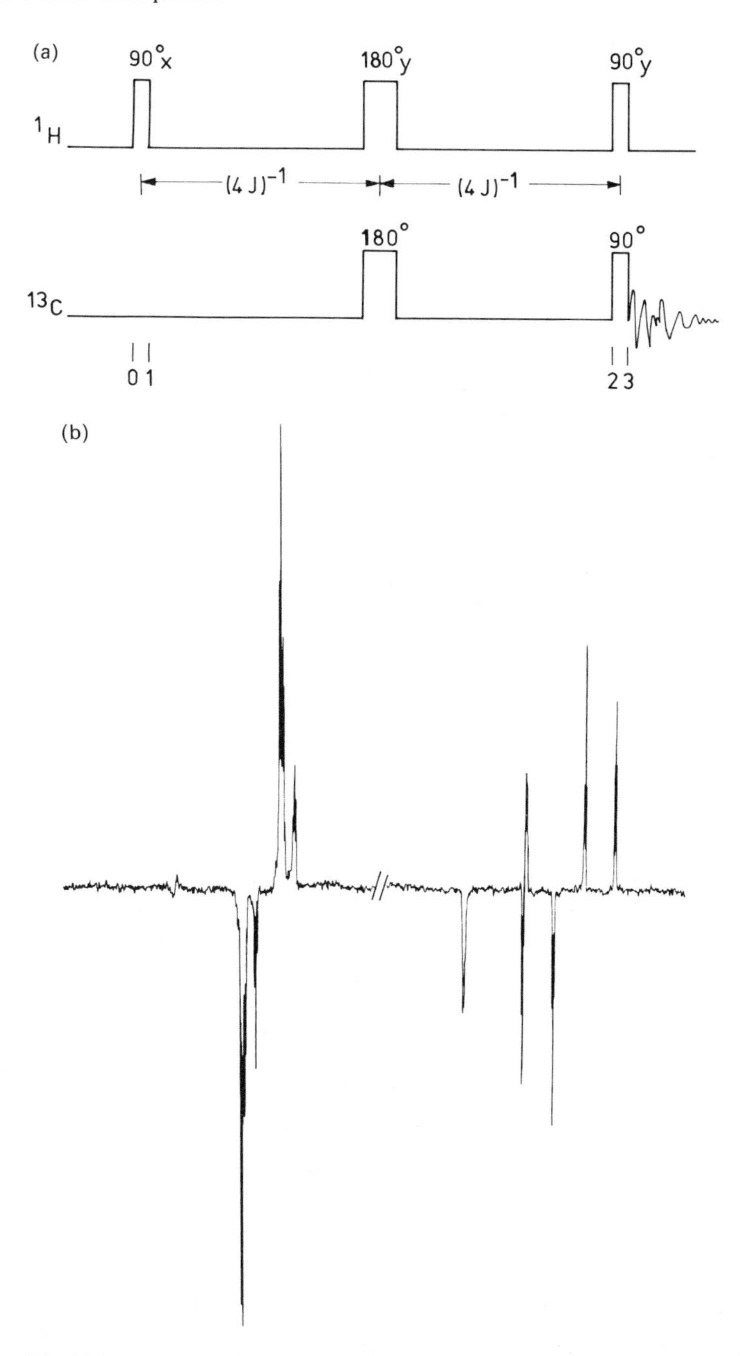

Figure 3.2. (a) INEPT pulse sequence. The duration between points 1 and 2 is denoted τ_1 in Eq. (8). (b) A 22.64-MHz ^{13}C INEPT spectrum of ethyl benzene. Note the occurrence of multiplet anomalies and the absence of phase anomalies. $\tau_1 = 3.846$ ms; $^1J_{CH}(\mathrm{al}) \simeq 126$ Hz, $^1J_{CH}(\mathrm{ar}) \simeq 160$Hz.

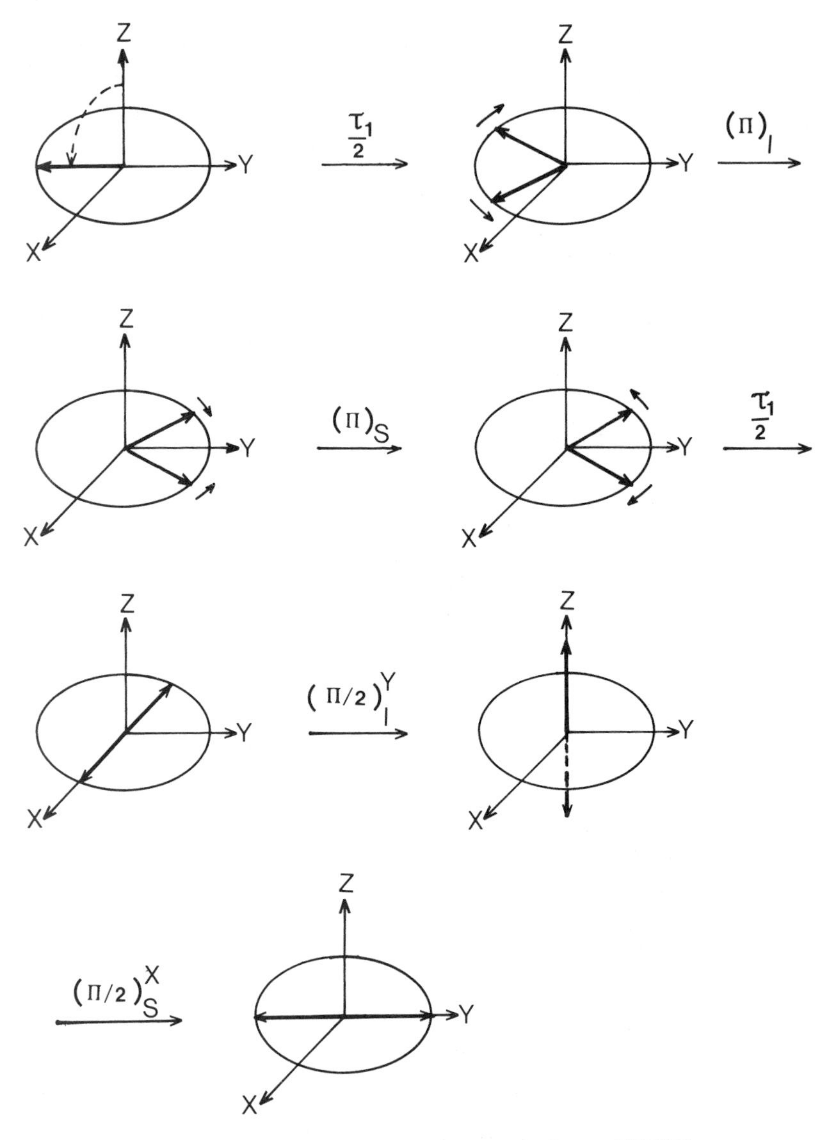

Figure 3.3. Vector picture of the sequence of events during an INEPT sequence.

In accordance with the discussion in Chapter 1, the anti-phase magnetization term $2I_zS_y$ is observable only if the multiplet structure is resolved; if the splitting due to J is made to vanish by decoupling during acquisition, in particular, zero signal results.

For an I_NS system, disregarding the natural S polarization, INEPT with phase cycling gives:

$$\begin{aligned}
\sigma_0 &= \sum_i I_{iz} \\
\sigma_1 &= -\sum_i I_{iy} \\
\sigma_2 &= -c_1 \sum_i I_{iy} + 2s_1 S_z \sum_i I_{ix} \\
\sigma_3 &= -c_1 \sum_i I_{iy} + 2s_1 \sum_i I_{iz} S_y
\end{aligned} \tag{11}$$

The S signal arises from the second term of σ_3. Following the results in Chapter 1, INEPT clearly gives spectra that do not involve any phase anomalies; i.e., there is no admixture of absorption and dispersion. In accordance with the results collected in Table 1.1, the multiplet patterns for the I_2S and I_3S systems are evidently $1/2(-1, 0, 1)$ and $1/8(-3, -3, 3, 3)$. Note especially that the central component of the triplet is missing. These INEPT multiplets are to be compared with the natural multiplets, which are, respectively, $1/4(1, 2, 1)$ and $1/8(1, 3, 3, 1)$. Signal enhancement factors for IS, I_2S, and I_3S systems are, respectively, 1,1 and 1.5 in units of $\gamma_{\text{inv}}/\gamma_{\text{obs}}$, taking into account the absolute intensities. In summary, the INEPT sequence gives anti-phase multiplets that exhibit multiplet anomalies and do not permit decoupled acquisition. There is no *net* polarization transfer (PT) *across* the multiplet. Notice also that the extent of PT depends on the factor $2 \sin \pi J \tau_1$: no single choice of τ_1 can insure optimal PT across the entire spectrum, because of variations in the value of J. This, together with the inherent unequal enhancement of different I_NS spin systems, constitutes the so-called intensity anomaly.

Should one desire to get in-phase multiplets that permit decoupled acquisition, further evolution under spin–spin coupling for a period τ_2 seconds prior to signal acquisition at the end of the five-pulse sequence suggests itself as the appropriate strategy. The modified sequence is displayed in Fig. 3.4 and insures that chemical shift evolution does not occur during τ_2. The refocusing of multiplet components may be pictured as in Fig. 3.5 and is described by the equation of motion of S_yI_z under weak coupling (the last member of Eq. 1.62). The S signal is now of the form:

$$\begin{aligned}
\sigma_{IS} &= s_1(2c_J I_z S_y - s_J S_x) \\
\sigma_{I_2S} &= s_1(2c_{2J}(I_{1z} + I_{2z})S_y - s_{2J}(1 + 4I_{1z}I_{2z})S_x) \\
\sigma_{I_3S} &= s_1[\tfrac{1}{2}(3c_{3J} + c_J)(I_{1z} + I_{2z} + I_{3z})S_y \\
&\quad + 6(c_{3J} - c_J)I_{1z}I_{2z}I_{3z}S_y \\
&\quad - \tfrac{3}{4}(s_{3J} + s_J)S_x \\
&\quad - (3s_{3J} - s_J)(I_{1z}I_{2z} + I_{1z}I_{3z} + I_{2z}I_{3z})S_x]
\end{aligned} \tag{12}$$

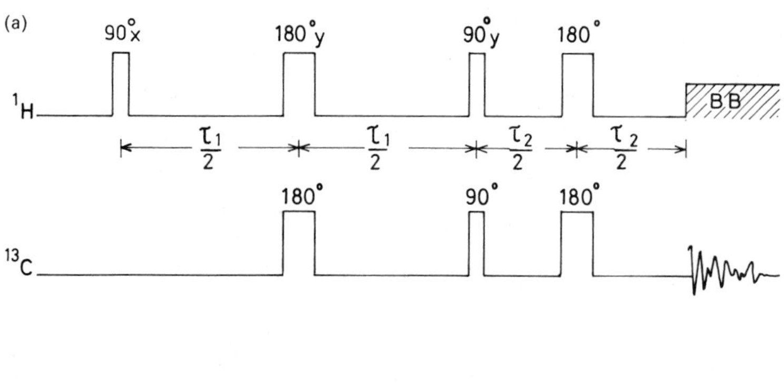

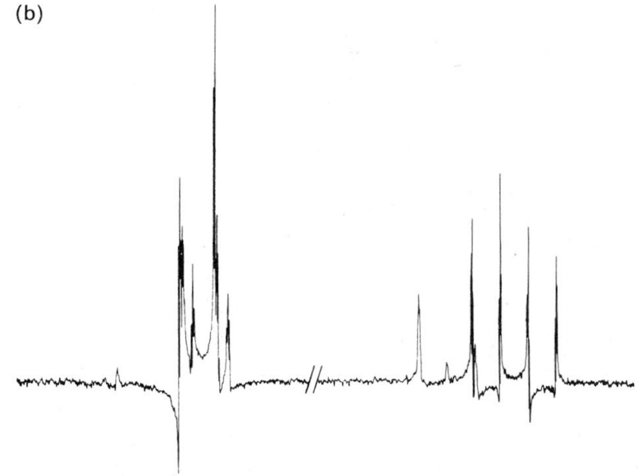

Figure 3.4. (a) A RINEPT pulse sequence (decoupling optional). (b) A 22.64-MHz ^{13}C RINEPT spectrum of ethyl benzene; note the occurrence of phase anomalies ($\tau_1 = 3.846$ ms, $\tau_2 = 1.924$ ms).

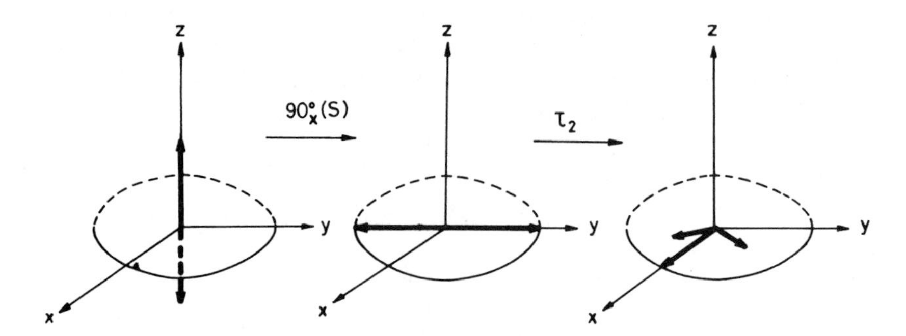

Figure 3.5. Vector picture of refocusing under J during τ_2 in the RINEPT sequence. A $(-1, -1, 1, 1)$ quartet refocuses as shown for $\tau_2 = 1/(6J)$; the phase anomaly may be noticed, the inner components having excursed by $\pm\pi/6$ radians during this period, whereas the outer components have refocused along the x axis.

In the above expressions, $c_{nJ} = \cos n\pi J\tau_2$ and $s_{nJ} = \sin n\pi J\tau_2$; terms corresponding to I-spin magnetization remaining after the PT step have been omitted. To allow maximal refocusing, the s_{nJ} terms should dominate the c_{nJ}'s. For the IS system, this implies $\tau_1 = \tau_2$, whereas for I_2S, $\tau_1 = 2\tau_2$. From the expressions given above for the form of the acquired S signal, it is clear that the refocused INEPT experiment, RINEPT, suffers in general from intensity, phase, and multiplet anomalies; phase anomalies occur owing to incomplete refocusing and cannot be altogether avoided by a suitable choice of τ_2, except for IS and I_2S systems; the multiplet anomalies propagate themselves from the INEPT stage. Decoupling during acquisition, however, gets rid of both phase and multiplet anomalies, because these anomaly terms involve exclusively anti-phase magnetizations.

Because the refocusing time for I_NS systems depends on N, we have a means of distinguishing between various I_NS spin systems and recording partial spectra with signals from groups with a fixed value of N alone. In other words, spectral editing may be achieved by using τ_2 as the parameter of discrimination. For example, with the choice $\tau_2 = (2J)^{-1}$, followed by acquisition with decoupling, only IS spin systems appear in the spectrum. With $\tau_2 = (4J)^{-1}$, on the other hand, IS, I_2S, and I_3S spin systems all appear in the spectrum with the same phase, although the IS and I_3S signals are attenuated to 0.707 of their normal intensities. When $\tau_2 = 3(4J)^{-1}$, the IS and I_3S signals occur with the same phase and attenuation characteristics as for $\tau_2 = (4J)^{-1}$, whereas I_2S signals are inverted, without any attenuation. Suitable linear combinations of RINEPT spectra with decoupling for these three values of τ_2 therefore lead to subspectra of IS, I_2S, and I_3S spin systems, respectively. However, the spectra obtained by such editing suffer in practice from "crosstalk" of one type of I_NS spin system into the subspectrum of another, because of the spread in J values that is encountered in practice in molecular species, rendering it impossible to select a universal value of τ_2.

The expressions for RINEPT signals for various I_NS systems given above suggest a means of getting rid of the anti-phase magnetization terms that lead to phase and multiplet anomalies in I-coupled RINEPT spectra. The application of a 90° pulse to the I spins just prior to acquisition has the desired effect, because this pulse converts the antiphase magnetizations to unobservable multiple-quantum coherences. This improved sequence, which allows one to record coupled polarization transfer spectra with the desired sensitivity improvement and also retains the natural multiplet patterns, is called INEPT$^+$ and is represented in Fig. 3.6. It must be noted, however, that in the process of "cleaning up" the coupled spectrum, INEPT$^+$ leads in general to a loss of enhancement: for the IS, I_2S, and I_3S systems enhancement factors are 1,1 and ≈ 1.15 repectively, in units of $\gamma_{\mathrm{inv}}/\gamma_{\mathrm{obs}}$. In comparison with the INEPT situation, there is an approximately 23% loss in enhancement with INEPT$^+$ for the I_3S system, with optimal $\tau_2 \simeq 1/5J$. The form of the acquired S signal with INEPT$^+$ is given in general by:

$$\sigma_{I_NS} = -Ns_1 s_J c_J^{N-1} S_x \tag{13}$$

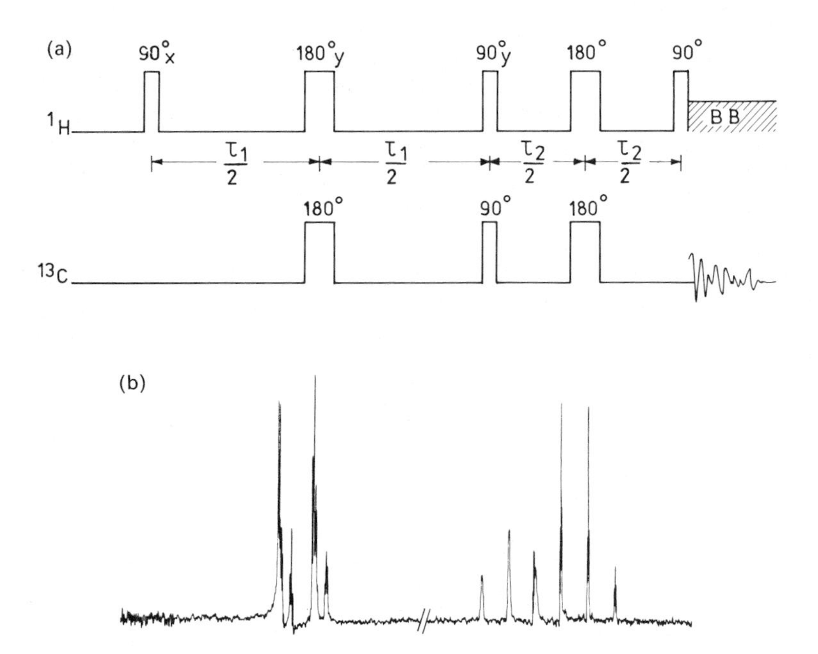

Figure 3.6. (a) An INEPT$^+$ pulse sequence (decoupling optional). (b) A 22.64-MHz ^{13}C INEPT$^+$ spectrum of ethyl benzene; $\tau_1 = 3.846$ ms, $\tau_2 = 1.924$ ms. Note the restoration of natural multiplets, although τ_2 has been set well off the optimum.

INEPT-Type Polarization Transfer Involving Nuclei With Spin >1/2. The idea of polarization transfer between a pair of spins by an SPI-type process is a general one that requires only the existence of resolved couplings between the pair. In particular, the process should work whatever be the value of the spin quantum number of the nuclei involved. Apart from PT from ^{1}H to spin $>1/2$ nuclei, there is a peculiar situation that has made PT from spin 1 to spin 1/2 worthwhile. This is the case of PT from ^{2}H to ^{13}C. This experiment is performed not so much for sensitivity improvement of ^{13}C spectra, but as a substitute for ^{13}C isotope labeling by taking recourse instead to ^{2}H labeling, which is performed chemically far more readily. By transferring polarization from ^{2}H to the attached ^{13}C in a normal INEPT-type experiment, one is effectively looking at the spectrum of only those carbons in the molecule that are bonded to ^{2}H, i.e., carbon atoms at specific sites in the molecule that have been selectively deuterated. The ratio $\gamma(^2\text{H})/\gamma(^{13}\text{C})$ is only about 0.6; however, signal averaging with an INEPT-type experiment demands delays between scans that are determined by the T_1 of the polarization *source* nuclei and *not* that of the observed nuclei; in the case of ^{2}H to ^{13}C PT, this factor is especially favourable because ^{2}H is a quadrupole nucleus and typically has short T_1's: this permits, typically, 10 times the number of scans per unit time in the ^{13}C–[^{2}H] case, as compared to the ^{13}C–[^{1}H] situation. In practice, ^{13}C spectra involving PT from ^{2}H achieve signal to noise ratios that are comparable to proton-decoupled ^{13}C spectra including NOE, in the same measurement time.

The form of the signal of the spin-1/2 nucleus when polarization is transferred to it from spin-1 nuclei by the INEPT pulse sequence is given by:

$$\sigma_{I_N S} = 2s_1 \sum_i I_{iz} S_y \tag{14}$$

This expression is identical to the corresponding expression for spin-1/2 to spin-1/2 PT. This is so because in the context of PT, only evolution under spin–spin coupling is a function of the spin quantum number of the nuclei involved, whereas evolution under rf pulses and offsets/chemical shifts is not (see Chapter 1). In the INEPT sequence, there is only one time period, τ_1, which involves evolution of transverse magnetization under spin–spin coupling. This evolution, however, is a function of the spin quantum number of the coupled nucleus only and *does not* depend on the spin quantum number of the nuclei whose transverse magnetization is subject to such evolution. As long as the evolution is under coupling to a spin-1/2 nucleus, therefore, it does not matter at all what the spin quantum number is of the polarization source nuclei whose transverse magnetization evolves under spin–spin coupling during τ_1 in the INEPT sequence (see Chapter 1). Hence the identity of expressions for INEPT to a spin-1/2 nucleus in $I_N S$ systems, no matter what the spin of the source nuclei I may be.

For ^2H to ^{13}C INEPT the enhancement factors are therefore 4/3, 16/9, and 20/9 for C^2H, C^2H_2, and C^2H_3 systems, respectively, in units of $\gamma(^2H)/\gamma(^{13}C)$ (see Table 1.1). It may be noted that these factors are all higher than for the corresponding $I_N S$ systems with $I = S = 1/2$.

The RINEPT strategy once again leads to the possibility of acquisition with decoupling. The S spin transverse magnetization evolves during τ_2 under coupling to spin-1 nuclei, however, and this evolution is of course different from that in the case with $I = S = 1/2$: the evolution frequencies are 0, $\pm J$, ..., $\pm NJ$. For the IS system, the multiplet anomaly persists with RINEPT, with the central line of the triplet missing; phase anomalies appear with misset τ_2. With the I_2S system, however, where the S multiplet exhibits two numerically different, nonzero frequencies, RINEPT exhibits phase anomalies regardless of the choice of τ_2, and of course the multiplet anomalies of INEPT persist.

The form of the RINEPT signals for IS and I_2S systems is given explicitly by:

$$\begin{aligned} \sigma_{IS} &= 2s_1(c_{2J} I_z S_y - s_{2J} I_z^2 S_x) \\ \sigma_{I_2S} &= 2s_1[c_{2J}(I_{1z} + I_{2z}) + (c_{4J} - c_{2J}) I_{1z} I_{2z}(I_{1z} + I_{2z})] S_y \\ &\quad - 2s_1[s_{4J} I_{1z} I_{2z} + s_{2J}(I_{1z}^2 + I_{2z}^2) + 2s_{2J}(c_{2J} - 1) I_{1z}^2 I_{2z}^2] S_x \end{aligned} \tag{15}$$

The idea of purging RINEPT spectra by an INEPT$^+$ strategy is only partially successful, because the 90° purge pulse on the I spins just prior to acquisition of the S signal suppresses phase anomalies, while multiplet anomalies persist.

For example:

$$\begin{aligned}\sigma_{IS} &= -2s_1 s_{2J} I_y^2 S_x \\ \sigma_{I_2S} &= -2s_1 s_{2J}[(I_{1y}^2 + I_{2y}^2) + 2(c_{2J} - 1)I_{1y}^2 I_{2y}^2]S_x\end{aligned} \tag{16}$$

In cleaning up the phases in the I_2S system, however INEPT$^+$ leads to a loss of signal enhancement. The multiplet patterns corresponding to the operators in the above expressions are listed in Table 1.1.

In contrast to PT from spin-1 to spin-1/2 nuclei, PT from spin-1/2 to spin-1 nuclei proceeds as for spin-1/2 to spin-1/2 PT with the INEPT, RINEPT, and INEPT$^+$ sequences, except that the creation of the J-ordered state of the source spin is achieved in half the time because the precession frequencies under coupling to spin-1 nuclei are 0, $\pm J$.

It may be mentioned here that in order to avoid loss in enhancement arising from imperfections in the π pulses during τ_2, one may avoid the π pulse pair during τ_2 altogether. Instead, one may resort to decoupling delayed by τ_2 seconds after commencement of acquisition, which in turn is triggered immediately after the $\pi/2$ pulse on the observe spins.

Sinept. The INEPT class of experiments involves at least two 90° pulses on the polarization source ("decoupler") channel; it is to be noted that the first two 90° pulses are in fact in quadrature. This requirement arises from the suppression of decoupler channel chemical shifts, which is necessary to make the PT chemical shift insensitive. Spectrometers that are not equipped with a phase shifter on the decoupler channel, however, can still be employed to gain at least partially the benefits of PT, albeit with a partial sacrifice of the chemical shift insensitivity. A pulse sequence of this kind, including refocusing to enable decoupling, is given in Fig. 3.7. The sequence gives rise to PT at stage 3, which is proportional to $-2\sum_i I_{iz}S_y \sin(\Delta_H\tau_1)\sin(\pi J\tau_1)$ and is termed SINEPT because of its sine dependence on the proton chemical shift, Δ_H radians per second. Optimal τ_1 is once again $(2J)^{-1}$, and PT is maximal for $\nu_H = (J/2)(2n + 1)$, $n = 0, \pm 1, \pm 2, \ldots$. Acquiring scans alternately with and without an initial π pulse on the observe spins achieves the phase cycling

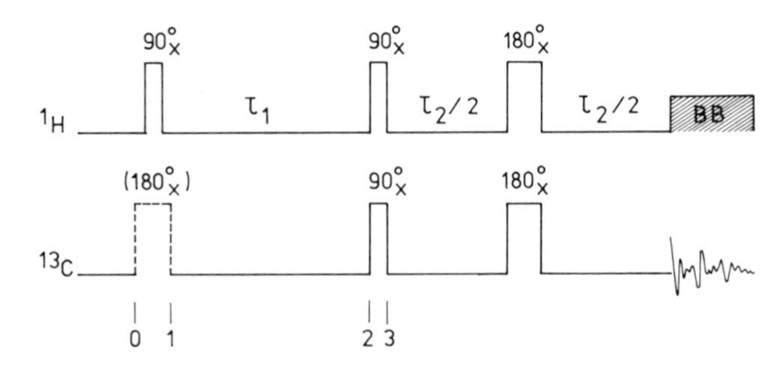

Figure 3.7. A SINEPT pulse sequence with refocusing (decoupling optional).

necessary to suppress signals from the natural polarization of the observe spins.

It may be noted also that if the S spin signals are acquired as a function of τ_1, the set of acquired S signals, that carry the I spin chemical shift modulation information as $\sin(\Delta_1 \tau_1)$ may be Fourier transformed with respect to τ_1 to extract the I spin chemical shifts. Note that it has been possible thus to record the I spin spectrum (including I–I couplings) with a receiver tuned for the Larmor frequency of the S spins! Indeed, this forms the basis of two-dimensional (2D) heteronuclear (shift) correlation spectroscopy, wherein the acquired signals are Fourier transformed once with respect to the acquisition time parameter and a second time with respect to τ_1. In this way the sensitivity enhanced S spectrum is obtained along one of the axes and the I spectrum along the other. The details of this class of 2D correlation experiments are discussed in Chapter 4, which deals with two-dimensional NMR methods.

Decoupler Calibration

There is a question of some practical importance that we must address next in the context of such PT experiments. This involves the need to calibrate pulse flip angles on the polarization source channel, which is the decoupler channel. It turns out that this is most conveniently performed by monitoring the sink spin signal while the flip angle on the decoupler channel is varied, employing the simple pulse sequence shown in Fig. 3.8.

Transverse magnetization of the observed spins is created by a $\pi/2$ pulse, preceded if necessary by broadband decoupling to establish NOE's. A θ pulse is applied on the decoupler channel after free evolution for a period τ and the observed spin signal is then acquired without decoupling. When protons are the "decoupled" nuclei, for example, zero signal results for $\tau = (2J)^{-1}$ and $\theta = \pi/2$, whereas an antiphase multiplet results for this choice of τ with $\theta \neq \pi/2$. An accurate null method is thus available for decoupler calibration. In order to elucidate the way the sequence works and also to characterize the

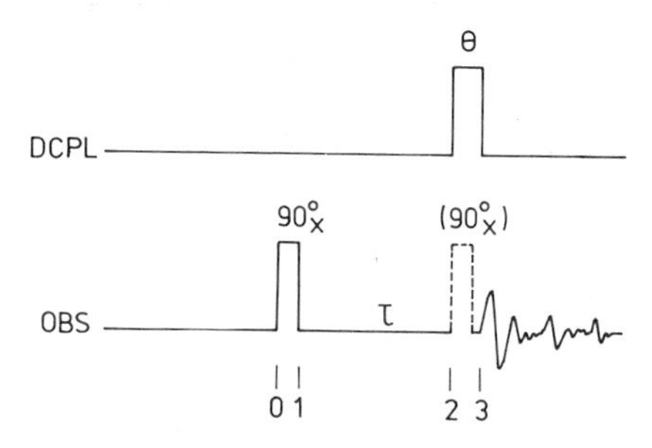

Figure 3.8. Pulse sequence for decoupler calibration. The second 90°_x pulse on the observe channel is to be employed only if the decoupler channel nucleus has spin 1. Noise decoupling prior to the first 90°_x pulse may be performed to take advantage of the NOE.

effect of the θ pulse, which has wide applications in PT and spectral editing, this simple pulse sequence is analyzed now in explicit detail.

To start with, we shall deal with the situation where both nuclei involved have spin 1/2. In this case, the density operator at strategic points in the sequence may be discussed as follows. At thermal equilibrium, ignoring the I-spin term, which does not contribute to the observed S multiplet,

$$\sigma_0 = S_z$$

Following the 90^0_x pulse on the S spins,

$$\sigma_1 = -S_y \tag{17}$$

After free evolution during τ, with the S spins on resonance:

$$\begin{aligned}
\sigma_{2,IS} &= -c_J S_y + 2s_J I_z S_x \\
\sigma_{2,I_2S} &= -c_J^2 S_y + 2c_J s_J (I_{1z} + I_{2z}) S_x + 4s_J^2 I_{1z} I_{2z} S_y \\
\sigma_{2,I_3S} &= -c_J^3 S_y + 2c_J^2 s_J \operatorname{Sym}(I_{iz}) S_x + 4c_J s_J^2 \operatorname{Sym}(I_{iz} I_{jz}) S_y \\
&\quad - 8s_J^3 S_x \prod_i I_{iz}
\end{aligned} \tag{18}$$

with:

$$c_J = \cos \pi J\tau \text{ and } s_J = \sin \pi J\tau.$$

For the definition of Sym, see footnote of Table 1.1B.

Following the θ_x pulse on the I spins,

$$\begin{aligned}
\sigma_{3,IS} &= -c_J S_y + 2s_J S_x (I_z c_\theta - I_y s_\theta) \\
\sigma_{3,I_2S} &= -c_J^2 S_y + 2c_J s_J S_x [(I_{1z} + I_{2z}) c_\theta - (I_{1y} + I_{2y}) s_\theta] \\
&\quad + 4s_J^2 S_y (I_{1z} c_\theta - I_{1y} s_\theta)(I_{2z} c_\theta - I_{2y} s_\theta) \\
\sigma_{3,I_3S} &= -c_J^3 S_y + 2c_J^2 s_J \operatorname{Sym}(I_{iz} c_\theta - I_{iy} s_\theta) S_x \\
&\quad + 4c_J s_J^2 \operatorname{Sym}[(I_{iz} c_\theta - I_{iy} s_\theta)(I_{jz} c_\theta - I_{jy} s_\theta)] S_y \\
&\quad - 8s_J^3 S_x \prod_i (I_{iz} c_\theta - I_{iy} s_\theta)
\end{aligned} \tag{19}$$

with $c_\theta = \cos\theta$ and $s_\theta = \sin\theta$.

The observable part of σ_3, acquiring the S signal, is given by:

$$\begin{aligned}
\sigma_{3,IS,\text{obs}} &= -c_J S_y + 2s_J c_\theta I_z S_x \\
\sigma_{3,I_2S,\text{obs}} &= -c_J^2 S_y + 2c_J s_J c_\theta (I_{1z} + I_{2z}) S_x + 4s_J^2 c_\theta^2 I_{1z} I_{2z} S_y \\
\sigma_{3,I_3S,\text{obs}} &= -c_J^3 S_y + 2c_J^2 s_J c_\theta \operatorname{Sym}(I_{iz}) S_x + 4c_J s_J^2 c_\theta^2 \operatorname{Sym}(I_{iz} I_{jz}) S_y \\
&\quad - 8s_J^3 c_\theta^3 S_x \prod_i I_{iz}
\end{aligned} \tag{20}$$

Terms describing multiple-quantum coherences, such as $s_\theta^2 \operatorname{Sym}(I_{iy} I_{jy}) S_y$ have been omitted in the above because they are unobservable. If τ is set to $(2J)^{-1}$, we have:

$$\sigma_{3,IS,\text{obs}} = 2c_\theta I_z S_x$$
$$\sigma_{3,I_2S,\text{obs}} = 4c_\theta^2 I_{1z} I_{2z} S_y \tag{21}$$
$$\sigma_{3,I_3S,\text{obs}} = -8c_\theta^3 I_{1z} I_{2z} I_{3z} S_x$$

From the above expressions, it is clear that the observed S signal is an anti-phase multiplet, with intensity patterns as given in Table 1.1: $1/2(-1, 1)$, $(1/4, -1/2, 1/4)$, and $(+1/8, -3/8, +3/8, -1/8)$, respectively, for IS, I_2S, and I_3S spin systems. It is to be noted that no PT is involved, and the total absolute intensities in the anti-phase multiplet equal the equilibrium values. Note also that there are no phase anomalies when τ is optimally set. Clearly, all the multiplets vanish when θ is set to 90°. This simple feature is employed to calibrate decoupler channel flip angles, as indicated above. The sequence creates unobservable multiple quantum coherences (MQC's) and "signed" magnetization leading to anti-phase multiplets (the S magnetization is labeled or signed with the I spin state) in various proportions depending on the value of θ. For the optimal choice of $\tau = (2J)^{-1}$, for instance, all magnetization in the I_NS system is trapped as unobservable MQC's for certain values of θ. The sequence therefore functions as a primitive multiple-quantum trap. Note the $\cos^N \theta$ dependence of the signals for I_NS systems, under optimal conditions.

Can the sequence in Fig. 3.8 also be employed to calibrate flip angles for spin-1 decouplers, e.g., for ^{2}H to ^{13}C PT experiments? It turns out that this is possible but induces complicated multiplet patterns that exhibit phase anomalies and do not vanish identically for any one setting of θ, for any I_NS system. This circumstance may be readily appreciated by considering the IS system, with the S spins on resonance. In this case, we find:

$$\sigma_{3,\text{obs}} = -S_y - (c_{2J} - 1)S_y(c_\theta^2 I_z^2 + s_\theta^2 I_y^2) + s_{2J} c_\theta I_z S_x \tag{22}$$

If at this point a second 90_x° pulse is given to the S spins "simultaneously" with the θ_x pulse on the I spins, we have the simpler expression:

$$\sigma_{4,\text{obs}} = s_{2J} c_\theta I_z S_x \tag{23}$$

which is of the same form as for the case treated earlier with $I = 1/2$. Optimal $\tau = (4J)^{-1}$. Phase anomalies are suppressed, and the resulting anti-phase multiplet vanishes for $\theta = \pi/2$. In the process, only the one pulse has been retained on the decoupler channel that is to be calibrated, and the total duration of the sequence is unaffected. In this simple situation of on-resonance S spins, we have taken advantage of the fact that the complications arise from the component of the S multiplet that corresponds to $m_I = 0$. This component does not evolve under J and therefore does not acquire the "signature" of the state of the coupled I spins; it therefore eludes the multiple-quantum trap altogether. Sending this component back to the $-z$ axis by a second 90° pulse that is in phase with the first insures that all detected magnetization has indeed passed the multiple-quantum trap, allowing a clean calibration of θ, as for the $I = 1/2$ case, for all I_NS systems. It should be noted, however, that this simple on-resonance strategy entails partial loss of S spin magnetization and also throws the $m_I = \pm 2$ components of the S multiplet in I_2S and I_3S systems to the z axis, when τ is chosen to equal $(4J)^{-1}$.

Spectral Editing by Multiple-Quantum Traps (SEMUT)

The $\cos^N \theta$ dependence of the observed signal for I_NS systems noted earlier for the case $I = S = 1/2$ with the primitive trap sequence suggests the possibility of displaying spectra for various I_NS systems ($N = 0, 1, 2, 3, \ldots$) individually. In order to permit spectrum acquisition with decoupling, the primitive trap sequence is modified by adding a 180° refocusing pulse on the S spins, which is applied "synchronously" with the θ pulse on I, followed by a refocusing period τ, at the end of which acquisition commences under decoupling. This is known as the SEMUT sequence (spectral editing by multiple-quantum trap) and is represented in Fig. 3.9.

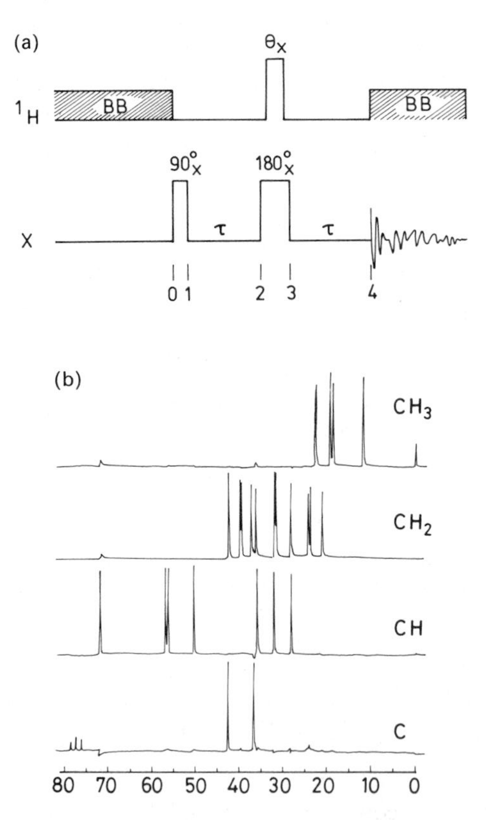

Figure 3.9. (a) A SEMUT pulse sequence. (b) A 25.1-MHz SEMUT ^{13}C spectrum of the aliphatic region of cholesterol; $\tau = 3.9$ ms. [Reproduced by permission. H. Bildsøe, S. Dønstrup, H.J. Jakobsen, and O.W. Sørensen, *J. Magn. Reson.*, **53**, 154 (1983), copyright 1983, Academic Press, New York]

Continuing with the analysis of the primitive trap sequence encountered earlier and dropping terms that contain $\cos \pi J\tau$ (which are close to zero for τ well "matched" to J), anti-phase terms at the point of acquisition (which are unobservable owing to the decoupling), and MQC's, we have for I_NS systems with even N:

$$\begin{aligned} \sigma_3 &= (-)^{N/2} 2^N s_J^N c_\theta^N S_y \prod_{i=1}^{N} I_{iz} \\ \sigma_4 &= s_J^{2N} c_\theta^N S_y \end{aligned} \tag{24}$$

With odd N, on the other hand, we have:

$$\begin{aligned} \sigma_3 &= (-)^{(N-1)/2} 2^N s_J^N c_\theta^N S_x \prod_{i=1}^{N} I_{iz} \\ \sigma_4 &= s_J^{2N} c_\theta^N S_y \end{aligned} \tag{25}$$

Therefore, σ_4 has the same general form for all I_NS systems. In the above, the effect of the 180°_x pulse has been taken care of by suppressing chemical shift evolution and changing the sign of S_y in σ_3 for even N.

With imperfect "matching" of τ to J, such that $c_J \neq 0$, σ_4 takes the form:

$$\sigma_4 = (c_J^2 + s_J^2 c_\theta)^N S_y \tag{26}$$

Like the primitive trap sequence, SEMUT does not involve any PT from the I to the S spins.

Considering CH_N systems with optimum τ, the CH_N intensities are given relative to unit intensity of the normal decoupled ^{13}C spectrum by:

$$\begin{aligned} S(\mathrm{C}) &= 1 \\ S(\mathrm{CH}) &= \cos\theta \\ S(\mathrm{CH_2}) &= \cos^2\theta = \frac{1}{2}(1 + \cos 2\theta) \\ S(\mathrm{CH_3}) &= \cos^3\theta = \frac{1}{4}(\cos 3\theta + 3\cos\theta) \end{aligned} \tag{27}$$

These expressions are identical to the SEFT case with θ put equal to $2\pi J\tau$ (see Chapter 2). Spectral editing with SEMUT may be performed by acquiring spectra for four different values of θ: 0°, 180°, φ°, 180° − φ°. This φ should be chosen to maximize the intensity difference between CH and CH_3, i.e., close to the magic angle, 54°44′. Calling the spectra with these four values of θ A, B, C, D, respectively, the linear combinations $X = (A + B)$ and $U = (C + D)$ give the C/CH_2 subspectra, whereas the combinations $Y = A - B$ and $V = C - D$ give the CH/CH_3 subspectra. Complete separation of the four subspectra is achieved by the combinations $U - c_\varphi^2 X$(C), $X - U$(CH_2), $V - c_\varphi^3 Y$(CH) and $c_\varphi Y - V$(CH_3).

How does SEMUT behave with the unavoidable "mismatch" between τ

and J? This may be examined noting that for nonoptimal τ, representing transverse S operators S_x or S_y generically as S_t, we have:

$$\sigma_2 \sim s_J^N 2^N S_t \prod_i I_{iz} \tag{28}$$

This evolves into such terms as:

$$s_J^{N-1} c_J 2^{N-1} S_t \prod_{i=1}^{N-1} I_{iz} \tag{29}$$

which has a flip angle dependence c_θ^{N-1}, mimicking a CH_{N-1} system. This product operator, after refocusing, finally contributes to the S signal a term proportional to:

$$c_\theta^{N-1} s_J^{2(N-1)} c_J^2 \tag{30}$$

Thus, J crosstalk occurs from $I_N S$ systems with higher values of N to the subspectra of those with lower values by dropping the signature of one or more of the I spins. Crosstalk is thus always in the "downwards" direction. The severest crosstalk of an $I_N S$ system is clearly into the $I_{N-1} S$ subspectrum.

Crosstalk problems are less severe with SEMUT than with SEFT. SEMUT also has the advantage over RINEPT and SEFT of not using a variable time delay, using instead a proton pulse of variable flip angle θ to achieve a separation of the subspectra. Note also that no phase shifting is involved on the decoupler channel.

Depending on the subspectrum (C, CH, or CH_2) considered, crosstalk from the next "higher" spin system ranges from 15 to 30% for a 25% variation in the value of J.

A generalized SEMUT sequence, called SEMUT GL, reduces the crosstalk problems significantly. This pulse sequence is given in Fig. 3.10. It involves a "purging sandwich" 90°–(τ_2/2)–180°–(τ_2/2)–90°, and reduces J crosstalk to about ±2% for 30% variation in J with appropriate choice of τ_1, τ_2, and τ_3. The SEMUT GL sequence functions essentially like SEMUT for ideal timing,

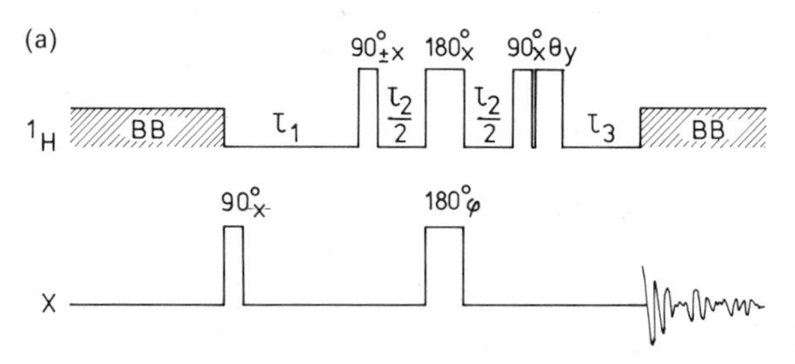

Figure 3.10. (a) A SEMUT GL pulse sequence. (b) A 75-MHz ^{13}C SEMUT GL spectrum of an ~1 : 1 mixture of brucine and 2-bromothiazole in $CDCl_3$ (τ_1 = 3.86 ms, τ_2 = 2.70 ms, τ_3 = 3.17 ms). [Reproduced by permission. O.W. Sørensen, S. Dønstrup, H. Bildsøe, and H.J. Jakobsen, *J. Magn. Reson.*, **55**, 347 (1983), copyright 1983, Academic Press, New York]

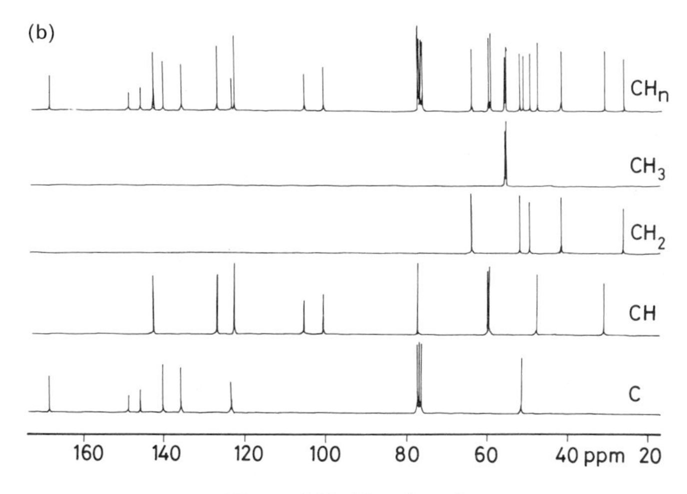

Figure 3.10. (Continued)

since $(N + 1)$ spin coherence in the $I_N S$ spin system does not evolve under J_{IS} during τ_2. For nonideal τ, however, J crosstalk contributions giving rise to N spin coherence after the first proton 90° pulse evolve during τ_2 and acquire a $\cos \pi J \tau_2$ coefficient, which results in attenuation of crosstalk. For CH_N systems with $125 < J < 225$ Hz, a suitable set of τ values is 3.79 ms, 2.87 ms, and 2.30 ms for minimizing crosstalk. The subspectra are generated as with SEMUT, except that the introduction of the 180° phase shift on the first or second proton 90° pulse allows one to work with all four values of θ less than 90°. Four suitable spectra would be $A(+x, 0°)$, $B(-x, 0°)$, $C(+x, 55°)$, $D(-x, 55°)$ where the phase of the first/second proton 90° pulse and the θ-pulse flip angle are shown as arguments of the four spectra.

PT Using Pulse Sequences with Variable Flip Angles θ and φ

In the SEMUT sequence, the θ pulse serves as a multiple quantum trap that helps discriminate between $I_N S$ systems of different multiplicities N.

An important feature to note in the SEMUT experiment is that the signed S magnetization, $\prod_i I_{iz} S_t$ recovered after the θ pulse is refocused in the following τ interval, $(2J)^{-1}$ regardless of the value of N. This is in contrast to the behavior in RINEPT where the signed S magnetization, $\sum I_{iz} S_t$, requires different time intervals for refocusing, depending on N; this is in general different from the INEPT time interval. This difference between SEMUT and RINEPT arises, clearly, from the different I-spin signatures (product and sum) in the two cases. For example,

$$\prod_{i=1}^{N} I_{iz} S_t \rightarrow s_J^N S_t + \cdots$$

whereas

$$\sum_{i=1}^{N} I_{iz}S_t \rightarrow s_{NJ}S_t + \cdots \tag{31}$$

The SEMUT strategy creates S-spin transverse magnetization, sends it through the trap, and recovers a refocused S signal that is labeled by N via a $\cos^N \theta$ term.

DEPT Class of Sequences

If one should create instead I-spin transverse magnetization, allow it to be signed by the S spin state by free precession under coupling, convert the signature into a transverse S-spin component [creating heteronuclear double-quantum coherences (DQC's) and zero-quantum coherences (ZQC's)] by an rf pulse, acquire $(N - 1)$ I-spin signatures under coupling, send the signed DQC's and ZQC's through a trap, allow refocusing under coupling, and acquire the S signal, one will have achieved the following:

1. With INEPT-type phase cycling, the resulting S signals will have originated solely from the I-spin polarization; i.e., PT will have been achieved.
2. The signals will carry a $\sin\theta \cos^{N-1}\theta$ dependence on the θ-pulse flip angle owing to the exit of signed heteronuclear ZQC's and DQC's as signed S magnetization.
3. The S signals will be the "refocused" versions, permitting decoupling during acquisition, with a refocusing period $(2J)^{-1}$ (independent of N), which is equal to the initial evolution period of the I magnetization.

It may be noted also that, as for refocusing, the time interval for acquiring $(N - 1)$ I-spin signatures on the heteronuclear zero- and double-quantum coherences, i.e., $\prod_i I_{iz} I_{jx} S_t$, can also be set independent of N: this in fact is equal to $(2J)^{-1}$, which is the I-spin evolution and S-spin refocusing times as well. The resulting highly symmetrical pulse sequence for PT has been called the distortionless enhancement by polarization transfer sequence (DEPT). This sequence is represented in Fig. 3.11.

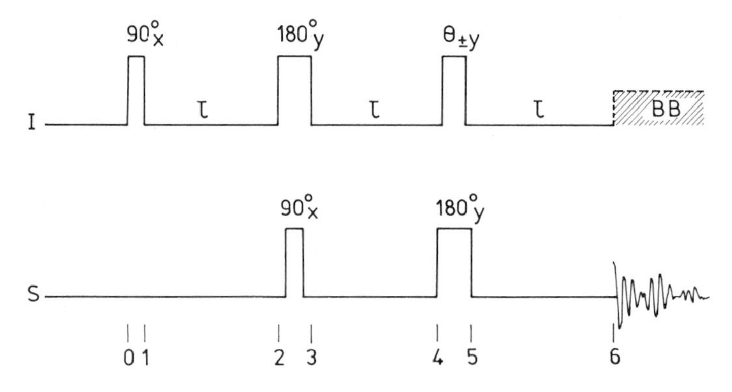

Figure 3.11. A DEPT pulse sequence (decoupling optional).

A detailed analysis of the signals arising from the DEPT sequence may be carried out as four INEPT. For the I_2S system for instance, we have:

$$\begin{aligned}
\sigma_0 &= I_{1z} + I_{2z} \\
\sigma_1 &= -(I_{1y} + I_{2y}) \\
\sigma_2 &= -c_J(I_{1y} + I_{2y}) + 2s_J(I_{1x} + I_{2x})S_z \\
\sigma_3 &= -c_J(I_{1y} + I_{2y}) + 2s_J(I_{1x} + I_{2x})S_y \\
\sigma_4 &= 2s_J c_J(I_{1x} + I_{2x})S_y - 4s_J^2(I_{1x}I_{2z} + I_{1z}I_{2x})S_x
\end{aligned} \tag{32}$$

ignoring the first term of σ_3, which leads to no PT.

$$\begin{aligned}
\sigma_5 = &-2s_J c_J s_\theta(I_{1z} + I_{2z})S_y + 2s_J c_J c_\theta(I_{1x} + I_{2x})S_y \\
&- 4s_J^2 s_{2\theta} I_{1z}I_{2z}S_x + 4s_J^2 s_{2\theta} I_{1x}I_{2x}S_x \\
&+ 4s_J^2 c_{2\theta}(I_{1x}I_{2z} + I_{1z}I_{2x})S_x
\end{aligned} \tag{33}$$

$$\begin{aligned}
\sigma_6 = &-2s_J c_J^3 s_\theta(I_{1z} + I_{2z})S_y + 2s_J^3 c_J s_\theta(I_{1z} + I_{2z})S_y \\
&+ 2s_J^2 c_J^2 s_\theta S_x + 8s_J^2 c_J^2 s_\theta I_{1z}I_{2z}S_x \\
&+ 2s_J c_J^2 c_\theta(I_{1x} + I_{2x})S_y - 4s_J^2 c_J c_\theta(I_{1x}I_{2z} + I_{1z}I_{2x})S_x \\
&- 4s_J^2 c_J^2 s_{2\theta} I_{1z}I_{2z}S_x - 2s_J^3 c_J s_{2\theta}(I_{1z} + I_{2z})S_y \\
&+ s_J^4 s_{2\theta} S_x + 4s_J^2 s_{2\theta} I_{1x}I_{2x}S_x \\
&+ 4s_J^2 c_J c_{2\theta}(I_{1x}I_{2z} + I_{1z}I_{2x})S_x \\
&+ 2s_J^3 c_{2\theta}(I_{1x} + I_{2x})S_y
\end{aligned} \tag{34}$$

The above expression for σ_6 does not include the effect of chemical shift evolution of the I spins during the final τ period; when this is included, all operators I_{ix} in σ_6 are to be replaced with $(I_{ix} \cos 2\pi\delta t + I_{iy} \sin 2\pi\delta t)$. It may be noted that the fifth, sixth, tenth, eleventh, and twelfth terms of σ_6 represent multiple-quantum coherences (MQC's), which do not contribute to the observed signal. This remains the case when the chemical shift evolution of the I spins during the final τ period is considered, as well. The only term that survives for $\tau = (2J)^{-1}$, i.e., $s_J = 1$, $c_J = 0$ is the ninth, $s_{2\theta}S_x$. With $\tau \neq (2J)^{-1}$, however, the first, second, and eighth terms constitute phase anomalies; the fourth and seventh terms constitute multiplet anomalies; the third term has a $\sin\theta$ dependence, which mimics the response of an IS spin system and so constitutes the crosstalk of the I_2S system into the IS subspectrum. DEPT therefore leads to distortionless spectra only in the event $\tau = (2J)^{-1}$. In any event, phase and multiplet anomalies disappear, as usual, on decoupling, whereas intensity anomalies and crosstalk problems remain.

If a 180° pulse is given to the I spins just prior to acquisition on alternate scans and the resulting FID's are coadded, the phase anomaly terms are suppressed in I-coupled spectra, owing to the sign change of $(I_{1z} + I_{2z})S_y$

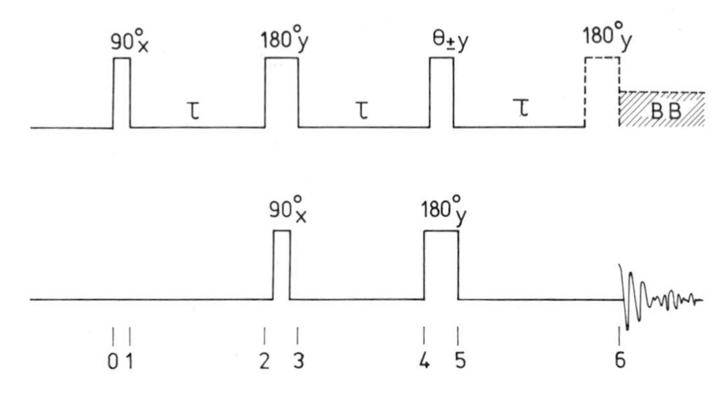

Figure 3.12. A DEPT$^+$ pulse sequence (decoupling optional).

resulting from the 180° pulse; this sequence, termed DEPT$^+$, is sketched in Fig. 3.12. DEPT$^+$ clearly leaves the multiplet anomalies intact because these involve operators of the kind $I_{1z}I_{2z}S_x$, which are invariant to the final 180° pulse on the I spins.

If a 90° pulse were applied to the I spins just prior to acquiring the S signal, the terms in σ_6 responsible for phase and multiplet anomalies would all be suppressed. However, owing to the chemical shift evolution of I_{ix} in the last τ period, MQC's would be partly converted to observable phase and multiplet anomaly contributions. This INEPT$^+$ type of strategy would work only if the I-spin chemical shift evolution during the last τ period were refocused, followed by the application of a 90°_x pulse to the I spins just before signal acquisition. A pulse sequence that achieves this is represented in Fig. 3.13 and is termed the DEPT^{++} sequence.

DEPT^{++} suppresses both phase and multiplet anomalies in I-coupled spectra but still leaves unsolved the intensity anomaly and spectral crosstalk problems. Summarizing, several observations about the DEPT sequences are in order.

For spectrometers not equipped with a phase shifter on the decoupler channel, a modified DEPT sequence may be employed to derive at least partially the benefits of PT, the motivation and modus operandi regarding this sequence being akin to those with SINEPT. This sequence may be represented as in Fig. 3.14. The initial 180°_x pulse on the S spins is applied on alternate scans to null the signal from the natural S-spin polarization upon adding the resulting FID's.

Like RINEPT and INEPT$^+$, the DEPT sequence also exhibit a loss in enhancement already for the I_3S system, compared with INEPT. Unlike RINEPT and INEPT$^+$, however, this 23% loss for I_3S has occurred already right after the PT step, i.e., after the θ pulse. In this sense, the polarization transferred by DEPT is intrinsically less than or equal to PT with INEPT, owing to the occurrence of

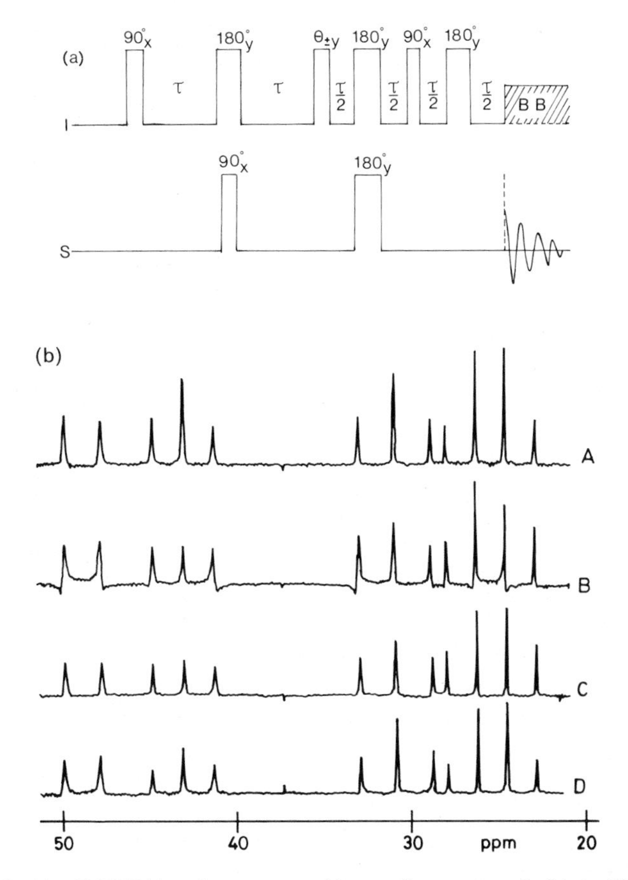

Figure 3.13. (a) a DEPT^{++} pulse sequence (decoupling optional). (b) the 75-MHz ^{13}C spectra of 1, 3-dibromobutane. (A) DEPT spectrum, $\tau = 3.58$ ms; $\theta = 45°$. (B) DEPT spectrum, $\tau = 1.79$ ms; $\theta = 45°$. (C) DEPT^{+} spectrum, $\tau = 1.79$ ms; $\theta = 45°$. (D) DEPT^{++} spectrum, $\tau = 1.79$ ms; $\theta = 135°$. [Reproduced by permission. O.W. Sørensen and R.R. Ernst, *J. Magn. Reson.*, **51**, 477 (1983), copyright 1983, Academic Press, New York]

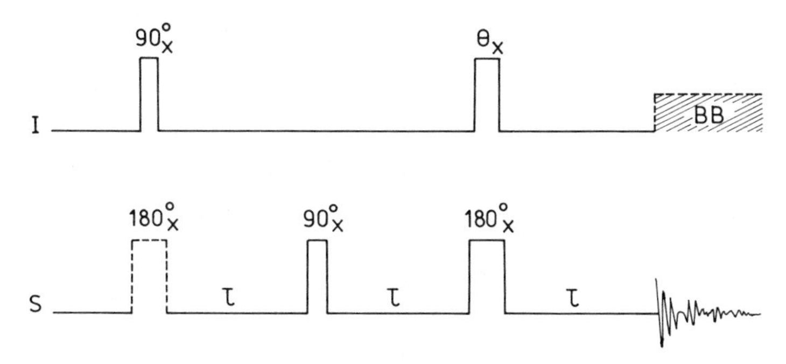

Figure 3.14. DEPT without a phase shifter (optional decoupling).

$$N 2^N s_\theta c_\theta^{N-1} \prod_{i=1}^{N} I_{iz} S_t \tag{35}$$

rather than

$$2\left(\sum_i I_{iz}\right) S_t$$

For I_4S systems, the loss in enhancement is about 13%.

$INEPT^+$ and $DEPT^{++}$ are qualitatively equally efficient because they both suppress phase and multiplet anomalies regardless of the spread in J values and lead to the same enhancement factors. Subspectral editing by $INEPT^+$ requires linear combination of spectra recorded with three values of τ_2, e.g., $1/(4J)$, $2/(4J)$, and $3/(4J)$, corresponding to $\pi J \tau_2 = 45°, 90°$, and $135°$; in an exact correspondence, DEPT editing requires linear combination of spectra recorded with three values of θ, e.g., 45°, 90°, and 135°. However, intensity anomalies under the optimal conditions of $\pi J \tau_2 = \theta = 54°$ are a much stronger function of variations in J for $INEPT^+$ than for $DEPT^{++}$. So also, subspectral editing by $DEPT^{++}$ is less prone to crosstalk than is the case with $INEPT^+$.

In fact the J crosstalk characteristics of DEPT spectral editing are identical to those of SEMUT, which, because DEPT is essentially SEMUT including PT, is to be expected. However, DEPT exhibits less sensitivity to missettings of the θ pulse than does SEMUT and has better signal to noise ratios in the subspectra.

It may also be mentioned that an extension of DEPT to DEPT GL, along the lines of SEMUT GL, reduces J crosstalk problems in DEPT substantially. The DEPT GL sequences is represented in Fig. 3.15, with phase cycling of the S spin π pulse to suppress refocusing errors.

The INEPT and DEPT classes of sequences may be related by observing that the inclusion of the purging sandwich $90°_y$–$(\tau/2)$–$180°_x$–$(\tau/2)$–θ_y in the RINEPT sequence leads to a sequence, termed INEPT GL, that functions

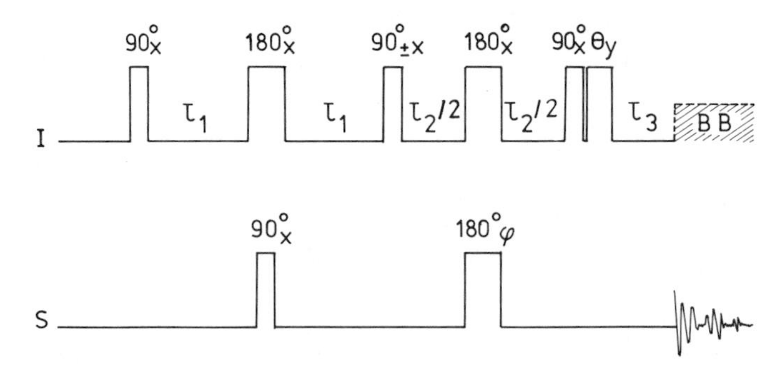

Figure 3.15. DEPT GL pulse sequence.

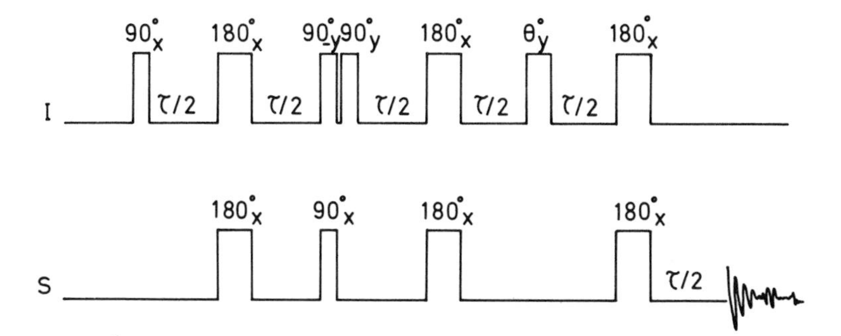

Figure 3.16. An INEPT GL pulse sequence.

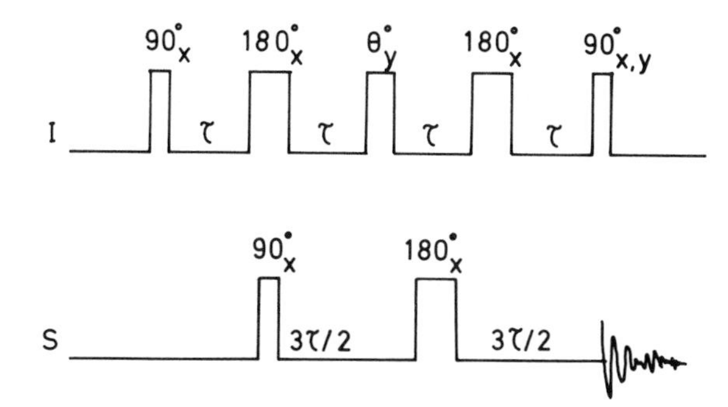

Figure 3.17. MODEPT pulse sequence.

identically to DEPT. The INEPT GL sequence is shown in Fig. 3.16. The third and fourth pulses on the proton channel can be consolidated and hence omitted. Consolidating the 180° pulses on both channels and introducing a 90° purge pulse on the I spins generates a sequence termed MODEPT, which functions in the same fashion as DEPT^{++}, while having the same duration and one less pulse than DEPT^{++}, MODEPT is represented in Fig. 3.17.

How does the basic DEPT idea work when PT is not between spin-1/2 nuclei but involves spin-1 $\leftrightarrow$ spin-1/2 situations?

In particular, we examine ^{2}H to ^{13}C PT by DEPT strategies. Upon carrying through an analysis of the motion of the IS spin system when subject to a DEPT sequence, we find for the form of the observed I-coupled S signal:

$$\begin{aligned}\sigma = & -2s_J c_J s_\theta (c_{2J} S_y I_z - s_{2J} S_x I_z^2) \\ & + 2s_J^2 s_{2\theta}(c_J^2 S_x I_x^2 - \tfrac{1}{2} s_{2J} S_y I_z) \\ & - 2s_J^2 s_{2\theta}(c_{2J} S_x I_z^2 + i s_J^2 S_x I_y [I_x, I_z]_+)\end{aligned} \tag{36}$$

For:

$$\tau = (2J)^{-1}$$

$$\sigma = -2s_{2\theta}(-S_x I_z^2 + iS_x I_y[I_x, I_z]_+) \tag{37}$$

$$= 2s_{2\theta}(S_x I_z^2 + S_x(I_x^2 - I_z^2)) = 2s_{2\theta} S_x I_x^2$$

Optimum θ is 45°, and the resultant signal is a (1/3)(1, 2, 1) triplet, in accordance with Table 1.1. The observable S signal arises in this case solely from the evolution of heteronuclear zero- and double-quantum coherence (ZQC and DQC) under J_{IS}, a situation in direct contrast to IS DEPT for $I = S = 1/2$, where the ZQC and DQC do not evolve under J. For $\tau \neq (2J)^{-1}$, phase anomalies appear in the signal. The DEPT$^+$ and DEPT^{++} sequences suppress phase anomalies. For DEPT^{++}, the signal takes the form:

$$\sigma = s_{2J}^2 s_\theta S_x I_y^2 + \tfrac{1}{2} s_{2J}^2 s_{2\theta} S_x I_x^2 - 2s_J^2 c_{2J} s_{2\theta} S_x I_y^2 \tag{38}$$

The intensity pattern is still a (1/3)(1, 2, 1) triplet. Apart from this multiplet anomaly, the occurrence of $\sin\theta$ as well as $\sin 2\theta$ dependences is to be noted.

In the reverse situation of PT from spin-1/2 to spin-1, DEPT does not transfer any polarization for $\tau = (2J)^{-1}$. However, with $\tau = (4J)^{-1}$, PT does occur and the resulting signal is given by:

$$\sigma = \tfrac{1}{4} s_{2J}^2 s_\theta I_x - s_{2J} c_J^2 s_\theta I_y S_z = \tfrac{1}{4} s_\theta (I_x - 2I_y S_z) \tag{39}$$

The result is a strongly phase-distorted doublet. The DEPT$^+$ and DEPT^{++} sequences suppress the phase anomaly, giving rise to a (1/8)(1, 1) doublet; optimal θ is 90°. In this case, evolution of heteronuclear ZQC and DQC under J has no observable effect.

Universal Polarization Transfer (UPT)

In the context of PT from N nuclei of arbitrary spin to M nuclei of arbitrary spin, a generalized version of DEPT has been developed, and is called the universal polarization transfer sequence (UPT). This sequence is given in Fig. 3.18. The standard UPT strategy of setting $\tau = (2J)^{-1}$ makes excellent sense for PT measurements on $I_N S_M$ systems with $I = S = 1/2$. The UPT sequence may be viewed as involving two multiple-quantum traps, one each on the S and I spin channels, so that the emerging S signal is labeled with sinusoids to powers up to M and N of ϕ and θ, respectively.

The UPT sequence has been employed to edit completely the ^{13}C spectra of CD_nH_m groups ($n = 1, 2, 3$; $m = 0, 1, 2$) in conjunction with gated proton decoupling (Fig. 3.18). In this case, $\phi = \pi/2$. Discrimination between CD, CD_2, and CD_3 groups is achieved by running three spectra with $\theta = \pi/6$, $\pi/3$, and $5\pi/12$. The complete editing is achieved by performing six different experiments as shown in Table 3.1.

The UPT sequence also leads to I-coupled S signals that exhibit phase and multiplet anomalies. A UPT$^+$ strategy—*à la* DEPT$^+$—would succeed in removing phase anomalies.

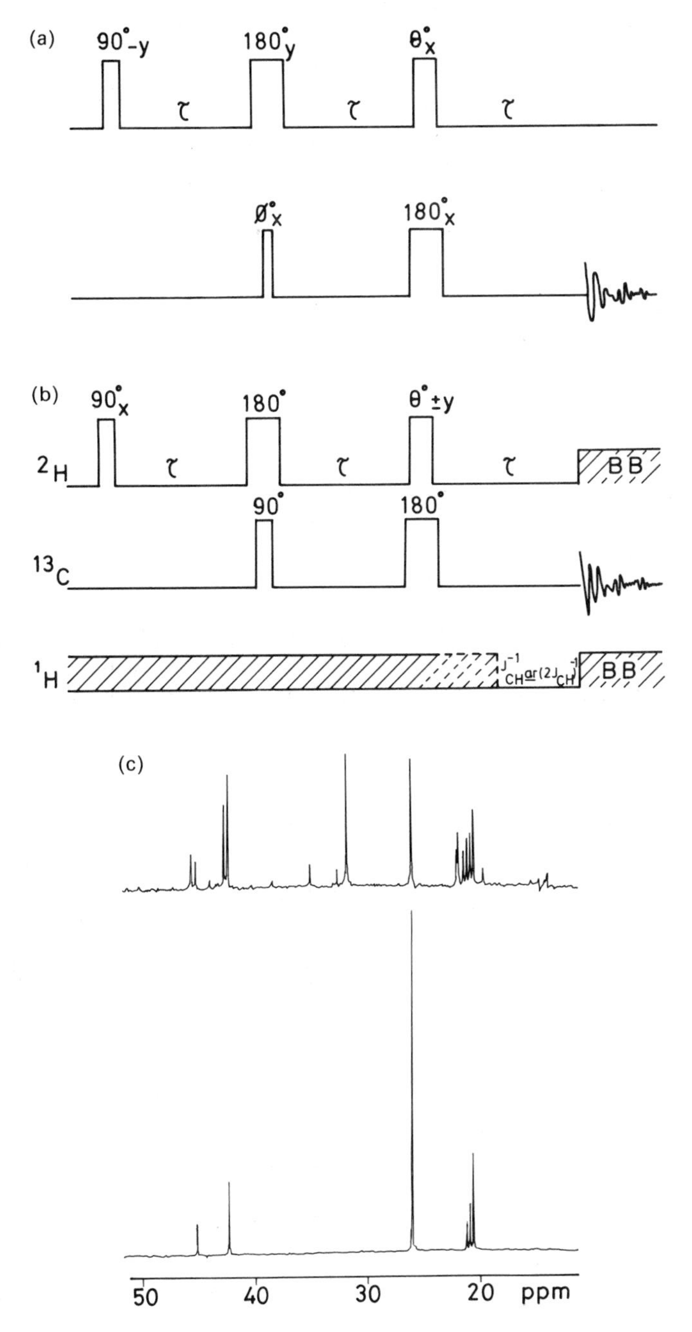

Figure 3.18. (a) UPT pulse sequence. (b) Pulse sequence for editing CD_nH_m spectra. (c) Upper trace: ^{13}C NMR of a mixture of $C_6H_5CH_3$, $C_6H_5CDH_2$, $C_6H_5CD_2H$, $C_6H_5CD_3$, $(C_6H_5)_2CH_2$, $(C_6H_5)_2CHD$, $(C_6H_5)_2CH—CH_3$, $(C_6H_5)_2CD—CH_3$, and C_6D_{12}. Lower trace: UPT spectrum with ^2D to ^{13}C PT. The deuterated groups of $(C_6H_5)_2CDCH_3$, $(C_6H_5)_2CDH$, C_6D_{12}, $C_6H_5CDH_2$, $C_6H_5CD_2H$, and $C_6H_5CD_3$ appear in order, from left to right. $\theta = \pi/6$; $\phi = \pi/2$.

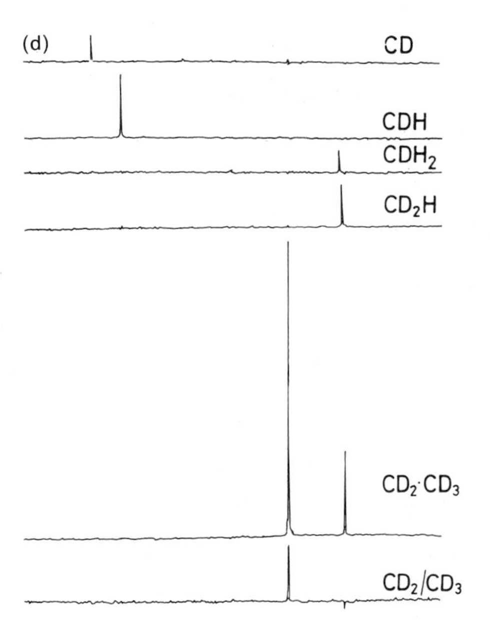

Figure 3.18. (d) The CD_nH_m subspectra of the mixture obtained with the sequence of (b), $n \neq 0$. [Parts (a)–(d) reproduced by permission. M.R. Bendall, D.T. Pegg, J.R. Wesener, and H. Günther, *J. Magn. Reson.*, **59**, 223 (1984), copyright 1984, Academic Press, New York]

Table 3.1. Parameters for Complete Editing of CD_nH_m Spectra

Experiment	θ	^1H Decoupling	Fractions of Maximum Signal Intensity					
			CD	CDH	CDH_2	CD_2H	CD_2	CD_3
1	$\pi/6$	Not gated	0.866	0.866	0.866	0.984	0.984	0.838
2	$\pi/6$	Off for $(J_{CH})^{-1}$	0.866	−0.866	0.866	−0.984	0.984	0.838
3	$\pi/3$	Not gated	0.866	0.866	0.866	0	0	0
4	$\pi/3$	Off for $(J_{CH})^{-1}$	0.866	−0.866	0.866	0	0	0
5	$\pi/3$	Off for $(2J_{CH})^{-1}$	0.866	0	0	0	0	0
6	$5\pi/12$	Off for $(2J_{CH})^{-1}$	0.5	0	0	0	−0.208	0.065

Rotating-Frame PT

The PT methods discussed so far in the heteronuclear context are all operative under the weak coupling Hamiltonian that survives after truncation in the doubly rotating frame. As such, these sequences may be classified as laboratory-frame PT sequences. As we noted, the immediate result of the PT step in these situations is always a signed magnetization leading to anti-phase multiplets; a further refocusing period after the PT stage is necessary to get net PT, leading to in-phase multiplets, although in general with residual anomalies that can be "cleaned up" with varying degrees of success by different purge strategies.

We shall here examine heteronuclear PT methods that are operative under a strong coupling Hamiltonian, such as may arise in the doubly rotating frame under conditions of matched spin locking.

These methods are consequently referred to as rotating-frame PT sequences. We shall also consider the strongly coupled homonuclear situation that arises under spin locking.

As shown in Chapter 1, such sequences lead to net PT but include the usual anomalies.

J Cross-Polarization. We shall first deal with the J cross-polarization (JCP) process introduced in Chapter 1. The pulse sequence employed is represented in Fig. 3.19.

Transverse magnetization of the I spins is created by a 90°_y pulse which is then spin locked by a long rf pulse, $(SL)_x$. Simultaneously, a long rf pulse is also applied to the S spins, such that the ratio of the two rf magnetic fields, B_{1I}/B_{1S}, equals γ_S/γ_I; this situation is termed Hartmann–Hahn matching.

Under these conditions, we have:

$$\gamma_I B_{1I} = \gamma_S B_{1S} \tag{40}$$

and the I and S spin systems have identical Larmor precession frequencies in the doubly rotating frame, when both spins are on resonance. As discussed in Chapter 1, the coupling Hamiltonian now takes on the form:

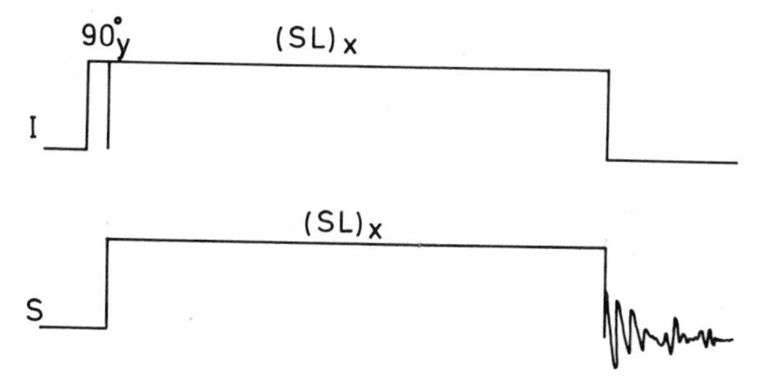

Figure 3.19. JCP pulse sequence (decoupling optional).

$$\mathscr{H} = \frac{hJ}{2}(I_z S_z + I_y S_y) \tag{41}$$

if B_{1I}, $B_{1S} \gg J$. Polarization transfer occurs between the I and S spins, and the natural S spin polarization may be suppressed by alternating the phase of the 90° pulse on the I spins and subtracting the FID's resulting in the 90°_{-y} experiment from those in the 90°_{+y} experiment, a strategy that may in this context be referred to loosely as spin–temperature alternation.

After a mixing or contact period of τ seconds during which both the long rf pulses are on, the resulting S signal is given by:

$$\sigma = \frac{1}{2}S_x(1 - \cos \pi J\tau) - I_z S_y \sin \pi J\tau \tag{42}$$

for an IS spin system with $I = S = 1/2$. For an I_2S spin system, on the other hand, we find

$$\sigma = \frac{1}{2}(\mathbf{I}^2 - I_x^2)S_x(1 - \cos \sqrt{2}\pi J\tau) - \frac{1}{\sqrt{2}} I_z S_y \sin \sqrt{2}\pi J\tau \tag{43}$$

where we have employed the composite particle approach, describing the two equivalent spins 1/2 as a composite spin-1 particle; treating the two I spins individually, we get the equivalent result:

$$\begin{aligned}\sigma = &\frac{1}{2}[1 + 2(I_{1z}I_{2z} + I_{1y}I_{2y})]S_x(1 - \cos \sqrt{2}\pi J\tau) \\ &- \frac{1}{\sqrt{2}}(I_{1z} + I_{2z})S_y \sin \sqrt{2}\pi J\tau\end{aligned} \tag{44}$$

The above results, which follow from the equations of motion given in Chapter 1 (Eqs. 1.69 and 1.70), clearly indicate that during the course of the contact period in the JCP process, there is an oscillatory transfer of part of the I spin polarization to the S spins as an in-phase multiplet term, as well as an anti-phase phase anomaly term, which is of course unobservable if acquisition is carried out under decoupling. The relative weightage of these terms is a function of the contact period. Regardless of the duration of contact τ, the phase anomalies can be suppressed by a 90°_x purge pulse on the I spins just prior to signal acquisition, provided mismatch of the rf fields is not high. This sequence called the "phase-corrected JCP" sequence, PCJCP, is represented in Fig. 3.20.

For the IS system, the frequency of PT between the I and S spins is $J/2$ Hertz. For a contact period of $\tau = (J)^{-1}$, it is clear from Eq. (1.69) that the I-spin polarization is completely transferred to in-phase S-spin transverse magnetization, and there are no phase anomalies. The maximum efficiency of transfer is therefore 1. The enhancement factor for the IS system is identical to RINEPT and DEPT; the duration of the sequence is equal to that of

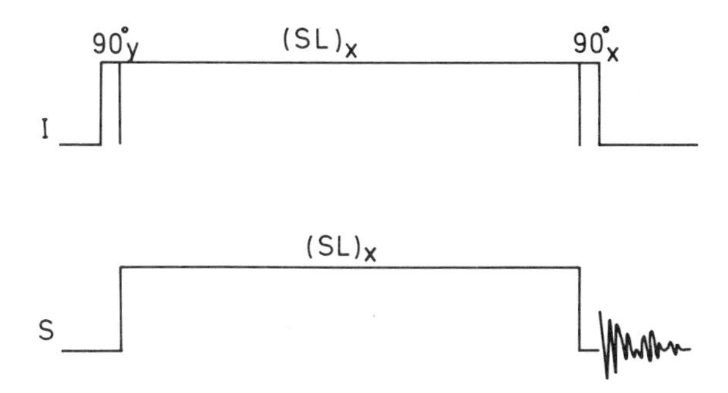

Figure 3.20. A PCJCP pulse sequence.

RINEPT, and is two thirds that of the DEPT sequence. Either phase of the purge pulse suppresses the phase anomaly, leading to identical results.

For the I_2S system, the frequency of polarization transfer between the I and S spins is 0.707 J. For a contact period of $\tau = (1.414J)^{-1}$, it follows from Eq. (1.70) that half the I-spin polarization is transferred to in-phase S-spin transverse magnetization, and there are no phase anomalies. The maximum efficiency of transfer is thus one half. The enhancement factor for the I_2S system is once again identical with that for RINEPT and DEPT; the duration of the sequence, however, is shorter than both RINEPT $[(1.333J)^{-1}]$ and DEPT $[(0.667J)^{-1}]$. Whereas a 90°_x purge pulse on the I spins suppresses phase anomalies and gives rise to a (3/8, 1/4, 3/8) triplet, a 90°_y purge pulse again suppresses phase anomalies and leads to a (1/4, 1/2, 1/4) triplet in accordance with the results in Table 1.1. This modified sequence, which may be called corrected J cross polarization (CJCP), is represented in Fig. 3.21.

It is interesting to note that for the I_2S system the CJCP sequence gives rise to the natural multiplet pattern. At the same time it is also to be noted that for the IS system with $I = 1$, $S = 1/2$, coadding two FID's of the PCJCP variety and one of the CJCP kind, one generates a (4/3)(1, 1, 1) triplet, which is the natural multiplet pattern for this system. This behavior of rotating-frame PT is to be contrasted with the laboratory-frame methods discussed earlier, where the (1/3)(1, 2, 1) multiplet resulting from each PT FID represents a multiplet anomaly that cannot be removed by INEPT, DEPT, or any of their variants. This difference between laboratory- and rotating-frame PT methods arises because the latter gives rise to different multiplet patterns with equally efficient suppression of phase anomalies, with purge pulses shifted in phase by 90°; whereas, RINEPT gives the same multiplet pattern for either phase of the purge pulse, and DEPT allows purging with only one phase of the purge pulse. The circumstance that rotating-frame PT requires less time than the laboratory-frame methods and allows natural multiplet patterns to be generated are factors in its favor compared to laboratory-frame PT, when dealing with spin-1 to spin-1/2 PT.

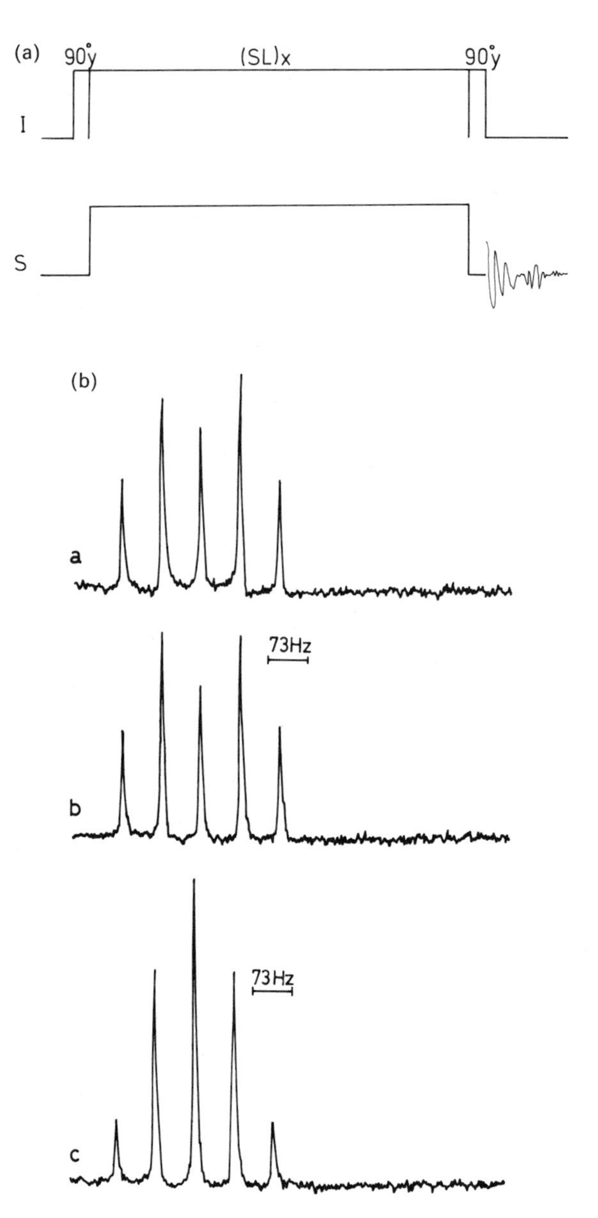

Figure, 3.21. (a) A CJCP pulse sequence. (b) The JCP, PCJCP, and CJCP spectra of a saturated acidified aqueous solution of 95.5% ^{15}N-enriched NH_4Cl, operating close to resonance, with rf fields of ca. 34.5 kHz. [Reproduced by permission. N. Chandrakumar, *J. Magn. Reson.*, **63**, 202 (1985), copyright 1985, Academic Press, New York]

The rotating-frame PT behavior of I_NS systems with $N > 2$ is more complicated. More than one frequency of PT is involved. For instance, the I-decoupled S signals are given by:

$$\begin{aligned}
&\left[\frac{1}{2}\sin^2\frac{(\pi J\tau)}{2} + \frac{1}{2}\sin^2\frac{(\sqrt{3}\pi J\tau)}{2} + \frac{1}{4}\sin^2(\pi J\tau)\right] \quad \text{for } N = 3 \\
&\left[\frac{3}{4}\sin^2\frac{(\pi J\tau)}{\sqrt{2}} + \frac{1}{4}\sin^2(\pi J\tau) + \frac{1}{4}\sin^2\frac{(\sqrt{6}\pi J\tau)}{2}\right] \quad \text{for } N = 4
\end{aligned} \tag{45}$$

In I-coupled S signal acquisition, different components of the multiplet exhibit different PT frequencies. For example, the following equations of evolution apply for the in-phase multiplet components. For $N = 3$:

$$\begin{aligned}
&\frac{1}{8}\sin^2\frac{(\sqrt{3}\pi J\tau)}{2} + \frac{3}{32}\sin^2(\pi J\tau) \qquad m_I = \pm\frac{3}{2} \\
&\frac{1}{4}\sin^2\frac{(\pi J\tau)}{2} + \frac{1}{8}\sin^2\frac{(\sqrt{3}\pi J\tau)}{2} + \frac{1}{32}\sin^2(\pi J\tau) \qquad m_I = \pm\frac{1}{2}
\end{aligned} \tag{46}$$

For the phase anomaly terms on the other hand, we have, for $N = 3$:

$$\begin{aligned}
&\pm\frac{\sqrt{3}}{32}\sin(\sqrt{3}\pi J\tau) \pm \frac{3}{64}\sin(2\pi J\tau) \qquad m_I = \pm\frac{3}{2} \\
&\pm\frac{1}{8}\sin(\pi J\tau) \pm \frac{\sqrt{3}}{32}\sin(\sqrt{3}\pi J\tau) \mp \frac{1}{64}\sin(2\pi J\tau) \qquad m_I = \pm\frac{1}{2}
\end{aligned} \tag{47}$$

Once again, the PCJCP sequence gets rid of the phase anomalies, as does CJCP, while leading to altered multiplet intensities; in both cases, the intensity ratios within the multiplet are a function of the contact period owing to differences in the time evolution of the different multiplet components, as seen above. Experimentally, the long rf pulses in Figs. 3.19 to 3.21 may be replaced on both channels by a train of 180° pulses at the rate of a few kilohertz.

In the case of off-resonance JCP experiments, Hartmann–Hahn matching refers to the equality of the effective rf fields:

$$(\Delta\omega_I^2 + \omega_{1I}^2)^{1/2} = (\Delta\omega_S^2 + \omega_{1S}^2)^{1/2} \tag{48}$$

In the presence of Hartmann-Hahn mismatch and off-resonance irradiation, the effective coherence transfer frequencies are increased and polarization transfer efficiencies are lowered. To appreciate the effects of mismatch somewhat more clearly, we compute the motion of a two-spin-1/2 system with both spins on resonance, but subject to resonant rf fields whose amplitudes do not satisfy the Hartmann–Hahn condition. Transforming to a synchronized doubly rotating frame rotating with frequency ω about the rf axis x, we have:

$$\frac{2\pi}{h}\mathscr{H} = \frac{\Delta}{2}(I_x - S_x) + \pi J(I_yS_y + I_zS_z) \tag{49}$$

where the rf amplitudes in radians per second are given by:

$$\omega_{1I} = \omega + \frac{\Delta}{2}, \qquad \omega_{1S} = \omega - \frac{\Delta}{2} \tag{50}$$

so that the mismatch is Δ radians per second. The two terms of the above Hamiltonian do not commute. Reexpressing it as:

$$\frac{2\pi}{h}\mathscr{H} = \pi J[k(I_x - S_x) + (I_y S_y + I_z S_z)] \tag{51}$$

where $k = \Delta/2\pi J$, we find the following equations of motion, employing the Hausdorff formula:

$$(I_x + S_x) \rightarrow (I_x + S_x) \tag{52}$$

$$\begin{aligned}(I_x - S_x) \rightarrow & \frac{1}{(4k^2 + 1)}(I_x - S_x)[\cos \pi J(4k^2 + 1)^{1/2} t + 4k^2] \\ & - \frac{4k}{(4k^2 + 1)}(I_y S_y + I_z S_z)[\cos \pi J(4k^2 + 1)^{1/2} t - 1] \\ & + \frac{2}{(4k^2 + 1)^{1/2}}(I_y S_z - I_z S_y) \sin \pi J(4k^2 + 1)^{1/2} t\end{aligned} \tag{53}$$

$$\begin{aligned}(I_y S_y + I_z S_z) \rightarrow & \frac{4k^2}{(4k^2 + 1)^{1/2}}(I_y S_y + I_z S_z)[\cos \pi J(4k^2 + 1)^{1/2} t + (1/4k^2)] \\ & - \frac{k}{(4k^2 + 1)}(I_x - S_x)[\cos \pi J(4k^2 + 1)^{1/2} t - 1] \\ & - \frac{2k}{(4k^2 + 1)^{1/2}}(I_y S_z - I_z S_y) \sin \pi J(4k^2 + 1)^{1/2} t\end{aligned} \tag{54}$$

$$\begin{aligned}(I_y S_z - I_z S_y) \rightarrow & (I_y S_z - I_z S_y) \cos \pi J(4k^2 + 1)^{1/2} t \\ & + \frac{1}{(4k^2 + 1)^{1/2}}[2k(I_y S_y + I_z S_z) \\ & - \tfrac{1}{2}(I_x - S_x)] \sin \pi J(4k^2 + 1)^{1/2} t\end{aligned} \tag{55}$$

Several features in the above equations of motion are worth noting. The equations correctly specify the behavior of the system under matched conditions, upon setting $k = 0$. In the presence of mismatch ($k \neq 0$), the added feature is the appearance of zz order and zero- and double-quantum coherences in the evolution of $(I_x - S_x)$, as well as in that of the anti-phase magnetization $(I_y S_z - I_z S_y)$. The coherence order therefore is not conserved, because the rf is "enabled" in this frame of reference owing to the mismatch. The term $(I_x + S_x)$ is a constant of the motion whether or not mismatch exists, because:

$$[I_x + S_x, I_x - S_x] = 0$$

and:

$$[I_x + S_x, I_zS_z + I_yS_y] = 0 \tag{56}$$

Also, $(I_yS_y + I_zS_z)$ is conserved when the mismatch parameter $k = 0$.

Finally, when $k \gg 1$, the evolution frequencies are:

$$\pi J(4k^2 + 1)^{1/2} = 2\pi Jk = \Delta \text{ rad/s} \tag{57}$$

When $\Delta t = \pm\pi/2$, therefore,

$$\begin{aligned}(I_x - S_x) &\rightarrow (I_x - S_x)\\ (I_yS_z - I_zS_y) &\rightarrow \pm(I_yS_y + I_zS_z)\\ (I_yS_y + I_zS_z) &\rightarrow \mp(I_yS_z - I_zS_y)\end{aligned} \tag{58}$$

In other words, "90°" evolution under strong mismatch leaves $(I_x - S_x)$ invariant and carries $(I_yS_z - I_zS_y)$ and $(I_yS_y + I_zS_z)$ into each other as by a rotation by 90° in a space spanned by these two operators.

From the above we find:

$$\begin{aligned}I_x \rightarrow &\frac{1}{2(4k^2+1)} I_x[(8k^2+1) + \cos \pi J(4k^2+1)^{1/2}t]\\ &+ \frac{1}{2(4k^2+1)} S_x[1 - \cos \pi J(4k^2+1)^{1/2}t]\\ &+ \frac{2k}{(4k^2+1)}(I_yS_y + I_zS_z)[1 - \cos \pi J(4k^2+1)^{1/2}t]\\ &+ \frac{1}{(4k^2+1)^{1/2}}(I_yS_z - I_zS_y) \sin \pi J(4k^2+1)^{1/2}t\end{aligned} \tag{59}$$

For a mismatch amounting to half the coupling, i.e., $\Delta = \pi J$, so that $k = 1/2$, we have:

$$\begin{aligned}I_x \rightarrow &\frac{1}{4}(3 + \cos \sqrt{2}\pi Jt)I_x\\ &+ \frac{1}{4}(1 - \cos \sqrt{2}\pi Jt)S_x\\ &+ \frac{1}{2}(1 - \cos \sqrt{2}\pi Jt)(I_yS_y + I_zS_z)\\ &+ \frac{1}{\sqrt{2}} \sin \sqrt{2}\pi Jt(I_yS_z - I_zS_y)\end{aligned} \tag{60}$$

which mimics the behavior of an I_2S system under matched conditions in terms of the efficiency and frequency of PT. Efficiency of transfer, in particular, has dropped to 1/2. For polarization transfer to be at least half optimal, therefore, the Hartmann–Hahn mismatch should be less than or equal to half of the coupling constant. With $J = 100$ Hz, at $\omega_1/2\pi = 3$ kHz, for instance, polarization transfer drops by more than 50% of the maximum outside a frequency range of about 1.1 kHz. This implies that the chemical shift window over which PT drops by less than 50% is about 1100 Hz, corresponding to about 48 ppm for ^{13}C at 2.1T. The practical utility of JCP is also limited by its sensitivity to the magnitude of J and the number N of equivalent source spins I. A strategy to alleviate both these problems involves refocusing, in a sequence called RJCP, shown in Fig. 3.22.

The functioning of RJCP may be qualitatively understood on the basis of

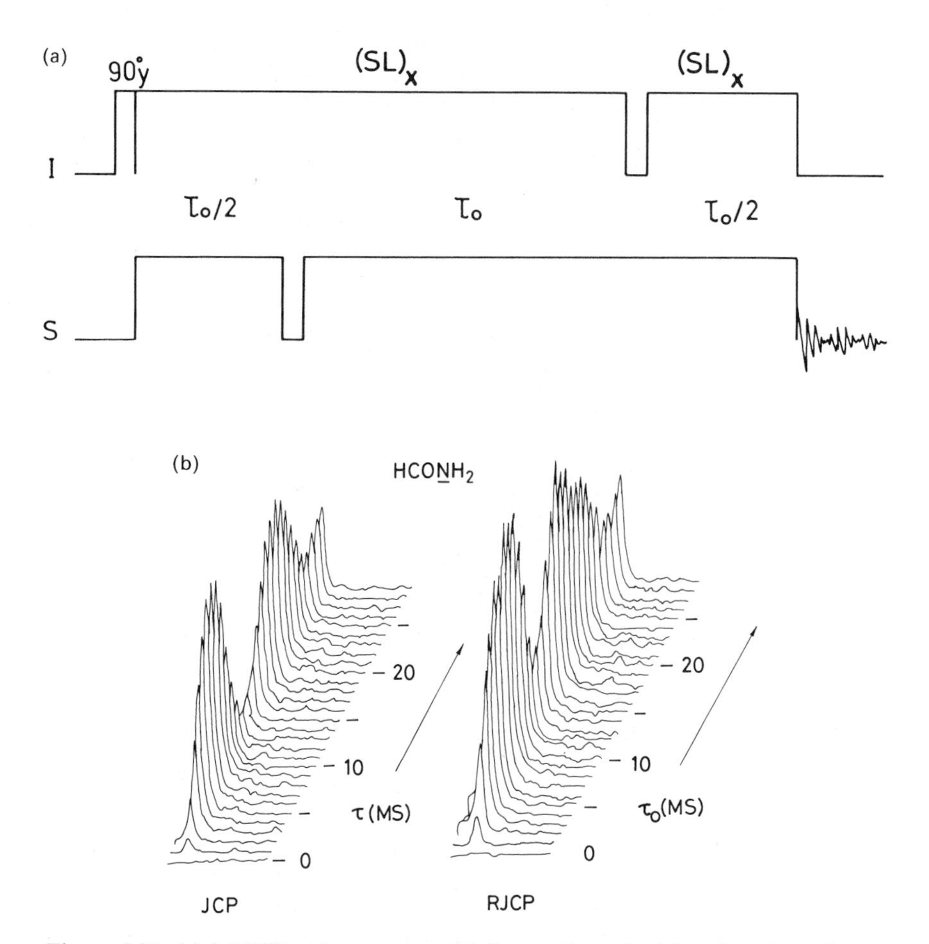

Figure, 3.22. (a) A RJCP pulse sequence. (b) Comparison of mixing time dependence of ^{15}N JCP and RJCP spectra of $HCONH_2$. [Reproduced by permission. G.C. Chingas, A.N. Garroway, R.D. Bertrand, and W.B. Moniz, *J. Magn. Reson.*, **35**, 283 (1979), copyright 1979, Academic Press, New York].

the fact that cross-polarization is rigorously analogous to spin inversion in the laboratory frame: the J coupling plays the role of the "observation" rf field, while Hartmann–Hahn mismatch is equivalent to resonance offsets. The problem that is faced in JCP is thus analogous to achieving spin inversion in the presence of resonance offset and rf inhomogeneity; RJCP achieves spin inversion by blanking the rf on each channel for a duration corresponding to a $\pi/2$ pulse, timed as shown in Fig. 3.22. We shall see in Chapter 7 that the spin inversion problem in the laboratory frame is solved in an entirely analogous fashion. We get a clearer view of the functioning of RJCP by calculating the equation of motion of a two-spin-1/2 system, which is subject to matched, resonant, double irradiation. For purposes of this analysis, the RJCP sequence may be viewed as comprising five time periods. During the first, third, and fifth periods, evolution is under matched, resonant rf fields, whereas during the second period mismatch is induced by blanking one rf channel. During the fourth period, mismatch of the opposite sign is induced by blanking the other rf channel. Mismatch during these two periods is in fact $\pm\omega$, which is much larger than J and is operative for time t such that $\omega t = \pi/2$.

Employing the equations of motion derived earlier for evolution under mismatch, setting $k = 0$ during the odd time periods and $k \gg 1$ during the even time periods, we find:

$$
\begin{aligned}
\frac{1}{2}(I_x - S_x) &\xrightarrow{T=\tau_0/2} \frac{1}{2}c_J(I_x - S_x) + s_J(I_yS_z - I_zS_y) \\
&\xrightarrow[(\omega t=\pi/2)]{+\omega} \frac{1}{2}c_J(I_x - S_x) + s_J(I_yS_y + I_zS_z) \\
&\xrightarrow{2T=\tau_0} \frac{1}{2}c_Jc_{2J}(I_x - S_x) + s_J(I_yS_y + I_zS_z) + c_Js_{2J}(I_yS_z - I_zS_y) \\
&\xrightarrow[(-\omega t=-\pi/2)]{-\omega} \frac{1}{2}c_Jc_{2J}(I_x - S_x) - c_Js_{2J}(I_yS_y + I_zS_z) + s_J(I_yS_z - I_zS_y) \\
&\xrightarrow{T=\tau_0/2} \left(\frac{1}{2}c_J^2c_{2J} - \frac{1}{2}s_J^2\right)(I_x - S_x) \\
&\qquad - c_Js_{2J}(I_yS_y + I_zS_z) \\
&\qquad + \left(\frac{1}{4}s_{4J} + \frac{1}{2}s_{2J}\right)(I_yS_z - I_zS_y)
\end{aligned} \tag{61}
$$

$$
(I_x + S_x) \xrightarrow{T=\tau_0/2} \xrightarrow[(\omega t=\pi/2)]{+\omega} \xrightarrow{2T=\tau_0} \xrightarrow[(-\omega t=-\pi/2)]{-\omega} \xrightarrow{T=\tau_0/2} (I_x + S_x) \tag{62}
$$

In the above, $c_{nJ} = \cos n\pi J\mathrm{T}$ and $s_{nJ} = \sin n\pi J\mathrm{T}$. Consolidating, we have at the end of the RJCP sequence:

$$I_x \rightarrow \frac{1}{2} I_x \left(\frac{3}{4} + \frac{1}{4} c_{4J} + c_{2J} \right) + \frac{1}{2} S_x \left(\frac{5}{4} - \frac{1}{4} c_{4J} - c_{2J} \right)$$
$$+ \frac{1}{2} (I_y S_z - I_z S_y) \left(\frac{1}{2} s_{4J} + s_{2J} \right) - \frac{1}{2} (I_y S_y + I_z S_z)(s_J + s_{3J}) \tag{63}$$

This equation of motion is to be compared with that under JCP for a similar period $4T$:

$$I_x \rightarrow \frac{1}{2} I_x (1 + c_{4J}) + \frac{1}{2} S_x (1 - c_{4J}) + (I_y S_z - I_z S_y) s_{4J} \tag{64}$$

The S signal acquired under decoupling after RJCP therefore exhibits a plateau at its maximum value when viewed as a function of T, owing to the superposition of the two harmonically related frequencies. After JCP, on the other hand, the signal is a single sinusoid, at the higher of the two frequencies, viz., $J/2$.

In other words, the value of JT is not critical with RJCP; for a given choice of T, for instance, RJCP gives maximum enhancement over a range of J values.

It may be noted also that the order of blanking of the two channels is immaterial. If the order were reversed with respect to the previous discussion, this would change the sign of s_J in steps 2 and 3 but restore it in step 4; in other words, the sign of $(I_y S_z - I_z S_y)$ at stage 4 would remain unchanged; the sign of the $(I_y S_y + I_z S_z)$ term, however, would be changed in steps 4 and 5.

It is clear also that blanking the same channel twice would be a disaster: this would lead to an S_x coefficient of $(1/8)(1 - c_{4J})$, which would involve exactly the same time dependence as in JCP but with only 25% of the PT efficiency. Single blanking, on the other hand, leads to an S_x coefficient of $(1/2)(1 - (1/2)c_{4J} - (1/2)c_{2J})$, which is not a very efficient "refocusing."

The total duration of the RJCP sequence is about twice that of JCP; for IS spin systems, enhancement is 94% of the optimum for PT frequencies lying in a range 0.7 to 1.3 of $J_0/2$, where τ_0 has been set equal to $(J_0)^{-1}$. This is to be compared with the 75% enhancement achieved with JCP under similar conditions.

In $I_N S$ systems where the Hartmann–Hahn condition is met, the S polarization resulting at $2\tau_0$ from the RJCP sequence is obtained upon substituting $(1 - \cos^4(2\pi\nu\tau_0))$ for $\sin^2(2\pi\nu\tau)$ in the corresponding JCP expressions, ν being the respective frequencies of oscillatory polarization transfer. In general, RJCP enables about 75% of optimal enhancement to be obtained with four times the mismatch and timing tolerance exhibited by JCP for a similar enhancement. Purge strategies work with RJCP as they do for JCP.

When the sample is spun in a crossed coil probe tuned separately for the I and S spins (A and X nuclei), or when sample heating is allowed under intense Hartmann–Hahn matched rf fields is a single coil probe tuned to both A and X nuclei, cross-polarization proceeds in a "quasi-stochastic" manner, leading to about 50% of the normal JCP enhancements and also averaging out the time dependences.

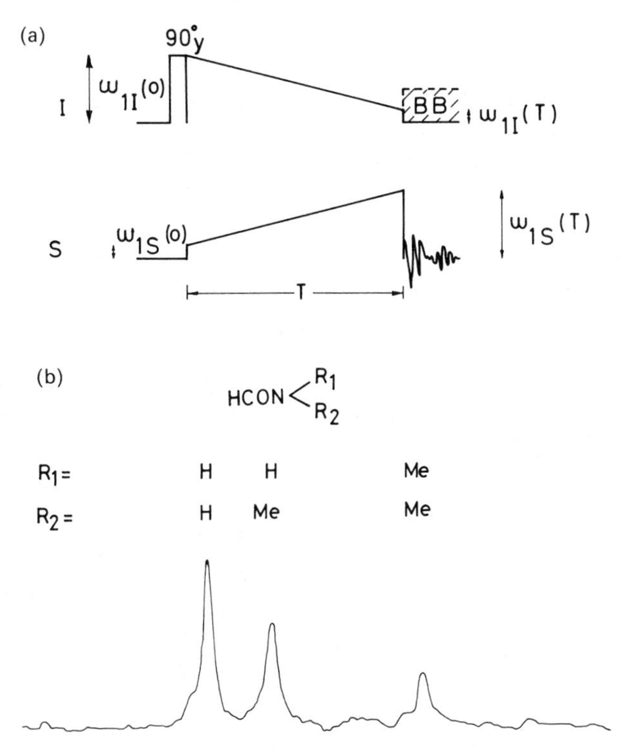

Figure 3.23. AJCP sequence. (b) The 10-MHz ^{15}N AJCP spectrum of an equivolume mixture of formamide, N-methylformamide and N, N-dimethyl formamide ($J_{NH} = 89$, 94, and 15 Hz, respectively). Fields ramped in 400 ms between 650 and 800 Hz. [Reproduced by permission. G.C. Chingas, A.N. Garroway, W.B. Moniz and R.D. Bertrand, *J. Amer. Chem. Soc.*, **102**, 2526 (1980); copyright (1980) American Chemical Society]

Adiabatic J Cross-Polarization. An altogether different J cross-polarization strategy, called adiabatic JCP (AJCP), enables full PT to be achieved for all J values above a lower cutoff threshold, over the full chemical shift range. This strategy involves a crossing of the Hartmann–Hahn condition, which requires that at least one of the rf channels be ramped. The method works provided the relevant relaxation times or exchange times are much longer than the period of the ramp. The experiment is analogous to adiabatic rapid passage for spin inversion.

The AJCP pulse sequence is shown in Fig. 3.23. After the I spins are spin locked as usual, the experiment commences with rf fields well off Hartmann–Hahn match: the matching condition is passed through by gradually decreasing the I- and increasing the S-spin rf powers. In this case, the Hamiltonian $\mathscr{H}$ is varied slowly in a time $\sim J^{-1}$, and the density operator commutes at all times with $\mathscr{H}$. Consequently the experiments can be described in terms of the diagonal elements of σ alone, i.e., in terms of the relative populations of the energy levels. Polarization is transferred in the rotating frame as the Hartmann–Hahn condition is crossed, and at the end the S spins contain the spin-lock polarization. Polarization transfer occurs, provided the coupling is at least equal to the square root of the rate of crossing $\Delta\nu_1$ ($= \Delta\omega_1/2\pi$) Hertz per second of the Hartmann–Hahn condition:

$$J \geqslant (\Delta\nu_1)^{1/2} = (\nu_{1I}(0) - \nu_{1S}(0) + \nu_{1S}(T) - \nu_{1I}(T)/T)^{1/2} \qquad (65)$$

Adiabaticity implies really the limit $\Delta\nu_1/J^2 \rightarrow 0$, but in practice it suffices if this ratio is unity. The detailed rf profile is uncritical, even allowing one of the rf fields to be held constant provided the other has enough range to cross the matching condition and the crossing is sufficiently slow. For ^1H to ^{13}C PT at 2.1 T, for instance, the respective chemical shift ranges correspond to 1–5 kHz. If the observe transmitter is placed at the center of the ^{13}C spectral window, this results in a ramp period $T \simeq 350$ ms. The use of trapezoidal ramps on the rf channels allows one to strike tradeoffs among chemical shift range, ramp duration, and minimal J coupling. Under these conditions, weaker couplings can be employed to cross-polarize in shorter times than is the case with "triangular ramps," where $\nu_{1S}(0) = \nu_{1I}(T) = 0$. Indeed, with trapezoidal ramps, AJCP has been employed successfully to achieve PT when J is as low as 15–20 Hz. Couplings below this level appear out of reach.

However, workable chemical shift ranges for the two species are reduced with trapezoidal ramps:

$$\begin{aligned} \Delta\nu_S &= (\nu_{1I}^2(0) - \nu_{1S}^2(0))^{1/2} \\ \Delta\nu_I &= (\nu_{1S}^2(T) - \nu_{1I}^2(T))^{1/2} \end{aligned} \qquad (66)$$

The mismatch parameters $\Delta\nu_1(t) = \nu_{1I}(t) - \nu_{1S}(t)$ must also satisfy the condition:

$$|\Delta\nu_1(t)| \gg |J_{\max}/2| \qquad \text{for } t = 0, \text{ as well as } t = T$$

The S-spin polarization produced by AJCP on an I_NS system is given by:

$$2\left[1 - \frac{(2m)!}{4^m(m!)^2}\right]\left|\frac{\gamma_I}{\gamma_S}\right| \qquad (67)$$

where $m = (1/2)(N + 1)$ if N is odd or $(1/2)N$ if N is even.

Isotropic Mixing. A more recent proposal for PT in the rotating frame is by the strategy of isotropic mixing. In this case, the heteronuclear spin systems are not allowed to evolve under their respective chemical shifts; at the same time, the weak coupling between them in the doubly rotating frame is made to take on strong coupling character in the average Hamiltonian sense by simultaneous identical pulsing on both channels. This may be accomplished by a SHRIMP sequence (scalar heteronuclear recoupled interactions by multiple pulse), shown in Fig. 3.24. In liquid-state PT with this method, the average Hamiltonian operative in the doubly rotating frame is of the form:

$$\overline{\mathscr{H}} = \tfrac{1}{3}J\mathbf{I}\cdot\mathbf{S} \qquad (68)$$

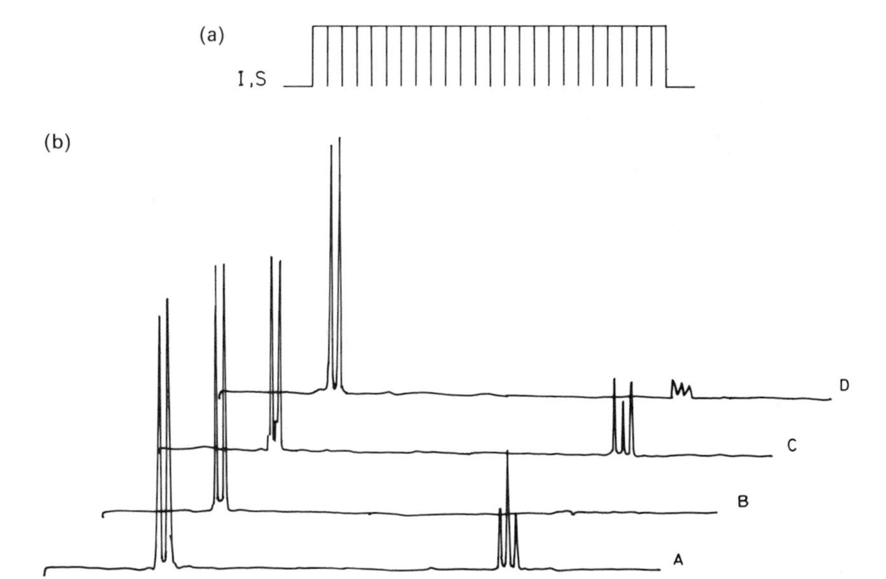

Figure 3.24. (a) An example of a SHRIMP sequence. Both I and S spins are subject to identical rf pulse sequences, a train of 90° pulses of phase $\bar{y}$, x, $\bar{y}$, $\bar{y}$, x, $\bar{y}$, y, x, y, y, x, y, $\bar{y}$, $\bar{x}$, $\bar{y}$, $\bar{y}$, $\bar{x}$, $\bar{y}$, y, $\bar{x}$, y, y, $\bar{x}$, y. ($\bar{i}$ indicates a 90° pulse of phase $-i$). (b) Homonuclear proton isotropic mixing, achieved by a rapid train of 180° pulses applied to a sample of propargyl bromide at 90 MHz. (A) Normal spectrum ($J = 2.62$ Hz). (B) Spectrum recorded following presaturation of high-field triplet. (C) Spectrum resulting on isotropic mixing for 127 ms following presaturation as in (B). Note the (1.5, 1, 1.5) triplet. (D) Isotropic mixing period of 254 ms, resulting in no transfer. [Reproduced by permission. N. Chandrakumar and S. Subramanian, *J. Magn. Reson.*, **62**, 332 (1985), copyright (1985), Academic Press, New York]

which is isotropic in spin operator space. A mixing Hamiltonian of this form gives rise to isotropic mixing.

The details of PT under isotropic mixing may be worked out from the equations of motion given in Chapter 1 [Eqs. (1.75) and (1.76)]. For the IS spin system ($I = S = 1/2$), the S-spin signal after a mixing period of τ seconds following the creation of I_x transverse magnetization is given by:

$$\sigma = \tfrac{1}{2}S_x(1 - \cos(2\pi J\tau/3)) - I_z S_y \sin(2\pi J\tau/3) \tag{69}$$

For the $I_s S$ system, on the other hand, we find for the S signal:

$$\sigma = \tfrac{4}{9}(\mathbf{I}^2 - I_x^2)S_x(1 - \cos\pi J\tau) - \tfrac{2}{3}I_z S_y \sin\pi J\tau \tag{70}$$

employing the composite particle approach for the two equivalent I spins. Treating the two I spins individually we have the equivalent result:

$$\begin{aligned}\sigma = {}& \tfrac{4}{9}(1 + 2(I_{1z}I_{2z} + I_{1y}I_{2y}))S_x(1 - \cos\pi J\tau) \\ & - \tfrac{2}{3}(I_{1z} + I_{2z})S_y \sin\pi J\tau\end{aligned} \tag{71}$$

In general, therefore, the consequence of isotropic mixing is an oscillatory transfer of part of the I spin polarization to the S spins, partly as in-phase S magnetization, and partly as anti-phase S magnetization, in varying proportions as a function of the mixing time. The PT frequency for the IS spin-1/2 system is $J/3$, which is to be compared with $J/2$ for the Hartmann–Hahn case. Efficiency of transfer is unity, as for JCP. For the I_2S spin-1/2 system, on the other hand, the PT frequency is $J/2$, compared to $J/\sqrt{2}$ for the Hartmann–Hahn experiment. Efficiency of transfer is 4/9, as against 1/2 for JCP. Isotropic mixing offers the same interesting possibility of recovering natural multiplets in spin-1 to spin-1/2 PT, since its behavior with respect to purging is obviously very similar to that of JCP.

For I_NS_M systems with N, $M \geqslant 2$, more than one frequency of PT is active; however, the states connected by this evolution are characterized by $\Delta(I_z + S_z) = 0$. This property arises because

$$[I_i + S_i, \mathbf{I}\cdot\mathbf{S}] = 0 \qquad \text{for } i = x, y, \text{ or } z \tag{72}$$

so that the quantum numbers $(I_i + S_i)$ are conserved by isotropic mixing. This is to be contrasted with the JCP processes, where only the net spin-locked component is a constant of the motion:

$$[I_x + S_x, I_yS_y + I_zS_z] = 0 \tag{73}$$

The periodic transfer of magnetization, component for component, between two spin-1/2 particles under isotropic mixing may be appreciated in a general sense by recalling the operator identity established by Dirac, which shows that the scalar coupling between two spins-1/2, I_1 and I_2, behaves in the same way as a permutation P_{12} of the two spins:

$$\mathbf{I}_1\cdot\mathbf{I}_2 = \tfrac{1}{4}(2P_{12} - 1) \tag{74}$$

In the case of homonuclear spin locking isotropic mixing also results, as discussed in Chapter 1. However, in this instance the effective coupling constant is the true J value and not $J/3$ as for the heteronuclear situation. For spin-1/2 systems, therefore, coherence transfer under isotropic mixing proceeds at a frequency of J Hertz for the IS system, and $3J/2$ Hertz for the I_2S system. The zero-quantum evolution processes occur at frequencies that are higher than the multiplet splittings observed in the normal single-quantum NMR spectra. In the homonuclear situation, therefore, PT can occur directly between spins that do not exhibit a resolved coupling, provided the true T_2 is favorable when compared to the inverse of the PT frequency.

In the heteronuclear situation, isotropic mixing under the unscaled coupling Hamiltonian occurs if the spins are placed in zero field during the mixing period, although the spin state is prepared and the NMR signal is detected in the Zeeman field. Effective heteronuclear P–N couplings have been "scaled up" in this fashion, as shown in Fig. 3.25.

Employing the equation of motion [Eq. (1.74)] of a two-spin-1/2 system under isotropic mixing, we find for the longitudinal component of the I-spin

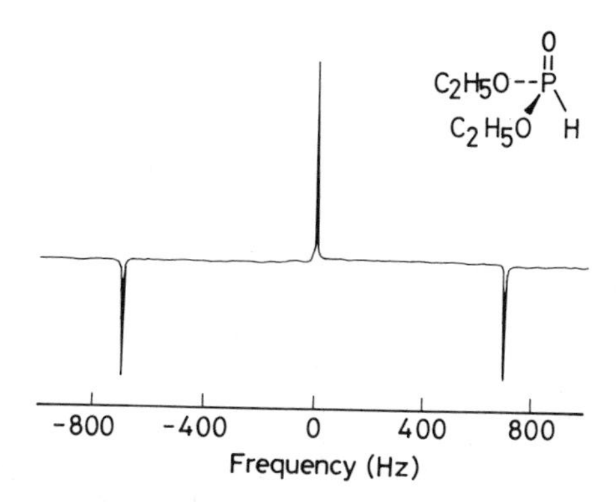

Figure 3.25. Zero-field ^{31}P spectrum of diethyl phosphite. [Reproduced by permission. D.B. Zax, A. Bielecki, K.W. Zilm, and A. Pines, *Chem. Phys. Lett.*, **106**, 550 (1984), North-Holland Physics Publishing, Amsterdam]

magnetization evolving under isotropic mixing for a period τ:

$$I_z \rightarrow \tfrac{1}{2}I_z(1 + \cos 2\pi J\tau) + \tfrac{1}{2}S_z(1 - \cos 2\pi J\tau) + (I_x S_y - I_y S_x)\sin 2\pi J\tau \tag{75}$$

This process therefore creates zero-quantum coherence (of phase y) from z order.

In contrast to J-cross polarization, evolution under isotropic mixing in AX_N systems (N arbitrary) is perfectly periodic, i.e., the various evolution frequencies are rationalmultiples of each other.

Coherence Transfer

In the discussion of laboratory-frame polarization transfer we observed that the first two pulses on the source spin channel that have a flip angle less than 180° are to be in quadrature with each other. This phase relationship is to be satisfied in order to maximize polarization transfer. In general, when chemical shifts are refocused, a pair of nonselective 90° pulses in quadrature with each other produce an odd-order coherence if the time interval between them is on the order of the inverse of the coupling between the spins.

In the SINEPT experiment we retained the chemical shift evolution and obtained with a pair of in-phase 90° pulses PT proportional to the sine of the chemical shift evolution. At the same time, this pair of in-phase 90° pulses produced even-order coherence, in this case weighted by the cosine of the chemical shift evolution. This is again a result of quite general validity: with the chemical shifts refocused, a pair of in-phase 90° pulses produce even-order coherences in the spin system.

The study of zero and higher order coherences, odd or even, reveals much significant information about the spin system. As shown in Chapter 1, however, only single-quantum coherences are observable in the FT NMR experiment. Other orders of coherence may be created by a suitable train of pulses, and allowed to evolve freely, followed by conversion to observable single-quantum coherences. By systematically varying the period of evolution of the multiple-quantum coherences (MQC's) prior to their conversion and observation, one can trace the evolution of MQC's as given by the resulting modulation of the observed single quantum coherences. Fourier transformation of the observed FID's, followed by a second Fourier transform of the resulting "interferograms," then leads to the multiple-quantum spectrum along the second frequency axis, whereas the normal single-quantum spectrum is displayed along the first. This strategy of two-dimensional spectroscopy to study MQC's is treated in Chapter 5.

In this chapter, however, we discuss some applications of MQC's in simplifying or cleaning up the spectrum of single-quantum coherences of the spin system. We have already employed the idea of a multiple-quantum trap in spectral editing and in polarization transfer; the purge strategies in both laboratory- and rotating-frame PT sequences also rely on the conversion of phase and multiplet anomaly components of the density operator into unobservable MQC's, thereby cleaning up the PT spectra.

We shall focus our attention now on some applications of MQC's in distinguishing between spin systems.

Double-Quantum Filtering

Single-spin systems that do not have any coupling partners can be distinguished from groups of coupled spins by having both their transverse magnetizations enter a double-quantum filter, which allows only the signals of coupled spins to pass. Features in a spectrum may thus be identified as arising from isolated spins or from coupled spins.

We shall discuss in detail an important application of this idea, named INADEQUATE (incredible natural abundance double-quantum transfer experiment), which is employed to measure ^{13}C–^{13}C couplings at natural abundance and so to derive the carbon connectivity in molecules. The pulse sequence is shown in Fig. 3.26.

Δ is a short time interval of a few microseconds. Proton noise decoupling is applied right through. The motion of the spins may be followed by the usual approach. At equilibrium, considering a two-spin system,

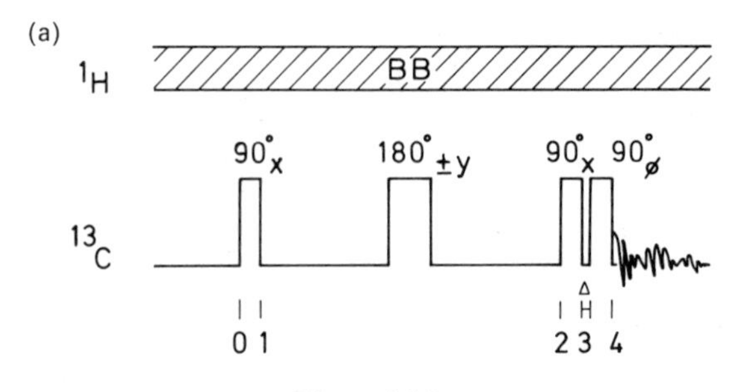

Figure 3.26.

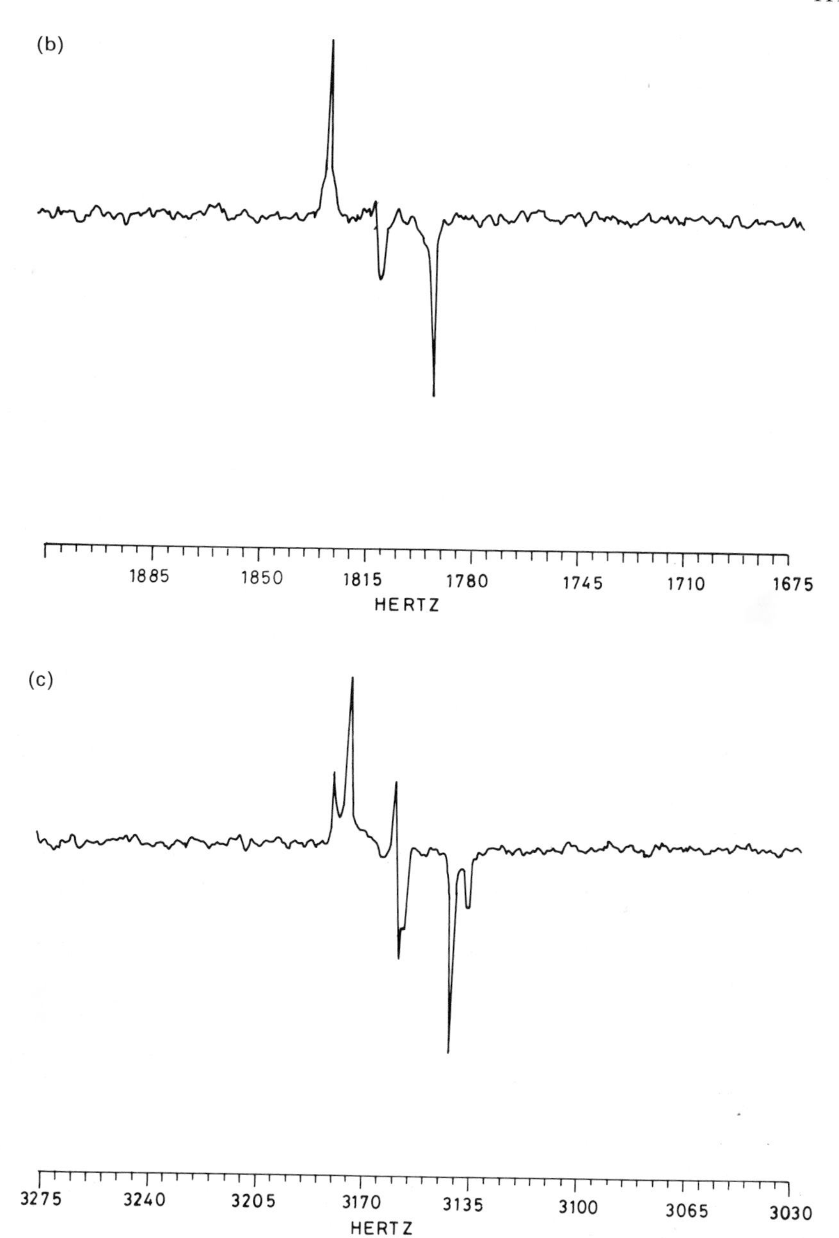

Figure 3.26. (a) INADEQUATE pulse sequence. The duration between points 1 and 2 is τ sec. (b) The 100-MHz ^{13}C INADEQUATE spectrum ethylbenzene, methyl carbon. (c) The 100-MHz ^{13}C INADEQUATE spectrum of ethyl benzene, methylene carbon.

$$\sigma_0 = I_{1z} + I_{2z} \tag{76}$$

After the first pulse, we have:

$$\sigma_1 = -(I_{1y} + I_{2y}) \tag{77}$$

During τ the spin system evolves under the mutual couplings:

$$\sigma_2 = -c_J(I_{1y} + I_{2y}) + 2s_J(I_{1x}I_{2z} + I_{1z}I_{2x}) \tag{78}$$

Following the second 90° pulse, we find:

$$\sigma_3 = -c_J(I_{1z} + I_{2z}) - 2s_J(I_{1x}I_{2y} + I_{1y}I_{2x}) \tag{79}$$

The second term of σ_3 describes a state of pure double-quantum coherence (DQC) (see Chapter 1). Depending on the phase ϕ of the "read" pulse we find:

$$\begin{aligned} \sigma_4 &= c_J(I_{1y} + I_{2y}) - 2s_J(I_{1x}I_{2z} + I_{1z}I_{2x}) && \text{for } \phi = x \\ &= -c_J(I_{1x} + I_{2x}) + 2s_J(I_{1z}I_{2y} + I_{1y}I_{2z}) && \text{for } \phi = y \\ &= -c_J(I_{1y} + I_{2y}) + 2s_J(I_{1x}I_{2z} + I_{1z}I_{2x}) && \text{for } \phi = -x \\ &= c_J(I_{1x} + I_{2x}) - 2s_J(I_{1z}I_{2y} + I_{1y}I_{2z}) && \text{for } \phi = -y. \end{aligned} \tag{80}$$

The signed magnetization component leading to anti-phase doublets is maximized when $\tau = (2n + 1)/2J$.

The unsigned component of the magnetization behaves as if it were subject to a 180° pulse prior to the read pulse, and, as a function of ϕ, its phase cycles in the characteristic fashion of a single-quantum coherence. The signed component, however, cycles in the opposite sense.

Arranging a "receiver phase" cycling (actually, cycling of the data routing to the computer in a quadrature detection system), which is in the same sense as the signed magnetization, one reinforces these signals while canceling the unsigned magnetizations. It may be noted that the roughly 200-fold stronger signal from molecules with a single ^{13}C behaves, of course, in the same fashion as the unsigned magnetization, so these signals are filtered out as well.

The pulse sequence with phase cycling therefore behaves like a double-quantum filter which allows only signals from coupled spin systems to emerge. It may be noted that at least a two-spin-1/2 system is required to create DQC and thus pass the double-quantum filter. However, spin systems with more than two particles also pass the filter. In the context of ^{13}C INADEQUATE, molecules with two ^{13}C nuclei are 1 in 100 of those with a single ^{13}C, whereas those with three are another 100-fold rarer. Taking into account the sensitivity factor, therefore, INADEQUATE functions in effect as a two-spin filter, i.e., rejects signals from all except two-spin systems. (We shall see later that when sensitivities are comparable, the realization of an N-spin filter is trickier and less satisfactory than that of an N-quantum filter.)

The effects of pulse imperfections and imbalances in the quadrature channels of the receiver are "cycled out" by a 32-step phase cycle: the four-step cycle discussed above is augmented by phase alternation of the 180° pulse and finally by CYCLOPS (see Chapter 7).

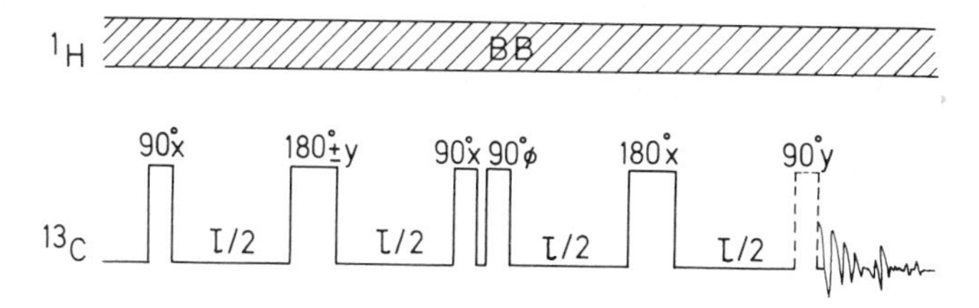

Figure 3.27. Symmetrical excitation detection sequence for INADEQUATE, generating in-phase multiplets. σ_5 in Eq. (82) denotes the density operator at the end of the last $\tau/2$ period.

Equivalent to the basic four-step cycle, the phase ϕ of the DQC excitation sandwich 90°_ϕ–$\tau/2$–180°_ϕ–$\tau/2$–90°_ϕ with respect to that of the read pulse is varied as $\phi = k\pi/2$ ($k = 0, 1, 2, 3$), leading to a phase shift of the DQC by 2ϕ (see Chapter 1). Alternatively, this phase cycling may be replaced by a θ pulse of variable flip angle ($\theta. = k\pi/2$) to give the INADEQUATE sequence:

$$90^\circ_x\text{–}\tau/2\text{–}180^\circ_x\text{–}\tau/2\text{–}\theta_y\text{–}90^\circ_x \qquad (81)$$

The FID's are alternately added and subtracted as θ (or ϕ) is cycled through 0°, 90°, 180°, and 270°.

As usual, the anti-phase doublets resulting from INADEQUATE may be refocused to give in-phase signals by extending the pulse sequence as shown in Fig. 3.27. This gives, for instance:

$$\sigma_5 = 2(I_{1x}I_{2z} + I_{1z}I_{2x})s_J c_J + s_J^2(I_{1y} + I_{2y}) \qquad (82)$$

The sequence of Fig. 3.27 is a symmetrical excitation/detection scheme, which, upon signal averaging with variable τ, results in uniform excitation/detection of the DQC's, independent of the magnitude of J. We have for this case:

$$\bar{\sigma}_5 = \tfrac{1}{2}(I_{1y} + I_{2y}) \qquad (83)$$

The final 90°_y pulse on alternate scans gets rid of phase anomalies regardless of the choice of τ, i.e., the relation between τ and $(J)^{-1}$. Alternatively a 45°_y purge pulse may be applied on each scan.

It may be mentioned that experiments involving coherence transfers to higher order coherences are strongly sensitive to rf inhomogenity and resonance offset effects. The INADEQUATE experiment performs better in practice when these effects are compensated by composite pulses (see Chapter 7).

It is clear that INADEQUATE spectra arise only from two-spin systems which can be either weakly coupled (AX spin systems) or strongly coupled (AB spin systems). The efficiency of creation of DQC's as a function of τ depends on the δ/J ratio. For δ/J down to about $\sqrt{3}$, $\tau = (2J)^{-1}$ is the best choice; for stronger coupling, $\tau = 3(2J)^{-1}$ is a better bet although this implies a threefold sensitivity to "mismatch" between τ and J^{-1}.

The INADEQUATE experiment clearly has very poor sensitivity because it selectively detects molecules with two ^{13}C nuclei, suppressing the much stronger signals from those with only one. In order to improve the sensitivity somewhat, it is possible to perform an INEPT–INADEQUATE experiment. This technique affords a modest improvement—by a factor of 2 to 3—in sensitivity in a given measurement time over the normal INADEQUATE experiment with NOE, especially since the sequence repetition rate is governed by the proton T_1's. The pulse sequence is set out in Fig. 3.28.

The phases of the rf pulses a to d and the receiver phase are cycled as given in Table 3.2. The performance of the sequence may be appreciated from the spectra in Fig. 3.29. Spectral editing techniques, such as SEMUT or DEPT or their GL versions, may be combined with INADEQUATE to yield the SEMINA experiment, which gives INADEQUATE subspectra of CH_n–CH_m fragments for various multiplicities ($n + m$). An especially useful SEMINA strategy is to generate two INADEQUATE subspectra depending on whether

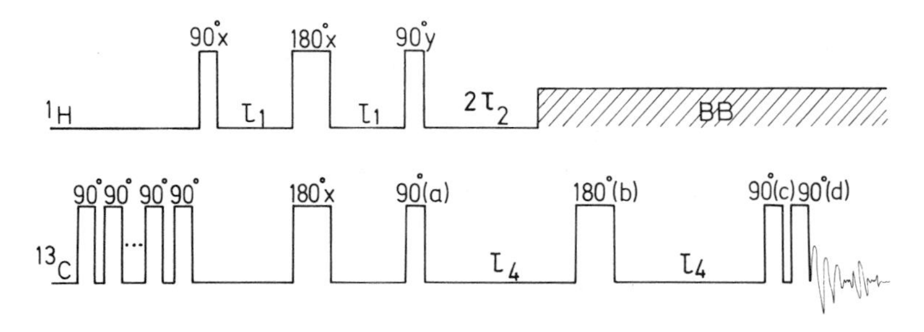

Figure 3.28. Pulse sequence for INADEQUATE preceded by INEPT for sensitivity enhancement.

Table 3.2. Phase Cycling for INEPT–INADEQUATE

Group	Run	(*a*)	(*b*)	(*c*)	(*d*)	(*e*)*
A	1–4	−Y	X	X	X	X
	5–8	−Y	X	X	Y	−Y
	9–12	−Y	X	X	−X	−X
	13–16	−Y	X	X	−Y	Y
B	17–20	−Y	X	−X	X	−X
	21–24	−Y	X	−X	Y	Y
	25–28	−Y	X	−X	−X	X
	29–32	−Y	X	−X	−Y	−Y
C	33–36	−Y	Y	X	X	−X
	37–40	−Y	Y	X	Y	Y
	41–44	−Y	Y	X	−X	X
	45–48	−Y	Y	X	−Y	−Y
D	49–52	−Y	Y	−X	X	X
	53–56	−Y	Y	−X	Y	−Y
	57–60	−Y	Y	−X	−X	−X
	61–64	−Y	Y	−X	−Y	Y

* Receiver phase

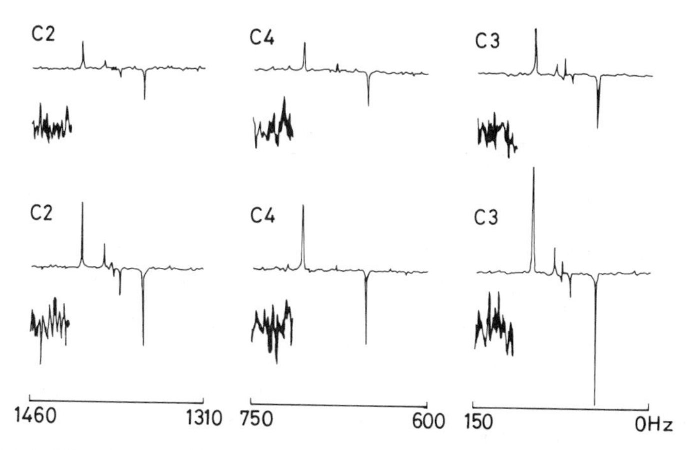

Figure 3.29. The 50-MHz ^{13}C INEPT–INADEQUATE spectrum of pyridine (lower trace) compared with the normal NOE–INADEQUATE spectrum (upper trace). [Reproduced by permission. O.W. Sørensen, R. Freeman, T. Frenkiel, T.H. Mareci, and R. Schuck, *J. Magn. Reson.*, **46**, 180 (1982), copyright 1982, Academic Press, New York]

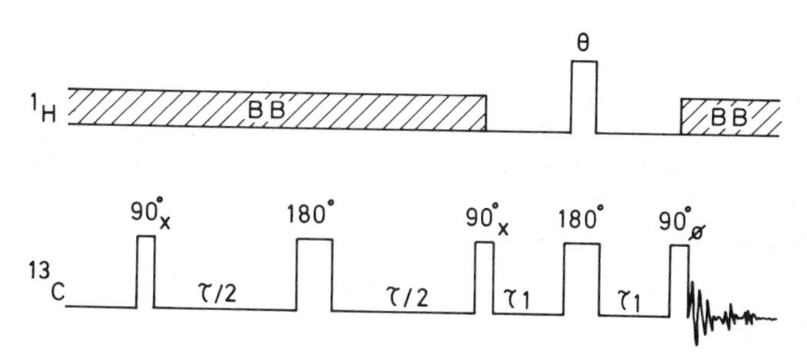

Figure 3.30. SEMINA sequence.

$(n + m)$ is odd or even. A suitable pulse sequence is shown in Fig. 3.30. Two INADEQUATE subexperiments are run for equal durations with $\theta = 0°$ and 180°. The first subexperiment ($\theta = 0°$) gives the normal, anti-phase INADEQUATE spectrum, whereas in the second ($\theta = 180°$), all anti-phase CH_n–CH_m doublets with odd $(n + m)$ appear inverted. The sum and difference of these two spectra thus yield the desired INADEQUATE subspectra. SEMINA is a useful assignment aid, especially when different ^{13}C–^{13}C subsystems give overlapping INADEQUATE responses. An example is shown in Fig. 3.31.

We now mention a related double-quantum filter technique termed DOUBTFUL (double quantum transitions for finding unresolved lines), which has been developed to identify two-spin systems with a specific double-quantum frequency. This method employs the same pulse sequence as on the

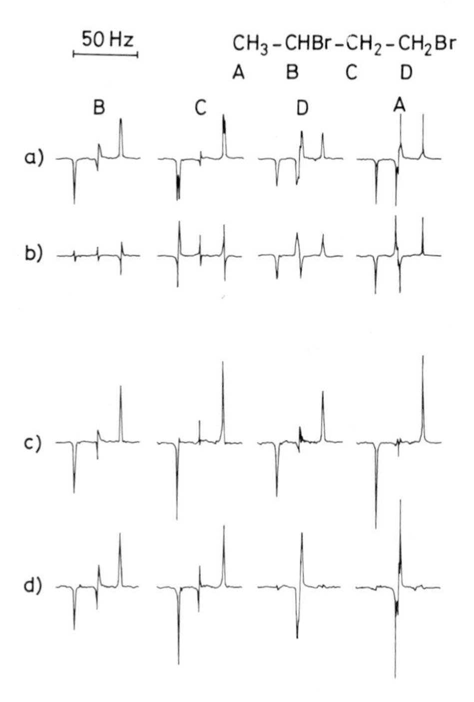

Figure 3.31. The ^{13}C SEMINA spectra of 1,3-dibromobutane at 75.45 MHz, $\tau = 121.6$ ms. (a) $\theta = 0°$; (b) $\theta = 180°$; (c) $n + m$ even subspectrum obtained by adding (a) and (b); (d) $n + m$ odd subspectrum obtained by subtracting (b) from (a). [Reproduced by permission. O.W. Sørensen, U.B. Sørensen, and H.J. Jakobsen, *J. Magn. Reson.*, **59**, 332 (1984), copyright 1984, Academic Press, New York]

^{13}C channel of Fig. 3.26, but the short period Δ is replaced by a variable MQC evolution period t_1. During t_1, the MQC's precess in the rotating frame at the algebraic sum of the offsets of the two coupled spins [Eqs. (1.53) to (1.54)]. If the normal phase cycling of the double-quantum filter is augmented by signal averaging as a function of t_1, all components interfere destructively except those for which the transmitter is placed so as to render the algebraic sum of the offsets zero, i.e., at the center of the AX/AB spectrum, so that the acquired signal is unmodulated as a function of t_1. Figure 3.32 gives an example of the application of the DOUBTFUL technique.

Multiple-Quantum Filters and Spin Filters. Our discussion has so far centered on double-quantum filters. The same general principles may be extended to multiple-quantum filters with a phase cycling of the excitation sandwich in steps of $\phi = k\pi/p$, p being the "order" of the filter and k taking on the values 0, 1, 2, 3, ..., $(2p - 1)$. The general symmetrical sequence for uniform excitation/detection of p-quantum coherence is shown in Fig. 3.33. For

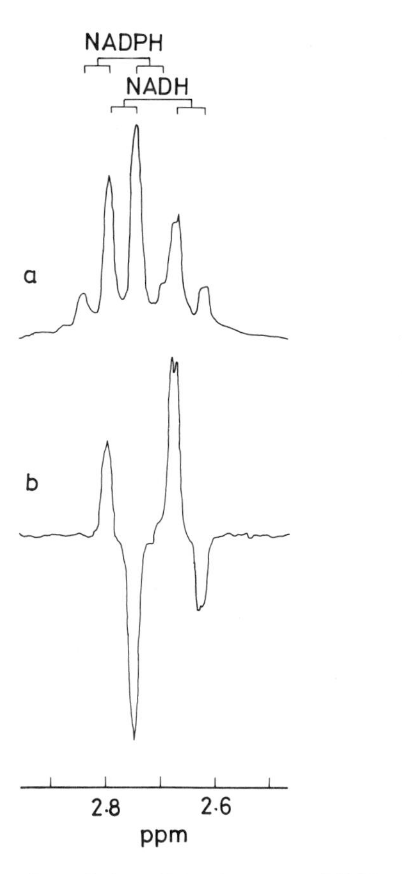

Figure 3.32. Trace (a) A portion of the 360-MHz proton NMR spectrum of equimolar β-nicotinamide adenine dinucleotide (NADH), and NADH c-2′-phosphate (NADPH), in D_2O. Trace (b) DOUBTFUL spectrum; transmitter placed at the center of the two NADH $C(4)H$ doublets; $\tau = 12.5$ ms; seven spectra ($t_1 = 0, 6, 12, 18, 24, 30$, and 36 ms) coadded. [Reprinted with permission from P.J. Hore, E.R.P. Zuiderweg, K. Nicolay, K. Dijkstra, and R. Kaptein, *J. Amer. Chem. Soc.*, **104**, 4286 (1982), copyright 1982, American Chemical Society]

odd-order coherences $\psi = 90°$ and $\psi = 0°$ for even-order coherences. This sequence, upon τ averaging, suppresses all magnetizations save those that originate from and end up on the same spin. The τ averaging may be improved by removing the refocusing pulses, allowing averaging under chemical shift precession during τ rather than under precession due to couplings. The resulting shift dependence of the desired magnetization components may be avoided by subtracting two experiments with the phase of both the second and third 90° pulses incremented together by (1) 0° and (2) 90° for each step of the ϕ and ψ phase cycle. The final purge pulse restores the natural multiplet patterns if applied on each scan; applied on alternate scans, it improves the

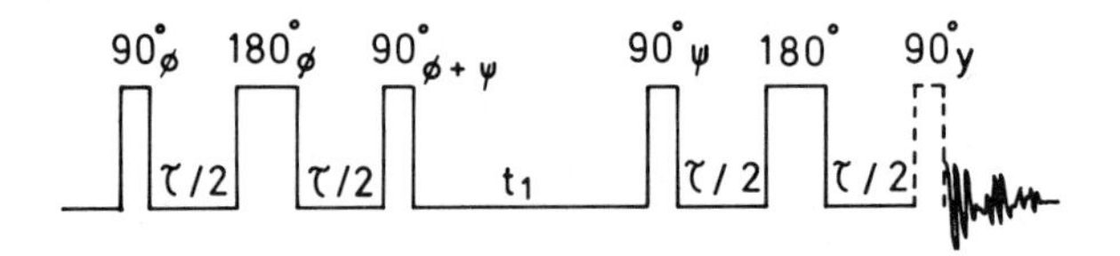

Figure 3.33. General pulse sequence for symmetric excitation/detection of p-quantum coherence.

suppression of antiphase magnetization. For a triple-quantum filter, for instance, the required phase shifts of the excitation sandwich are 0, 60°, 120°, 180°, 240°, and 300°. One may in principle translate these phase shifts into pulses with variable flip angles $k\pi/p$, whose phase is y or x depending on whether p is even or odd, and alternately add and subtract the FID's as the flip angle is incremented cyclically. The pulse sequence $90^\circ_x-\tau/2-180^\circ_x-\tau/2-\beta_y-90^\circ_x$, for instance, creates pure p-quantum coherence (p: even) when β is cycled through $k\pi/p$ and the FID's alternately added and subtracted.

Similarly, the sequence $90^\circ_x-\tau/2-180^\circ_x-\tau/2-\beta_x-90^\circ_y$ creates pure p-quantum coherence for odd p. The one-dimensional filters thus take the form $90^\circ_x-\tau/2-180^\circ_x-\tau/2-\beta_{y,x}-\tau/2-180^\circ_x-\tau/2-(90^\circ_y)$, where the β pulse must have phase y for even-quantum filters, and otherwise phase x. However, in practice, the replacement of phase shifts with variable flip angle pulses is subject to large errors depending on the offset, rf inhomogeneity, etc., so the method is perhaps not of universal practical value.

The application of a double-quantum filter is illustrated in Fig. 3.34 with a mixture of one-, two-, and three-spin systems.

The selection of a specific isolated N-spin system is also possible by an

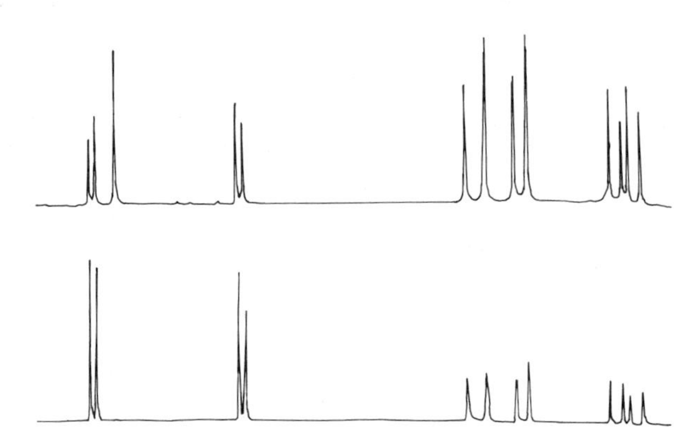

Figure 3.34. Double-quantum filter applied to a mixture of benzene (A_6), 2,3-dibromothiophene (AX) and acrylonitrile (AMX). Top trace: normal 300-MHz proton spectrum. Lower trace: filtered spectrum. τ was varied in 8-ms steps from 8 to 736 ms. [Reproduced by permission. O.W. Sørensen, M.H. Levitt and R.R. Ernst, *J. Magn. Reson*, **55**, 104 (1983), copyright 1983, Academic Press, New York]

extension of the idea of p-quantum filtering in analogy to the DOUBTFUL strategy. For instance, the following sequence may be used to excite and detect triple-quantum coherence selectively:

$$90^\circ_x - \tau/2 - 180^\circ_x - \tau/2 - \beta_x - 90^\circ_y - \tau_1 - 90^\circ_y - \text{Acquisition } (\theta) \qquad (84)$$

Here, β takes the values 0°, 60°, 120°, 180°, 240°, and 300°, and the receiver phase θ is alternately y and $-y$. Pulse imperfections may be compensated by cycling together in 90° steps the phases of the receiver as well as all five pulses. The transmitter is positioned such that the triple-quantum coherence (3QC) of interest does not precess during t_1; signal averaging with different values of t_1 then tends to cancel the oscillatory responses of the other 3QC's while the selected 3QC adds coherently. Figure 3.35 gives an example of the application of this technique.

It is clear that in general a p-quantum filter behaves as a highpass filter in the spin number domain, passing signals from all N-spin systems with $N \geqslant p$, although the responses for $N > p$ are attenuated compared to the $N = p$ situation. An attempt to design a bandpass filter in the spin number domain leads to the idea of the p-spin filter, which passes signals only from p-spin systems. A pulse sequence that functions in this fashion and its application to a mixture of p-spin systems with $p = 1$, 2, and 3 are shown in Figs. 3.36 and 3.37.

The p-spin filter exploits the fact that p-quantum coherences of p-spin

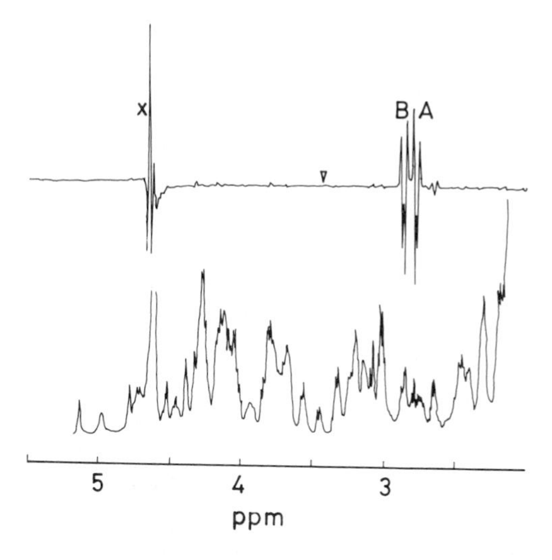

Figure 3.35. The 500-MHz proton NMR of 6.2 mM lac repressor headpiece (residues 1–51) in D_2O. Lower trace: normal spectrum. Upper trace: 3QC filter and t_1 averaging to pick up the ABX fragment alone ($\tau = 40$ ms; $t_1 = 0, 1.25, 2.5, \ldots, 25$ ms). The transmitter position is indicated by the arrow. [Reproduced by permission. P.J. Hore, R.M. Scheek, and R. Kaptein, *J. Magn. Reson.*, **52**, 339 (1983), copyright 1983, Academic Press, New York]

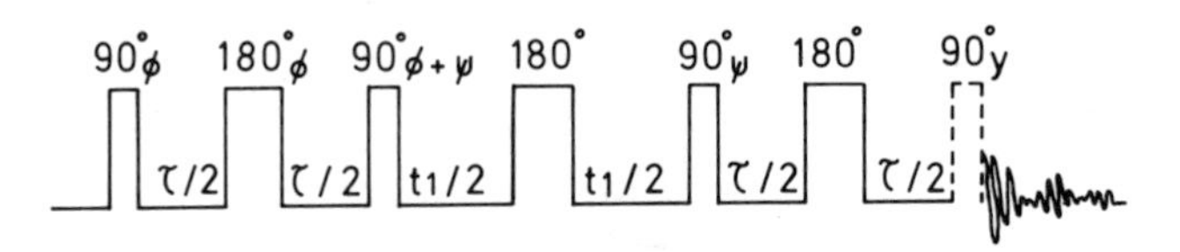

Figure 3.36. A p-spin filter.

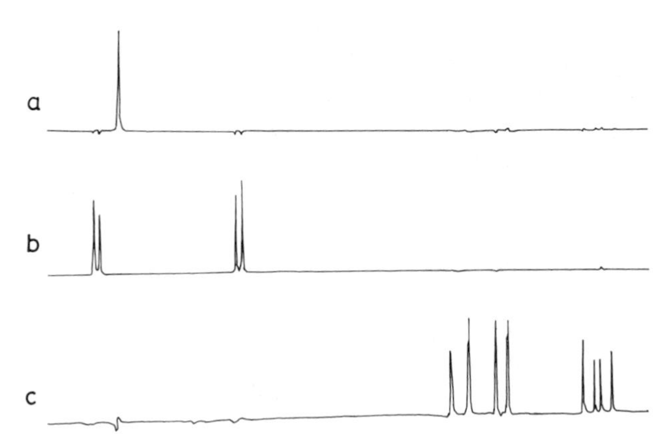

Figure 3.37. One-, two-, and three-spin filtered 300-MHz proton spectra of a mixture of benzene, 2,3-dibromothiophene, and acrylonitrile. For trace (a) (one-spin filter), spectra with 78 values of t_1 from 32 to 648 ms have been coadded; for trace (b) (two-spin filter), $\tau = 80$ ms, and t_1 has been varied from 32 to 824 ms in 100 steps; for trace (c) (three-spin filter), τ was varied from 32 to 648 ms in 78 steps; $t_1 = 0$. [Reproduced by permission. O.W. Sørensen, M.H. Levitt, and R.R. Ernst, *J. Magn. Reson.*, **55**, 104 (1983), copyright 1983, Academic Press, New York]

systems do not evolve under coupling, while p-quantum coherences of N-spin systems ($N > p$) do, under coupling to the remaining ($N - p$) passive spins. The period t_1 during which the MQC's evolve (which is a vanishingly small time interval in one-dimensional p-quantum filters) is therefore made variable, and the resulting FID's are averaged to try to eliminate the responses of spin systems with $N > p$. However, it should be stressed that this strategy is not entirely satisfactory, since, in general, an N-spin system with $N > p$ will pass the p-spin filter whenever one of its p-quantum transitions has an unmodulated multiple-quantum multiplet component.

Pulse sequences that seek to identify spin systems of the type A_nX_m not on the basis of the number of coupled spins alone as in the above cases, but also on the basis of the topography of the coupling network are being developed. Such sequences employ symmetrical excitation/detection schemes and generate MQC's with a higher efficiency than the sequences discussed so far. Two such sequences for A_nX systems, for n odd and even, respectively, are shown in Fig. 3.38.

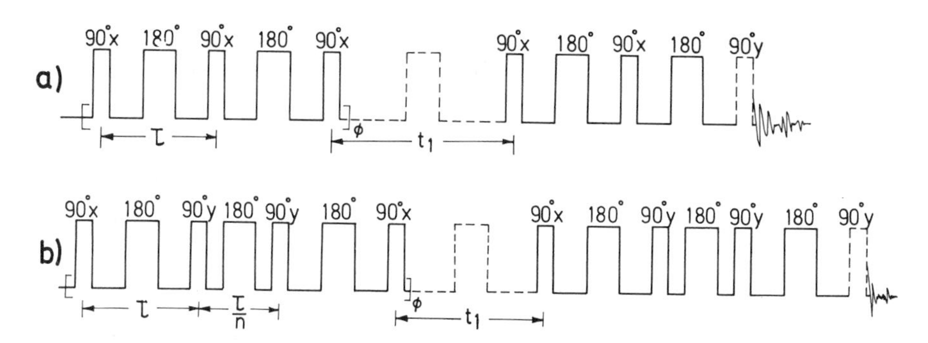

Figure 3.38. Pulse sequences for recognition of A_nX spin systems. (a) n : even; (b) n : odd.

In these sequences, ϕ is cycled through $k\pi/(n+1)$ $(k = 0, 1, 2, \ldots, 2n+1)$, while $\tau = (2J)^{-1}$. Signal averaging is performed with a range of t_1 values, $0 < t_1 < J^{-1}$ as usual and the final purging may be performed with a y pulse of variable flip angle when anti-phase, quadrature magnetization components experience a modulation and therefore lend themselves to suppression, unlike the in-phase y-magnetization components. It may further be noted that considering four-spin systems, for instance, the sequence of Fig. 3.38b passes only signals of A_3X systems and not those of A_2X_2, whereas the same sequence with the pulses of phase y replaced with x-phase pulses in the excitation sandwich passes A_2X_2, but not A_3X. Some experimental results using these sequences are shown in Fig. 3.39.

It may be noted, however, that such sequences work well only for weak coupling and τ well matched with $(2J)^{-1}$; they also lead to poor sensitivities owing to T_2 effects during the long sequences.

z Filters

Finally, we briefly discuss the idea of z filters, which are less sensitive to transverse relaxation effects and work well as a purge strategy for suppressing phase and multiplet anomalies, when added just prior to acquisition. The z filter is represented in Fig. 3.40.

The first 90° pulse throws the desired component of the transverse magnetization onto the z axis, while converting the undesired components into MQC's and multiple spin order. The MQC's are suppressed by coadding results of a number of experiments with different MQC precession periods τ_z. The final 90° pulse reconverts the z components that are retained during τ_z into transverse magnetization components and unobservable MQC's. The filter efficiency may be improved by using a four-step phase cycle with 90° increments (such as CYCLOPS) on the second pulse and receiver reference phase. Figure 3.41 shows the effect of applying a z filter to a simple spin-echo multiplet selection sequence.

Less gratifying performance, however, is encountered with a spread of J and/or T_2.

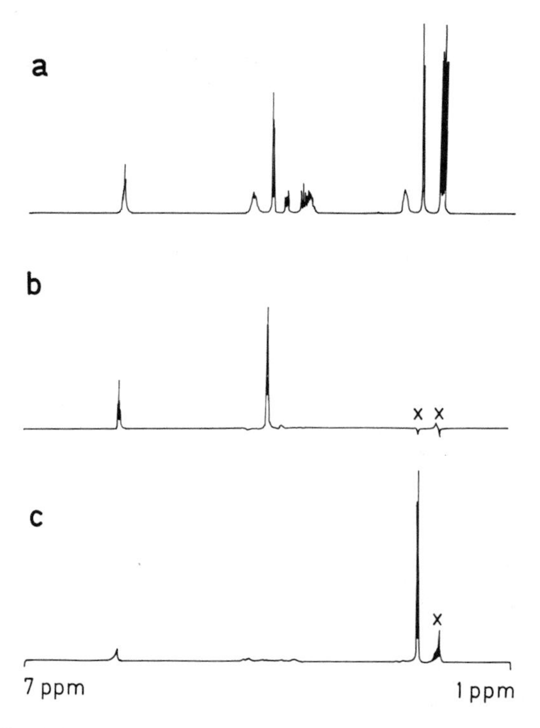

Figure 3.39. The 300-MHz proton spectra of an equimolar mixture of 1,1-dichloroethane (A_3X), 1,1,2-trichloroethane (A_2X), 1,3-dibromobutane (A_3MPQ "XY"), and 1,2-dibromopropane (A_3MPQ). (a) Conventional spectrum. (b) Suppression of all but the A_2X resonances, employing the sequence of Fig. 3.38a. (c) Suppression of all but the A_3X resonances, employing the sequence of Fig. 3.38b. Incompletely suppressed peaks are marked by ×. [Reproduced by permission. M.H. Levitt and R.R. Ernst, *Chem. Phys. Lett.*, **100**, 119 (1983), North-Holland Physics Publishing, Amsterdam]

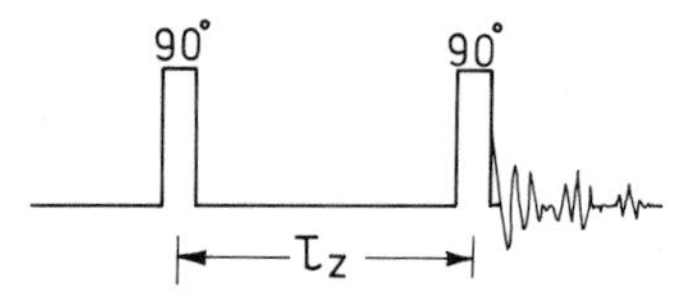

Figure 3.40. z-filter pulse sequence.

Figure 3.41. (a) The 300-MHz proton spin-echo spectrum (0.4–4.5 ppm) of an equimolar mixture of isobutanol, isopropanol, 2-butanone, and 2-butanol. Top trace: with the $90^\circ_x-\frac{\tau}{2}-180^\circ_\phi-\frac{\tau}{2}$ sequence ($\tau = 156$ ms). Bottom trace: with z filter employing 10 τ_z values (0, 18, 36, ... ms). (b) Top trace: single-pulse spectrum of the mixture of (a). Middle trace: additive combination of top trace spectrum with z-filtered spectrum of (a), giving the spectrum exclusively of protons with an odd number of coupling partners. Bottom trace: Subtractive combination of top trace spectrum with z-filtered spectrum of (a). This gives the spectrum of only those protons with an even number of coupling partners. [Reproduced by permission. O.W. Sørensen, M. Rance, and R.R. Ernst, *J. Magn. Reson.*, **56**, 527 (1984), copyright 1984, Academic Press, New York]

(a)

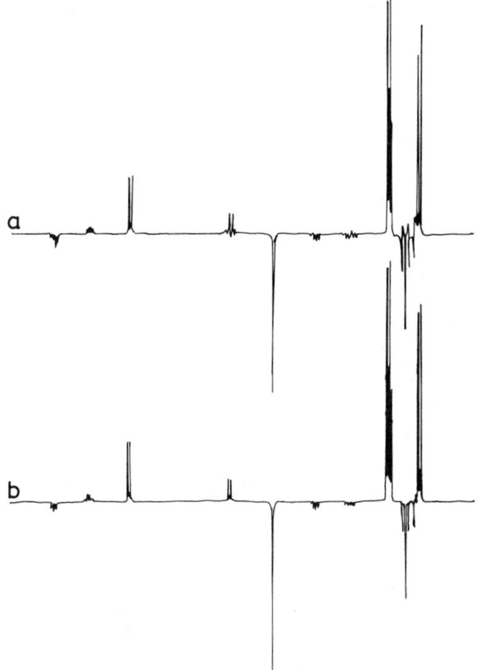
a
b

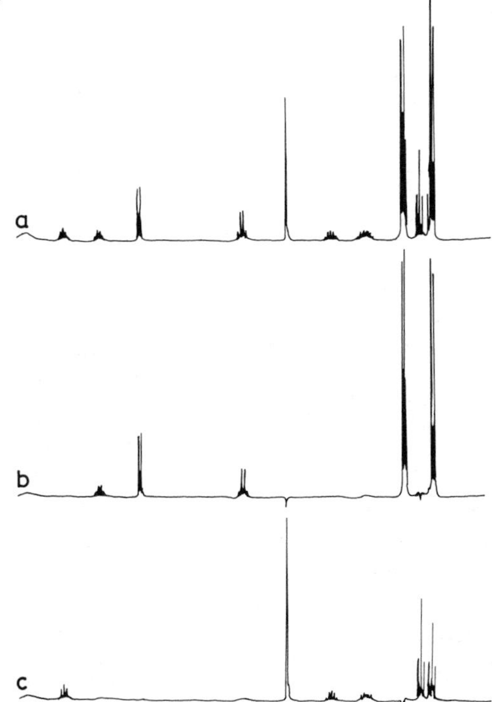
(b)
a
b
c

CHAPTER 4

Two-Dimensional Experiments in Liquids

Introduction

The high-resolution NMR of complex molecules containing several chemically shifted nuclei with their associated spin–spin couplings can give rise to a fairly complicated frequency spectrum. The analysis of the spectrum obtained in a continuous wave (CW) mode or by simple one-dimensional pulse FT mode can be fraught with ambiguities so that unequivocal assignments may not be possible, especially in the case of such complex biomolecules as proteins and nucleic acids. To analyze these latter, one can resort to techniques mentioned in Chapter 2, such as selective, off-resonance, and broadband decoupling. There is no guarantee, however, that these will lead to a complete analysis. For example, in off-resonance decoupling, in which it may be possible to scale down multiplet separation to reduce intermultiplet overlap, there is obviously a practical lower bound to the relevant chemical shift difference, below which overlap is unavoidable, even with uniform J scaling methods.

In this connection it is interesting to ponder for a moment on the time evolution of a complex spin system. The free induction decay (FID) is a complex time evolution of various circularly polarized transverse components at their respective chemical shifts (depending on the offset) which are further modulated by hetero- or homonuclear spin–spin coupling. The single-quantum coherences evolve in the rotating frame with their offset frequencies modulated by other zero-field terms in the NMR Hamiltonian; other coherences will have precessional frequencies depending on the order (Chapter 1). Additionally, the various coherences are interlinked via direct and cross-relaxation processes. Suppose it is possible to monitor the time evolution of the system instead of a function of a single time variable, but as a function of two successive periods t_1 and t_2 such that we are able to "tailor" the evolution under t_1 and t_2 in a predetermined way (say in period t_1 only scalar couplings are allowed to operate on the spin system, whereas in period t_2 both chemical shifts and spin couplings are allowed to operate). Then the various modulation frequencies present in the two time regions can be looked at separately, and in the process we get an unraveling of the complex pattern that would otherwise be obtained in a simple NMR spectrum. This concept of disentangling a complex spectrum by projecting out different spectra by suitably dressing the Hamiltonian has developed into an entirely new class of exciting experiments in high-resolution NMR of liquids. We shall not attempt to enumerate all such two-dimensional (2D) experiments; we restrict our attention to the fundamental principles underlying the 2D philosophy and include representative examples of 2D chemical shift correlation through homo- and heteronuclear scalar coupling, 2D homonuclear and heteronuclear J-resolved spectroscopy, 2D cross-relaxation, and exchange spectroscopy. There are a number of other 2D experiments involving multiple-quantum coherences as well as 2D experiments in solids. These are discussed, respectively, in Chapters 5 and 6.

Principles of 2D Spectroscopy

In order to understand the basic aspects of 2D spectroscopy, it is best to start with a schematic of the time evolution of the system under suitable pulse sequences and to divide the total region into four time realms; namely, the preparation period, the evolution period, the mixing period, and the detection period (Fig. 4.1).

During the preparation period a nonequilibrium state of the system is created by applying suitable pulses. The preparation can be with or without broadband decoupling in the heteronuclear case. It may be a selective or nonselective $\pi/2$ or θ pulse. It could be a pair of closely spaced $\pi/2$ pulses, etc.

During the evolution period t_1 (this variable is important because whatever modulation that takes place in this period is going to be "carried over" to further periods of evolution) the system is allowed to evolve under suitably tailored Hamiltonians. During this period the spin system may freely evolve or may be subject to interrupted evolution under suitable pulses.

In the third period of mixing, which is not mandatory for all 2D experiments, one can employ suitable pulses for effecting coherence transfer, or monitor cross-relaxation or chemical exchange, etc.

The fourth period, t_2, which corresponds to the acquisition of the final response of the observe spins, contains all the information. This period is usually free from any pulses of the observed spin, although broadband decoupling of the hetero spins can be applied during acquisition. This period,

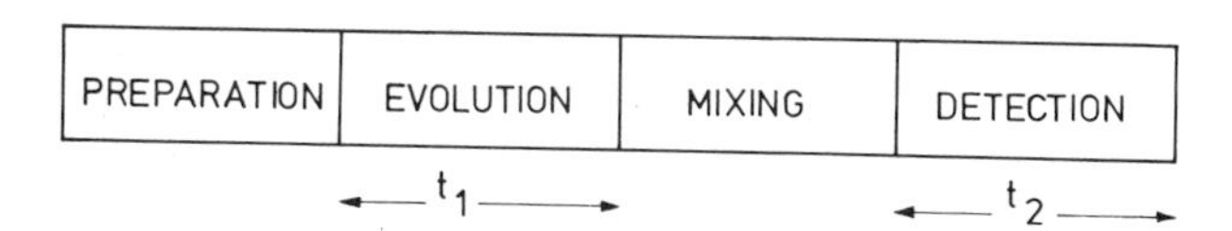

Figure 4.1. Timing sequence for 2D NMR spectroscopy.

however, can be tailored in a number of ways when working with solids. The important point to remember is that the total time of evolution from preparation to the start of acquisition is such that it is much less than the spin–spin relaxation time, so that the spin system "remembers" its history and retains the phase memories.

Double Fourier Transformation. In a 2D experiment one acquires a signal as a function of the detection period t_2 for a progressively incremented evolution period t_1. Note that t_1 is a "virtual" time variable, the receiver being entirely inactive during this period. "Quadrature detection" and "filtration" in t_1 are therefore conceptually—and in practice—quite different from the corresponding processes in t_2, as we shall see later. The sampling rate in the t_2 period determines the spectral width in the corresponding frequency dimension. The sensitivity of the time-domain signal can be increased by repeating the acquisition a number of times for a given t_1 and coherently adding the resulting time responses. This signal, which is labeled $S(t_1, t_2)$, is treated as a normal FID. It is suitably weighted and Fourier transformed (FT):

$$S(t_1, t_2) \xrightarrow{FT} S(t_1, F_2) \tag{1}$$

where $S(t_1, F_2)$ is a spectrum in F_2 dimension as a function of t_1. The resulting frequency spectrum, however, will be a phase- or amplitude- (or both) modulated version of the normal spectrum (see Fig. 4.2). Thus the first Fourier

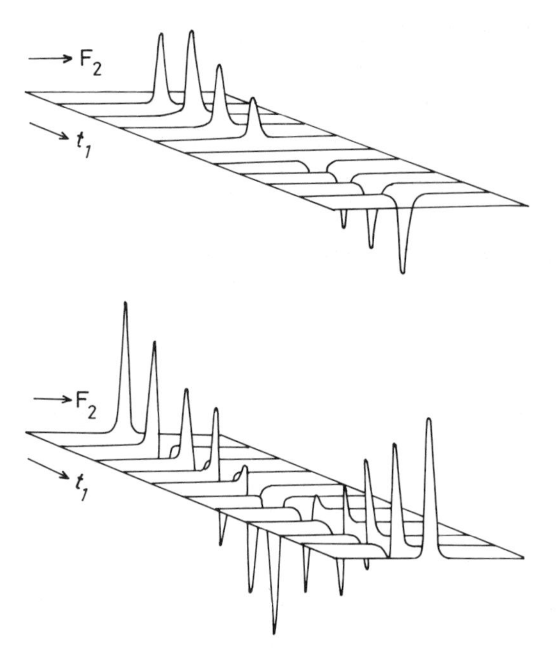

Figure 4.2. Single FT of $S(t_1, F_2)$ leading to interferograms $S(t_1, t_2)$ showing amplitude (top) and phase modulation along the t_1 axis.

transform contains the phase/amplitude modulation as a function of the evolution period t_1 in the form of sine and cosine transforms S^c, S^s. The parameter t_1 is then incremented, keeping in mind the range of frequencies in which one is interested in the next frequency dimension F_1. A 2-ms increment of t_1 corresponds to a frequency spread of 250 Hz in the F_1 dimension. Each successive FID is stored in the computer, and as soon as the core is full it is transferred to a disk. In the normal FT mode the input data are scaled by binary shifts to use the dynamic range of the computer effectively. Similar scaling also occurs in the fast Fourier transform algorithm. Such scaling factors have to be carefully kept track of in 2D experiments, because all the spectra (for say nt_1 increments) should be scaled by the same factor. Generally, this is done by storing the scale factor of each of the n spectra so that after the first Fourier transformation it can generate the $S(t_1, F_2)$ data matrix by a uniform scale factor depending on the highest signal. These spectra form the rows of the $S(t_1, F_2)$ matrix and each column corresponds to variation of a given frequency component as a function of time t_1 and is referred to as an "interferogram." It contains specific information depending upon the type of Hamiltonian that has been operative in the period t_1. The columns of the $S(t_1, F_2)$ matrix are subject to long-term instabilities of the spectrometer and hence usually weighted by an exponentially decaying function and apodized before Fourier transformation (Appendix 3).

Although the data matrix $S(t_1, F_2)$ is stored as a series of spectra $S(F_2)$, the second Fourier transform must be done on the series of interferograms $S(t_1)$, which are not conveniently retrievable in a short time. The following description is perhaps typical when a computer of only limited CPU memory is available. Consider a square matrix of 1024×1024 data points of $S(t_1, F_2)$. We can transfer the first 128 spectra of $S(t_1, F_2)$ into the core, which contains the first 128 points of all 1024 interferograms. However, the computer core is insufficient to handle 1024 full length interferograms. The remaining seven data blocks have to be read in sequentially until all the 1024 points of the first 128 interferograms have been loaded. This is then Fourier transformed and stored in another portion of the disk corresponding to 128 traces of $S(F_1, F_2)$. The whole sequence is repeated in blocks of 128 until all the interferograms are transformed and the complete two-dimensional $S(F_1, F_2)$ is ready. If additional zero filling of the interferogram is done the new matrix $S(F_1, F_2)$ becomes correspondingly larger. This sequential loading, transposition, and Fourier Transformation typically takes a few minutes to hours depending on the size of the data matrix and speed of the data processing. It is not necessary that the data matrix be square. In fact, in practice, in the F_1 dimension the size of data matrix is usually several factors smaller and the time required for 2D transformation is much less. In some special cases, when the normal spectrum is well known, it is enough that only interferograms corresponding to regions in the F_2 dimension where NMR resonances are present be fed in for the second Fourier transformation.

Thus the first Fourier transformation generates the sine and cosine parts

$S^c(t_1, F_2)$ and $S^s(t_1, F_2)$, which, upon the second Fourier Transformation, yield four parts $S^{cc}(F_1, F_2)$, $S^{cs}(F_1, F_2)$, $S^{sc}(F_1, F_2)$, and $S^{ss}(F_1, F_2)$. If in the t_1 dimension there are phase rotations of either sign (for example, in J-resolved spectroscopy, vide infra), then all the four components are to be retained to distinguish positive and negative frequencies in the F_1 axis. No doubt, therefore, that the adjustment of the resonance lineshapes to absorption/dispersion modes is quite complicated in a 2D spectrum and is particularly so in a phase-modulated spectrum, because the weightage of all the four different parts, S^{cc}, S^{cs}, S^{sc}, and S^{ss}, are to be controlled. Usually, a small portion of the 2D spectrum is displayed on the oscilloscope and, using interactive software, one can adjust the phases to any level of accuracy. A cross-section of a typical 2D response parallel to frequency axes at the center of the line gives a pure absorption mode lineshape, whereas parallel cross-sections on either side show rapid change in lineshape, for example, through negative absorption–dispersion–positive absorption. This is the so-called phase-twisted lineshape. A typical phase-twisted lineshape and the general schematics of handling the data matrix for double Fourier transformation are shown in Figs. 4.3 and 4.4. The 2D spectra are presented by stacking the spectra one beside the other by displacements along the x and y axes, to produce a three-dimensional effect. The perspective is further enhanced by the so-called "whitewash" plot routine, in which peaks with lower intensities appearing "behind" another peak are not plotted.

Phase and Amplitude Modulation. As has been pointed out earlier the signal that is acquired in the detection period has "memory" of what happened in the t_1 period. This can affect the phase, the amplitude, or both of the final signal. Phase-modulation information can give the signs of modulation fre-

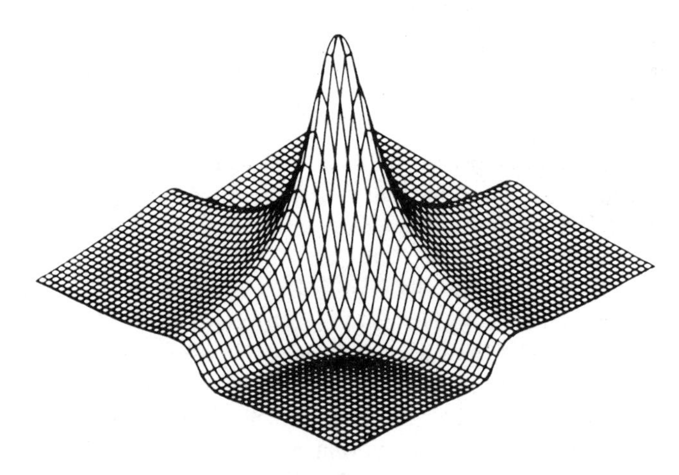

Figure 4.3. Typical phase-twisted lineshape obtained in many 2D NMR experiments. The lineshape typically goes through dispersion to absorption to negative dispersion as it passes through resonance.

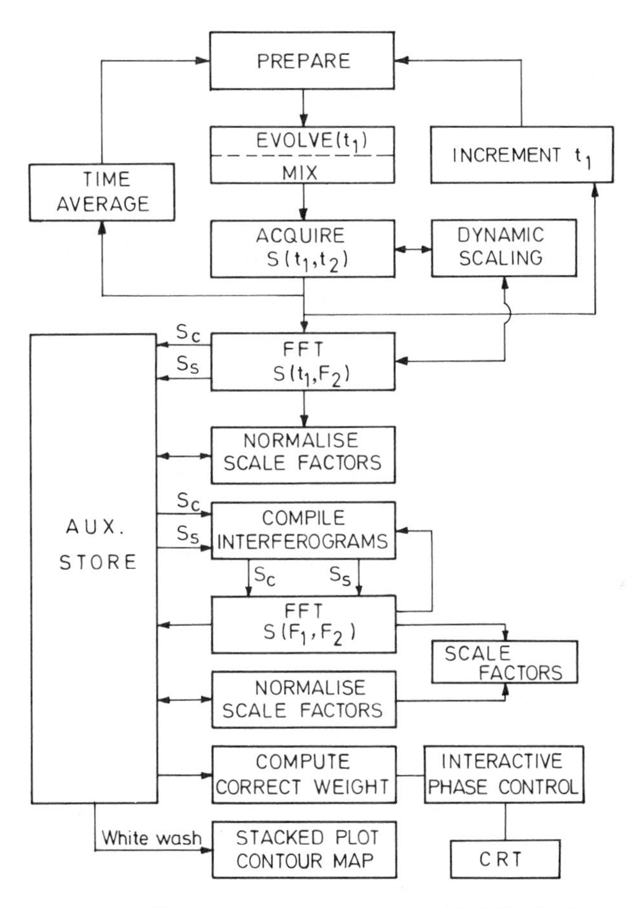

Figure 4.4. Data handling and manipulation in a typical 2D NMR experiment.

quency in the t_1 period. Consider that the transverse magnetization is modulated at frequency $+\Omega_1(2\pi F_1)$ during t_1 and at frequency $+\Omega_2(2\pi F_2)$ in the t_2 period. The acquired signal $S(t_1, t_2)$ inclusive of the spin–spin relaxation time in both the periods [$T_2^{(1)}$ and $T_2^{(2)}$, respectively] is:

$$S(t_1, t_2) \sim \exp(i\Omega_1 t_1)\exp(i\Omega_2 t_2)\exp(-t_1/T_2^{(1)})\exp(-t_2/T_2^{(2)})$$

$$S(F_1, F_2) \sim \int_0^\infty \exp(-i\omega_1 t_1)\,dt_1 \int_0^\infty S(t_1, t_2)\exp(-i\omega_2 t_2)\,dt_2 \qquad (2)$$

$$S(t_1, F_2) \sim \exp(i\Omega_1 t_1)[A_2(F_2) + iD_2(F_2)]\exp(-t_1/T_2^{(1)})$$

where $A_2(F_2)$ and $D_2(F_2)$ are the absorption and dispersion signals of the spectrum in the F_2 domain. Depending on the sign of Ω_1 the phase twist is either right handed or left handed. A further complex Fourier transformation leads to the 2D signal. Thus the real part of the 2D spectrum is a complex superposition of two-dimensional absorption and two-dimensional dispersion, giving the so-called "phase-twisted" lineshape, which cannot be made into pure absorption or dispersion by any normal phase adjustment.

$$\begin{aligned} S(F_1, F_2) \sim\ & A_1(\Omega_1)A_2(\Omega_2) \\ & - D_1(\Omega_1)D_2(\Omega_2) \\ & + iA_1(\Omega_1)D_2(\Omega_2) \\ & + iD_1(\Omega_1)A_2(\Omega_2) \end{aligned} \tag{3}$$

If, in addition, the reverse precession signal is present, corresponding to phase modulation during t_1 at $-\Omega_1$, then:

$$S(t_1, t_2) \sim \exp(-i\Omega_1 t_1)\exp(i\Omega_2 t_2)\exp(-t_1/T_2^{(1)})\exp(-t_2/T_2^{(2)}); \tag{4}$$

Adding the two signals prior to Fourier transformation gives, for the time-domain signals,

$$S(t_1, t_2) \sim 2\cos\Omega_1 t_1 \exp(i\Omega_2 t_2)\exp(-t_1/T_2^{(1)})\exp(-t_2/T_2^{(2)}); \tag{5}$$

which corresponds to pure amplitude modulation. This does not reveal the sign of Ω_1 but does allow pure lineshape in the 2D spectrum.

Consider the case where the modulation in t_1 time occurs at a single frequency F_1 but this frequency is present in both a positive and negative sense. The signal $S(t_1, t_2)$ is first Fourier transformed to produce the sine and cosine transforms in the F_2 domain

$$S(t_1, t_2) \rightarrow S^{c}(t_1, F_2) + S^{s}(t_1, F_2) \tag{6}$$

Assuming that a single frequency is operative in the F_2 domain these two spectra are the absorption–dispersion pair and one can adjust the phase detector to detect either the absorption or the dispersion mode. The lines in this spectra are subjected to a positive and negative twist when viewed along the t_1 axis and, since positive and negative twists are superimposed on the same signal at a given frequency, the lines apparently are only amplitude modulated. A second Fourier transform of the sine and cosine signals mentioned above along the t_1 direction leads to the 2D signals:

$$S^{c}(t_1, F_2) \rightarrow S^{cc}(F_1, F_2) + S^{sc}(F_1, F_2)$$

and:

$$S^{s}(t_1, F_2) \rightarrow S^{cs}(F_1, F_2) + S^{ss}(F_1, F_2) \tag{7}$$

It can be shown that:

$$\begin{aligned} S^{cc}(F_1, F_2) - S^{ss}(F_1, F_2) &\sim \sin(\theta_1 + \theta_2) \\ S^{cs}(F_1, F_2) + S^{sc}(F_1, F_2) &\sim \cos(\theta_1 + \theta_2) \end{aligned} \tag{8}$$

whereas the aliased signal at $(-F_1^{\circ}, F_2^{\circ})$ is completely suppressed, and:

$$\bar{S}^{cc}(F_1, F_2) + \bar{S}^{ss}(F_1, F_2) \sim \sin(\theta_2 - \theta_1)$$
$$\bar{S}^{cs}(F_1, F_2) - \bar{S}^{sc}(F_1, F_2) \sim \cos(\theta_2 - \theta_1) \tag{9}$$

where $\bar{S}^{cc}$, for example, is the two-dimensional double cosine transform of a time domain signal that has been modulated during t_1 with a frequency $-F_1^\circ$. Thus we can distinguish the positive and negative frequencies in the F_1 domain. Here, $\cos\theta_1$ and $\sin\theta_2$ refer to positive absorption and a dispersion that is negative at the low-frequency side.

By looking at the nature of the signals one can infer the lineshape that will result upon projection onto F_1 and F_2 axes. We have, of course, neglected so far the possible phase errors that are introduced by not acquiring the signal at the correct "0" of the t_2 period (starting from the exact focused point of an echo sequence, for example). Also, we may not have a $S(t_1, F_2)$ for $t_1 = 0$. Add to these problems arising in setting the phase of the detector exactly in quadrature to the transmitter. These phase errors can be corrected by the correct linear combinations of the S^{cc}, S^{cs}, S^{sc}, and S^{ss} in an interactive way with the computer.

In order to simplify the presentation one can also plot the spectra in absolute value mode $(u^2 + v^2)^{1/2}$ so that in the positive quadrant:

$$S^P = \{(S^{cc} - S^{ss})^2 + (S^{cs} + S^{sc})^2\}^{1/2}$$

and in the negative quadrant:

$$S^N = \{(\bar{S}^{cc} + \bar{S}^{ss})^2 + (\bar{S}^{cs} - \bar{S}^{sc})^2\}^{1/2} \tag{10}$$

In such a presentation the cross-section along $\Delta F_1 = 0$ and $\Delta F_2 = 0$ (at the respective frequencies in the two domains) both give lineshapes identical to the corresponding one-dimensional NMR.

We said that if the interferograms corresponding to positive and negative modulation in the t_1 domain were added prior to Fourier transformation the phase information would be lost, because:

$$\exp(i\phi) + \exp(-i\phi) = 2\cos\phi \tag{11}$$

and amplitude modulation results. For example, in looking at ^{13}C resonance, if we allow the spin–spin coupling to evolve during time t_1 and decouple the ^1H by broadband decoupling during the acquisition time t_2, this is precisely equal to pure amplitude modulation. The time signal is then:

$$S(t_1, t_2) \sim \cos\Omega_1 t_1 \exp(i\Omega_2 t_2)\exp(-t_1/T_2^{(1)})\exp(-t_2/T_2^{(2)})$$
$$S(t_1, F_2) \sim \cos\Omega_1 t_1 [A_2(\Omega_2) + iD_2(\Omega_2)]\exp(-t_1/T_2^{(1)}) \tag{12}$$

An example for such a modulation is given in Fig. 4.2. The second Fourier transformation with respect to t_1 gives (S^{sc}, S^{cc}):

$$S(F_1, F_2) \sim A_2(\Omega_2)[A_1(\Omega_1) + A_1(-\Omega_1) + iD_1(\Omega_1) + iD_1(-\Omega_1)] \tag{13}$$

It is possible to set the phase detector to get the absorption mode signal in both dimensions and one has the advantage of resolving closely spaced lines. In fact in this case one can do a real cosine Fourier transformation with respect to t_1 of the correctly phased cosine transformed signal with respect to t_2.

In some situations the amplitude of the signal at t_2 can be phase modulated at frequency F_P and amplitude modulated at a frequency F_A such that the time-domain signal can be represented as:

$$S(t_1, t_2) \sim \cos 2\pi F_A t_1 \exp(i2\pi F_P t_1) \exp(i2\pi F_2 t_2) \tag{14}$$

excluding the relaxation terms. This can be rewritten as:

$$\begin{aligned} S(t_1, t_2) \sim \exp(it_1 2\pi(F_P - F_A)) \exp(i2\pi F_2 t_2) \\ + \exp(it_1 2\pi(F_P + F_A)) \exp(i2\pi F_2 t_2) \end{aligned} \tag{15}$$

Thus, the complex double Fourier transformation should give phase-twisted lines at $(F_P - F_A)$, F_2 and $(F_P + F_A)$, F_2.

Display Cosmetics of 2D Spectra. As we have seen the 2D FT spectrum consisting of signals as a function of two frequencies, F_1 and F_2 (depending on the frequencies under which the spin system evolves in the t_1 and t_2 durations) and contains a complicated mixture of absorption and dispersion lineshapes. The question now arises as to how to represent the 2D spectrum in specific situations. We shall very briefly outline some of the techniques that are employed to present 2D high-resolution NMR data.

In the phase-sensitive mode if one has a possible absorption–absorption lineshape in the two frequency domains, when correctly phased one gets spectra that have the linewidth of the normal spectra. This makes it possible to distinguish between positive and negative intensities. Also it is possible to get spectra parallel to any one of the frequency axes and the spectra can always be phase adjusted for a pure absorption mode, without affecting the rest of the spectra. This is possible even if the general lineshape is phase twisted.

Nonoverlapping Lorentzian lines do give lineshapes corresponding to Lorentzian absorption. When large numbers of absorption and dispersion lines overlap, the dispersion overlap is subtractive, whereas absorption overlap is additive and severe lineshape distortion can occur. In fact in the absolute value mode the width at half-height will be 1.732 times larger than the corresponding absorption mode, and, at the base of the line this factor reaches 10. In spite of this the absolute value mode is one of the most frequently used ways of representing 2D spectra.

In presenting 2D spectra, we have already mentioned the stacked plot method using the so-called "whitewash" routine. Although it gives a good 3D perspective, peaks smaller in intensity that come immediately "behind" a certain intense line are masked. Also the plotting of the stacked plot for large data matrix can be quite time consuming.

An alternate method is to project the various peaks as in contour plots of altitudes in a geographical map. Thus the lowest signal amplitude L can be taken as a basis so that all peaks are projected down onto their bases at heights

L, $2L$, $3L$, etc. Therefore the higher amplitude peaks have a larger number of rings. A cursory look then should give some idea of the amplitude, the bigger "blobs" corresponding to more intense resonances. When intense lines occur close together these contours can overlap and also, in the process, mask weak lines. However, the time taken to obtain a contour plot is much less than that for a stacked plot.

In some of the 2D spectra information regarding certain correlations may be obtained by simply taking typical spectra at specific frequencies parallel to or making a definite angle to the F_1 and F_2 axes. These correspond to the so-called cross-sections and can be plotted as normal spectra and information can be extracted.

Finally, it is also useful to project the 2D spectra by integrating the intensity of cross-section perpendicular to F_1 or F_2 axis. This gives spectra characteristic of the Hamiltonian operative in the evolution or detection period, so that simplified spectra can be obtained (vide infra).

The signals can also be subjected to one of several digital filters to improve either sensitivity or resolution of the 2D peaks. To a high degree of accuracy it can be shown that a filter function $g(t_1, t_2)$ to be applied to the acquired signal $S(t_1, t_2)$ can in general be applied in sequence, i.e.,

$$S(t_1, t_2)g(t_1, t_2) = S(t_1, t_2)g_1(t_1)g_2(t_2) \tag{16}$$

Then the digital filtering to smooth out spectra in the 2D scheme can go as follows:

$$\begin{aligned} S(t_1, t_2)g(t_2) \overset{FT}{\rightarrow} S(t_1, \omega_2) \overset{TR}{\rightarrow} S(\omega_2, t_1) \cdot g(t_1) \\ \overset{FT}{\rightarrow} S(\omega_1, \omega_2) \end{aligned} \tag{17}$$

The acquired signal is usually weighted by an exponential function $g(t)$ given by:

$$g(t) = \exp(t/T_w) \tag{18}$$

where depending upon the sign of T_w, positive or negative, one will artificially lengthen or shorten the FID corresponding to resolution or sensitivity enhancement. The lineshape of pure absorption will still be Lorentzian. The optimal value should be the spin–spin relaxation time of the system, if known. Sometimes, the use of only the first 60–70% of the weighted FID with a large zero filling improves the resolution. The width at half maximum height is $1/\pi T_2'$, where:

$$(T_2')^{-1} = (T_2^*)^{-1} + (T_w)^{-1} \tag{19}$$

while the base is nearly 10 times the above width. A 2D exponentially weighted absorption signal has a four- cornered starlike appearance when contour projections are made. However, it will be difficult to select a unique exponential that will be optimal if the linewidths in the spectra vary.

In case the sampling in either time domain is limited by limited data storage facilities, a Gaussian weighting can be done to avoid truncation and to produce better resolution or sensitivity enhancement. Here $g(t)$ is given by:

$$g(t) = \exp(t/T_{w_1})\exp(-t^2/T_{w_2}) \tag{20}$$

where T_{w_1} is set close to T_2 and T_{w_2} can have a large positive value. Whereas the sensitivity enhancement achieved is almost the same as in exponential weighting there is an advantage of much narrower width near the base. This so-called Lorentzian to Gaussian transformation should be useful in cases where the lineshapes of unresolved multiplets resemble Gaussian rather than Lorentzian lineshapes.

In the pseudo-echo-transformation (PE) technique the time domain signal is first multiplied by an increasing exponential (as in exponential filtering for resolution enhancement). Then the resultant time-domain response is multiplied by a decreasing exponential or Gaussian function of time on either side of the midpoint, say t_m, so that the total filter function can be represented as:

$$g(t) = \exp(t/T_2)\exp(-|t - t_m|/T_{PE})$$

or:

$$g(t) = \exp(t/T_2)\exp(-(t - t_m)^2/T_{PE}) \tag{21}$$

The constant T_{PE} must be chosen carefully; for Gaussian echo transform the optimal value is approximately one fourth of or less than the duration of the FID and for the exponential echo filter is about one eighth of the duration of the FID. The pseudo-Gaussian echo is the better of the two because it gives a much narrower base in the absolute value mode and produces 2D peaks that give circular contour maps centered around the line position.

In the convolution difference filtration (CDF) the time-domain signal is first weighted with a fast decaying exponential filter, which is supposed to leave only broad resonance components. The same is then filtered from the original FID corresponding to the filter function:

$$g(t) = 1 - \exp(-t/T_{\text{CDF}}) \tag{22}$$

The resulting signal, which is now purged of broad background signals, can be weighted again by functions that are mentioned above for sensitivity/resolution enhancement. Because in this way one is taking a difference between two Lorentzian lines, the dispersive components at large offsets from the center are eliminated and long tailing is suppressed at the expense of some sensitivity. If we have a mixture of absorption and dispersion peaks in a 2D spectrum then the dispersion signals can be largely reduced by convolution difference and any absorption peaks close to a broad dispersion peak can be brought to light. The various lineshapes mentioned above, their absolute value modes, 2D stacked plots, and the corresponding contour plots are shown in Fig. 4.5.

Nomenclature of Various 2D Resonances. If the Hamiltonians under which the evolution is taking place in the t_1 and t_2 domains are identical, the modula-

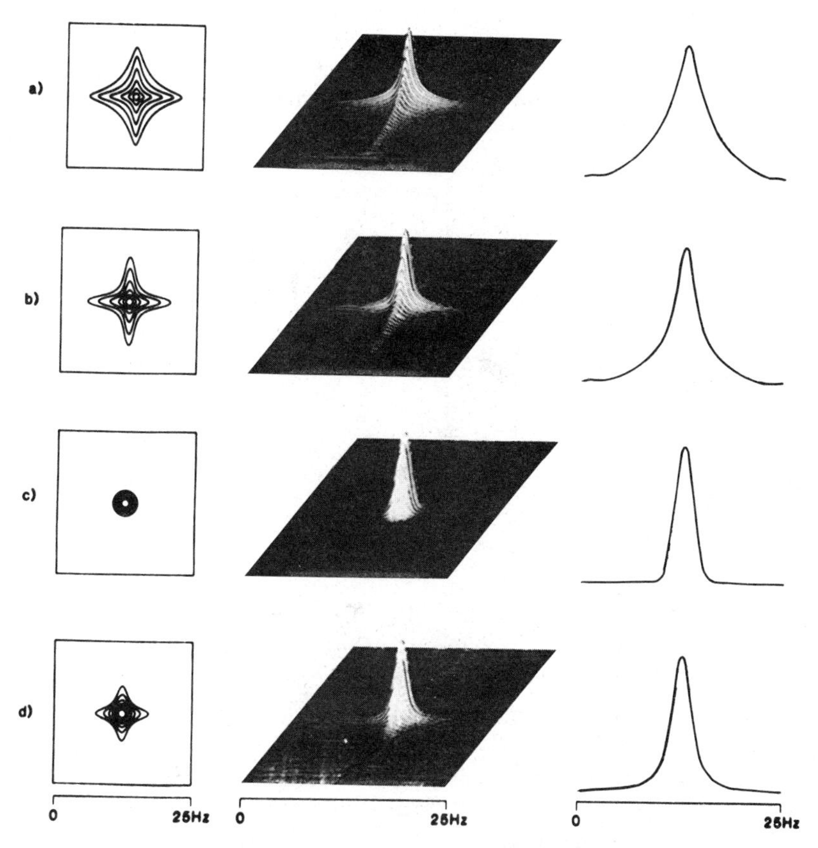

Figure 4.5. Computer simulation of typical absolute value mode 2D lineshapes: contour plots, stacked plots, and cross-section at exact resonance are shown for (a) exponential filter, (b) Lorentzian to Gaussian transformation, (c) pseudo-Gaussian echo transformation, and (d) convolution difference filter. [Reproduced by permission. A. Bax, *Two Dimensional Nuclear Magnetic Resonance in Liquids*, Delft University Press, D. Reidel Publishing Company, Dordrecht-Holland]

tions and the frequencies present in both dimensions will be identical. Hence, in the 2D spectrum one sees only diagonal peaks (in the absence of any mixing period), inclined at an angle of 45° to either frequency axes.

It is possible that in an IS system in the t_1 duration, I spins are evolving under their chemical shifts and under I–S coupling and by suitable pulses one can transfer the magnetization corresponding to I spin single quantum coherence to S spin coherence which evolves under S spin chemical shift and I–S coupling in the t_2 domain. Such coherences that evolve at some frequency Ω_1 in t_1 and later evolve at Ω_2 in t_2 give rise to cross-peaks in the 2D spectrum at the point (Ω_1, Ω_2) (Fig. 4.6).

If however, during the t_1 period some z magnetization is restored by T_1

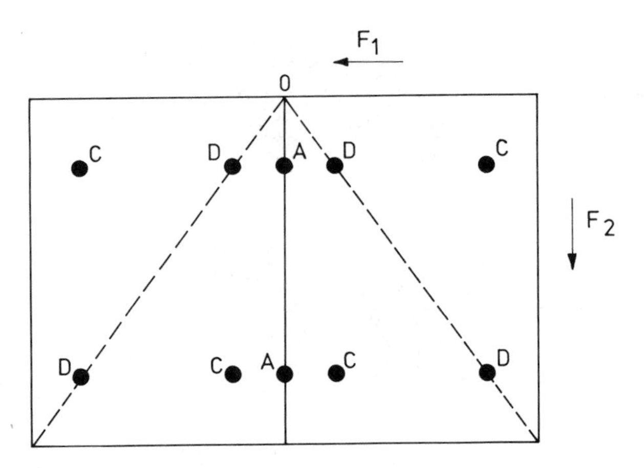

Figure 4.6. Schematic diagram showing a contour plot of diagonal (D), cross (C), and axial (A) peaks in a 2D NMR spectrum.

processes (corresponding to diagonal elements of the density matrix) and evolves in t_2 period at frequency Ω_2, then the corresponding peaks will occur along the line at $\Omega_1 = 0$. These are called axial peaks.

The way in which the lineshapes and intensities are predicted is exactly analogous to the normal one-dimensional situation: the product operator corresponding to single quantum coherences will be either in-phase or anti-phase resonances (I_y, I_x, I_yS_z, I_xS_z, etc.) and, depending on the sinusoids appearing in t_1 and t_2 domains, they will occur as absorption or dispersion. For example, a product operator:

$$\cos\omega_1 t_1 \cos\delta_1 t_1 \sin\omega_2 t_2 \cos\delta_2 t_2 I_{1x}S_z \tag{23a}$$

will correspond (when the phase of the detector is along the x axis) to an in-phase absorption when viewed parallel to the F_1 axis and to an anti-phase dispersion when viewed along the F_2 axis. Another operator product is, say,

$$-\cos\omega(I_1)t_1 \sin\delta t_1 \sin\omega(I_2)t_2 \sin\delta t_2 I_{1y}I_{2z} \tag{23b}$$

where $\omega(I_1)$ is the frequency characteristic of spin I_1 and $\omega(I_2)$ that of spin I_2. This will give four cross-peaks at $\omega(I_1) \pm \delta$, $\omega(I_2) \pm \delta$, of which the lineshape along F_1 axis will be anti-phase dispersion (anti-phase dispersion because product of cosine and sine), whereas the lineshapes along F_2 axis will be in-phase absorption.

Apart from the diagonal, axial, and cross-peaks (which can be predicted by an analysis of the evolution of the density matrix, assuming perfect phases and ideal flip angles) additional artifacts do appear in 2D spectra. "Ghost" lines appear, for example, when an imperfect 180° refocusing pulse is applied during the evolution period. This imperfect pulse, although focusing most of the transverse magnetization, will convert some magnetization into longitudinal order and leave some unaffected. Therefore, a portion of the magnetization would have evolved under chemical shift. For the very same reason the magnetic field inhomogeneities also are not eliminated, giving rise to a set of ghost lines in the 2D spectra. These appear parallel to the main spectrum, being displaced along the F_1 axis by the chemical shifts.

"Phantom" lines appear arising from the flip angle errors in the pulses. Thus an echo sequence $90^\circ-\tau-180^\circ-\tau$, when the pulses are weaker than they ought to be, leaves some longitudinal magnetization in the z direction after the first pulse. Although transverse magnetization is refocused by the 180° pulse, this imperfect pulse can create a small amount of fresh transverse magnetization that can dephase for a time τ after the pulse. Thus, during the evolution a small portion of the magnetization evolves under chemical shift for a period $t_1/2$ and spin–spin coupling, whereas the major portion evolves only under coupling. This produces the so-called "phantom" lines, which are offset from the normal line by half the chemical shift along the F_1 axis.

One common property of ghost and phantom lines is that they are subject to the field inhomogeneities and hence by shortening T_2^* they may largely be eliminated.

J-Resolved 2D Spectroscopy

Heteronuclear J Spectroscopy. In high-resolution NMR of liquids the two most important parameters are the chemical shift δ and the scalar coupling constant J. In large molecules (especially in the ^{1}H-coupled ^{13}C NMR spectra, for example) because of considerable overlap of the multiplets from different chemically shifted nuclei, spectral assignments become formidably difficult. In 2D J spectroscopy these two principal parameters are resolved along the two axes of the 2D spectra. Typically one of the axes contains information of spin–spin coupling only, and along the other axis chemical shift dispersion and the spin–spin coupling or chemical shift alone may be present. The basic methodology is to use the idea of a $90^\circ-\tau-180^\circ-\tau$ echo sequence in the t_1 period so that chemical shifts are focused. While evolution under spin–spin coupling is unaffected by the nonselective π pulse for homonuclear situation, in heteronuclear coupled systems the evolution under J is also focused. This can be avoided, for example, by applying π pulses to both spins, thereby restoring J modulation in the t_1 period.

A simple technique in 2D J-resolved heteronuclear spectroscopy is the so-called gated decoupler method, which is entirely analogous to the SEFT experiment (Chapter 2) except for a systematic variation of t_1. Here the idea is to resolve the various ^{13}C multiplets using the heteronuclear coupling as the spreading parameter into the F_1 dimension. The pulse sequence used is shown in Fig. 4.7. The broadband noise decoupler is on for some time prior to the 90°_x pulse on ^{13}C to give rise to NOE and is switched off during the second τ period after the 180°_x pulse. During the acquisition the ^{1}H's are decoupled.

The ^{13}C chemical shifts are focused at the end of the t_1 period and because

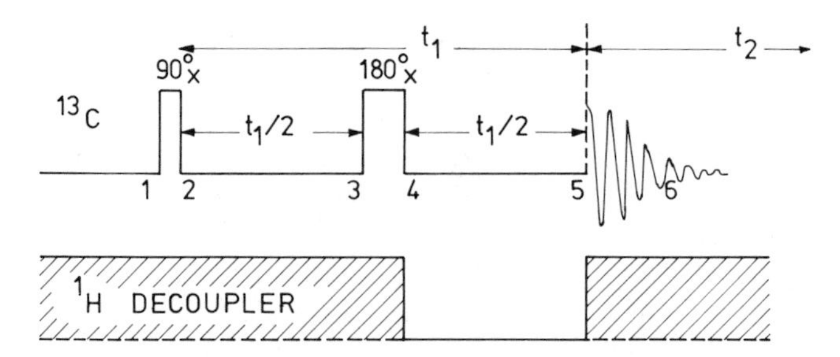

Figure 4.7. Pulse sequence for 2D heteronuclear J-resolved spectroscopy by the gated decoupler method.

the decoupler is off during the second half of the t_1 period the evolution under the spin–spin coupling produces a net modulation that corresponds to scaling all C–H couplings by a factor of 0.5. Looking at the density matrix at various points,

$$\begin{aligned} \sigma_1 &= I_z \\ \sigma_2 &= -I_y \\ \sigma_3 &= -I_y \end{aligned} \tag{24}$$

neglecting T_2, T_1, and evolution under spin–spin coupling and chemical shift (which will be focused at the end of t_1),

$$\begin{aligned} \sigma_4 &= I_y \\ \sigma_5 &= \cos(\pi J t_1/2) I_y - 2\sin(\pi J t_1/2) I_x S_z \end{aligned} \tag{25}$$

When the decoupler is switched on at the start of the acquisition the second term vanishes, leading to:

$$\sigma_5 = \cos\left(\frac{\pi J t_1}{2}\right) I_y \tag{26}$$

which evolves freely into:

$$\sigma_6 = \cos\left(\frac{\pi J t_1}{2}\right)\cos(2\pi\delta t_2) I_y - \cos\left(\frac{\pi J t_1}{2}\right)\sin(2\pi\delta t_2) I_x \tag{27}$$

With the detector referenced along the y axis this leads to a pure absorption mode in both the F_1 and F_2 dimensions, with F_2 containing chemical shifts only (decoupled spectrum) whereas in the F_1 dimension the spin–spin couplings are scaled by a factor of 0.5.

Projection of the 2D spectra onto the F_2 axis produces a normal broadband proton-decoupled spectrum, whereas projection on the F_1 axis gives the J spectrum, consisting only of scaled multiplets symmetrically around zero frequency. If each of the J multiplets is rotated about its center of gravity by

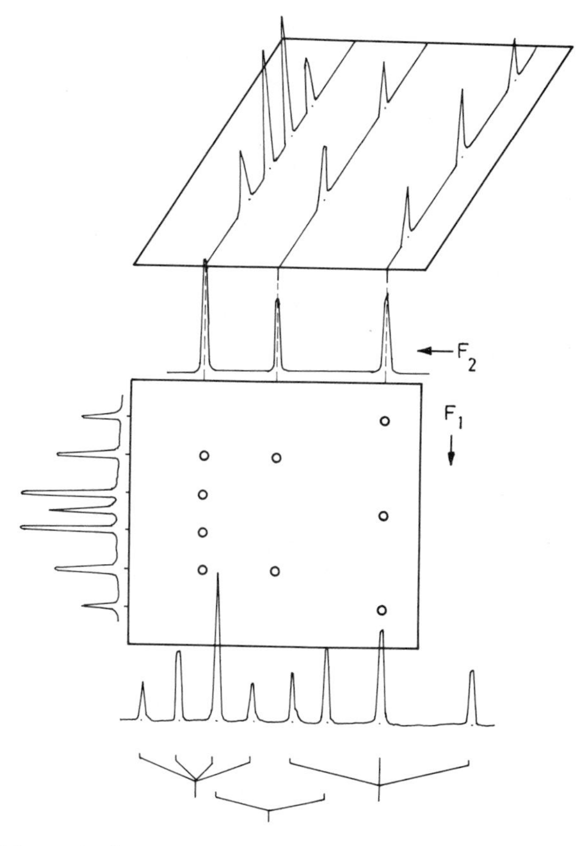

Figure 4.8. Schematic diagram of a 2D heteronuclear *J*-resolved spectrum. On the top, cross-sections parallel to the F_1 axis show separated multiplets with their *J* scaled by a factor 0.5. The bottom figure shows a projection onto the F_2 axis, giving ^1H-decoupled ^{13}C spectrum, and a projection on to the F_1 axis, which gives the *J* spectrum (with reduced splitting). The normal ^1H-coupled spectrum is represented schematically at the bottom.

90° and the resulting pattern is projected onto the F_2 axis, one obtains the fully coupled spectrum with normal chemical shifts and uniformly scaled spin–spin coupling (see Fig. 4.8). Thus one gets undistorted coupled multiplets without any prior knowledge of the various spin–spin coupling constants. This is in contrast to results using DEPT and INEPT, although in the latter situation subsequent purging does take care of the anomalies (Chapter 3).

In the proton-flip experiment, which is a simple modification of the above procedure, the following pulse sequence is used (Fig. 4.9). Here the difference is that the spin–spin coupling is allowed to evolve for the full t_1 period so that unscaled heteronuclear spin–spin coupling appears on the F_1 axis, while on the F_2 axis the observe nuclei chemical shifts appear. While chemical shifts of

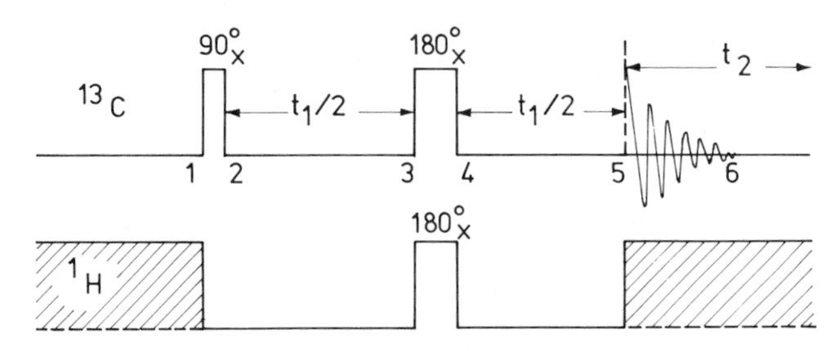

Figure 4.9. The pulse sequence for 2D heteronuclear J-resolved spectroscopy where the broadband decoupling during the first half of the t_1 period (see Fig. 4.7) is replaced by a 180° pulse at the middle of the t_1 period, in the so-called proton-flip experiment.

^{13}C isochromats are refocused by the ^{13}C π pulse, precession under J coupling, which is reversed by this pulse, is rereversed by the ^{1}H (π) pulse. Hence, evolution under heteronuclear J coupling proceeds fully for the entire t_1 period (Fig. 4.9). Thus, for $I = {}^{13}\mathrm{C}$, $S = {}^{1}\mathrm{H}$,

$$\begin{aligned}
\sigma_1 &= I_z \\
\sigma_2 &= -I_y \\
\sigma_3 &= -\cos\left(\frac{\pi J t_1}{2}\right) I_y + 2\sin\left(\frac{\pi J t_1}{2}\right) I_x S_z \\
\sigma_4 &= \cos(\pi J t_1/2) I_y - 2\sin(\pi J t_1/2) I_x S_z \\
\sigma_5 &= \cos(\pi J t_1) I_y - 2\sin(\pi J t_1) I_x S_z
\end{aligned} \tag{28}$$

Evolution under broadband decoupling cancels the second term so that at the beginning of the free evolution:

$$\sigma_5 = \cos(\pi J t_1) I_y \tag{29}$$

and this evolves into:

$$\sigma_6 = \cos(\pi J t_1)\cos(2\pi\delta t_2) I_y - \cos(\pi J t_1)\sin(2\pi\delta t_2) I_x \tag{30}$$

Thus, in the F_2 axis we get only resonances at the respective chemical shifts and along the F_1 axis only spin–spin coupling, $(2I + 1)$ number of lines, centered around zero frequency. In both dimensions pure absorption lineshapes result. In the F_1 dimension natural linewidths are obtained, whereas in F_2, the linewidth corresponds to T_2^*, affected by both field inhomogeneities and diffusion effects.

A two-dimensional J-resolved spectrum of cholesterol is shown in Fig. 4.10. It is obvious how easy and straightforward the analysis is.

When there is strong coupling between protons due to second-order effects, the multiplet structure around the chemical shift frequency may not be symmetrical so that J modulation in the t_1 duration does not take place at exactly

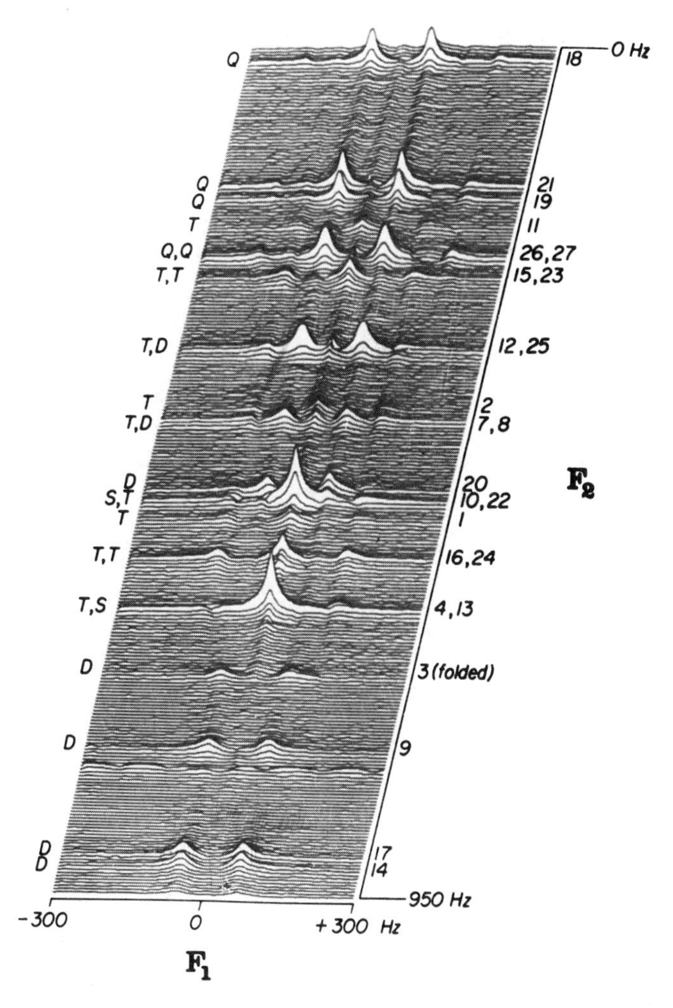

Figure 4.10. The heteronuclear J-resolved 2D spectrum of cholesterol. Q, T, D, and S stand for quartets (CH_3), triplets (CH_2), doublets (CH), and singlets (C). [Reproduced by permission. D.L. Turner, and R. Freeman, *J. Magn. Reson.*, **29**, 587 (1978), copyright 1978, Academic Press, New York]

the same frequency of either sign. This leads to independent phase modulations, corresponding to different spin multiplets in the asymmetrical conventional spectrum. Although the modulation present in the t_1 period produces a faithful J spectrum along the F_1 axis, no additional frequencies appear. This is in contrast to the homonuclear 2D J-resolved spectra of strongly coupled systems (*vide infra*). The 2D lineshape, however, will be one of phase-twisted absorption. A cross-section parallel to F_1 axis corresponding to a given carbon chemical shift can nevertheless be pulled out and adjusted for pure phase.

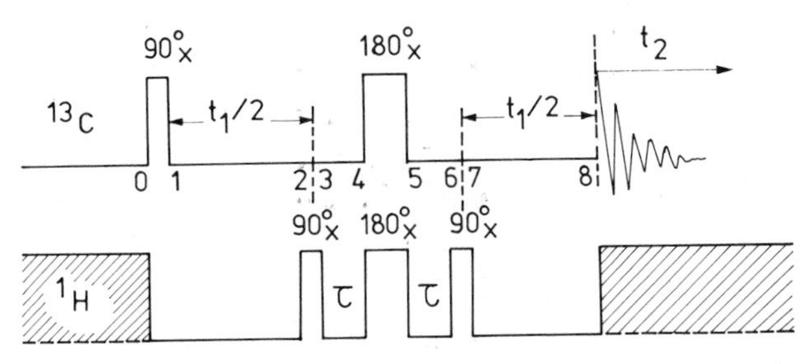

Figure 4.11. Pulse sequence for discriminating long-range and short-range spin–spin coupling in heteronuclear J-resolved 2D spectroscopy by using the BIRD sandwich.

Another interesting variation of the proton-flip experiment is to tailor the evolution using the so-called bilinear rotation decoupling (BIRD) sequence by introducing a pair of 90°_x pulses or a 90°_x and a 90°_{-x} pulse symmetrically about the π pulse in the proton channel with a gap of $(2J_{CH})^{-1}$. It can be shown that such a pulse sandwich is capable of discriminating short-range and long-range ^{13}C–^{1}H coupling constants. Considering the pulse sequence (Fig. 4.11), the introduction of the 90°_x–τ–180°_x–τ–$90^\circ_{\pm x}$ pulse can be used to flip the spin orientations selectively of either directly bound or distant coupled protons. The principle relies on the fact that while the directly bonded $^{1}J_{C\text{–}H}$ is on the order of 100 Hz, distant $^{n}J_{C\text{–}H}$ is less than 10 Hz, so that for short intervals of time much less than $(J)^{-1}$ we may ignore J evolution of ^{13}C under long-range coupling. Considering first the short-range protons with coupling constant J and $\tau = (2J)^{-1}$:

$$\begin{aligned} \sigma_0 &= I_z \\ \sigma_1 &= -I_y \\ \sigma_2 &= -c_{J/2} I_y + 2 s_{J/2} I_x S_z \end{aligned} \tag{31}$$

with a 90°_x pulse on the proton channel,

$$\sigma_3 = -c_{J/2} I_y - 2 s_{J/2} I_x S_y \tag{32}$$

With evolution under scalar coupling for $\tau = (2J)^{-1}$:

$$\sigma_4 = 2 c_{J/2} I_x S_z - 2 s_{J/2} I_x S_y \tag{33}$$

With an 180°_x pulse on both S and I spin channels:

$$\sigma_5 = -2 c_{J/2} I_x S_z + 2 s_{J/2} I_x S_y \tag{34}$$

With evolution under scalar coupling for $\tau = (2J)^{-1}$:

$$\sigma_6 = -c_{J/2} I_y + 2 s_{J/2} I_x S_y \tag{35}$$

Consider applying the 90° pulse with $+x$ phase on proton channel.

We then have at the start of acquisition:

$$\sigma_8 = -c_{J/2}^2 I_y + 2 c_{J/2} s_{J/2} I_x S_z + 2 s_{J/2} c_{J/2} I_x S_z + s_{J/2}^2 I_y \tag{36}$$

Under decoupling of ^{1}H, the modulation that is present will be:

$$\sigma_8' = -c_J I_y \tag{37}$$

and this will evolve under chemical shift during the period t_2.

If, on the other hand, the second 90° pulse has been a $-x$ pulse, then,

$$\sigma_7 = -c_{J/2} I_y - 2s_{J/2} I_x S_z \tag{38}$$

Evolving under spin–spin coupling during $t_1/2$ leads to

$$\begin{aligned}\sigma_8 &= -c_{J/2}^2 I_y + 2c_{J/2}s_{J/2} I_x S_z - 2c_{J/2}s_{J/2} I_x S_z - s_{J/2}^2 I_y \\ &= -I_y\end{aligned} \tag{39}$$

corresponding to cancellation of the J modulation in the t_1 period. Thus the 90°_x–τ–180°_x–τ–90°_{-x} pulse sandwich corresponds to no net rotation of the directly bonded ^{1}H spins. Hence J modulation averages to zero.

The effect of a J-selective (BIRD) sequence may perhaps be more transparent on viewing its propagator. For a weakly coupled spin-1/2 IS system, for example, the sequence $\frac{\pi}{2}(I, y) - \tau - \pi(I, S, y) - \tau - \frac{\pi}{2}(I, y)$ has the following propagator:

$$U = \exp(-4\pi i J I_x S_z \tau)\exp(-2\pi i I_y)\exp(-i\pi S_y) \tag{40}$$

which leads to

$$\begin{aligned} UI_xU^\dagger &= I_x, \quad \tau \text{ arbitrary} \\ UI_yU^\dagger &= -I_y, \quad \tau = (2J)^{-1} \\ UI_zU^\dagger &= -I_z, \quad \tau = (2J)^{-1} \\ UI_yU^\dagger &= I_y, \quad \text{if } J = 0 \\ UI_zU^\dagger &= I_z, \quad \text{if } J = 0 \end{aligned} \tag{41}$$

On the other hand, the sequence $\frac{\pi}{2}(I, x) - \tau - \pi(I, S, x) - \tau - \frac{\pi}{2}(I, -x)$ has as propagator:

$$U = \exp(-4\pi i J I_y S_z \tau)\exp(-i\pi I_x)\exp(-i\pi S_x), \tag{42}$$

leading to

$$\begin{aligned} UI_xU^\dagger &= -I_x, \quad \tau = (2J)^{-1} \\ UI_yU^\dagger &= -I_y, \quad \tau \text{ arbitrary} \\ UI_zU^\dagger &= I_z, \quad \tau = (2J)^{-1} \\ UI_xU^\dagger &= I_x, \quad \text{if } J = 0 \\ UI_zU^\dagger &= -I_z, \quad \text{if } J = 0 \end{aligned} \tag{43}$$

In considering the long-range ^{1}H coupling we assume that evolution during the period 2τ between the two pulses for these isochromats is so slow that this can be neglected. Under these circumstances, 90°_x–τ–180°_x–τ–90°_{+x} corresponds to a 360° rotation of protons, leading to no flip of the protons at all, whereas the 90°_x–τ–180°_x–τ–90°_{-x} acts like a simple π pulse and causes J modulation to persist in the t_1 period. Thus, the second 90° pulse with x phase gives a 2D spectrum in which only short-range (large-magnitude) spin–spin couplings are present in the F_1 axis. Here, because the J modulation is symmetrical about zero frequency, multiple folding via aliasing can be allowed, yielding symmetrical lines. This allows sampling rates of 30–40 Hz, giving a frequency range of ± 15 to ± 20 Hz in the F_1 direction. Here the actual coupling constant is an integer multiple of the frequency range in F_1 plus the measured separation in the F_1 direction. The same sampling rate can be kept for the 90°_x–τ–180°_x–τ–90°_{-x} situation, in which only long-range (small-J) couplings are present in the F_1 axis. This way one can take cross-sections of the maps corresponding to different chemically shifted carbons and read off the short-range/long-range coupling constants (See Fig. 4.12).

Homoscalar J-Resolved 2D Spectroscopy. These experiments are also based on the 90°_x–τ–$180^\circ_{x(y)}$–τ–echo sequence (Fig. 4.13). In the weakly coupled homonuclear situation, whereas the π pulse in the middle of evolution refocuses chemical shifts, spin–spin couplings evolve unaffected because all bilinear

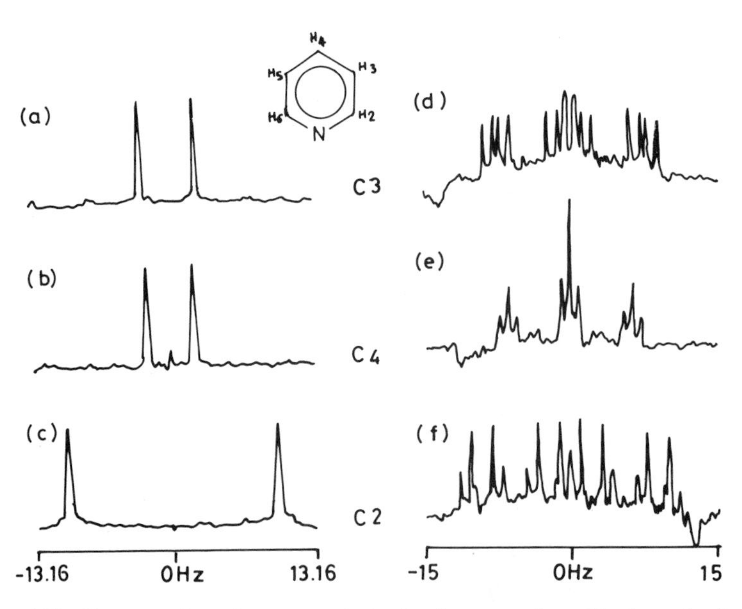

Figure 4.12. Cross-sections from a semiselective heteronuclear 2D J-resolved spectrum where long-range and short-range ^{13}C–^{1}H couplings are distinguished. The one-bond $^1J_{CH}$ are given in a, b, and c, using the BIRD sandwich 90_x–τ–180_x–τ–90_x, whereas the long-range (small J) couplings given in d, e, and f are obtained using a 90_x–τ–180_x–τ–90_{-x} sandwich. A 30-Hz sampling rate in both experiments leads to aliasing in a, b, and c. [Reproduced by permission. A. Bax, *J. Magn. Reson.*, **52**, 330 (1983), copyright 1983, Academic Press, New York]

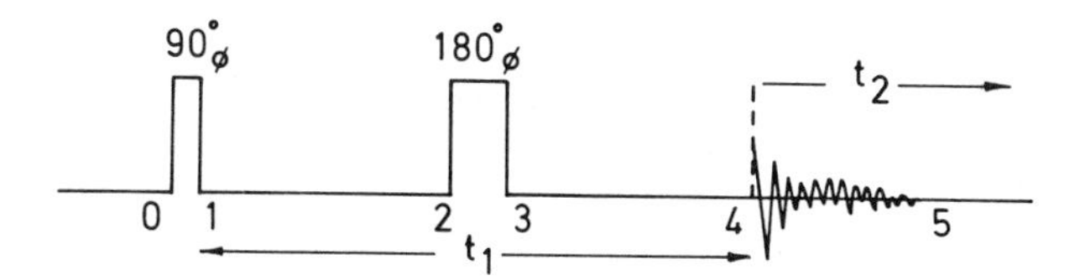

Figure 4.13. Pulse sequence for homoscalar J-resolved 2D NMR.

terms in the Hamiltonian are not affected. Thus in the t_1 period, J modulation alone is operative, while in the t_2 period both J and chemical shifts are operative. Consider the evolution of magnetization of a particular spin-I homoscalar coupled to another spin S with a coupling constant J Hertz:

$$\begin{aligned}
\sigma_0 &= I_z \\
\sigma_1 &= -I_y \\
\sigma_4 &= -c_J(t_1)I_y + 2s_J(t_1)I_xS_z \\
\sigma_5 &= -c_J(t_1)c_{2\delta}(t_2)c_J(t_2)I_y + 2s_J(t_1)c_{2\delta}(t_2)c_J(t_2)I_xS_z \\
&\quad + 2c_J(t_1)c_{2\delta}(t_2)s_J(t_2)I_xS_z + s_J(t_1)c_{2\delta}(t_2)s_J(t_2)I_y \\
&\quad + c_J(t_1)s_{2\delta}(t_2)c_J(t_2)I_x + 2s_J(t_1)s_{2\delta}(t_2)c_J(t_2)I_yS_z \\
&\quad + 2c_J(t_1)s_{2\delta}(t_2)s_J(t_2)I_yS_z - s_J(t_1)s_{2\delta}(t_2)s_J(t_2)I_x
\end{aligned} \tag{44}$$

For the detector phase set to the $-y$ axis only the first, fourth, sixth, and seventh terms contribute and the lineshape, as can be seen, is a mixture of absorption and dispersion. While only spin–spin couplings are present in the F_1 dimension both spin–spin coupling and chemical shifts are present in the F_2, so that spectra occur at coordinates corresponding to $(m_SJ, \delta \pm m_SJ)$. If the scale is identical in both frequency axes one should get multiplets belonging to a given proton occurring at an angle of 45° to the F_2 axis (Fig. 4.14).

The spectral lineshapes in both dimensions consist of linear combinations of absorption and dispersion and the resulting phase-twisted line cannot be adjusted for pure phase by any phase-correction scheme.

Let us digress for a moment as to how to manipulate 2D spectra to obtain projections of features that lie on a cross-section, making angle in the range of 0–$\pi/2$ (in this case it is $\pi/4$) with one of the axes. Considering the double Fourier transformed signal,

$$S(\omega_1, \omega_2) = \int_{-\infty}^{+\infty} dt_1 \exp(-i\omega_1 t_1) \int_{-\infty}^{+\infty} dt_2 \exp(-i\omega_2 t_2) S(t_1, t_2) \tag{45}$$

for a given cross-section $c(t)$ in the time domain described by $s(t\cos\phi, t\sin\phi)$, ϕ being the angle between the direction of the cross-section and the positive t_1 axis, we can define the corresponding cross-section, $P(\omega)$, in the frequency

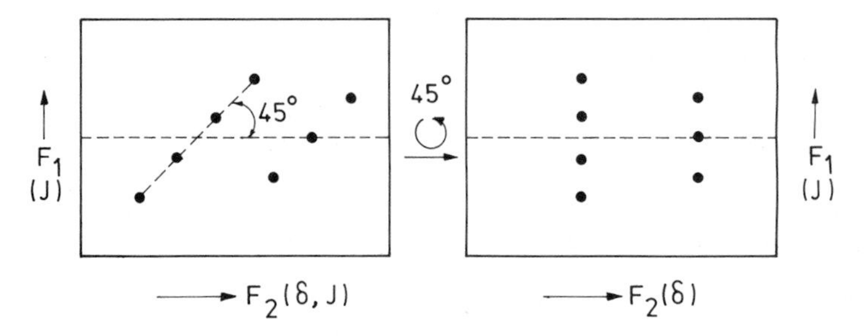

Figure 4.14. Schematic representation of contour plots of 2D homonuclear J-resolved spectrum of an A_3X_2 system. Note that the multiplet pattern makes an angle 45° to both F_1 and F_2 axes. An absolute value mode projection at 45° to F_1 gives the plot on the right. Here projection onto F_1 will give the J spectrum, while onto F_2 it gives the "broad-band-decoupled" ^{1}H spectrum.

domain at the same angle by:

$$P(\omega) = \int_{-\infty}^{+\infty} dx(-x \sin\phi + \omega\cos\phi, x\cos\phi + \omega\sin\phi) \tag{46}$$

$c(t)$ and $P(\omega)$ form a FT pair such that:

$$P(\omega) = \mathrm{FT}[c(t)]$$

and

$$c(t) = (\mathrm{FT})^{-1}[P(\omega)] \tag{47}$$

In the above expression x is the direction of $S(\omega_1, \omega_2)$. Because $S(t_1, t_2)$ is zero for $t_1 < 0-$ and/or $t_2 < 0$ (causality principle) it has nonzero value only in the first quadrant. Therefore, the Fourier transform of either $c(t)$ or $P(\omega)$ for the range $0 < \phi < \pi/2$ is zero except for $t = 0$. Thus, the projection of the homonuclear J-resolved 2D spectrum, corresponding to a cross-section parallel to the J multiplets that occur at an angle of 45° to ω_1 and ω_2 axes onto the line that makes 135° with ω_1 axis, are identically zero. In fact, looking at the phase-twisted lineshape, one can see that projection along the J multiplets leads to cancelation of negative dispersive and positive absorptive contributions (see Fig. 4.15). Therefore, all projections must be attempted only in the absolute value mode. Once this is done, then a 45° tilted projection onto the F_1 axis produces a J spectrum related to natural linewidths, and one onto the F_2 axis gives the so-called broadband homonuclear decoupled spectrum. The 2D spectrum itself can be presented in the tilted form. To do this, one has to interpolate the 2D data. We represent the frequency signal matrix as:

$$S(n, m) = S(n\Delta_1, m\Delta_2) \tag{48}$$

where the size of matrix is $n \times m$ and Δ_1 and Δ_2 are the frequency increment per point in the two domains. The interpolated values parallel to ω_1 axis for

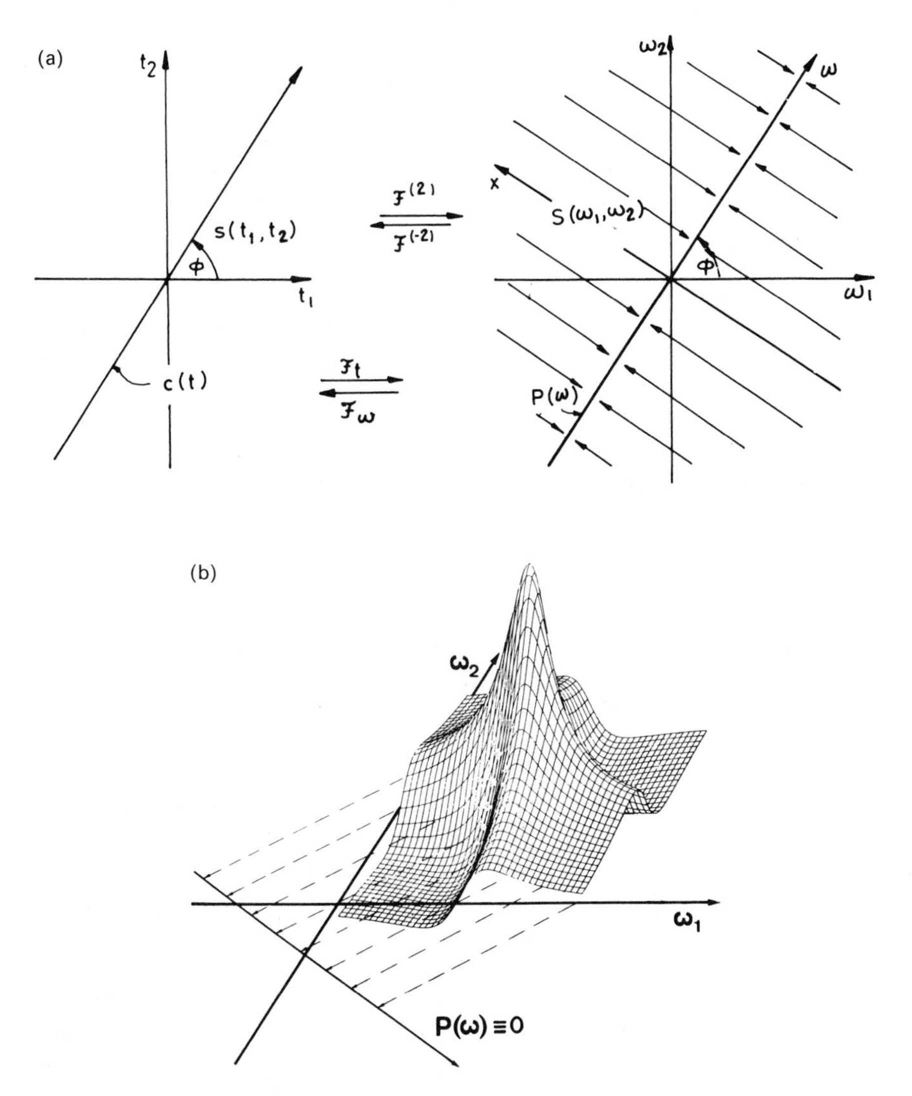

Figure 4.15. (a) Representation of the projection cross-section theorem. $S(t_1, t_2)$ and $S(\omega_1, \omega_2)$ represent time-domain and frequency-domain signals, respectively. $c(t)$ is a cross-section in the time domain, with $P(\omega)$ representing the corresponding projection in the frequency domain. The direction of projection is X, perpendicular to $P(\omega)$. (b) Superposition of absorptive and dispersive 2D peaks. Projection $P(\omega)$ obtained by projecting along lines with $\phi = \pi/4$ onto lines with $\phi = 3\pi/4$ vanishes due to exact cancelation of absorptive and dispersive contributions. [(a) and (b) are reproduced with permission. K. Nagayama, P. Bachmann, K. Wüthrich, and R.R. Ernst, *J. Magn. Reson.*, **31**, 133 (1978), copyright 1978, Academic Press, New York]

a constant ω_2 can be written:

$$S(n', m') = S(n'\Delta_1, m'\Delta_2 + n'\Delta_1 \tan 2\phi) \tag{49}$$

where ϕ is the tilt angle. The tilt angle depends on Δ_1 and Δ_2; for homonuclear J-resolved 2D spectra it is 45° when $\Delta_1 = \Delta_2$. Interpolation likewise can be done along the ω_1 axis. Such "tilt-corrected" spectra appear similar to the 2D heteronuclear J-resolved spectra discussed in the previous section. Moreover, one can apply different tilt angle corrections and introduce a uniform J scaling in the F_2 direction. Because the linewidth in the F_2 direction is subject to magnetic field inhomogeneities one can also use resolution enhancement weighting prior to Fourier transformation. Excellent resolution of various closely spaced, chemically equivalent protons is possible and J-resolved homonuclear spectroscopy can be used to unravel and assign ^1H spectra of fairly complicated systems, as long as there are no strong couplings that can lead to extra resonances (*vide infra*). An illustrative example of a 2D J-resolved spectrum of bovine pancreatic trypsin inhibitor (BPTI) is shown in Fig. 4.16. Additional improvement in resolution in the t_1 domain can be accomplished by taking advantage of the fact that during t_1 only the true spin–spin relaxation is operative and perhaps some broadening due to diffusion. To get the full advantage, one should extend the t_1 period upto six to seven times T_2. This will lead to a limitation on the t_2 period. However, Lorentzian to Gaussian weighting in the t_1 and Lorentzian weighting in the t_2 period can be used to increase resolution in the F_1 axis and to avoid truncation effects on the F_2 axis, respectively.

Diffusion effects can be reduced in the t_1 domain by applying a series of pulses instead of a single π pulse. There are other variations of the homonuclear 2D J-resolved spectroscopy based on the basic echo sequence. Instead of applying $[90^\circ_x - \frac{t_1}{2} - 180^\circ_x - \frac{t_1}{2} - \mathrm{Acq} - t_2]$ n times and collecting $S(t_1, t_2)$, all the necessary data matrix of the same can be generated by a single Carr–Purcell–Meiboom–Gill (CPMG) echo sequence by acquiring the signals during the windows between the pulses as shown in Fig. 4.17. Time t_2 is always the duration between the echo maximum and the next π pulse. The sampling in the t_2 period is limited to constant interval T and hence limited in resolution.

In another modification of the homonuclear J-resolved 2D spectroscopy, a weak rf pulse affecting only a very small range (covering approximately 200–300 Hz) corresponding to a flip angle π is applied at the end of the evolution period for every value of t_1. The resulting $S(t_1, t_2)$, with and without the weak pulse, is coadded (see Fig. 4.18). Because the selective pulse will affect only one of the partners in coupling, it will not change the modulation frequency at t_1 but will change the phase. Coadding two signals such as this for a series of t_1 results in amplitude modulation so that it is possible to record 2D spectra in the absorption mode.

A more general strategy to get rid of phase twists in homonuclear 2D

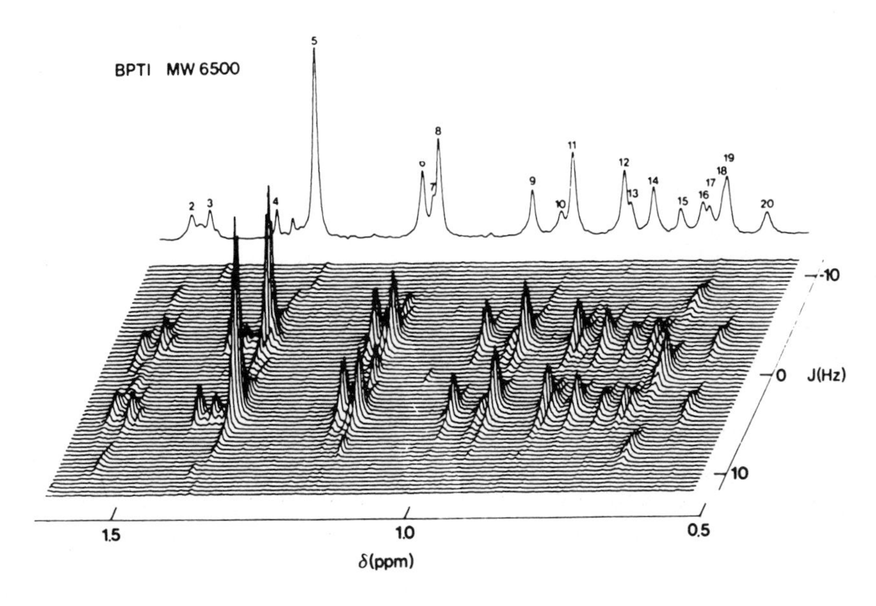

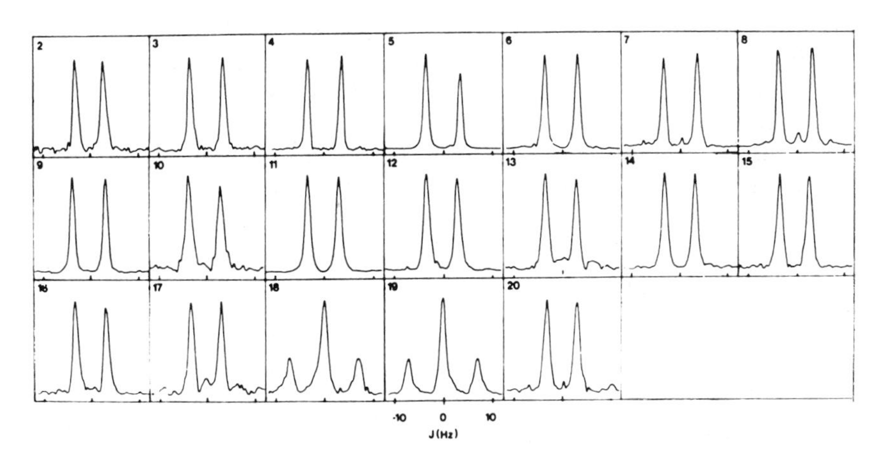

Figure 4.16. A 400-MHz homonuclear *J*-resolved spectrum of bovine pancreatic trypsin inhibitor (BPTI). The top figure shows the 2D spectrum and in the bottom, cross-sections corresponding to various chemically shifted protons show clearly the homonuclear multiplets. [Ref. Bruker Report, **2**, 4 (1979), reproduced by permission, Spectrospin AG, Zürich]

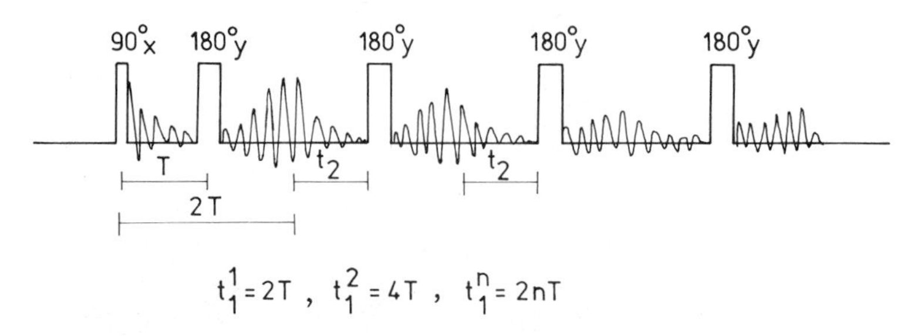

Figure 4.17. Pulse sequence for getting $S(t_1, t_2)$ in a one-shot experiment using the CPMG echo sequence.

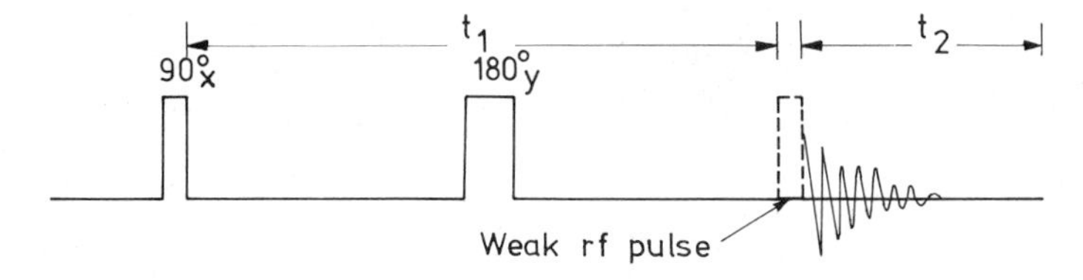

Figure 4.18. Pulse sequence for getting homonuclear 2D J-resolved spectra in the absorption mode. For each value of t_1, $S(t_1, t_2)$ is generated by coadding signals with and without the weak rf pulse prior to acquisition.

J-resolved spectra with a fair measure of success employs the pulse sequence shown in (Fig. 4.19). Addition or subtraction of the FID's from these two sequences converts the phase modulation into amplitude modulation; on working with the resultant of both addition and subtraction, the sign information in F_1 is also recovered. These results follow at once on noting that for a two-spin 1/2 system:

$$\begin{aligned} \sigma_0 &= I_{1z} + I_{2z} \\ \sigma_1 &= -(I_{1y} + I_{2y}) \\ \sigma_2 &= -c_J(I_{1y} + I_{2y}) + 2s_J(I_{1x}I_{2z} + I_{1z}I_{2x}) \end{aligned} \tag{50}$$

With a 90°_y pulse just prior to acquisition, on the other hand, we find:

$$\sigma_2 = -c_J(I_{1y} + I_{2y}) - 2s_J(\mathrm{I}_{1x}I_{2z} + I_{1z}I_{2x}) \tag{51}$$

It should be noted, however, that the final 90°_y pulse creates odd-order MQC's, such as triple-quantum coherence in systems of three or more spins. This leads to residual phase modulation because the single-quantum signal is correspondingly reduced in intensity in the experiment including the 90°_y pulse.

Multiple-Quantum Filtered 2D Resolved Spectroscopy. Two-dimensional resolved spectra may be modified by suppressing the strong singlet responses of "isolated" spins, allowing the multiplet patterns of close-lying coupled spin systems to be thrown in relief.

The idea of multiple quantum filtering introduced in Chapter 3 is employed to this end. The filter is introduced before the commencement of the evolution

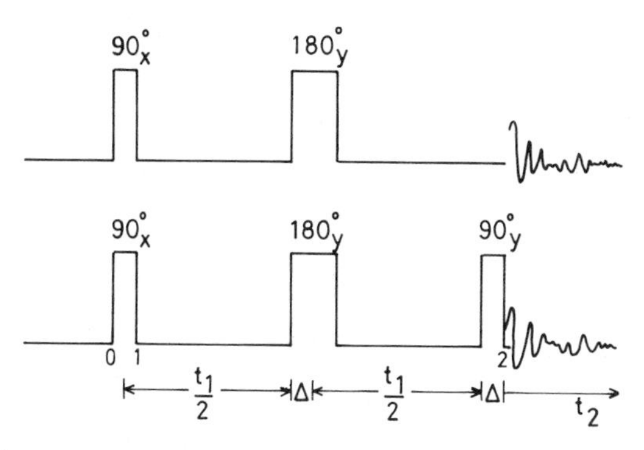

Figure 4.19. Pulse sequence to get rid of phase twists in homonuclear 2D J-resolved spectra by the introduction of a $\pi/2$ pulse in every alternate scan.

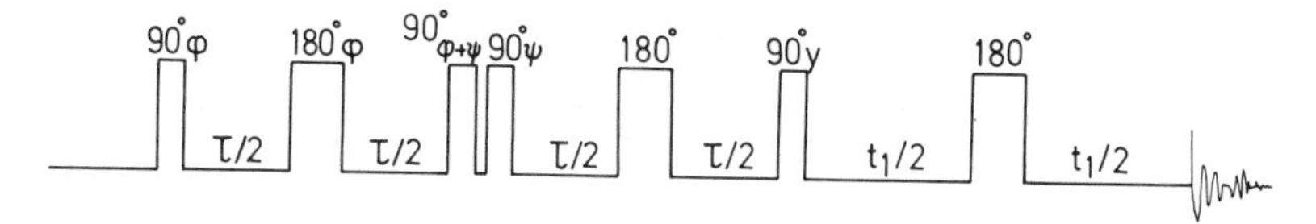

Figure 4.20. Pulse sequence for MQ-filtered, 2D J-resolved spectra.

period t_1, as shown in Fig. 4.20. Phase cycling is carried out as discussed in Chapter 3. The symmetrical excitation/detection strategy is used in this homonuclear experiment to insure uniform excitation efficiency and undistorted multiplets.

The performance of this multiple-quantum filtered homonuclear J, δ-resolved experiment is exemplified in Fig. 4.21, which employs double-quantum filtering to suppress the seven strong N-methyl resonances as well as the strong residual solvent resonance in a sample of cyclosporin A dissolved in benzene. Suppression of the N-methyl resonances permits in this case the clear identification of two other resonances that otherwise are hidden.

Presence of Strong Homonuclear Coupling. In homonuclear J-resolved 2D spectra in the presence of strong coupling, additional resonances occur. This can be understood in terms of the eigenstates and the corresponding single-quantum coherences before and after the refocusing π pulse.

Using second-order perturbation theory we can write the eigenfunctions in a AB NMR system as (see Fig. 4.22):

$$
\begin{aligned}
|1) &= |\alpha\alpha) \\
|2) &= \cos\theta\,|\alpha\beta) + \sin\theta\,|\beta\alpha) \\
|3) &= \cos\theta\,|\beta\alpha) - \sin\theta\,|\alpha\beta) \\
|4) &= |\beta\beta)
\end{aligned}
\tag{52}
$$

The presence of strong coupling terms modifies the otherwise uncoupled

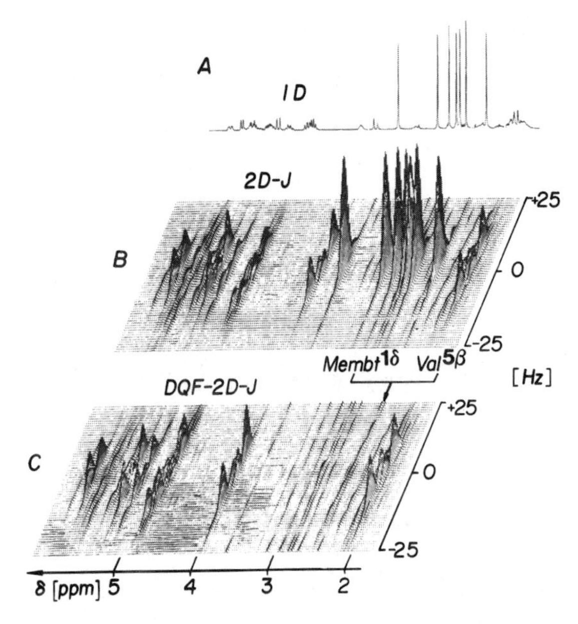

Figure 4.21. A 300-MHz double-quantum filtered, homonuclear, 2D J-resolved spectrum of cyclosporin A in-a benzene-d_6 (A) Conventional 1D spectrum between 2 and 6 ppm. (B) Stacked plot of 2D J-resolved spectrum. (C) Stacked plot of 2QF–2D J spectrum using the pulse sequence given in Fig. 4.20; 64 t_1 increments have been used. [Reproduced by permission. H. Kessler, H. Oschkinat, O.W. Sørensen, H. Kogler, and R.R. Ernst, *J. Magn. Reson.*, **55**, 329 (1983), copyright 1983, Academic Press, New York]

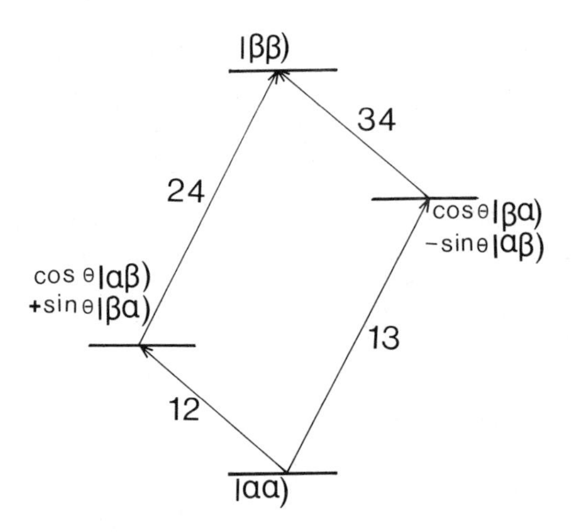

Figure 4.22. First-order corrected wave-functions and schematic energy level scheme for an AB NMR system.

states $|\alpha\beta)$ and $|\beta\alpha)$ into linear combinations of each other, the mixing coefficient being proportional to the relative magnitudes of J and Δ, such that $\tan 2\theta = J/\Delta$, where Δ is the difference between the chemical shifts of A and B. In the presence of strong coupling, the various single quantum coherences, such as (12), (13), (24), and (34), will be evolving at their respective frequencies. Let us consider the coherence (12) evolving at the frequency ω_{12}, corresponding to a coherent superposition of levels $|1)$ and $|2)$. This will evolve at the frequency ω_{12} for a time t_1. The 180°_y pulse modifies the states $|1)$ and $|2)$ to new states given by:

$$\begin{aligned} |\alpha\alpha) &\xrightarrow{180^\circ_y} |\beta\beta) \\ \cos\theta\,|\alpha\beta) + \sin\theta\,|\beta\alpha) &\xrightarrow{180^\circ_y} \cos\theta\,|\beta\alpha) + \sin\theta\,|\alpha\beta) \end{aligned} \tag{53}$$

Thus the states that are created can be considered as:

$$\begin{aligned} &|1) \to |4) \\ &|2) \to \sin 2\theta\,|2) + \cos 2\theta\,|3) \end{aligned} \tag{54}$$

so that the coherence (12) splits into two other coherences evolving at frequencies ω_{24} and ω_{34} with amplitudes $\sin(2\theta)$ and $\cos(2\theta)$. In the absence of strong coupling the coherence (12) evolving at frequency ω_{12} would after the 180°_x pulse have evolved as coherence (34) at the frequency ω_{34}. Therefore, in the case of strong coupling there is an additional frequency evolving with a phase difference $(\omega_{24} - \omega_{12})t_1/2$. Hence additional modulation at this frequency will be present in the t_1 domain, leading to a cross-peak in the eventual 2D spectrum at the coordinates $(\omega_{24} - \omega_{12})/2, \omega_{24}$. Similar cross peaks also occur at $(\omega_{34} - \omega_{13})/2, \omega_{34}$; $(\omega_{12} - \omega_{24})/2, \omega_{12}$; and $(\omega_{31} - \omega_{34})/2, \omega_{31}$. Simple vector pictures, corresponding to strong coupling effects, are shown in Fig. 4.23. Four additional peaks will appear at the coordinates indicated above.

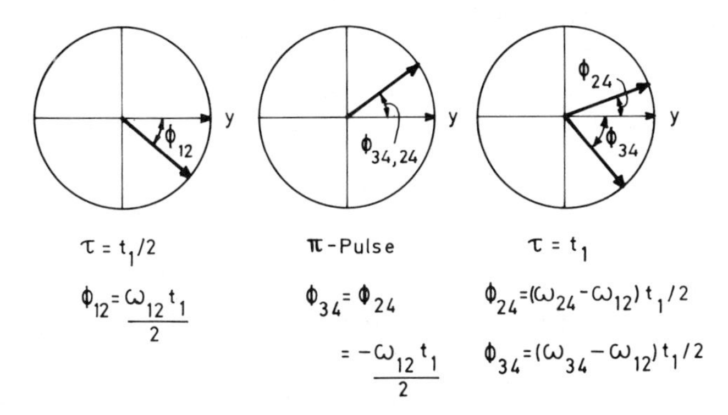

Figure 4.23. Vector picture showing the effect of strong coupling in homonuclear 2D *J*-resolved spectrum. The figure shows the relative phases of single-quantum coherence (12) before the pulse, after the π pulse, and the consequent splitting of this into coherences (24) and (34) at the end of the evolution period.

Two-Dimensional Correlation Spectroscopy

Consider the effect of two 90°_x pulses separated by an interval τ on a system of two weakly coupled spin-$\frac{1}{2}$ nuclei. We find after the second pulse:

$$\begin{aligned}\sigma = &-c_J(c_{2\delta_I}I_z + c_{2\delta_S}S_z) - 2s_J(c_{2\delta_I}I_xS_y + c_{2\delta_S}I_yS_x) \\ &+ c_J(s_{2\delta_I}I_x + s_{2\delta_S}S_x) - 2s_J(s_{2\delta_I}I_zS_y + s_{2\delta_S}I_yS_z)\end{aligned} \tag{55}$$

This density operator is composed of four types of terms:

1. Frequency-labeled longitudinal magnetization
2. Zero- and double-quantum coherences
3. In-phase transverse magnetizations
4. Anti-phase transverse magnetizations labeled with the frequencies of the "passive" spin

All the major categories of two- and three-pulse 2D correlation experiments may be characterized immediately, depending on the above density operator terms they track.

Terms 3 and 4 are responsible for the signals in a 2D correlation spectroscopy (COSY) experiment, where acquisition commences after the second pulse.

Term 1 may be used to follow cross-relaxation and exchange processes, which result in exchange of I_z with S_z during a mixing time τ_m; a subsequent 90° read pulse followed by signal acquisition permits the correlation of frequency labels during the evolution and detection periods, by 2D NOESY (nuclear Overhauser effect correlation spectroscopy).

Term 2 may be employed to monitor the evolution of multiple-quantum coherence (MQC's) during a free precession interval t_1; a subsequent 90° read pulse followed by signal acquisition permits the correlation of MQ frequencies with single-quantum frequencies. (Note that if a 180° refocusing pulse were employed at the middle of the interval between the two 90° pulses, pure double-quantum coherences would have been created after the second pulse.)

If the second pulse has a flip angle $\theta \neq (2n + 1)\pi/2$, longitudinal two-spin order is created also. Migration of such two-spin order during a mixing time τ_m may be monitored by a subsequent read pulse with flip angle $\theta \neq (2n + 1)\pi/2$.

Each of these 2D correlation experiments and their extensions is dealt with in some detail in this chapter, as well as in Chapter 5.

Chemical Shift Correlation Spectroscopy

In J-resolved spectroscopy we could unravel homonuclear or heteronuclear coupled high-resolution spectra by separating out different chemically shifted nuclei and could examine in detail the spin–spin coupling to a given nucleus without interference from multiplets belonging to another chemically shifted nucleus.

J-resolved spectroscopy is limited to separation of the multiplets with no correlation/identification of the chemical shift of the coupled nuclei. Two-dimensional chemical shift correlation just aims to achieve this via the hetero- and homonuclear scalar coupling (long range or short range), chemical exchange, cross-relaxation, and, in ordered liquids, dipolar coupling. These interactions are the most important link in some sort of polarization transfer from one spin to another and these manifest themselves as cross-peaks in the 2D spectra. Both frequency axes contain chemical shifts and they require, in addition to the preparation and evolution times, the mixing time mentioned in the introduction which brings about either coherent transfer of magnetization from one spin to another (homo- or heteronuclear scalar coupling in liquids or the dipolar coupling in partially ordered phases) providing the transfer pathway, or incoherent transfer via nuclear Overhauser effect brought about by dipolar cross-relaxation or chemical exchange. We shall first deal with 2D correlation spectroscopy through scalar coupling.

Heteroscalar Correlated 2D NMR in Liquids. In heteroscalar correlated 2D NMR spectroscopy one tries to unravel heteronuclear coupling networks or connectivities using pulse sequences similar to the polarization transfer schemes such as DEPT and INEPT (Chapter 3). As representative of heteronuclear systems we shall take ^{13}C and ^{1}H, which cover a majority of organic and biological systems. In these experiments usually ^{13}C will be the observe nuclei and polarization transfer from ^{1}H to ^{13}C will lead to, apart from establishing connectivities, sensitivity enhancement as a bonus. It is not mandatory that the less sensitive nuclei be observed. However, in the C–H coupling it is advantageous because $\gamma(^{1}H)/\gamma(^{13}C) \approx 4$ and spectral simplification can be achieved through broadband proton decoupling. This would be easier to accomplish than observing ^{1}H and decoupling ^{13}C, whose chemical shifts would be spread over hundreds of parts per million. Nevertheless, the larger dispersion of ^{13}C chemical shifts would warrant much higher sampling rate in t_2 and also larger number of FID's for each value of t_1 for S/N improvement. If, however, one wants to monitor ^{1}H in a C–H coupled situation, in order not to lose the sensitivity by transferring polarization from ^{13}C to ^{1}H, one can start the experiment by transferring maximum polarization from ^{1}H to ^{13}C and then go ahead with the specific pulse sequence.

The simplest of schemes in shift correlation spectroscopy consists in creating transverse magnetization of the abundant spin, letting it evolve under chemical shift and heteronuclear spin–spin coupling, thereby generating single-quantum coherences which are signed by the z components of the coupled rare spin. These signed components, corresponding to anti-phase magnetization, can be flipped back into two-spin zz order. Considering the low abundance of ^{13}C, one can, for all intents and purposes, assume subsequent carbon polarization to be governed by $\gamma(^{1}H)$. A $\pi/2$ pulse now at the frequency of the rare spin would produce transverse magnetization of the rare spin with enhanced sensitivity. Looking at the basic sequence (Fig. 4.24) and writing the density matrix in terms of product operators, with:

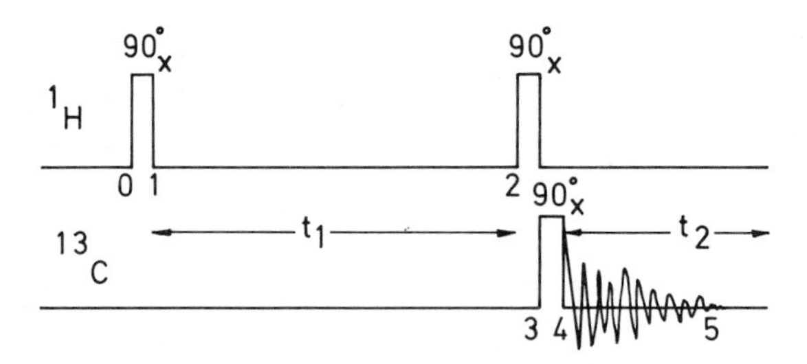

Figure 4.24. Pulse scheme for 2D heteronuclear shift correlation.

$$\begin{aligned} {}^1\mathrm{H} = I, \qquad & {}^{13}\mathrm{C} = S \\ \sigma_0 = I_z & \\ \sigma_4 = c_{2\delta_I}(t_1)[-c_J(t_1)I_z & - 2s_J(t_1)I_xS_y] \\ + s_{2\delta_I}(t_1)[c_J(t_1)I_x & - 2s_J(t_1)I_zS_y] \end{aligned} \tag{56}$$

The last term corresponds to observable S magnetization and evolves during the t_2 period:

$$\begin{aligned} -2s_{2\delta_I}(t_1)s_J(t_1)I_zS_y \to & -2s_{2\delta_I}(t_1)s_J(t_1)c_{2\delta_S}(t_2)c_J(t_2)I_zS_y \\ & + s_{2\delta_I}(t_1)s_J(t_1)c_{2\delta_S}(t_2)s_J(t_2)S_x \\ & + 2s_{2\delta_I}(t_1)s_J(t_1)s_{2\delta_S}(t_2)c_J(t_2)I_zS_x \\ & + s_{2\delta_I}(t_1)s_J(t_1)s_{2\delta_S}(t_2)s_J(t_2)S_y \end{aligned} \tag{57}$$

We can see that only the first and the last term contribute to the signal for $-y$ detection. Writing these two terms explicitly we get:

$$\begin{aligned} & -\frac{1}{2}\left\{\left[-\cos\left(\delta_I + \frac{J}{2}\right)t_1 + \cos\left(\delta_I - \frac{J}{2}\right)t_1\right]\right. \\ & \qquad \left.\times\left[\cos\left(\delta_S + \frac{J}{2}\right)t_2 + \cos\left(\delta_S - \frac{J}{2}\right)t_2\right]\right\} I_zS_y \\ & -\frac{1}{2}\left\{\left[\cos\left(\delta_I + \frac{J}{2}\right)t_1 - \cos\left(\delta_I - \frac{J}{2}\right)t_1\right]\right. \\ & \qquad \left.\times\left[-\cos\left(\delta_S + \frac{J}{2}\right)t_2 + \cos\left(\delta_S - \frac{J}{2}\right)t_2\right]\right\} S_y \end{aligned} \tag{58}$$

In the F_2 dimension, therefore, only S-spin chemical shifts and spin–spin coupling occur as anti-phase absorption, and in the F_1 dimension only I-spin chemical shifts and spin–spin coupling occur again as anti-phase absorption. The most important result is that the magnetization that evolves in t_2 with S-spin chemical shift and I–S coupling has been, in the t_1 duration, evolving with I-spin chemical shift and I–S coupling. Hence, cross-peaks (anti-phase absorption in shape) appear at coordinates ($\delta_I \pm (J/2)$, $\delta_S \pm (J/2)$). If in addition some transverse spin components are created by the S-spin $\pi/2$ pulse from single spin Zeeman order, axial peaks appear at coordinates $(0, \delta_S \pm J/2)$.

Therefore, there will always be cross-peaks between two nuclei that have a resolvable spin–spin coupling, and hence we can identify J connectivity. However, for each $^{13}C-^{1}H$, there will be four cross-peaks, for $^{13}C-^{1}H_3$, eight cross-peaks, etc.

We can simplify the above scheme by introducing proton broadband decoupling during acquisition. However, as we have seen, at the beginning of the detection period we see that the only observable S spin magnetization is anti-phase in nature (at $t_2 = 0$) and hence decoupling immediately after the mixing period leads to zero signal by the cancelation of the anti-phase signals. However, a time delay Δ between the $\pi/2$ S pulse and acquisition makes sure that the S-spin magnetization is finite even under decoupling conditions. Ideally the delay $\Delta = (2J)^{-1}$ (see RINEPT in Chapter 3). Interestingly any unmodulated magnetization that enters the signal in the t_2 period will just become anti-phase in the $(2J)^{-1}$ so that in this case the axial peaks at $(0, \delta_S \pm \frac{J}{2})$ will be absent. The scheme is shown in Fig. 4.25. Just before acquisition the y magnetization is given by:

$$\sigma_5 = s_{2\delta_I}(t_1) s_J(t_1) s_{2\delta_S}(\Delta) S_y \tag{59}$$

This evolves further only under S-spin chemical shift and the detected component will be

$$s_{2\delta_I}(t_1) s_J(t_1) s_{2\delta_S}(\Delta) c_{2\delta_S}(t_2) S_y$$

which is an absorption mode signal in the F_2 domain and anti-phase absorption in the F_1 domain. Cross-peaks appear at locations $(\delta_I \pm \frac{J}{2}, \delta_S)$ so that projections onto the F_2 axis in the absolute value mode should give the decoupled S spectrum and ones onto F_1 axis give the fully coupled I spectrum.

It is possible to simplify further by removing the spin–spin coupling from

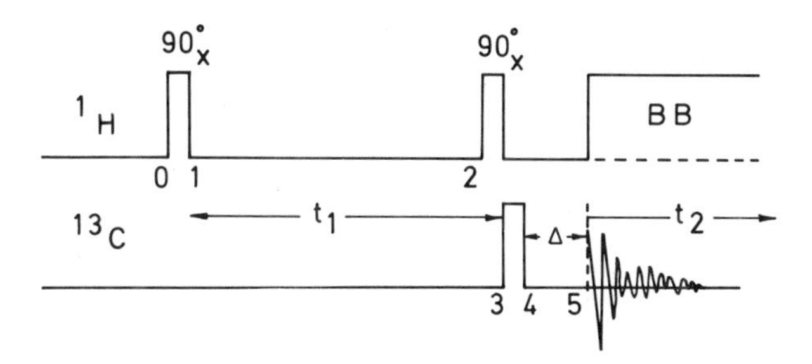

Figure 4.25. Pulse sequence for simplification of 2D heteronuclear shift correlation spectroscopy where heteronuclear J couplings are removed from the F_2 dimension by broadband decoupling. The delay Δ allows anti-phase ^{13}C multiplets to evolve so that magnetization will be finite under decoupling conditions.

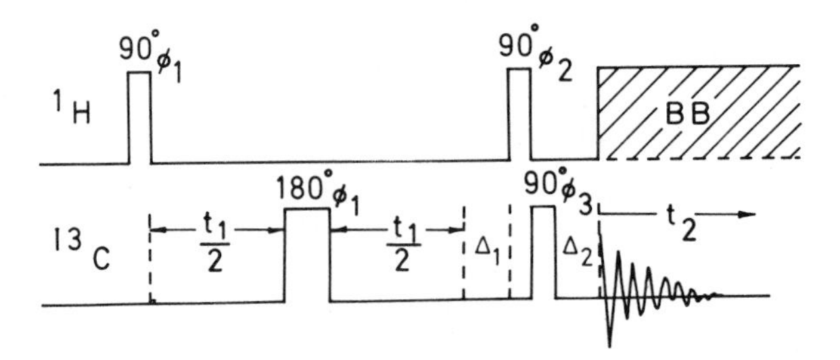

Figure 4.26. Pulse scheme for removing spin–spin coupling from either axis in heteronuclear 2D shift correlation spectra. σ_4 (Eq. (60)) is the state of the system at the end of the $90^\circ_{\phi_2}$ pulse.

either axis of the 2D spectrum by doing ^{13}C decoupling by a π pulse on this channel (this changes the sign of evolution under spin–spin coupling) and acquiring the ^{13}C signal after mixing and the delay mentioned above. However, if the carbon π pulse is applied exactly at the midpoint between the preparation and mixing period, we defeat the purpose of J correlation; anti-phase magnetization is not created at the end of the t_1 period, which is mandatory for polarization transfer (Chapter 3), which in turn is crucial for chemical shift correlation. The sequence (Fig. 4.26) leads to:

$$\sigma_0 = I_z$$

$$\begin{aligned}\sigma_4 = &-c_{2\delta_I}(t_1)c_{2\delta_I}(\Delta_1)c_J(\Delta_1)I_z + 2c_{2\delta_I}(t_1)c_{2\delta_I}(\Delta_1)s_J(\Delta_1)I_xS_z \\ &+ c_{2\delta_I}(t_1)s_{2\delta_I}(\Delta_1)c_J(\Delta_1)I_x + 2c_{2\delta_I}(t_1)s_{2\delta_I}(t_1)s_J(\Delta_1)I_zS_z \\ &+ s_{2\delta_I}(t_1)c_{2\delta_I}(\Delta_1)c_J(\Delta_1)I_x + 2s_{2\delta_I}(t_1)c_{2\delta_I}(\Delta_1)s_J(\Delta_1)I_zS_z \\ &+ s_{2\delta_I}(t_1)s_{2\delta_I}(\Delta_1)c_J(\Delta_1)I_z - 2s_{2\delta_I}(\Delta_1)s_{2\delta_I}(t_1)s_J(\Delta_1)I_xS_z\end{aligned} \quad (60)$$

Considering only single-quantum coherences involving S-spin magnetization that are created after the $\pi/2$ S pulse:

$$\sigma_5 = -2c_{2\delta_I}(t_1)s_{2\delta_I}(\Delta_1)\, s_J(\Delta_1)I_zS_y - 2s_{2\delta_I}(t_1)c_{2\delta_I}(\Delta_1)s_J(\Delta_1)I_zS_y \quad (61)$$

turns out to be anti-phase magnetization of the S spin, which when decoupled without a delay, vanishes. Allowing this to evolve for a further period of Δ_2 leads to:

$$\begin{aligned}\sigma_6 = &-2c_{2\delta_I}(t_1)s_{2\delta_I}(\Delta_1)s_J(\Delta_1)[(c_J(\Delta_2)I_zS_y - \tfrac{1}{2}s_J(\Delta_2)S_x)c_{2\delta_S}(\Delta_2) \\ &- (c_J(\Delta_2)I_zS_x + \tfrac{1}{2}s_J(\Delta_2)S_y)s_{2\delta_S}(\Delta_2)] \\ &- 2s_{2\delta_I}(t_1)c_{2\delta_I}(\Delta_1)s_J(\Delta_1)[(c_J(\Delta_2)I_zS_y - \tfrac{1}{2}s_J(\Delta_2)S_x)c_{2\delta_S}(\Delta_2) \\ &- (c_J(\Delta_2)I_zS_x + \tfrac{1}{2}s_J(\Delta_2)S_y)s_{2\delta_S}(\Delta_2)]\end{aligned} \quad (62)$$

Acquisition under decoupling of I spins and detecting with the detector phase as $-y$ the relevant part of the density operator corresponding to observable $S(t_1, t_2)$ is given by:

$$\sigma_{obs} = c_{2\delta_S}(t_2)c_{2\delta_S}(\Delta_2)s_J(\Delta_1)s_J(\Delta_2)[c_{2\delta_I}(t_1)s_{2\delta_I}(\Delta_1) + s_{2\delta_I}(t_1)c_{2\delta_I}(\Delta_1)]S_y \quad (63)$$

For a $^{13}C-^{1}H$ system if $\Delta_2 = \Delta_1 = (2J)^{-1} = \Delta$, this leads to:

$$\sigma_{obs} = c_{2\delta_S}(t_2)c_{2\delta_S}(\Delta)[c_{2\delta_I}(t_1)s_{2\delta_I}(\Delta) + s_{2\delta_I}(t_1)c_{2\delta_I}(\Delta)]S_y \quad (64)$$

Thus the double Fourier transform of $S(t_1, t_2)$ should give only chemical shift information on both axes. The lineshape along the F_2 axis is pure absorption, while along the F_1 axis it is a mixture of absorption and dispersion, depending on the delays Δ_1 and Δ_2. Thus the delays Δ_1 and Δ_2 play crucial roles; the first one creates anti-phase magnetization as a necessary prelude to polarization transfer, and the second Δ_2 converts anti-phase magnetization that would be unobservable under broadband decoupling into observable signal by allowing the anti-phase components to evolve further to produce a nonzero component along the detector phase. It is therefore obvious that the delays Δ_1 and Δ_2 cannot be chosen to be ideal for different I_nS groups with different spin–spin coupling. In the case of $^{13}C-^{1}H$ systems, the one-bond carbon–proton coupling is in the range 100–170 Hz (see SEFT, Chapter 2). If $\Delta_2 = (3J)^{-1}$ reasonable amplitudes are obtained for all types of carbons. Projection of the 2D spectrum onto the F_2 axis gives the conventional broadband decoupled ^{13}C spectrum, projection onto the F_1 gives the ^{1}H spectrum with any homonuclear coupling (Fig. 4.27).

Heteronuclear shift correlation spectra can be made quite efficient if we can reduce the sampling frequency in the t_1 duration by distinguishing positive and negative frequencies. That is, we have to manipulate the scheme to get pure phase modulation in the t_1 period by the following phase cycling. In this process we can also suppress axial peaks. Four experiments are performed with the phase settings shown in Table 4.1, and the resulting $S(t_1, t_2)$ are coadded prior to Fourier transformation.

There is one other modification of the heteronuclear 2D shift correlation spectroscopy that allows differentiating homonuclear J multiplets if these are attached to different chemically shifted heteronuclei. In other words, for example, it is possible to separate ^{1}H (I-spin) coupled multiplets of identically chemically shifted protons, if these are attached to different chemically shifted ^{13}C (S-spin) nuclei. The pulse sequence to be used is shown in Fig. 4.28. The experiment consists of allowing the ^{1}H transverse magnetization to evolve under only homonuclear coupling in the t_1 period by providing a π pulse at the middle of the t_1 period; this focuses both ^{1}H chemical shift evolution and $^{13}C-^{1}H$ heteronuclear coupling. At the end of the t_1 period both homo- and heteronuclear spin–spin coupling is allowed to operate by a $((\Delta_1/2) - \pi_S, \pi_I - (\Delta_1/2))$ pair of π pulses on both channels. This creates the necessary anti-phase magnetization still keeping out chemical shift evolution and magnetic field inhomogeneities effects. The antiphase magnetization at the end of the Δ_1 period is converted into two-spin zz order by a $\pi/2$ I pulse; this is followed by the $\pi/2$ pulse in the S channel, which produces transverse magnetization of S spins with polarization transfer. This anti-phase magnetization has to evolve for a period Δ_2, so that the S-spin signals do not vanish upon broadband

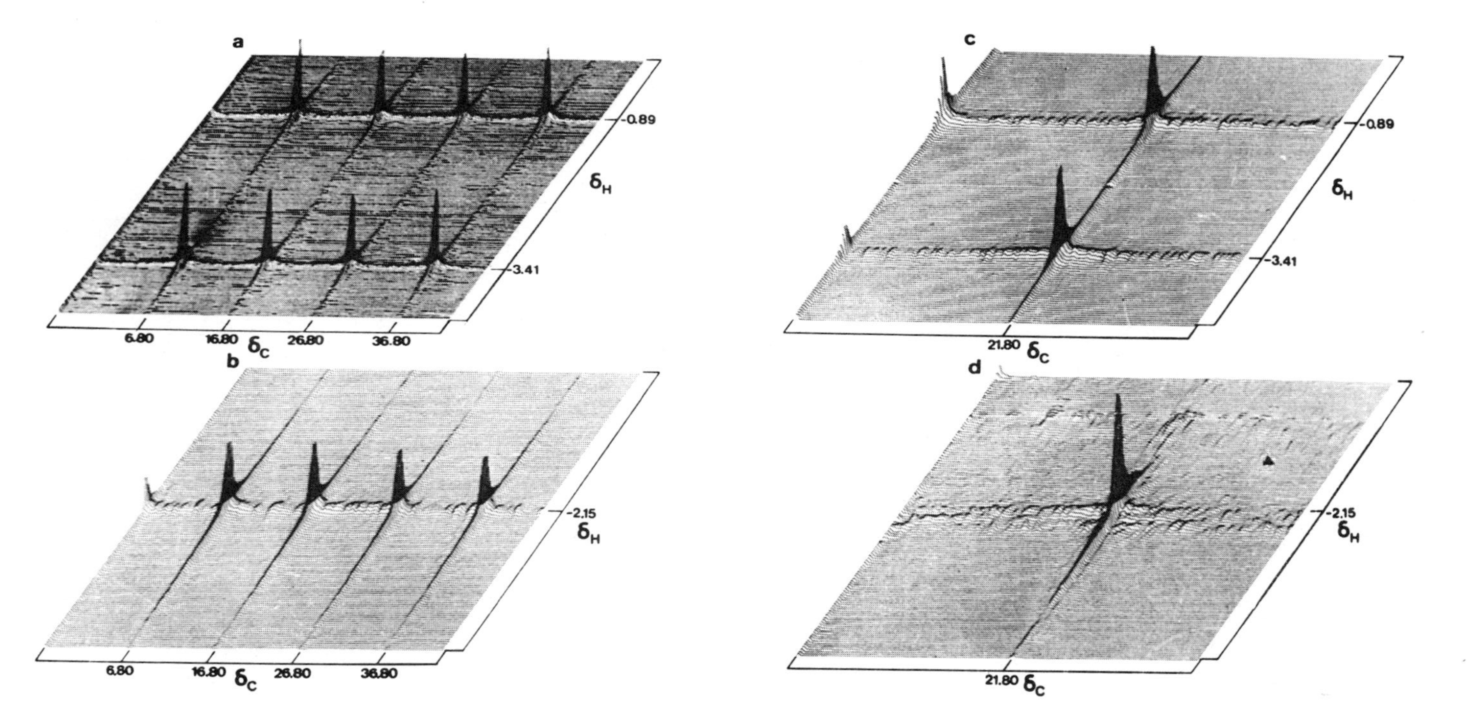

Figure 4.27. The 2D heteronuclear shift correlation spectra of methyl iodide (a) both δ and J are present in both axes. (b) only δ in F_1, and δ and J in the F_2 dimension; (c) δ and J in F_1, but δ only in the F_2 dimension; (d) only δ in both dimensions giving a peak at (δ_c, σ_H). [Reproduced with permission. A.A. Maudsley, L. Muller and R.R. Ernst, *J. Magn. Reson.*, **28**, 463 (1977), copyright 1977, Academic Press, New York]

Table 4.1. Phase Cycling for Heteronuclear COSY

Experiment	ϕ_1	ϕ_2	ϕ_3	Receiver Phase
1	x	x	x	x
2	x	y	x	$-y$
3	x	$-x$	x	$-x$
4	x	$-y$	x	y

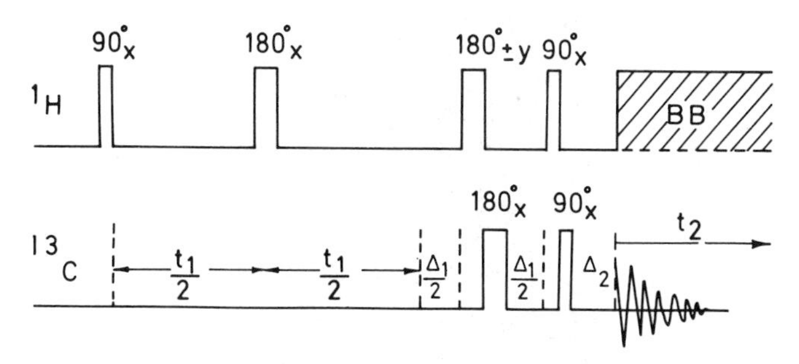

Figure 4.28. Modified 2D heteronuclear shift correlation pulse sequence to differentiate homonuclear J multiplets that are coupled to different chemically shifted heteronuclei.

I-spin decoupling. Except for small distortions introduced by Δ_1 and Δ_2 the 2D spectrum will have an absorption lineshape in both frequency dimensions. In the F_2 direction ^{13}C chemical shifts alone appear, while in the F_1 direction H–H spin–spin coupling multiplets appear, corresponding to a proton that is spin coupled to the carbon. The advantage is that proton multiplets that have identical chemical shifts can be separated as long as the carbons coupled to them have different chemical shifts, whereas in homonuclear J spectroscopy this is not possible (see Figs. 4.29 and 4.30).

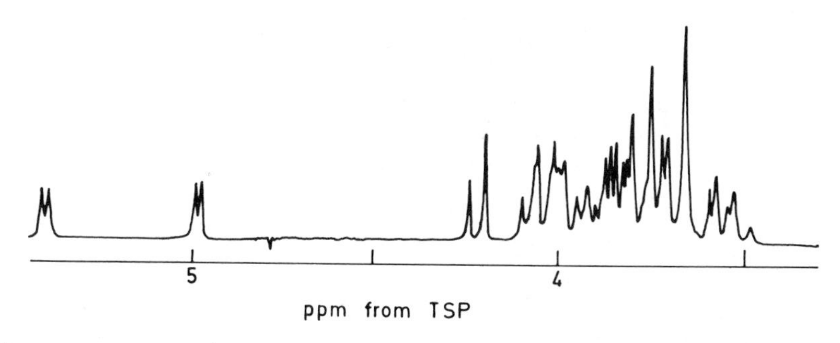

Figure 4.29. Normal ^{1}H spectrum of an aqueous solution of raffinose(II) at 200 MHz with solvent suppression. [Reprinted with permission. G.A. Morris, *J. Magn. Reson.*, **44**, 277 (1981), copyright 1981, Academic Press, New York]

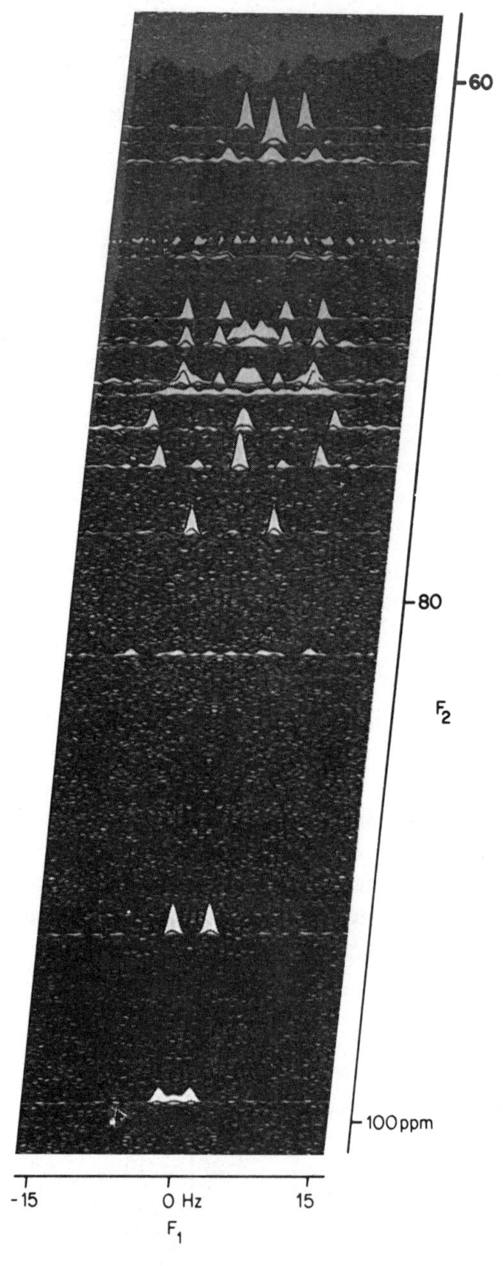

Figure 4.30. Indirect *J* spectrum of aqueous raffinose(II) showing excellent separation of very closely chemically similar protons brought about by the differences in the chemical shift of their coupled ^{13}C partners, a vast improvement over Fig. 4.29 for analysis. [Reprinted with permission. G.A. Morris, *J. Magn. Reson.*, **44**, 277 (1981), copyright 1981, Academic Press, New York]

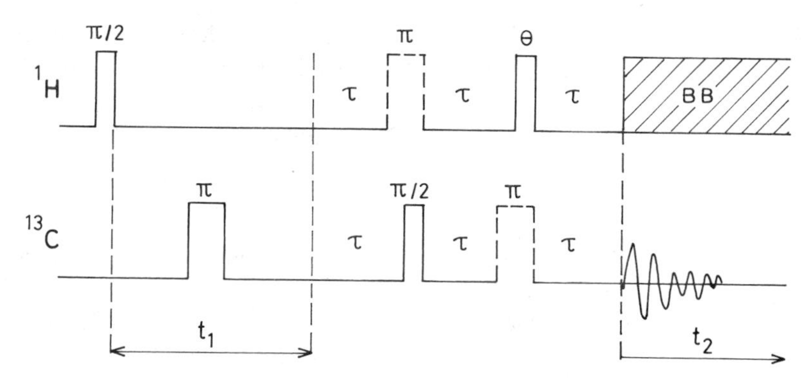

Figure 4.31. Pulse sequence for multiplet-seperated 2D heteronuclear shift correlation spectroscopy.

Multiplet Separated 2D Hetero COSY. It is also possible to modify the 2D heteronuclear correlation spectroscopy (COSY) in the following way to separate different types of coupling networks in the so-called multiplet-separated 2D hetero COSY. In the normal hetero COSY we suppress heteronuclear coupling in both dimensions to enhance resolution and sensitivity but in the process lose information such that CH, CH_2, CH_3 units can no longer be distinguished, unless a separate 2D J-resolved experiment is performed to retrieve this additional information. One can of course resort to one-dimensional spectral editing techniques, such as INEPT or DEPT and their improved versions (Chapter 3). The pulse scheme is given in Fig. 4.31. The first $\pi/2$ pulse in the proton channel creates ^1H coherences that evolve only under chemical shift during the period t_1 because of the presence of a π refocusing pulse halfway through the t_1 period in the ^{13}C channel. The mixing process in this scheme involves the DEPT sequence. The three delays τ in the mixing period are on the order of $(2J_{C\text{-}H})^{-1}$, where $J_{C\text{-}H}$ is the average one-bond C–H coupling. The polarization transfer depends on the flip angle θ of the proton pulse. For C–H the transferred magnetization is proportional to $\sin\theta$, for CH_2 to $\sin 2\theta$, and for CH_3 it is $\sin\theta + \sin 3\theta$. There are three ways in which one can use this flip angle dependence of polarization transfer.

One can perform 2D experiments and acquire data matrices for $\theta = \pi/4$, $\pi/2$, and $3\pi/4$ and linearly combine them either before or after FT to produce edited 2D spectra. Thus, subtraction of spectrum obtained with $\theta = \pi/4$ from that with $\theta = 3\pi/4$ eliminates all but CH_2 group magnetizations. When pulses are not perfect and the one-bond J_{CH} is spread over large values, editing is incomplete.

Alternatively, one can perform the experiment for $\theta = 3\pi/4$, in which case the resonances from CH_2 become inverted relative to those from CH and CH_3 and, if chemical shift dispersion is favorable, all three types can be distinguished. One must therefore present spectra in the phase-sensitive 2D mode and the refocusing pulses (given by dotted lines in the figure) are needed. Data must also be collected for $t_1 = 0$ to avoid frequency-dependent phase shifts.

A third version, similar to the accordion strategy, involves incrementing θ with t_1 such that:

$$\theta = \omega_\theta t_1 \tag{65}$$

Thus, apart from chemical shift modulation as a function of t_1, the magnetization will be modulated in addition as a function of $\sin(\omega_\theta t_1)$; i.e., CH resonances will be modulated as a function of $\sin(\omega_\theta t_1)$, whereas CH_2 will be modulated by $\sin(2\omega_\theta t_1)$ and CH_3 by $(\sin \omega_\theta t_1 + \sin(3\omega_\theta t_1))$. Thus the 2D peaks in correlation maps get split into sidebands in the F_1 dimension, two for CH and CH_2 and four for CH_3. If the minimum sideband separation ω_θ exceeds the estimated spread in the ^1H chemical shifts, the 2D peaks from different groups (CH, CH_2, CH_3) will be well separated from each other. This leads to distinct heteronuclear shift correlation maps in a single experiment. There is still a difficulty arising from the presence of CH_3 group resonances in CH regions; however, because these resonances have distinct chemical shifts and, in addition, methyl resonances will give four sidebands, identification is still possible. The standard phase cycling procedure is invoked to distinguish between positive and negative frequencies. The phases of the first proton pulse and the final ^{13}C pulse are cycled in steps of $\pi/2$ radians in opposite senses and the resulting signals coadded. This will also suppress signals from ^{13}C that did not originate from polarization transfer. To summarize, in this third version, the 2D spectra have resonances at:

$$\begin{aligned}
(F_1, F_2) &= (\delta_H \pm \omega_\theta, \delta_C) \text{ for CH} \\
(F_1, F_2) &= (\delta_H \pm 2\omega_\theta, \delta_C) \text{ for CH}_2 \\
(F_1, F_2) &= (\delta_H \pm \omega_\theta, \delta_C) \text{ and } (\delta_H \pm 3\omega_\theta, \delta_C) \text{ for CH}_3
\end{aligned} \tag{66}$$

Figure 4.32 shows an example of such a scheme.

Homoscalar Correlated 2D NMR in Liquids. This was the very first 2D NMR experiment to be suggested and carried out. The principle is to create transverse magnetization by a suitable nonselective pulse and allow single-quantum coherences of a given spin to evolve under chemical shift and spin–spin coupling. During the mixing period another nonselective pulse of arbitrary flip angle is applied, by which the signed single-quantum coherences of a given spin are converted either into unobservable zero-, double-, or multiple-quantum coherences involving the coupled spins, or into observable single-quantum coherences corresponding to one of the coupled partners. For example, in an IS situation, signed coherence such as $I_z S_y$, evolving under S-spin chemical shift and spin–spin coupling, will after a $\pi/2$ mixing pulse evolve as $-I_y S_z$ under I-spin chemical shift and spin–spin coupling, thereby establishing correlation. Such a coherence transfer is possible only when there is a resolved coupling between the spins involved and if a particular spin is

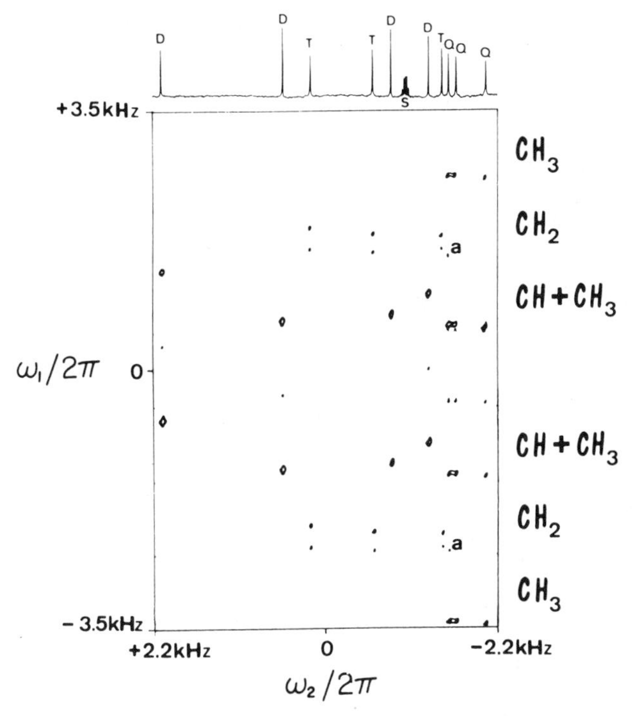

Figure 4.32. Multiplet-separated heteronuclear shift correlation spectrum of menthol obtained by incrementing θ (see Fig. 4.31) synchronously with the evolution period t_1. The ω_2 dimension corresponds to ^{13}C chemical shift, whereas in ω_1 the frequencies of coupled protons occur at $\delta_H \pm \omega_\theta$, $\pm 2\omega_\theta$, $\pm 3\omega_\theta$ depending on the nature of the multiplet (see text for details). On the top is given the ^1H broadband-decoupled ^{13}C spectrum. [Reproduced with permission. M.H. Levitt, O.W. Sørensen, and R.R. Ernst, *Chem. Phys. Lett.*, **94**, 540 (1983), North-Holland Physics Publishing, Amsterdam]

coupled to a large number of other spins, the magnetization will be aportioned. This particular scheme of 2D correlation spectroscopy, known as COSY, is of great significance in the analysis of complex high-resolution NMR spectra. It is possible, by observing cross-peaks, for example at frequencies

$$\left(\delta_I \pm \frac{J_{IS}}{2}, \delta_S \pm \frac{J_{IS}}{2}\right)$$

to go through with proton–proton connectivities and hence proximities in molecules, so that ideally one should be able to go from one end of the molecule to the other end by following up cross-peak connectivities. As a consequence, valuable information on molecular conformation will result. Because resolved J coupling is mandatory for establishing connectivities, different molecules in a given sample can be distinguished. (This may be of significance in actual biological samples where one can isolate impurity peaks.)

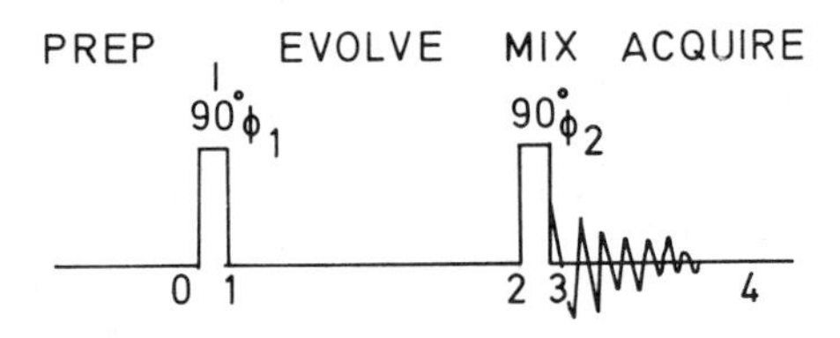

Figure 4.33. Pulse sequence for 2D homonuclear shift correlation spectroscopy, COSY.

Two-dimensional COSY has therefore become an important analytical and structural tool in the area of biological systems, such as proteins, nucleic acids, and polymers. The pulse sequence is given in Fig. 4.33.

Let us consider a pair of $(\pi/2)_x$ pulses applied to a homonuclear coupled $I_1 I_2$ system and look at the evolution of the density matrix in terms of the product operators:

$$\begin{aligned}\sigma_0 &= I_{1z} + I_{2z}\\ \sigma_3 &= -c_J(t_1)[c_{2\delta_1}(t_1)I_{1z} + c_{2\delta_2}(t_1)I_{2z}]\\ &\quad + 2s_J(t_1)[-c_{2\delta_1}(t_1)I_{1x}I_{2y} - c_{2\delta_2}(t_1)I_{1y}I_{2x}]\\ &\quad + c_J(t_1)[s_{2\delta_1}(t_1)I_{1x} + s_{2\delta_2}(t_1)I_{2x}]\\ &\quad - 2s_J(t_1)[s_{2\delta_1}(t_1)I_{1z}I_{2y} + s_{2\delta_2}(t_1)I_{1y}I_{2z}]\end{aligned} \tag{67}$$

The first term is unobservable Zeeman order, and the second term is unobservable zero- and double-quantum coherences involving the coupled spins. Therefore, during the detection period only the third and fourth terms will evolve. It is important to note that for the fourth term, if evolution of each of these coherences is under the chemical shift and spin–spin coupling of one of the partners in the t_1 period, it will be under the chemical shift and spin–spin coupling of the other partner in the t_2 period. During the t_2 period the acquired signal is represented by the density matrix:

$$\begin{aligned}\sigma_4 &= s_{2\delta_1}(t_1)c_J(t_1)c_{2\delta_1}(t_2)c_J(t_2)I_{1x} + s_{2\delta_2}(t_1)c_J(t_1)c_{2\delta_2}(t_2)c_J(t_2)I_{2x}\\ &\quad - 2s_{2\delta_1}(t_1)c_J(t_1)s_{2\delta_1}(t_2)s_J(t_2)I_{1x}I_{2z} - 2s_{2\delta_2}(t_1)c_J(t_1)s_{2\delta_2}(t_2)s_J(t_2)I_{1z}I_{2x}\\ &\quad + s_{2\delta_1}(t_1)s_J(t_1)c_{2\delta_2}(t_2)s_J(t_2)I_{2x} + s_{2\delta_2}(t_1)s_J(t_1)c_{2\delta_1}(t_2)s_J(t_2)I_{1x}\\ &\quad - 2s_{2\delta_1}(t_1)c_J(t_1)s_{2\delta_2}(t_2)c_J(t_2)I_{1z}I_{2x} - 2s_{2\delta_2}(t_1)c_J(t_1)s_{2\delta_1}(t_2)c_J(t_2)I_{2z}I_{1x}\\ &\quad + 2s_{2\delta_1}(t_1)c_J(t_1)c_{2\delta_1}(t_2)s_J(t_2)I_{1y}I_{2z} + 2s_{2\delta_2}(t_1)c_J(t_1)c_{2\delta_2}(t_2)s_J(t_2)I_{1z}I_{2y}\\ &\quad + s_{2\delta_1}(t_1)c_J(t_1)s_{2\delta_1}(t_2)c_J(t_2)I_{1y} + s_{2\delta_2}(t_1)c_J(t_1)s_{2\delta_2}(t_2)c_J(t_2)I_{2y}\\ &\quad - 2s_{2\delta_1}(t_1)s_J(t_1)c_{2\delta_2}(t_2)c_J(t_2)I_{1z}I_{2y} - 2s_{2\delta_2}(t_1)s_J(t_1)c_{2\delta_1}(t_2)c_J(t_2)I_{1y}I_{2z}\\ &\quad + s_{2\delta_2}(t_1)s_J(t_1)s_{2\delta_1}(t_2)s_J(t_2)I_{1y} + s_{2\delta_1}(t_1)s_J(t_1)s_{2\delta_2}(t_2)s_J(t_2)I_{2y}\end{aligned} \tag{68}$$

For $-y$ detection only the last eight terms will induce the signal. Of them, the first four terms correspond to single-quantum coherences, which evolve under identical chemical shifts in both the durations and hence produce diagonal peaks at $(\delta_i \pm \frac{J}{2}, \delta_i \pm \frac{J}{2})$. The last four terms produce cross-peaks since these correspond to the evolution of transferred coherences, which evolve under different chemical shifts in the evolution and detection periods and hence establish the J connectivities. They occur at frequencies $(\delta_i \pm \frac{J}{2}, \delta_j \pm \frac{J}{2})$. In all, for an AX system one gets eight diagonal peaks and eight cross-peaks. The first term, corresponding to the I_1 diagonal peaks, simplifies to:

$$\frac{1}{2}\left[\sin\left(\delta_1 + \frac{J}{2}\right)t_1 + \sin\left(\delta_1 - \frac{J}{2}\right)t_1\right] \times \left[\sin\left(\delta_1 + \frac{J}{2}\right)t_2 - \sin\left(\delta_1 - \frac{J}{2}\right)t_2\right] I_{1y}I_{2z} \tag{69}$$

In the F_2 direction this corresponds to in-phase dispersions (the anti-phase $\sin\alpha - \sin\beta$ dispersion is made in phase by the product operator $I_{1y}I_{2z}$). Likewise, in the t_1 period the $(\sin\alpha + \sin\beta)$ form predicts again in phase dispersion. The third term corresponding to I_1 diagonal peaks can be written:

$$\frac{1}{4}\left[\sin\left(\delta_1 + \frac{J}{2}\right)t_1 + \sin\left(\delta_1 - \frac{J}{2}\right)t_1\right] \times \left[\sin\left(\delta_1 + \frac{J}{2}\right)t_2 + \sin\left(\delta_1 - \frac{J}{2}\right)t_2\right] I_{1y} \tag{70}$$

again corresponding to in-phase dispersion on both dimensions. On the other hand, the fifth term, corresponding to the cross-peaks, simplifies to:

$$\frac{1}{2}\left[\cos\left(\delta_1 + \frac{J}{2}\right)t_1 - \cos\left(\delta_1 - \frac{J}{2}\right)t_1\right] \times \left[\cos\left(\delta_2 + \frac{J}{2}\right)t_2 + \cos\left(\delta_2 - \frac{J}{2}\right)t_2\right] I_{1z}I_{2y} \tag{71}$$

Thus in both frequency dimensions absorption lineshapes with anti-phase character appear. The same lineshape is predicted from the other three terms of the cross-peaks. Thus, all cross-peaks are anti-phase absorptions, while the diagonal peaks are in phase dispersion. This is represented in Fig. 4.34. It is therefore not possible to phase the spectrum for pure absorption.

Although it is too difficult to predict the exact intensities of cross-peaks, it is possible to generalize on this. Since in the 1D spectrum the intensity of a given chemically shifted peak is aportioned among its spin-coupled multiplet lines, in the corresponding 2D this is further distributed among all protons to which the polarization is transferred by the second $\pi/2$ pulse. Thus, in an AX system, the intensity of the one-dimensional AX quartet is now shared by eight diagonal and eight cross-peaks, not necessarily equally. If the transverse

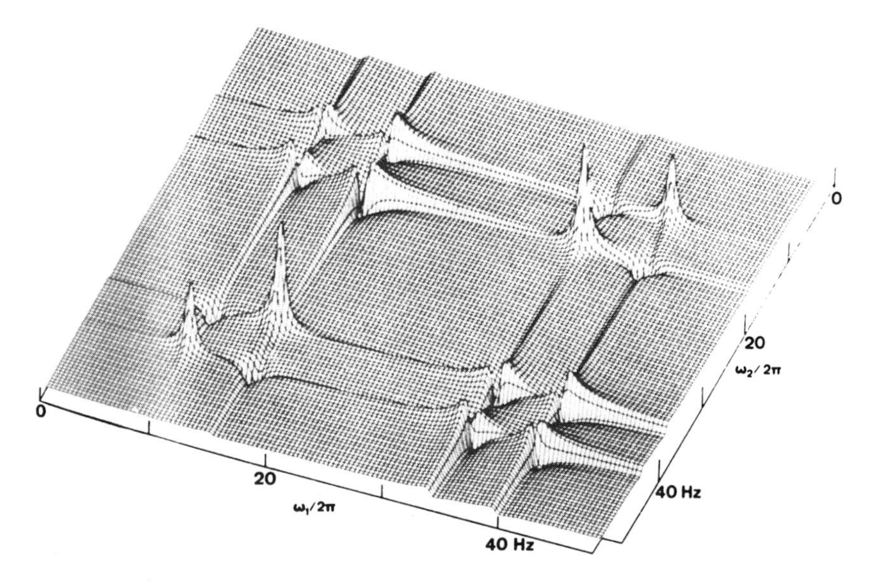

Figure 4.34. Theoretical COSY spectrum of 2,3-dibromothiophene at 60 MHz. Note that the diagonal peaks are in-phase dispersion, whereas the cross-peaks are anti-phase absorption in nature. [Reproduced with permission. W.P. Aue, E. Bartholdi, and R.R. Ernst, *J. Chem. Phys.*, **64**, 2229 (1976), copyright 1976, American Institute of Physics, New York]

relaxation times T_2^* are short compared to $(J_{AX})^{-1}$ then the transferred magnetization by the second $\pi/2$ pulse will be so low that the cross-peaks may be very low in intensity. Large coupling alone cannot insure strong cross-peaks.

In order to distinguish between positive and negative frequencies in the F_1 axis (this is important when placing the transmitter frequency somewhere in the middle of the spectrum), which enables sampling frequency to be reduced by a factor of 2, one must modify the experiment so as to produce phase modulation in the t_1 duration. This can be done by changing the phase of the second $\pi/2$ pulse from x to y and subtracting the resulting signal from the one with the $(\pi/2)x$ pulse. This converts the original amplitude modulation to phase modulation so that a complex 2D FT can distinguish between positive and negative frequencies. Additionally, any axial peaks $(0, \delta_I \pm J/2)$ arising from T_1 recovery during the t_1 period can also be removed by additional phase cycling. Thus one performs four experiments with the following setting and coadds the FID's before processing (Table 4.2).

To further combat errors in the quadrature detector system each experiment can be repeated four times by incrementing the phases by 90° for one full cycle including the detector phase. This will lead to 2D COSY spectra without axial peaks and monitor only coherence transfer echoes.

The coherence transfer echoes, which are a general phenomenon in 2D

Table 4.2. Phase Cycling for Homonuclear COSY

Experiment	$\pi/2$ Preparation Phase	$\pi/2$ or (θ) Mixing Phase	Detector Phase
1	x	x	x
2	x	y	$-x$
3	x	$-x$	x
4	x	$-y$	$-x$

experiments, can be defined as follows. In general any single-quantum coherence pq that is modulated at ω_{pq} can after the mixing pulse appear as another coherence rs, now modulated, say, at the frequency ω_{rs}. The transverse magnetization can thus be written:

$$M_{(pq)(rs)}(t_1, t_2) \sim \cos(\omega_{pq} t_1) \exp(i\omega_{rs} t_2) \tag{72}$$

where we have written the modulation in the t_1 period as an amplitude modulation. This can also be considered as the superposition of phase modulation at two identical frequencies of opposite signs. The two coherences pq and rs will have their characteristic frequencies ω_{pq}, ω_{rs} affected by the strength and the inhomogeneity of the field, but the ratio ω_{pq}/ω_{rs} is independent of both the field strength and its quality. Therefore, at a time $t_2/t_1 = \omega_{pq}/\omega_{rs}$ the phase of the detected components will be independent of field strength/inhomogeueities corresponding to the so-called N-type peak or "coherence transfer echo." This signal corresponds to modulation with negative phase during t_1, whereas the other component, which has been modulated with positive phase, will continue to decay without refocusing after coherence transfer. The corresponding signal is called the anti-echo, the so-called P-type peak.

It is better to acquire the coherence transfer echoes in 2D experiments since they will have high resolution. The above phase cycling therefore will detect only coherence transfer echoes and suppress axial peaks. To eliminate truncation errors one can use Gaussian weighting in the t_1 period. We have noted that the cross-peaks appear as anti-phase multiplets in either frequency axis. As such, in the absolute value mode of presentation at large distances from the center of these cross-peaks, the dispersion contribution cancels so that these peaks do not show tailing. However, if the peak separations in the cross-peaks are so low as to overlap with each other, partial cancelation of these does occur. To avoid such cancelation, fine digitization of the two time domains t_1 and t_2 is required. Thus the length of the samplings in both time axes must be at least of the order of $(J)^{-1}$.

By adding the time-domain signals corresponding to diagonal and cross-peaks in the AX system one can get:

$$S(t_1, t_2)_{\text{cross}} \sim s_J(t_1) s_J(t_2) e^{-2\pi i \delta_A t_1} e^{+2\pi i \delta_X t_2} e^{-t_1/T_2^{*(1)}} e^{-t_2/T_2^{*(2)}}$$

and:

$$S(t_1, t_2)_{\text{dia}} \sim c_J(t_1)c_J(t_2)e^{-2\pi i\delta_A t_1}e^{+2\pi i\delta_A t_2}e^{-t_1/T_2^{*(1)}}e^{-t_2/T_2^{*(2)}} \quad (73)$$

Here we have introduced spin–spin relaxation during both the time dimensions. Suppose we introduce weighting functions that increase with time and reach a maximum when $t = (2J)^{-1}$. This will favor cross-peaks at the expense of diagonal peaks. Thus a weighting function that modifies the signal to a Gaussian pseudo-echo, such as $g(t) = 1 - \exp(-t/(2J)^{-1})$, where J is the largest coupling in the molecule, (a) will lead to enhancement of cross peaks and (b) in the absolute value mode will suppress long tailing of the diagonal peaks. This brings to light cross-peaks that occur close to the diagonal peaks. Choosing identical lengths of the t_1 and t_2 period and using the pseudo-echo filter in both dimensions lead to 2D spectra that are mirror images about the diagonal. Again, one must remember that absence of cross-peaks does not imply absence of spin–spin coupling.

Effect of Flip Angle on the Nature of COSY Spectra. When the mixing pulse corresponds to a flip angle of $\pi/2$ coherence transfer takes place among all spins that are connected by nonvanishing coupling constants. However, it can be shown (see below) that at flip angles other than 90° (assuming that the Hamiltonian corresponds to weak coupling), magnetization transfer takes place to coherences that share a common energy level in a greater magnitude than to those which are unconnected. Thus it is possible to choose the flip angle to enhance intensities of cross-peaks corresponding to directly connected transitions.

Taking the weakly coupled three-spin system IST (see Fig. 4.35), let us concentrate on S-spin coherence corresponding to one of the outer transitions at the end of the t_1 period under coupling to both I and T spins.

The second (mixing) pulse, if applied along the x axis, transfers y compo-

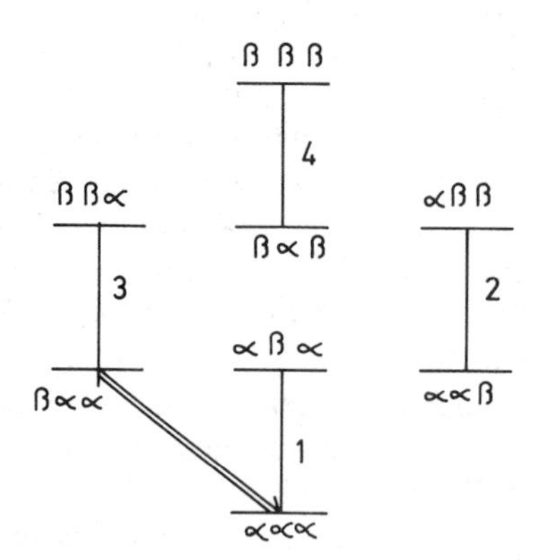

Figure 4.35. Energy levels diagram of an SIT system. The S spin coherence corresponding to the slanted arrow will be transfered to four I-spin and four S-spin coherences during the mixing. (1) and (3) are directly connected coherences; (2) and (4) are distant coherences.

nents of the S-spin magnetization into y components of the I and T spins and, if applied along the y axis, transfers x components of the S-spin magnetization into x components of the I and T spins. Considering the x magnetization of S spins at the end of the evolution period, the relevant product operators, corresponding to (see "2" in Fig. 4.33):

$$\sigma_2 \sim 4S_xI_zT_z + 2S_x(I_z + T_z) + S_x. \tag{74}$$

We can represent the above product operator in terms of single S transitions as in Fig. 4.36. Let us now apply a θ pulse about the y axis and look at only product operators which correspond to cross-peaks between S and I spins:

$$4S_xI_zT_z + 2S_x(I_z + T_z) + S_x \xrightarrow{\theta_y} -\cos\theta\sin^2\theta\, 4S_zI_xT_z - \sin^2\theta\, 2S_zI_x \tag{75}$$

The corresponding intensity patterns are shown in Fig. 4.36. Whereas the transitions labeled (1) and (3) are reinforced corresponding to a total intensity $\pm\sin^2\theta(1 + \cos\theta)$, those corresponding to (2) and (4) are reduced, resulting in an intensity of $\pm\sin^2\theta(1 - \cos\theta)$. In the energy level diagram of Fig. 4.35, the slanted arrow corresponds to S coherence under consideration prior to polarization transfer, and the vertical arrows are the I spin coherences that are created from the S spin by the θ pulse (mixing pulse). Of them, (1) and (3) share the levels corresponding to the S-spin coherence and come up with an intensity $\sim(1 + \cos\theta)$, whereas the coherences (2) and (4), which are not connected, come up with intensities $\sim(1 - \cos\theta)$. Hence, relative intensities of directly connected cross-peaks to that of unconnected ones will be equal for a $\pi/2$ mixing pulse; for a $\pi/4$ mixing pulse, in contrast, the ratio of directly connected peaks to those of indirect peaks is 5.8 times larger and the polarization transfer is also 50% as efficient as the $\pi/2$ pulse.

These connected cross-peak to unconnected cross-peak intensities can also be calculated by assuming the second nonselective pulse to be a cascade of transition-selective pulses. Arguments can be advanced to show that transfer of given coherence to unconnected coherences will go through hypothetical double- and zero-quantum pathways. We believe that the simple product operator formalism unambiguously demonstrates the aportioning of the transferred coherences in terms of the connected and "distant" coherences.

Similar arguments show that the intensities of the "parallel" transitions that corresponds to polarization transfer within the same multiplet also are subject to the flip angle effects of the second pulse. Thus, for the second pulse less than $\pi/2$, these "parallel" transitions lose intensity faster than the "connected" transitions, and some parallel transitions lose intensity faster than others.

It is also possible to infer the relative signs of coupling constants. For the second pulse of small flip angle, as shown earlier, directly connected transitions are more intense than the others. Thus, in an AMX system, under ideal conditions it is possible to wipe out eight of the 16 resonances in each cross-peak using a suitable small flip angle. This small flip angle situation is like selective decoupling in that for an AMX system, for example, polarization transfer is considerable only among directly connected transitions and only

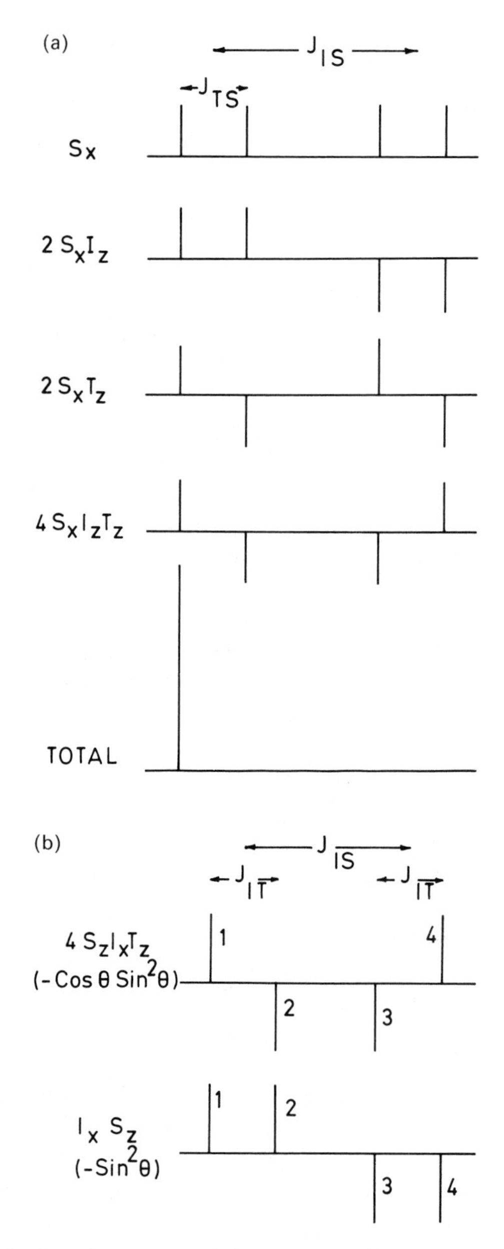

Figure 4.36. (a) S-spin coherences and the corresponding spectra which add up to a single transition. (b) Cross-peaks corresponding to I–S coherence transfer after the θ mixing pulse and the corresponding intensity patterns.

eight cross-peaks appear between any two resonances, instead of the 16 expected for a second $\pi/2$ pulse. Considering the cross-peaks for A and X resonances if AM and XM have coupling constants of the same sign, then (since J_{AM} and $J_{MX} > 0$) the displacement of the AX cross-peaks will be in the same direction. Therefore, the two sub-cross-multiplets, when connected center to center, will make an angle less than 45° to the main diagonal. If J_{AM} and J_{MX} are of opposite sign, in contrast, then the center of gravity of the sub-cross-multiplets will make an angle more than 45°. If J_{AM} or $J_{MX} \approx 0$ then such tilting effects will not be there (Fig. 4.37).

A J-selective (BIRD) sequence $(\frac{\pi}{2})_x(I) - (2J)^{-1} - \pi_x(I,S) - (2J)^{-1} - (\frac{\pi}{2})_{-x}(I)$ may be inserted in the middle of t_1, leading to a homonuclear broadband decoupled spectrum in F_1. Decoupling is in principle achieved for all but inequivalent geminal protons. In practice, however, such an experiment often leads to "artifacts" in F_1, which arise whenever relative shift of two protons matches $^1J_{CH}$, i.e., $(\nu_A - \nu_B) \sim J_{BX}$, A and B denoting protons and X being ^{13}C, with $J_{AX} \sim 0$.

Multiple-Quantum Filtered 2D Correlated Spectra. Two-dimensional correlated spectra may also be modified to advantage by suppressing strong responses from uncoupled spin systems, which are generally responsible for the overwhelming diagonal peaks such as that from the solvent. The idea is to insert a multiple-quantum filter to distinguish systems of coupled and isolated spins. A useful pulse sequence is shown in Fig. 4.38. The usual phase cycling is performed to achieve the desired p-quantum filtering (pQF). Quadrature detection in F_1 is insured by phase cycling both for $\psi = 0$ and $\psi = \pi/2$ and alternately adding and subtracting the signals ($\varphi = k\pi/p$, $k = 0, 1, \ldots, (2p-1)$).

In the weakly coupled situation the appearance of a diagonal peak in a pQF–COSY spectrum implies that the spin in question has resolved couplings to $(p-1)$ other spins. In contrast, the appearance of a cross-peak between two spins in pQF–COSY indicates that they are coupled not only to each other but also to a common set of at least $(p-2)$ other spins.

It may be noted that the filter has been inserted at the end of the evolution period, just prior to detection. This is in contrast to the corresponding experiment for 2D-resolved spectroscopy, where the filter is inserted before commencement of the evolution period, since filtering between the evolution and detection periods would cause undesired mixing, leading in this case to additional peaks.

When the double-quantum filtered (DQF) 2D correlation experiment is analyzed in detail, it can be shown that with appropriate phase cycling, an AX system leads to:

$$\sigma_{obs} = [(I_{kx}I_{lz} + I_{kz}I_{lx})\cos\Delta_k t_1 + (I_{kx}I_{lz} + I_{kz}I_{lx})\cos\Delta_l t_1]\sin\pi J_{kl}t_1 \quad (76)$$

This expression clearly indicates an anti-phase doublet structure for diagonal peaks in both dimensions. Together with the suppression of diagonal peaks from isolated spins, this feature results in the efficient suppression of the

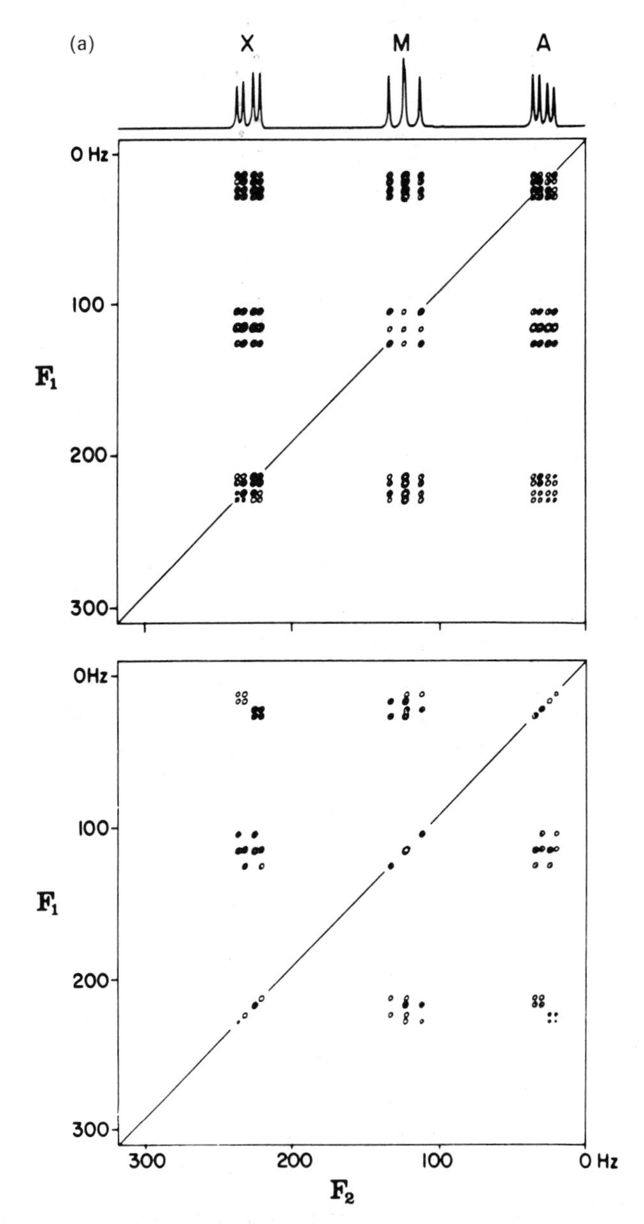

Figure 4.37. (a) Contour plots of COSY spectrum of 2,3-dibromopropionic acid (AMX system). The upper spectrum was obtained using a $\pi/2$ mixing pulse and the lower one with a $\pi/4$ mixing pulse. Only directly connected transitions are intense enough to show contours. (b) The COSY spectrum of a tricyclodecane derivative using a mixing pulse corresponding to a flip angle of $\pi/4$. The arrows indicate the "tilt" of the cross-peaks (see text) and the tilts are in opposite directions depending on the relative signs of the coupling constants. [Reproduced with permission. A. Bax and R. Freeman, *J. Magn. Reson.*, **44**, 542 (1981), copyright 1981, Academic Press, New York]

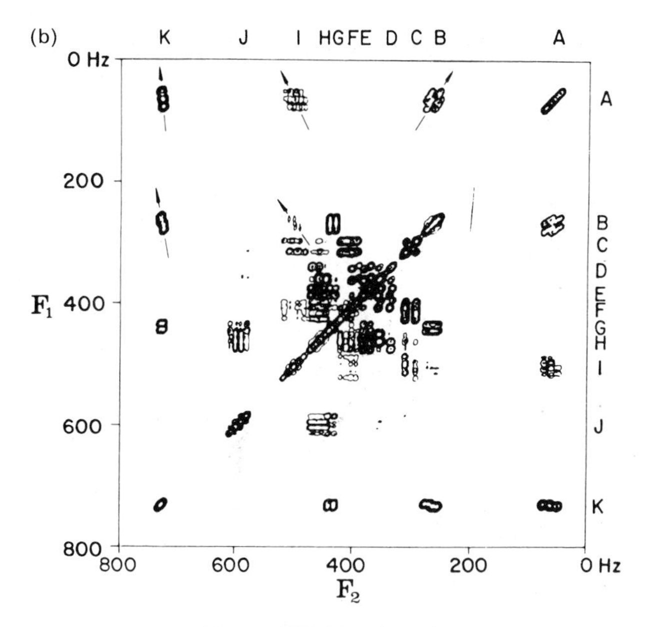

Figure 4.37. (Continued)

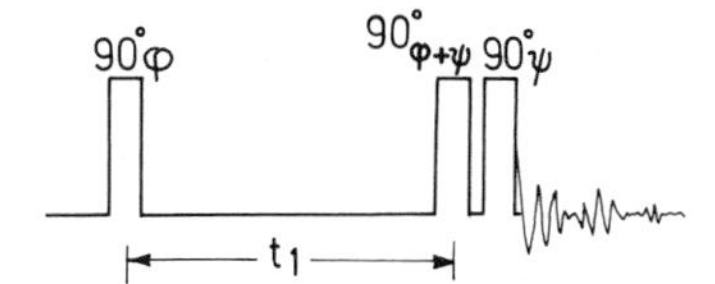

Figure 4.38. Pulse sequence for MQ-filtered COSY.

dominant diagonal ridge of conventional 2D experiments, permitting the identification of cross-peaks lying close to the diagonal. It may be noted that, in contrast to COSY, both diagonal and cross-peaks have the same lineshapes in the DQF–COSY situation. Although the cross-peaks in DQF–COSY have only one half the intensity compared to COSY, the effective sensitivities are comparable because the amount of t_1 noise and the size of the cross-peaks relative to the intensity of the diagonal dispersion tails are really the sensitivity-determining factors. The significant improvement that DQF–COSY represents over the normal COSY experiment is exemplified in Fig. 4.39.

As a further example, the implementation of a 4QF–COSY experiment proceeds on the basis of the following sequences:

$$90^\circ_\varphi - \tau - 90^\circ_\varphi - 90^\circ_0 - \text{Acquisition } (\psi)$$

and:

$$90^\circ_\varphi - \tau - 90^\circ_{\varphi+\pi/2} - 90^\circ_{\pi/2} - \text{Acquisition } (\psi + \pi) \tag{77}$$

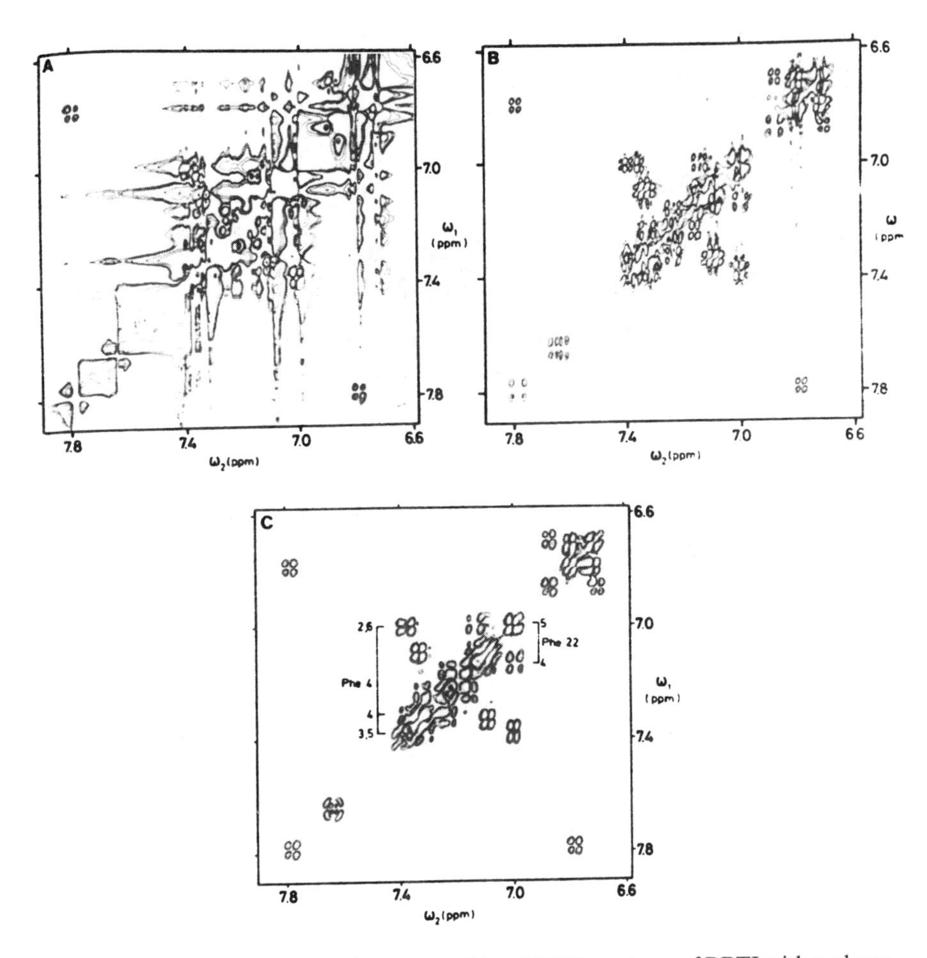

Figure 4.39. (A) Conventional phase-sensitive COSY spectrum of BPTI with a phase-shifted sinebell window filter function with a shift of $\pi/8$ along t_2 and $\pi/4$ along t_1. (B) Same as in (A) but with sinebell squared window function with a shift of $\pi/16$ along t_2 and $\pi/32$ along t_1. The dispersive tails of the diagonal peaks are considerably reduced. (C) DQF–COSY spectrum of BPTI showing much better resolution in the cross peaks. [Reproduced with permission. M. Rance, O.W. Sørensen, G. Bodenhausen, G. Wagner, R.R. Ernst and K. Wüthrich, *Biochem. Biophys. Res. Comm.*, **117**, 479 (1983), copyright 1983, Academic Press, New York]

where $\varphi = k\pi/4\,(k = 0, 1, 2, \ldots, 7)$ and ψ is alternated between 0 and π radians. This pair of excitation/detection schemes insures that when the responses are combined in quadrature the 4QC excitation is independent of the offset/chemical shift of the individual spins and results further in the selection of N-peaks (coherence transfer echoes). An example of the application of this strategy is shown in Fig. 4.40.

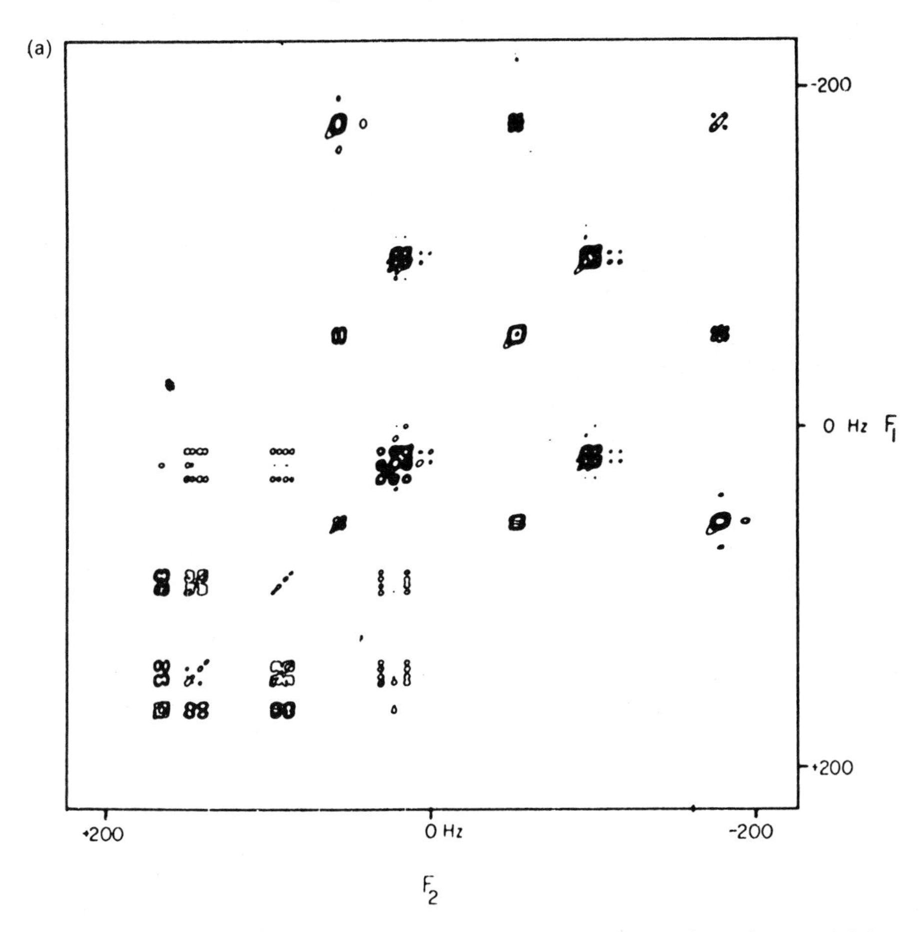

Figure 4.40. (a) A 200-MHz conventional COSY spectrum of a mixture of 2,3-dibromothiophene (AX), 2-furoic acid (AMX), and 1-bromo-3-nitrobenzene (AMQX). [Reproduced with permission. A.J. Shaka and R. Freeman, *J. Magn. Reson.*, **51**, 169 (1983), copyright 1983, Academic Press, New York]

Super COSY. One disadvantage in the normal COSY spectrum is that the diagonal peaks, being in phase dispersion, lead to long tails and broad contours in the absolute value mode. Often, if a matched filtering is not done, this leads to poor cross-peaks (where, because of the anti-phase absorptions, intensities tend to cancel if the cross-peaks occur close to each other). Thus, valuable information close to the diagonal will be masked by intense tailing of the phase-twisted autocorrelated lines. This can be circumvented by introducing finite delays, such as the one used in $^{13}C-^{1}H$ heteroshift correlation spectroscopy, with suitable π pulses interposed in between the delays to focus chemical shifts. Such a scheme, known as super COSY, gives 2D correlated spectra in which the diagonal peaks occur as pairs of anti-phase absorptions

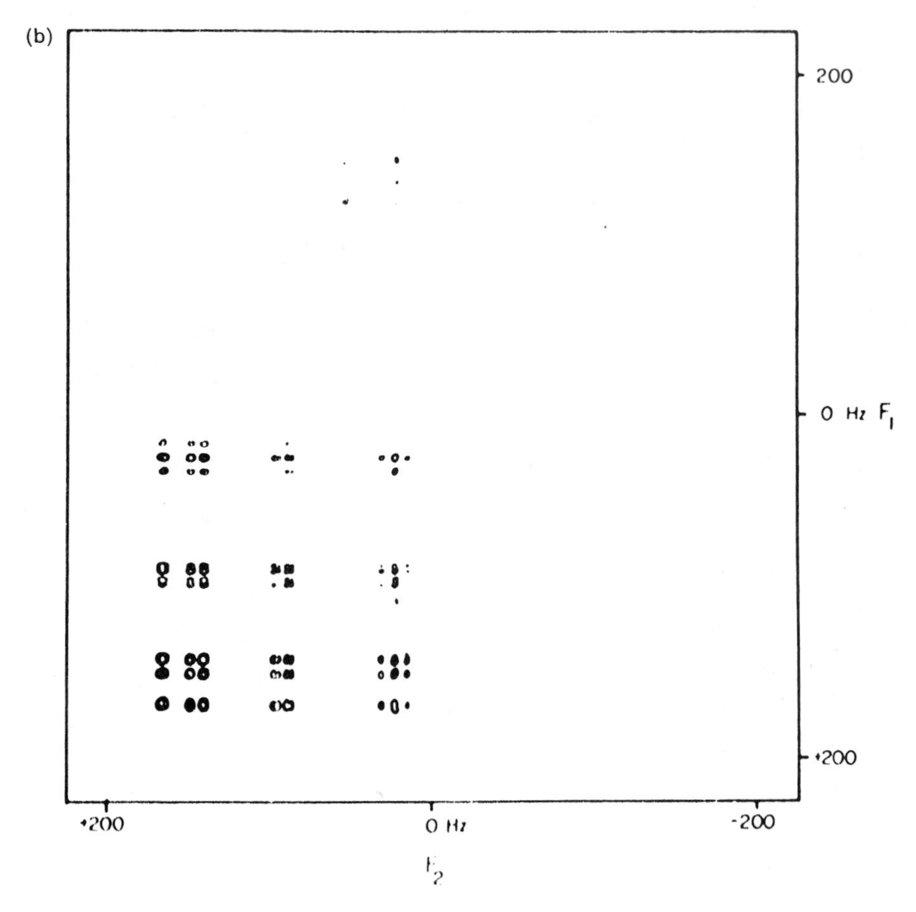

Figure 4.40. (b) A 200-MHz 4QF–COSY spectrum of the same system as in (a), where only the 1-bromo-3-nitrobenzene spectra are retained (AMQX alone show up). [Reprinted with permission. A.J. Shaka and R. Freeman, *J. Magn. Reson.*, **51**, 169 (1983), copyright 1983, Academic Press, New York]

and the cross-peaks occur as in phase dispersion (Fig. 4.41). Compared to COSY, super COSY allows for an evolution exclusively under homonuclear scalar coupling for an extra period of 2Δ before the mixing pulse, and for a similar evolution under coupling immediately after mixing and prior to acquisition. Looking at the density matrix in terms of product operators,

$$\begin{aligned}\sigma_0 &= I_{1z} + I_{2z}\\ \sigma_2 &= -c_J(t_1)[c_{2\delta_1}(t_1)I_{1y} - c_{2\delta_2}(t_1)I_{2y}] + 2c_{2\delta_1}(t_1)s_J(t_1)I_{1x}I_{2z}\\ &\quad + 2c_{2\delta_2}(t_1)s_J(t_1)I_{1z}I_{2x} + s_{2\delta_1}(t_1)c_J(t_1)I_{1x} + s_{2\delta_2}(t_1)c_J(t_1)I_{2x}\\ &\quad + 2s_{2\delta_1}(t_1)s_J(t_1)I_{1y}I_{2z} + 2s_{2\delta_2}(t_1)s_J(t_1)I_{1z}I_{2y}\end{aligned} \tag{78}$$

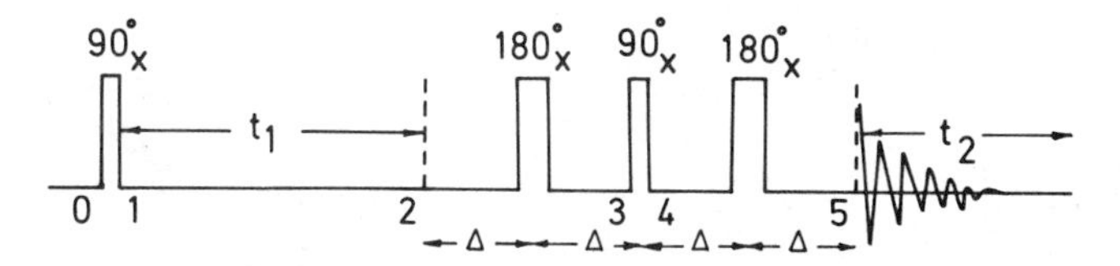

Figure 4.41. Pulse sequence for super COSY.

Evolving only under spin–spin coupling for a time $2\Delta = (1/2J)$,

$$\begin{aligned}\sigma_3 = {} & 2c_{2\delta_1}(t_1)c_J(t_1)I_{1x}I_{2z} + 2c_{2\delta_2}(t_1)c_J(t_1)I_{2x}I_{1z} \\ & + c_{2\delta_1}(t_1)s_J(t_1)I_{1y} + c_{2\delta_2}(t_1)s_J(t_1)I_{2y} \\ & + 2s_{2\delta_1}(t_1)c_J(t_1)I_{1y}I_{2z} + s_{2\delta_2}(t_1)c_J(t_1)I_{2y}I_{1z} \\ & - s_{2\delta_1}(t_1)s_J(t_1)I_{1x} - s_{2\delta_2}(t_1)s_J(t_1)I_{2x}\end{aligned} \tag{79}$$

After the $\pi/2$ x pulse,

$$\begin{aligned}\sigma_4 = {} & -2c_{2\delta_1}(t_1)c_J(t_1)I_{1x}I_{2y} - c_{2\delta_1}(t_1)c_J(t_1)I_{2x}I_{1y} \\ & + c_{2\delta_1}(t_1)s_J(t_1)I_{1z} + c_{2\delta_2}(t_1)s_J(t_1)I_{2z} \\ & - 2s_{2\delta_1}(t_1)c_J(t_1)I_{1z}I_{2y} + 2s_{2\delta_2}(t_1)c_J(t_1)I_{2z}I_{1y} \\ & - s_{2\delta_1}(t_1)s_J(t_1)I_{1x} - s_{2\delta_2}(t_1)s_J(t_1)I_{2x}\end{aligned} \tag{80}$$

Evolution again under spin–spin coupling for a period $2\Delta = (2J)^{-1}$ (only the last four terms evolve) gives:

$$\begin{aligned}\sigma_5 = {} & s_{2\delta_1}(t_1)c_J(t_1)I_{2x} - s_{2\delta_2}(t_1)c_J(t_1)I_{1x} \\ & - 2s_{2\delta_1}(t_1)s_J(t_1)I_{1y}I_{2z} - 2s_{2\delta_2}(t_1)s_J(t_1)I_{2y}I_{1z}\end{aligned} \tag{81}$$

The first two terms give rise to cross-peaks, whereas the last two terms give the diagonal peaks after evolving during period t_2. Comparison of this with the corresponding COSY density matrix immediately reveals that the cross-peaks and diagonal peaks have interchanged their phases and hence appear, respectively, as in phase dispersions and anti-phase absorptions in both dimensions. Apart from this advantage of emphasizing the cross-peaks, additional advantage of super COSY over COSY is the possible discrimination between short-range and long-range coupling by "tuning" the value of Δ. In addition, poorer frequency resolution will suppress diagonal peaks because of their anti-phase absorption character in super COSY. One can thus get away with a smaller data matrix (for an example see Fig. 4.42). However, the additional delay of 4Δ causes considerable loss of transverse magnetization, especially for large molecules.

The SECSY Sequence. In a modification of the homo COSY experiment, the mixing pulse can be advanced to the middle of the t_1 period. Then, during the t_1 period, those coherences which get transferred by the second $\pi/2$ pulse will

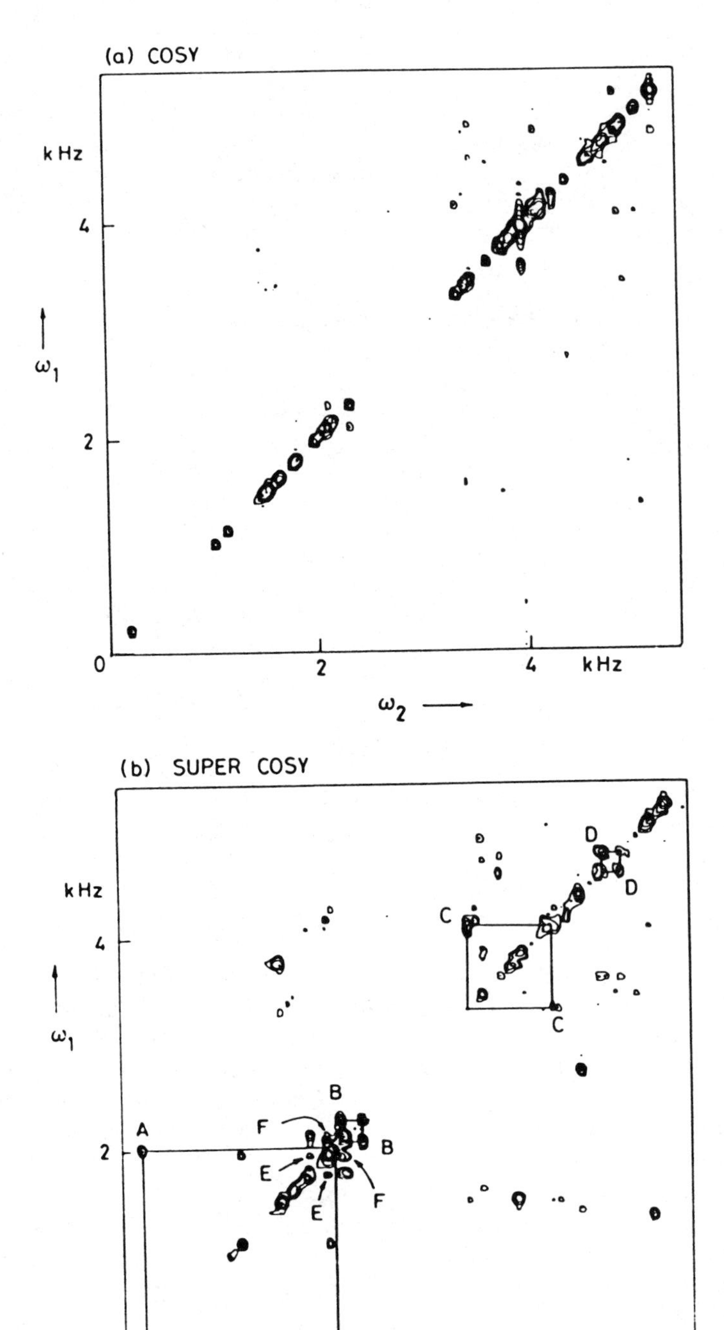
(a) COSY
kHz
4
2
ω1
0
2
4
kHz
ω2
(b) SUPER COSY
kHz
4
2
ω1
A
B
B
C
C
D
D
E
E
F
F
0
2
4
ω2

have evolved at one chemical shift frequency during the first $t_1/2$ period, whereas in the second $t_1/2$ period they will have evolved under the chemical shift of the transferred coherence. The overall modulation for a two-spin system therefore is $(\delta_1 - \delta_2)/2$ during the t_1 period and δ_1 or δ_2 in the t_2 period. To avoid foldover in the F_1 dimension, the sampling frequency should be larger than the largest difference in chemical shifts in the molecules. The resolution along the F_1 direction will be poorer than that in COSY because we have only "differences in δ" rather than δ itself. The so-called diagonal peaks will occur at zero frequency in the F_1 axis. The sensitivity of the SECSY experiment is poorer than that of COSY because much of the transferred coherence is lost between the mixing pulse and the start of the acquisition. The density operator at the start of acquisition is:

$$\begin{aligned}\sigma_4 = {} & 2s_{2\delta_1}(t_1/2)c_J(t_1/2)c_{2\delta_1}(t_1/2)s_J(t_1/2)I_{1y}I_{2z} \\ & + 2s_{2\delta_2}(t_1/2)c_J(t_1/2)c_{2\delta_2}(t_1/2)s_J(t_1/2)I_{1z}I_{2y} \\ & + s_{2\delta_1}(t_1/2)c_J(t_1/2)s_{2\delta_1}(t_1/2)c_J(t_1/2)I_{1y} \\ & - 2s_{2\delta_1}(t_1/2)s_J(t_1/2)c_{2\delta_1}(t_1/2)c_J(t_1/2)I_{1y}I_{2z} \\ & - 2s_{2\delta_1}(t_1/2)s_J(t_1/2)c_{2\delta_2}(t_1/2)c_J(t_1/2)I_{1z}I_{2y} \\ & + s_{2\delta_2}(t_1/2)c_J(t_1/2)s_{2\delta_1}(t_1/2)c_J(t_1/2)I_{2y} \\ & + s_{2\delta_2}(t_1/2)s_J(t_1/2)s_{2\delta_1}(t_1/2)s_J(t_1/2)I_{1y} \\ & + s_{2\delta_1}(t_1/2)s_J(t_1/2)s_{2\delta_2}(t_1/2)s_J(t_1/2)I_{2y}\end{aligned} \qquad (82)$$

This will further evolve uninterrupted in the t_2 period. The first four terms correspond to coherences that have not been transferred from one spin to another during the t_1 period and give rise to the diagonal peaks. The last four terms, however, correspond to coherences that evolve in the first $t_1/2$ period with one chemical shift, and in the second $t_1/2$ period with that of the spin to which coherence has been transferred through the J connectivity. The former coherences evolve into the t_2 period with no modulation, and the latter will have been modulated by the net differences in the chemical shifts between connected nuclei reduced by a factor of 2, because each partner evolves only for $t_1/2$. Therefore, in this sequence when the chemical shifts are focused by phase cycling for N-type peaks, as in the COSY scheme, in the t_1 period the "diagonal" peaks corresponding to COSY occur at zero frequency parallel to the F_2 axis, whereas cross-peaks occur at $(\delta_A, (\delta_A - \delta_B)/2)$ and $(\delta_B, (\delta_B - \delta_A)/2)$. To these positions the spin–spin coupling $\pm J/2$ must be added in both

◀ Figure 4.42. (a) COSY and (b) super COSY spectra of the decapeptide LHRH (*p*-Glu–His–Trp–Ser–Tyr–Gly–Leu–Arg–Pro–Gly–NH_2) at 500 MHz in a DMSO-d_6 solution. Note the considerable enhancement of cross-peaks and suppression of dispersive tails of diagonal peaks in super COSY. [Reprinted with permission. Anil Kumar, R.V. Hosur, and K. Chandrasekhar, *J. Magn. Reson.*, **60**, 143 (1984), copyright 1984, Academic Press, New York]

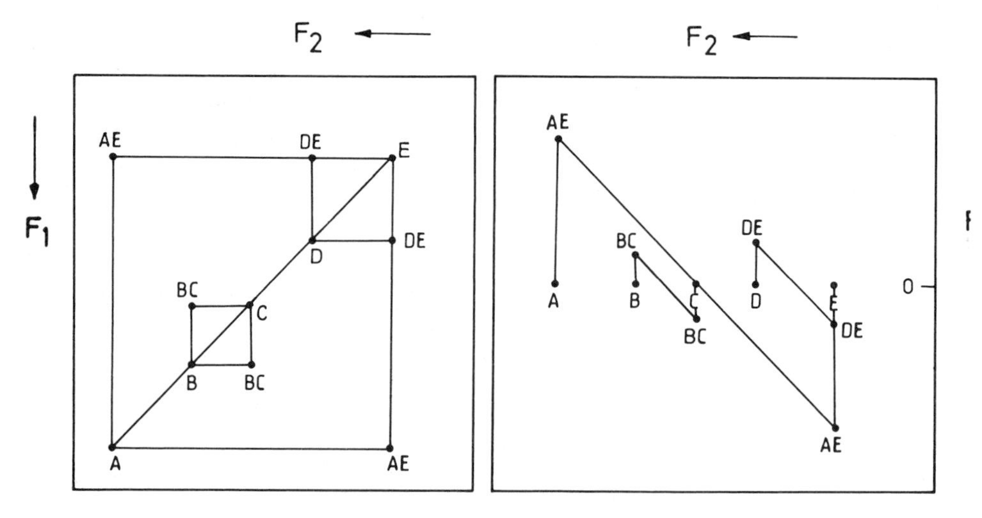

Figure 4.43. Schematic representation of COSY and SECSY format of an ABCDE system with nonvanishing coupling between A and E, B and C, and D and E.

dimensions; schematically the spectra appear as shown in Fig. 4.43. Since we have chemical shift differences scaled by a factor of 0.5 in the F_1 axis, the sampling frequency can be quite low. The disadvantage is the loss of resolution. If we have to double the resolution, the sampling time along the t_1 axis must be doubled compared to the COSY scheme. This is the so-called spin-echo correlated spectroscopy (SECSY).

***z*-Filtered SECSY Spectra.** Conventional SECSY experiments lead to phase-twisted lineshapes owing to the presence of both transverse magnetization components at the commencement of acquisition, making it necessary to record absolute value spectra; strong resolution enhancement must always be applied at the expense of sensitivity, in order to suppress the long dispersive tails. Also, SECSY cross-peaks are always 90° out of phase with the diagonal peaks, as in COSY.

The pulse sequence for z-filtered SECSY is shown in Fig. 4.44. z-filtered (zF) SECSY selects one of the in-phase magnetization components and permits phase-sensitive zF-SECSY spectra to be recorded. Diagonal and cross-peaks exhibit the same phase. In weakly coupled spin systems, the standard density operator treatment shows that all multiplets in F_2 are in phase, with their normal intensity patterns. In F_1, the diagonal peaks exhibit an in-phase triplet with splitting $(1/2)J_{AX}$ for each nucleus X coupled to the observed spin A, while the cross-peaks between A and X exhibit in F_1, a $-1:2:-1$ triplet with splitting $J_{AX}/2$ for the AX coupling and in-phase doublet structure with splittings $J_{AM}/2$ or $J_{RX}/2$ for all other couplings involving the spins A and X. The mixing pulse phase ψ is cycled through x, y, $-x$, $-y$ as in SECSY, alternately adding and subtracting the FID's. This cycle is repeated four times with the last pulse of the z filter being cycled through $\xi = x, y, -x, -y$, with

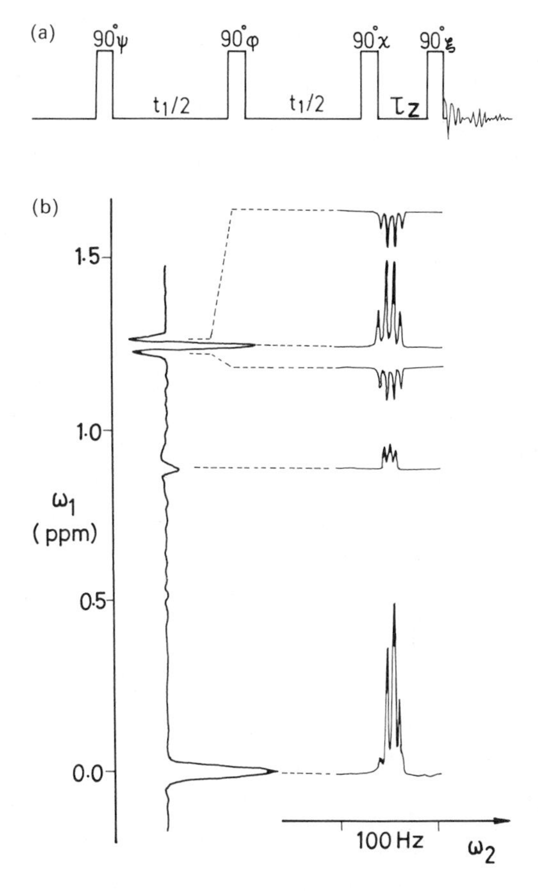

Figure 4.44. (a) Pulse sequence for z-filtered SECSY. (b) Sections from phase-sensitive 300-MHz z-filtered SECSY of a mixture of amino acids. The spectrum parallel to the ω_1 axis is a cross-section through one of the inner peaks of alanine quartet and touching the C_α–H resonance of glutamate. The corresponding multiplet structures parallel to the ω_2 dimension are shown beside each peak in the ω_1 section. The peak at $\omega_1 = 0$ (arbitrary) represents overlap of a triplet and a quartet from C_α–H protons of alanine and glutamate. The triplet at $\omega_1 = 0.9$ ppm results from glutamate α–β coupling and peak at 1.25 ppm arises from alanine α–β coupling. The spectra were recorded at 360 MHz. [Reproduced with permission. O.W. Sørensen, M. Rance and R.R. Ernst, *J. Magn. Reson.*, **56**, 527 (1984), copyright 1984, Academic Press, New York]

CYCLOPS data routing. The first pulse may be subjected to the time proportional phase incrementation (TPPI) strategy to achieve quadrature detection in F_1 (see Chapter 5).

It is possible to achieve good suppression of the dispersive signal components with just a few values of τ_z. Figure 4.44 shows sections from the zF-SECSY spectrum of a mixture of amino acids including alanine and glutamate in D_2O. In this case two values of τ_z, 10 and 30 ms, were adequate.

FOCSY. Finally, one other version of the homonuclear shift correlation spectroscopy that has been proposed is the so-called foldover correlation spectroscopy, FOCSY, which is simply a modification in the presentation of the data from a COSY experiment. The frequency spectrum is modified in such a way that the (F_1, F_2) in COSY is changed to $(F_1 - F_2, F_2)$, thereby transforming the COSY into a SECSY format so that the diagonal peaks (F_1, F_1) now occur horizontally at $(0, F_1)$. The advantage is that if the chemical shift frequencies are larger than one half of the sampling frequency in the t_1 axis, then in the F_1 axis of the FOCSY presentation the corresponding F_1 coordinate is raised or lowered by $2\pi\mu$ where μ is the sampling frequency in the t_1 domain. This avoids folding errors and presents correlation in a compact form. Apart from compacting the spectrum, FOCSY has no special advantage over COSY, although it has improved sensitivity compared to SECSY.

Homo- and Heteronuclear Relayed COSY

We have mentioned that for establishing connectivity between two spins, whether homo- or heteronuclear, in two-dimensional shift correlation spectroscopy the necessary condition is that the two spins have a resolvable spin–spin coupling. This nonvanishing coupling is responsible for the creation of signed single-quantum coherences represented by bilinear spin operators (such as $I_{1z}I_{2y}$ or $I_{1x}I_{2z}$), which, when subjected to a $(\pi/2)$ x or y pulse as the case may be, leads to polarization transfer and to cross-peaks in the 2D spectrum. It is possible to extend this coherence transfer in a relayed sense, say, in an AMX system, by a mixing period that is modified from the COSY sequence to include additional $(\pi/2)$ pulses, whereby coherence transfer can be achieved from A to M and then M to X in some sort of a relay under conditions where $J_{AM} \neq 0$, $J_{MX} \neq 0$ but $J_{AX} = 0$. The sequence is given in Fig. 4.45, where we have replaced the single $\pi/2$ mixing pulse of COSY by a 90°_x–τ_m–180°_x–τ_m–90°_x sandwich in which the two $\pi/2$ pulses are interposed with a π pulse to refocus chemical shifts. The above scheme is for a homonuclear two-step relay, where coherence would be relayed from A → M → X. It is conceivable to extend the number of mixing stages, and in the two-dimensional experiment one may vary t_1 and τ_m together in the so-called "accordion" fashion.

Since we have already seen the form of the density operators in a COSY experiment it is enough if we deal with only those components that are

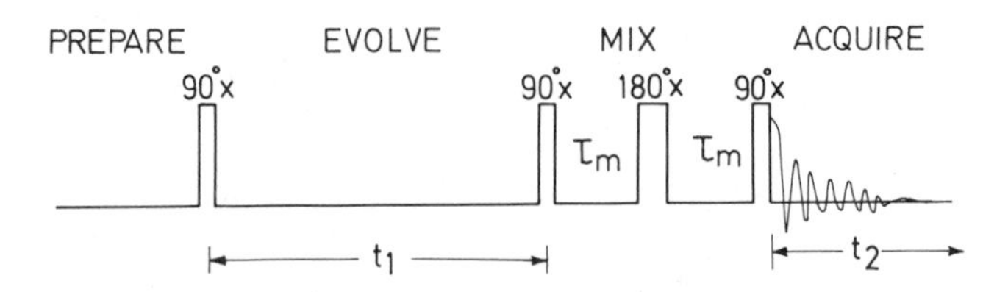

Figure 4.45. Pulse sequence for homonuclear relayed COSY.

responsible for relayed coherence transfer. Just before the mixing period the density matrix is as follows, where we label the AMX system as I_1, I_2, I_3 with $J_{I_1I_2} = J_{12} \neq 0$; $J_{I_2I_3} = J_{23} \neq 0$; $J_{I_1I_3} = J_{13} = 0$.

$$\begin{aligned}\sigma = &-c_{2\delta_1}(t_1)c_{J_{12}}(t_1)I_{1y} + c_{2\delta_1}(t_1)s_{J_{12}}(t_1)I_{1x}I_{2z} - c_{2\delta_2}(t_1)c_{J_{12}}(t_1)c_{J_{23}}(t_1)I_{2y} \\ &+ 2c_{2\delta_2}(t_1)c_{J_{12}}(t_1)s_{J_{23}}(t_1)I_{2x}I_{3z} + 2c_{2\delta_2}(t_1)s_{J_{12}}(t_1)c_{J_{13}}(t_1)I_{1z}I_{2x} \\ &+ 4c_{2\delta_2}(t_1)s_{J_{12}}(t_1)s_{J_{23}}(t_1)I_{1z}I_{2y}I_{3z} - c_{2\delta_3}(t_1)c_{J_{23}}(t_1)I_{3y} \\ &+ 2c_{2\delta_3}(t_1)s_{J_{23}}(t_1)I_{2z}I_{3x} + s_{2\delta_1}(t_1)c_{J_{12}}(t_1)I_{1x} + \underset{\sim\sim\sim}{2s_{2\delta_1}(t_1)s_{J_{12}}(t_1)I_{1y}I_{2z}} \\ &+ s_{2\delta_1}(t_1)c_{J_{12}}(t_1)c_{J_{23}}(t_1)I_{2x} + 2s_{2\delta_2}(t_1)c_{J_{12}}(t_1)s_{J_{23}}(t_1)I_{2y}I_{3z} \\ &+ 2s_{2\delta_2}(t_1)s_{J_{12}}(t_1)c_{J_{23}}(t_1)I_{2y}I_{1z} - 4s_{2\delta_2}(t_1)s_{J_{12}}(t_1)s_{J_{23}}(t_1)I_{2x}I_{1z}I_{3z} \\ &+ s_{2\delta_3}(t_1)c_{J_{13}}(t_1)I_{3x} + \underset{\sim\sim\sim}{2s_{2\delta_3}(t_1)s_{J_{13}}(t_1)I_{3y}I_{2z}}\end{aligned} \tag{83}$$

Concentrating only on terms that lead to coherence transfer in the mixing period, the $(\pi/2)_x$ pulse leads to:

$$\begin{aligned}I_{1y}I_{2z} &\rightarrow -I_{1z}I_{2y} \\ I_{3y}I_{2z} &\rightarrow -I_{3z}I_{2y}\end{aligned} \tag{84}$$

Thus normal COSY-type correlation is already established. However, these will evolve further for a period of $2\tau_m$ under the scalar couplings J_{12} and J_{23}:

$$\begin{aligned}-I_{1z}I_{2y} \xrightarrow{J_{12},J_{23}} &-c_{2J_{12}}(\tau_m)c_{2J_{23}}(\tau_m)I_{1z}I_{2y} \\ &+ 2c_{2J_{12}}(\tau_m)s_{2J_{23}}(\tau_m)I_{1z}I_{2x}I_{3z} + \tfrac{1}{2}s_{2J_{12}}(\tau_m)c_{2J_{23}}(\tau_m)I_{2x} \\ &+ \underset{\sim\sim\sim}{s_{2J_{12}}(\tau_m)s_{2J_{23}}(\tau_m)I_{2y}I_{3z}} \\ -I_{2y}I_{3z} \xrightarrow{J_{12},J_{23}} &-c_{2J_{23}}(\tau_m)c_{2J_{12}}(\tau_m)I_{2y}I_{3z} \\ &+ 2c_{2J_{23}}(\tau_m)s_{2J_{12}}(\tau_m)I_{2x}I_{1z}I_{3z} + \tfrac{1}{2}s_{2J_{23}}(\tau_m)c_{2J_{12}}(\tau_m)I_{2x} \\ &+ \underset{\sim\sim\sim}{s_{2J_{23}}(\tau_m)s_{2J_{12}}(\tau_m)I_{2y}I_{1z}}\end{aligned} \tag{85}$$

Again the last two terms have the correct form to produce coherence transfer by the third $\pi/2$ pulse, so that

$$\begin{aligned}s_{2J_{23}}(\tau_m)s_{2J_{12}}(\tau_m)[I_{2y}I_{3z} + I_{2y}I_{1z}] &\xrightarrow{(\pi/2)_x} \\ &- s_{2J_{23}}(\tau_m)s_{2J_{12}}(\tau_m)[I_{2z}I_{3y} + I_{2z}I_{1y}]\end{aligned} \tag{86}$$

Thus a complete connectivity of AMX is established. The extent of relayed transfer between two spins A and X via an intermediary spin M is proportional to $\sin(2\pi J_{AM}\tau_m)\sin(2\pi J_{MX}\tau_m)$. These will evolve in the t_2 period with their respective Larmor frequencies and spin–spin coupling. Thus, it is possible to establish proton networks in a given molecule by a 2D experiment. The only protons that will not be correlated are those which do not have a resolvable coupling to any other proton in the molecule (see Fig. 4.46 for an example).

The homonuclear relay that results from sufficiently long mixing times τ_m

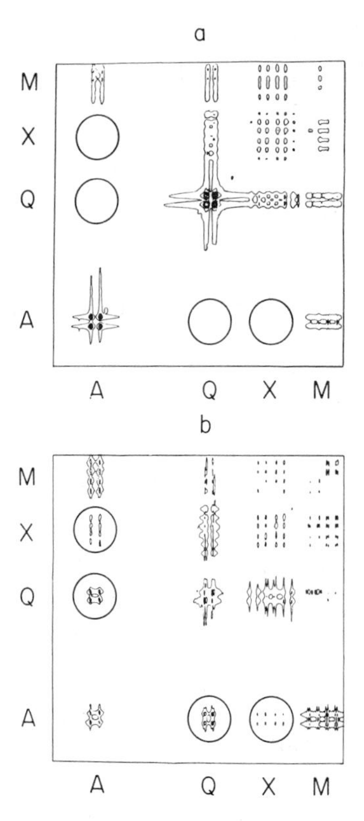

Figure 4.46. Two-dimensional relayed COSY of crotonaldehyde in 50% acetone-d_6. The top spectrum shows conventional COSY map showing the absence of connectivity (empty circles) between the aldehyde proton (A), the methyl group (Q), and the trans proton (X). The bottom spectrum corresponding to relayed COSY shows that all resonances belong to the same coupling network. The resonances A and Q are aliased because of the small spectral width of 250 Hz. [Reprinted with permission. G. Eich, G. Bodenhausen, and R.R. Ernst, *J. Amer. Chem. Soc.*, **104**, 3731 (1982); copyright (1982), American Chemical Society).]

under the mixing propagator generated by the 90°_x–τ_m–180°_x–τ_m–90°_x sequence fails to go through with short mixing times. Under these conditions however, a COSY spectrum results with a reduced number of peaks. This follows from the selection rules operative under the mixing propagator, which in this case causes bilinear rotation. Such a COSY spectrum, termed bilinear COSY, exhibits no diagonal or near-diagonal peaks and yields simplified cross-peak patterns that allow higher resolution and provide the relative signs of couplings on inspection, in the limit of short mixing times ($2\pi J\tau_m \ll 1$), and with N-type peak selection.

The same principle of relayed correlation can also be extended to heteronuclear shift correlation. The pulse scheme is basically the same as in relayed

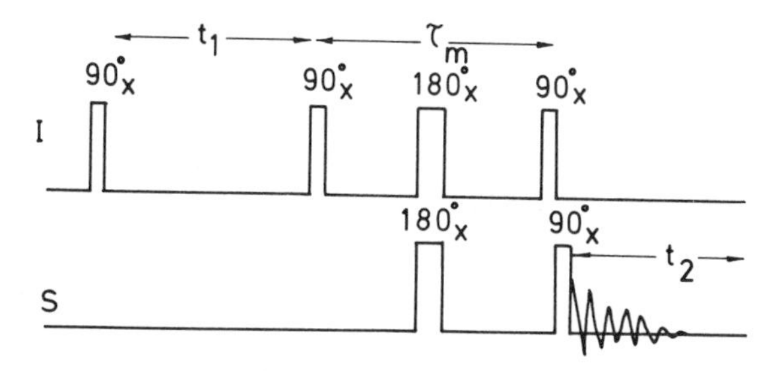

Figure 4.47. Pulse sequence for heteronuclear relayed coherence transfer 2D experiment.

COSY (see Fig. 4.47). Coherence transfer can be optimized by a previous knowledge of the consecutive couplings. A 180°_x pulse applied synchronously in both the channels at the middle of the mixing period will insure focusing of chemical shifts, while retaining the heteronuclear spin–spin coupling constant. Thus in a linear heteronuclear AMX system ($A = M = {}^1H, X = {}^{13}C$ or ${}^{31}P, J_{AM} \neq 0, J_{MX} \neq 0, J_{AX} = 0$) depending on the mixing time τ_m the transfer efficiency is proportional to $\sin(\pi J_{AM}\tau_m)\sin(\pi J_{MX}\tau_m)$ as in the previous case. It is not possible to have a unique τ_m that is effective for all sorts of relays in a molecule. In the 2D spectrum cross-peaks will appear among all the spins A, M, and X. To obtain similar connectivity information in a continuous wave (CW) method would require, in a heteronuclear system of three different spins, triple resonance experiments. Additional mixing periods can relay the coherence to farther neighbors. In a large molecule with a lengthy carbon skeleton, relayed heteronuclear COSY should yield for each ${}^{13}C$ at least three cross-peaks, one corresponding to directly bonded protons and two vicinal protons.

It is possible to modify the heteronuclear relayed 2D shift correlation spectroscopy so as to present only chemical shifts of the heteronuclear spins involved on both the frequency axes. Again, taking a linear AMX system, let $A = H$, $M = H$, and $X = {}^{13}C$. The pulse sequence is as shown in Fig. 4.48.

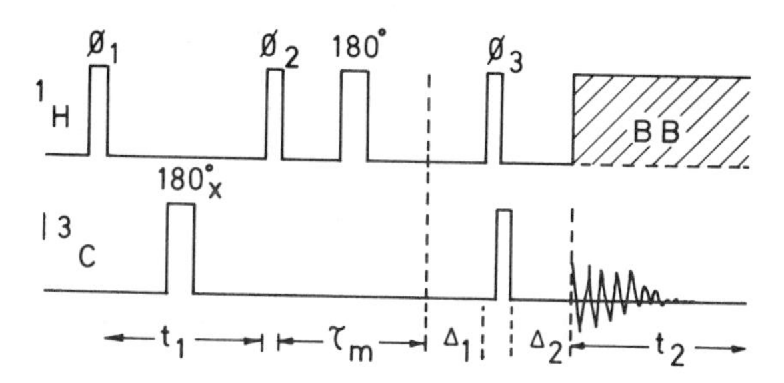

Figure 4.48. Pulse sequence for heteronuclear relayed coherence transfer 2D experiment with only chemical shifts of the nuclei in either dimension.

The first two pulses in the proton channel transfer coherence between the remote ^{1}H and neighbor ^{1}H which are J connected just as in the conventional COSY experiment. The presence of a π ^{13}C pulse in the middle of these two pulses refocuses heteronuclear spin–spin coupling, so that in the F_1 dimension only ^{1}H chemical shifts and homonuclear spin–spin couplings evolve. During the mixing period τ_m, the transferred coherences evolve under homo- and heteronuclear spin–spin coupling so that anti-phase magnetization is created wherein the coherence transferred from A to M is now signed by X. The two synchronous 90° pulses at the end of $\tau_m + \Delta_1$ transfer polarization from the M nucleus to X nucleus, thereby establishing remote (relayed) connectivity. The delays Δ_1 and Δ_2 provided flanking the mixing pulses make the coherence transfer fairly nonselective irrespective of the number of equivalent M nuclei and also enable acquiring ^{13}C with broadband ^{1}H decoupling.

Thus, the sequence correlates chemical shift of a given carbon to the directly connected protons and remotely connected protons. Also, in the F_2 axis only ^{13}C chemical shifts are present, so that in the 2D spectra a projection onto the F_2 axis gives a proton-decoupled carbon spectrum, and one onto the F_1 axis gives a normal ^{1}H spectrum. Thus, relayed heteronuclear COSY using additional mixing cycles and proper phase cycling procedures can indirectly lead to carbon–carbon connectivities. The only other method that gives directly bonded ^{13}C–^{13}C connectivities is the INADEQUATE scheme. Whereas INADEQUATE is limited to the 10^{-4} probability of ^{13}C–^{13}C pair occurrence, relayed heteronuclear coherence uses all the ^{13}C sites. However, while the INADEQUATE coherences lead to unambiguous ^{13}C–^{13}C connectivity there may be residual ambiguity in the present method. Nevertheless, adjacent ^{13}C nuclei in a molecule can be identified if they have two common frequencies in the F_1 dimension (Fig. 4.49).

Lowpass J-Filtered Relay Spectra. Heteronuclear relayed correlation spectra often have to be recorded with a short proton–proton transfer period to avoid loss of magnetization by relaxation. Under these conditions, the signals arising from direct polarization transfer through one-bond heteronuclear couplings $^1J_{\mathrm{XH}}$ could dominate the weaker signals arising from the relayed polarization transfer from "remote" protons. It is possible to distinguish "neighbor" cross-peaks from "remote" cross-peaks and selectively suppress the former by a pulse sequence which selectively removes all signals associated with couplings exceeding a (lower) threshold, $J_{\min}$. Such "lowpass J-filtering" as incorporated in a hetero relay sequence is shown in Fig. 4.50.

The lowpass J filter is inserted immediately after the first proton $(\pi/2)_x$ pulse and consists of a set of suitably spaced $\pi/2$ pulses on the X nuclei and a π pulse on ^{1}H synchronized with one of these if phase sensitive displays are required. Denoting remote and neighbor protons, respectively, as H_R and H_N, we have during the preparation delay τ_p following the first $(\pi/2)_x$ pulse:

$$-I_{Ny} \rightarrow -I_{Ny}\cos(\pi J_{\mathrm{CH}_N}\tau_p) + 2I_{Nx}S_z\sin(\pi J_{\mathrm{CH}_N}\tau_p) \tag{87}$$

The following $(\pi/2)$ $(S, \pm x)$ pulse converts the anti-phase term into hetero-

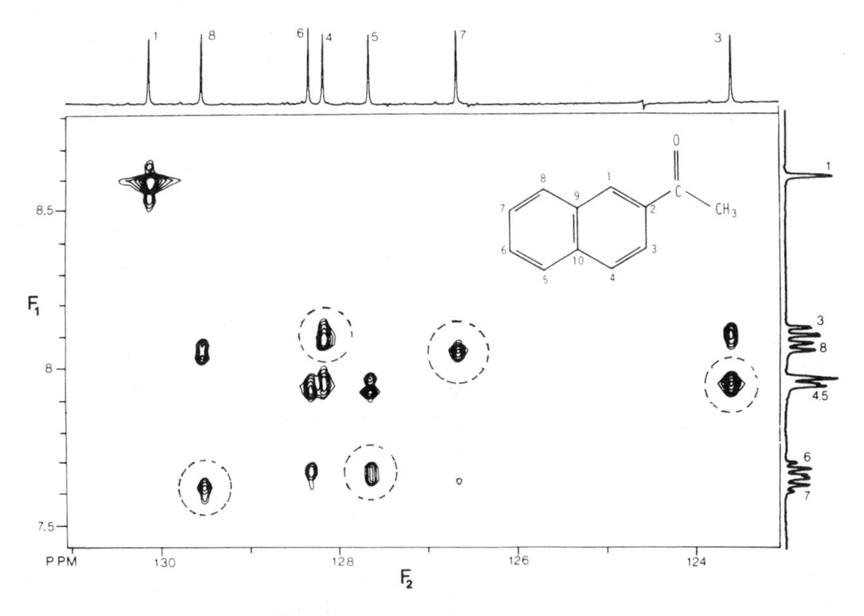

Figure 4.49. Relayed heteronuclear COSY spectrum of aromatic region of 2-acetonaphthalene in acetone-d_6. The peaks arising from relay are circled. [Reprinted with permission. A. Bax, *J. Magn. Reson.*, **53**, 149 (1983), copyright 1983, Academic Press, New York]

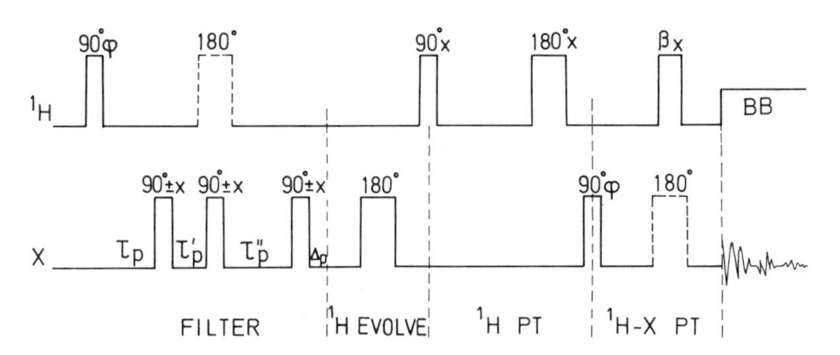

Figure 4.50. Pulse sequence for lowpass J-filtered 2D heteronuclear relayed correlation spectra.

nuclear MQC's, $\mp 2I_{Nx}S_y$. This filtering may be repeated with different intervals $\tau'_p, \tau''_p, \ldots$, in order to eliminate all neighbor peaks to an arbitrary degree in the presence of a wide range of coupling constants J_{CH_N}.

However, in order to take into account the reconversion of MQC's into observable magnetization by the successive stages of filtering with $\pi/2(S, x)$ pulses, the phase of each filter pulse must be alternated independently on successive scans and the resulting FID's added: 2^N transients are thus required with an N-stage filter. This is to be combined with the usual phase cycle of four steps for quadrature detection in F_1, leading to a 2^{N+2} step cycle. Provided all intervals τ_p remain short compared to J_{CH_R}, however, the magnetization of H_R is not trapped as MQC's.

The response of the lowpass J filter with N filter pulses is given by:

$$f = \prod_{k=1}^{N} \cos(\pi J_{CH} \tau_p^{(k)}) \tag{88}$$

The maximum permissible residual amplitude may be used as the criterion to determine numerically the optimum set of $\tau_p^{(k)}$'s, given the range of J to be suppressed and the number of filter stages N. Figure 4.51 displays normal and filtered relay spectra of equilibrated glucose with both α and β anomers. Three-stage filtering has been employed to suppress $135 \leqslant J \leqslant 175$ Hz. The filtered relay spectrum demonstrates clearly the possibility of discrimination between remote and neighbor peaks even when they overlap because of accidental chemical shift degeneracies: the test sample, in fact, features extensive overlaps in the ^1H spectrum.

Should a phase-sensitive display be desired, the optional π pulses and the delay $\Delta_p = \tau_p + \tau_p' - \tau_p''$ shown in Fig. 4.50 should all be included.

Lowpass J filtering may also be applied to other heteronuclear experiments and also to homonuclear two-spin systems, e.g., in INADEQUATE experiments, with π pulses inserted in the middle of each $\tau_p^{(k)}$ interval.

Coherence Transfer through Isotropic Mixing in Homonuclear COSY–Total Correlation Spectroscopy (TOCSY). So far we have seen that homonuclear and heteronuclear shift correlation in 2D spectroscopy is achieved via coherence transfer through scalar coupling and it is possible to modify the polarization transfer so as to relay coherence among consecutively coupled spins in the so-called relayed COSY. A particularly interesting scheme emerges from the phenomenon of coherence transfer via isotropic mixing that has been described in detail in Chapter 3.

The principle is that in a homonuclear situation, when the system is allowed to evolve under strong coupling, the spin system gets into a "collective mode" in which, depending on the number of spins involved, the coherence transfer takes place at a number of frequencies among the spins. Thus, in an AX system the coherence transfer will be at a frequency of J (as against a J modulation frequency $\pm J/2$) and in an AX_2 system it will be at a frequency of $(3/2)J$ (as against J modulation frequencies of $1/2J$, J). There are a number of ways of achieving isotropic mixing. Schematically, this is shown in Fig. 4.52.

During the evolution period single-quantum coherences evolve under weak scalar coupling. However, during the mixing period, since chemical shift evolution is arrested, the spins are "strongly coupled" and hence get into collective modes, whereby coherences (polarizations) are transferred back and forth among all spins (see Chapter 3) via multiple relays. It can be shown that:

$$\begin{aligned} I_{1x} \xrightarrow{JI_1 \cdot I_2} & \tfrac{1}{2} I_{1x}(1 + c_{2J_{12}}(\tau_m)) + \tfrac{1}{2} I_{2x}(1 - c_{2J_{12}}(\tau_m)) \\ & + (I_{1y} I_{2z} - I_{1z} I_{2y}) s_{2J_{12}}(\tau_m) \end{aligned} \tag{89}$$

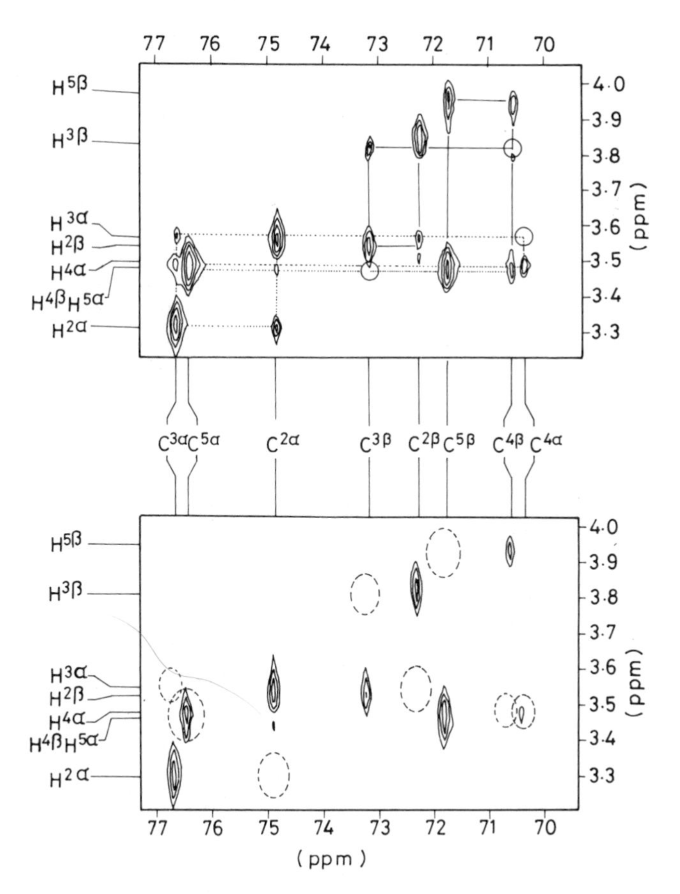

Figure 4.51. Partial contour plots of 300-MHz 2D heteronuclear relayed correlation spectra (in absolute value mode) of equilibrated glucose with α and β anomers. The top picture is the normal relayed 2D heteronuclear correlation spectrum; the bottom corresponds to the lowpass J-filtered analog. The suppressed neighbor peaks in the bottom picture are indicated by dotted contours. Relayed connectivities for the α and β anomers in the top figure are indicated by dotted and solid lines, respectively. [Reprinted by permission. H. Kogler, O.W. Sørensen, G. Bodenhausen, and R.R. Ernst, *J. Magn. Reson.*, **55**, 157 (1983), copyright 1983, Academic Press, New York]

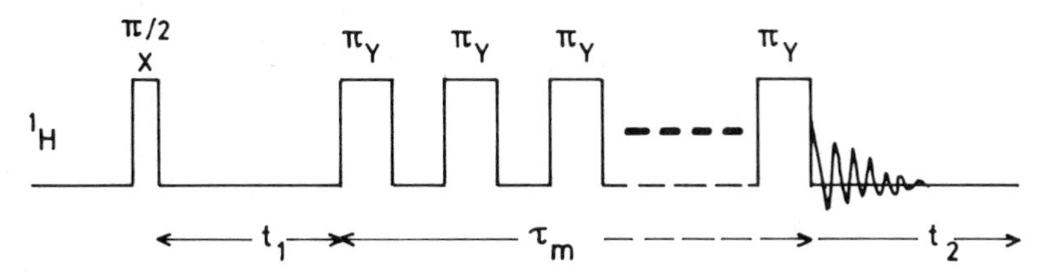

Figure 4.52. Pulse sequence for homonuclear isotropic mixing and 2D total correlation spectroscopy, TOCSY.

which for a mixing time of $\tau_m = (2J)^{-1}$ transfers coherence completely from I_{1x} to I_{2x}. If the period of mixing were $(J)^{-1}$ it would have first been transferred to I_2 and would have come back fully to I_1. Similar transfer can be understood among other coherences which transform as:

$$\begin{aligned} I_{1z}I_{2y} &\rightarrow I_{2z}I_{1y} \\ I_{1y} &\rightarrow I_{2y}, \quad \text{etc.} \end{aligned} \tag{90}$$

at the isotropic mixing frequency. During the isotropic mixing period both in-phase and anti-phase magnetizations tend to develop. In order to average out the out of phase components and to reinforce the in-phase components, a number of mixing periods are used and the resulting FID's co-added prior to Fourier transformation. Such a 2D version of homonuclear shift correlation is called TOCSY, total correlation spectroscopy. Also, depending upon the coupling constants, it is possible to emphasize cross-peaks at the expense of diagonal peaks or vice versa by suitably varying the mixing time (see Fig. 4.53).

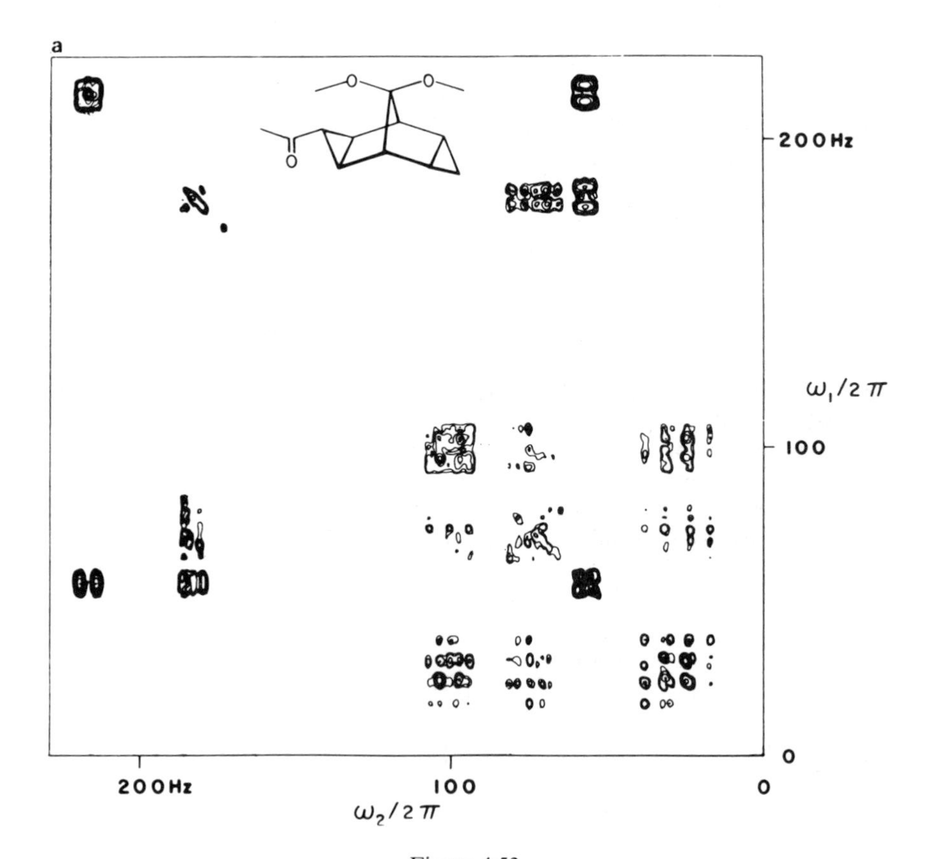

Figure 4.53.

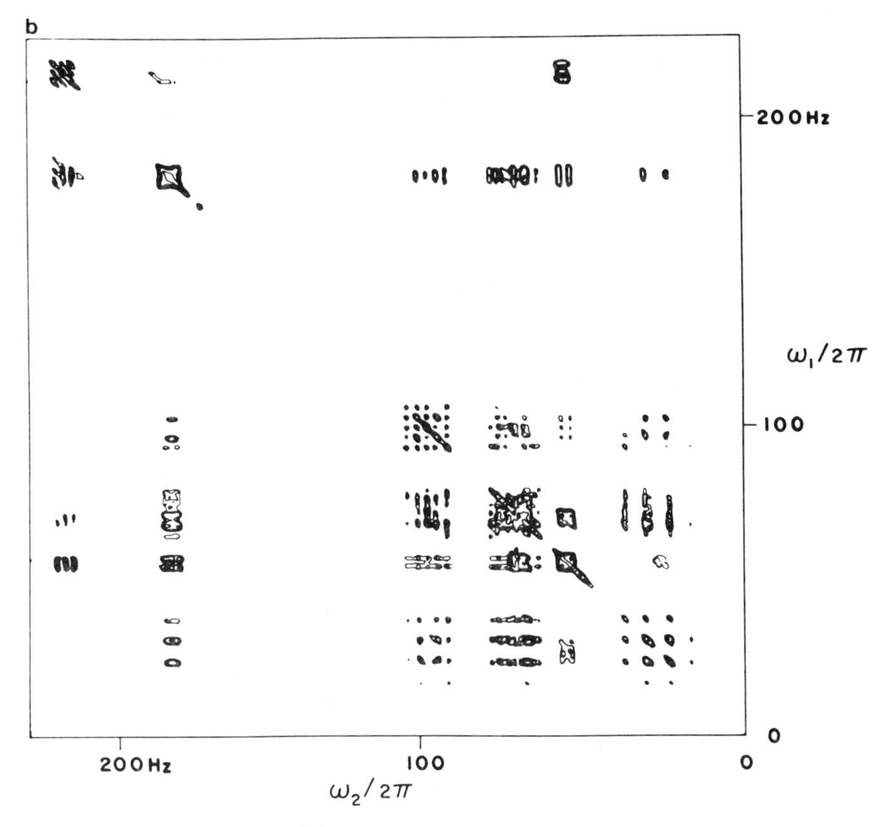

Figure 4.53. (a) The 2D COSY spectrum of 3-acetyl-9, 9′-dimethoxy-3-*exo*-7-*endo*-tetracyclo-[3.3.1.$0^{2.4}$.$0^{6.8}$]-nonane showing only J connectivity. (b) the 2D TOCSY spectrum of the compound of (a); connectivity between all types of protons that have a common coupling partner is established. [Reprinted with permission. L. Braunschweiler and R.R. Ernst, *J. Magn. Reson.*, **53**, 521 (1983), copyright 1983, Academic Press, New York]

Two-Dimensional Cross-Relaxation and Chemical Exchange (NOESY)

So far we have dealt with J-resolved and chemical shift correlation spectroscopy using the 2D Fourier transform methodology. In these methods we have seen correlation via nonvanishing scalar spin–spin coupling acting as a vehicle in transferring coherences among coupled spins, with variations such as relayed coherence transfer and isotropic mixing (TOCSY). There is yet an important realm of 2D correlation spectroscopy wherein internuclear transfer of magnetization takes place via incoherent processes such as the nuclear Overhauser effect (NOE) brought about by the matrix elements of homonuclear and heteronuclear dipolar Hamiltonian (see Appendix 4 and also Chapter 1). Incoherent transfer can also occur via chemical exchange, hindered internal rotation, or conformational interconversions among nuclei exchanging sites. The study of the dynamics of small molecules using temperature-dependent lineshape analysis in one-dimensional NMR spectroscopy has provided a wealth of information on the kinetics of exchange as well as the related thermodynamic parameters. Such dynamic processes, which are dominated by cross-relaxation, diffusion, and exchange, are governed by very similar principles and can be described by a master equation of a set of coupled differential equations. One important distinguishing feature between coherent and incoherent magnetization transfer processes is that, in the latter, there is no need of a "through-bond" connectivity, such as scalar coupling. It is enough if there is a process that can transfer information of one environment to the other via an exchange of locations or if the two nuclei involved have a finite through-space dipolar coupling. It is possible to establish this "through-space" proximity or "exchange" connectivity by double-resonance experiments, such as saturation transfer, selective population inversion, or studying the buildup of nuclear Overhauser enhancement in the one-dimensional sense. These experiments can be too tedious and often the results may also be difficult to evaluate. Thus, the coalescence of two resonances of chemically distinct nuclei, upon rapid exchange can proceed through line broadening to such an extent that complete and accurate followup of the lineshape throughout the temperature range may be rather difficult.

There are excellent and elegant 2D methods available that can unravel complicated intra- and intermolecular exchange pathways, correlate proximal nuclei in macromolecular folded backbones, and give extraordinary insight into the three-dimensional structure of large molecules in solution, which are not accessible for study by x-ray methods. Before we look into the two-dimensional NOE/exchange spectroscopy it is relevant to look into the various magnetization transfer pathways in coupled nuclear systems.

In the absence of scalar spin–spin coupling magnetization exchange in nonequilibrium situation can be adequately described by the Bloch equations, to which can be added expressions to take into account of chemical exchange and cross-relaxation. The density matrix at any time t is governed by:

$$\sigma(t) = -i[\mathscr{H}, \sigma(t)] - \hat{T}_R[\sigma(t) - \sigma(0)]\hat{E}\sigma \tag{91}$$

where the operator $\hat{T}_R$ drives the nonequilibrium density matrix to its diagonal form with the expected Boltzmann weightages and the operator $\hat{E}$ reorders basis functions corresponding to all the exchange processes taking place at the respective exchange frequencies, so that

$$\hat{E}\sigma = \sum_i v_i[E_i \sigma \bar{E}_i - \sigma] \tag{92}$$

where $(v_i)^{-1}$ is the lifetime in a given state i, and E_i suitably exchanges the labels of the related spin operators.

The equilibrium density operator commutes with the Hamiltonian and is not modified by the chemical exchange or relaxation. We shall examine the

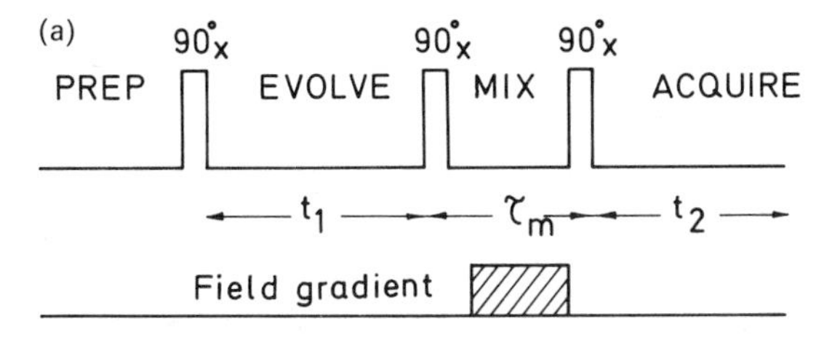

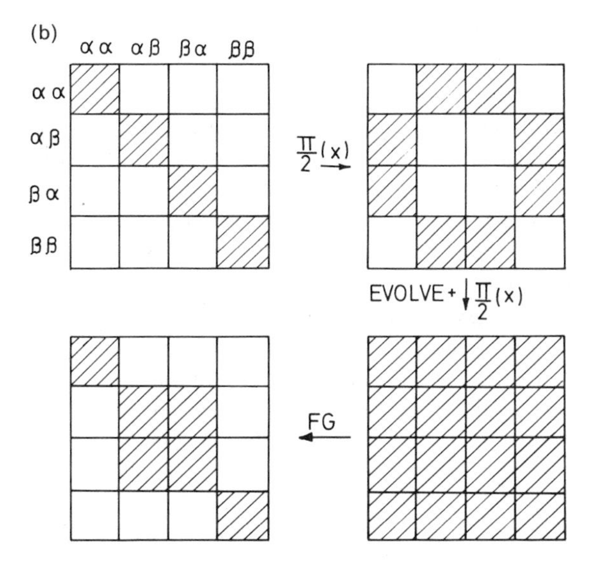

Figure 4.54. (a) Pulse sequence for 2D correlation via exchange or cross-relaxation—2D NOESY. (b) Representation of the density matrix of an AX system subject to NOSEY sequence. Hatched squares correspond to nonzero values for the corresponding density matrix element.

effect of exchange and cross-relaxation by 2D spectroscopy. The basic pulse sequence used to study either process is the same and consists of the usual preparation, evolution, mixing, and detection period, except that a magnetic field gradient pulse is applied at the beginning of the mixing period to destroy all transverse components of magnetization immediately following the second $\pi/2$ pulse (Fig. 4.54).

Following the first pulse, single-quantum coherences of all resonant spins are created in the transverse plane and these evolve under their chemical shifts. Those spins which have scalar coupling to others will additionally evolve under this coupling also.

$$I_z \xrightarrow{(\pi/2)_x} -I_y \xrightarrow{\delta} -c_{2\delta}(t_1)I_y + s_{2\delta}(t_1)I_x \tag{93}$$

and if scalar coupling to other spins is finite, then:

$$-c_{2\delta}(t_1)I_y + s_{2\delta}(t_1)I_x \xrightarrow{JI_zS_z} -c_{2\delta}(t_1)[c_J(t_1)I_y - 2s_J(t_1)I_xS_z] + s_{2\delta}(t_1)[c_J(t_1)I_x + 2s_J(t_1)I_yS_z] \tag{94}$$

The second pulse generates such components as I_z, I_x, I_xS_y, I_zS_y and the field gradient pulse destroys the transverse components corresponding to single- and double-quantum coherences, while the zero-quantum coherences and single-spin Zeeman order will be unaffected by the same. We represent these by a 4×4 density matrix, where shaded squares correspond to finite value for the corresponding element (see Fig. 4.54).

The evolution period, however, provides the frequency labeling for the various z-components. These nonequilibrium z-components can reach steady state via direct as well as cross-relaxation processes. We shall for the moment defer discussion of such T_1 processes and examine what happens if internal rotation and other molecular conformational changes are present that exchange sites of a particular number of nuclei. Thus, if two nuclei A and X in the system are interchanged by an intramolecular rotation process this is equivalent to changing the labels of A by X and X by A. Therefore, in the final detection period they will evolve under the chemical shift of the new environment, ultimately producing a 2D spectrum where cross-peaks will appear at the coordinates (Ω_A, Ω_X) and (Ω_X, Ω_A), establishing exchange correlation. To understand the relative intensities of cross- to diagonal peaks, let us assume that two sites A and B of a given molecule exchange at rates K_{AB} and K_{BA}:

$$\mathrm{A} \underset{K_{BA}}{\overset{K_{AB}}{\rightleftharpoons}} \mathrm{B} \tag{95}$$

governed by the kinetic matrix:

$$K = \begin{pmatrix} -K_{AA} & K_{BA} \\ K_{AB} & -K_{BB} \end{pmatrix} \tag{96}$$

The equilibrium constant K is given by:

$$K = \frac{K_{AB}}{\mathrm{MF_B}} = \frac{K_{BA}}{\mathrm{MF_A}} \tag{97}$$

where $\mathrm{MF_A}$ and $\mathrm{MF_B}$ are the mole fractions of site occupancies.

Considering a system of one type of spin exchanging between two sites with equal mole fraction and assuming a slow chemical exchange, i.e.,

$$K_{AB}, K_{BA} \ll |(\delta_A - \delta_B)| \tag{98}$$

the exchange matrix corresponding to the scrambling of the sites will be given by:

$$\mathscr{L} = \begin{pmatrix} P & -K/2 \\ -K/2 & P \end{pmatrix} \tag{99}$$

where

$$P = \frac{1}{T_{2A}} + \frac{1}{T_{1A}} + \frac{K}{2}$$
$$= \frac{1}{T_{2B}} + \frac{1}{T_{1B}} + \frac{K}{2} \tag{100}$$

Here T_{2A} and T_{1A} are the transverse and spin–lattice relaxation times and the dynamics of the mixing are governed by the differential equation:

$$\frac{dM_z}{dt} = \mathscr{L}(M_z - M_0) \tag{101}$$

Integration of the equation of motion leads to peak amplitudes given by:

$$A_{\text{diag}} = \tfrac{1}{4}\exp(-\tau_m/T_1)[1 + \exp(-K\tau_m)]$$
$$A_{\text{cross}} = \tfrac{1}{4}\exp(-\tau_m/T_1)[1 - \exp(-K\tau_m)] \tag{102}$$

The two cross-peaks and diagonal peaks are of equal intensity and they have the same phase irrespective of the mixing time and the rate of spin–lattice relaxation. In the above expression we have neglected spin–lattice relaxation in the evolution and the detection periods. When T_1 is long, for an infinite mixing time, all four peaks are of equal intensity. However, when T_1 is finite, then the cross-peak amplitude reaches a maximum at:

$$\tau_{\text{opt}} = \frac{1}{K}\ln\frac{(1/T_1) + K}{1/T_1} \tag{103}$$

with a cross-peak (C) to diagonal peak (D) ratio of:

$$\frac{\text{C}}{\text{D}}(\text{opt}\,\tau) = \frac{1}{(2/T_1) + K} \tag{104}$$

The two-site exchange case may be demonstrated by looking at 2D spectroscopy using the pulse sequence given above, in *N*, *N*-dimethyl acetamide (Fig. 4.55). The same sequence of motional broadening and the subsequent motional narrowing as in the normal CW spectra can be seen on the cross- and diagonal peaks. Careful analysis of the intensity of exchange peaks provides rates of exchange and certainly gives clues to various exchange pathways, although precise quantitative data are rather difficult to evaluate.

Let us now consider the establishment of Boltzmann distribution from non-equilibrium starting points through cross-relaxation. We shall for the present assume that the scalar coupling is absent. The cross-relaxation of any longitudinal component I_{kz} (z component of the kth spin) will depend on dipolar links between magnetic nuclei in the same molecule or to the external surroundings (intra- and intermolecular fluctuating dipolar interaction being the main mechanism for relaxation in spin-1/2 nuclei):

$$\frac{dI_{kz}}{dt} = -\boldsymbol{R}I_{kz} \tag{105}$$

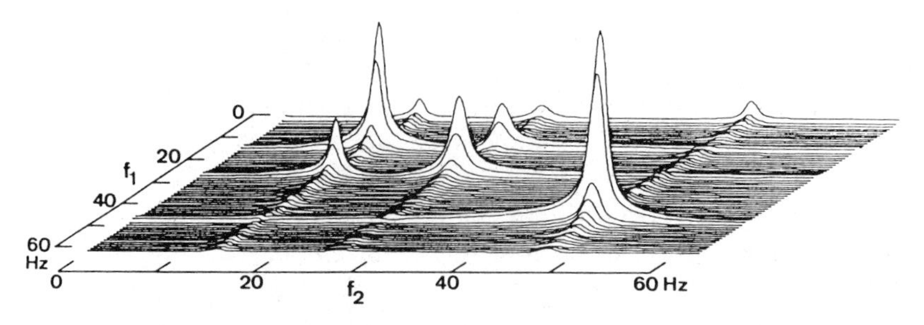

Figure 4.55. The 60-MHz proton 2D exchange spectrum of *N*,*N*-dimethylacetamide (DMA) at 30°C. Cross peaks can be seen between the two *N*-methyl groups. $\tau_m =$ 0.5 s. [Reproduced by permission. J. Jeener, P. Bachmann and R.R. Ernst, *J. Chem. Phys.*, **71**, 4546 (1979) copyright (1979) American Institute of Physics]

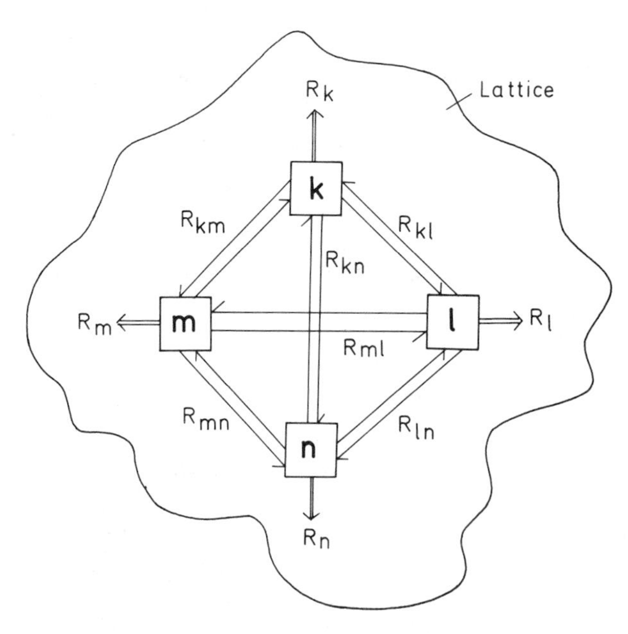

Figure 4.56. Relaxation network schematic for a group of four nuclei. R_{ij} represents cross-relaxation rates; R_i represents "leakage" relaxation rate to the surroundings.

Here $\boldsymbol{R}$ represents the relaxation matrix, containing rate constants R_{kl}, which in turn include cross-relaxation as well as external relaxation. The cross-relaxation rates between spins k and l are R_{kl} and relaxation by thermal link to the surroundings is R_k or R_l. This is schematically represented in Fig. 4.56.

Let us consider the three-pulse sequence that is used in the above exchange correlation. During the evolution period t_1 and just before the mixing period, each component is labeled by its chemical shift modulations:

$$\sigma(t_1) = -[1 + \cos(2\pi\delta_l t_1)\exp(-t_1/T_{2l})I_{lz}] \qquad (l = 1, \ldots, n) \tag{106}$$

During the mixing period the nonequilibrium Zeeman orders reach Boltzmann equilibrium:

$$\sigma(\tau_m) = \exp(-\boldsymbol{R}\tau_m)\sigma(t_1) \tag{107}$$

During the detection period t_2, the observed y magnetization (assuming that all the three $\pi/2$ pulses are applied along the $+x$ axis) is given by:

$$\begin{aligned}\sigma(t_1, \tau_m, t_2) = \sum_{k,l} &\cos(2\pi\delta_k t_2)\exp(-t_2/T_{2k})\exp(-\boldsymbol{R}\tau_m)_{kl} \\ &\times [1 + \cos(2\pi\delta_l t_1)\exp(-t_1/T_{2l}) - \delta_{kl}]\end{aligned} \tag{108}$$

where we have assumed cross-relaxation between two nuclei k and l coupled via dipolar interaction (δ_{kl} is the Kronecker delta). Because transfer of magnetization among spins in a state of Zeeman order is governed by both cross-relaxation and dissipation into other degrees of freedom (which provide the necessary sink for radiationless loss of magnetization), during the mixing period populations are driven toward Boltzmann equilibrium. This causes the nuclear Overhauser effect. The "untransferred" magnetization, in contrast, becomes transverse magnetization with the same frequency labels as in the t_1 period; upon 2D Fourier transformation, this gives the so-called auto peaks (diagonal peaks). Those which have acquired magnetization via "transfer" produce the cross-peaks in the 2D spectrum. Typically the signal corresponding to an exchange of magnetization between spins k and l is given by:

$$\begin{aligned}S(\omega_1, \tau_m, \omega_2) \sim &\frac{1}{2}\frac{T_{2k}^{-1}}{(\omega_2 - \omega_k)^2 + T_{2k}^{-2}}\exp(-\boldsymbol{R}\tau_m)_{kl} \\ &\times \frac{1}{2}\frac{T_{2l}^{-1}}{(\omega_1 - \omega_l)^2 + T_{2l}^{-2}}\end{aligned} \tag{109}$$

However, any residual Zeeman order that is left by an imperfect preparation pulse or those magnetizations that have recovered during the evolution process along the z axis and hence have no frequency labels during t_1 produce the so-called axial peaks at $(0, \omega_2)$. These axial peaks can become uncomfortably large for extended mixing periods and have to be removed by suitable phase cycling.

The integrated intensities (Int) of the peaks in 2D cross-relaxation spectra are proportional to the mixing coefficient and under ideal conditions should follow the NOE buildup (*vide infra*) rates,

$$\mathrm{Int}_{kl} \sim \exp(-\boldsymbol{R}\tau_m)_{kl} \tag{110}$$

For short mixing periods we can expand the exponential and truncate at a suitable point depending on the magnitude of τ_m,

$$\exp(-\boldsymbol{R}\tau_m) = 1 - \boldsymbol{R}\tau_m + \tfrac{1}{2}\boldsymbol{R}^2\tau_m^2 - \ldots \tag{111}$$

so that for short τ_m,

$$\mathrm{Int}_{kl} \sim (\delta_{kl} - R_{kl}\tau_m) \tag{112}$$

If we include the quadratic term,

$$\mathrm{Int}_{kl} \sim (\delta_{kl} - R_{kl}\tau_m) + \frac{1}{2}\sum_j (R_{kj}R_{jl}\tau_m^2) \tag{113}$$

Dipolar Cross-Relaxation. It will be pertinent here to examine cross-relaxation via the dipolar mechanism in an AB system. Let us take the homonuclear AB system with a small chemical shift difference so that $|\delta_A - \delta_B| \simeq 0$. Let us also assume that the direct relaxation processes T_{1A} and T_{1B} are equal. In the case of protons we can assume that the populations are equal i.e., $n_A = n_B$. In the AB energy level diagram (see Appendix 4) are indicated various probabilities of transitions among levels labeled by the conventional single-quantum transition probability W_1, the zero-quantum cross-relaxation probability W_0, and the double-quantum transition probability W_2. The mixing coefficients (a) for diagonal and cross-peaks are given by:

$$W_{1A} = W_{1B} = W_1$$

$$a_{AA}^{\tau_m} = a_{BB}^{\tau_m} = \frac{M_0}{4}\exp(-\boldsymbol{R}_L\tau_m)[1 + \exp(-\boldsymbol{R}_C\tau_m)]$$

and:

$$a_{AB}^{\tau_m} = a_{BA}^{\tau_m} = \frac{M_0}{4}\frac{(W_2 - W_0)}{|(W_2 - W_0)|}\exp(-\boldsymbol{R}_L\tau_m)[1 - \exp(-\boldsymbol{R}_C\tau_m)]$$

where:

$$\boldsymbol{R}_C = 2|(W_2 - W_0)|$$

and:

$$\boldsymbol{R}_L = R_1 + 2W_1 + W_0 + W_2 - |(W_2 - W_0)| \tag{114}$$

Here $\boldsymbol{R}_C$ is cross relaxation, $\boldsymbol{R}_L$ is relaxation *via* coupling to the lattice and includes R_1 which is the relaxation of the two nuclei by processes excluding the direct dipolar mechanism.

It is in this context that we want to differentiate between dipolar relaxation in the extreme narrowing limit, the so-called "white" spectral approximation and the spin diffusion limit, the so-called "black" spectral limit (see Appendix 4 for a general description of dipolar relaxation). In the white spectral limit, we can assume that the power spectrum has identical finite amplitude at all relevant frequencies such that the following equalities hold:

$$J(\omega_A) = J(\omega_B) = J(\omega_A - \omega_B) = J(\omega_A + \omega_B) \tag{115}$$

Comparing with the transition probabilities W_0, W_1, and W_2 given in Appendix 4,

$$J(\omega_A) = J(\omega_B) = J(\omega_B - \omega_A) = J(\omega_A + \omega_B) = 2\tau_c$$

and:

$$W_1 = \left[\frac{3}{20}\gamma^2 \frac{h^2}{4\pi^2} r_{\mathrm{AB}}^{-6}\right]\tau_c = \frac{3}{2} q\tau_c$$

where:

$$q = \frac{1}{10}\gamma^2 \frac{h^2}{4\pi^2} r_{\mathrm{AB}}^{-6}$$

$$W_2 = 6q\tau_c$$

and

$$W_0 = q\tau_c \tag{116}$$

Here q replaces the terms given in the square brackets. The rate constant for cross-relaxation involves W_2 and W_0 and is given by:

$$R_C = \frac{1}{T_{1\mathrm{AB}}} = W_2 - W_0 = 5q\tau_c \tag{117}$$

and for the independent single-spin relaxation,

$$\begin{aligned} R_{\mathrm{L}} &= T_1^{-1} + R_1 \\ &= R_1 + 2W_1 + W_0 + W_2 - |(W_2 - W_0)| \\ &= R_1 + 5q\tau_c \end{aligned} \tag{118}$$

While R_{C} exclusively exchanges magnetization between dipolar-coupled partners, R_{L} will distribute magnetization among dipolar-coupled partners as well as leak some magnetization onto the "lattice" as sort of a thermal dissipation to surroundings. What is interesting is that R_{L}, apart from being the external relaxation link, is supplemented by the AB dipolar interaction as well, which can contribute to magnetization loss (the term $5q\tau_c$ in R_{L}). Therefore, even if we can somehow damp R_1, the dipolar interaction between A and B itself is capable of attenuating the intensities of cross-peaks. In other words, while R_{C} conserves the total magnetization, R_{L} attenuates all the four peaks (two diagonal and two cross) uniformly. A plot of amplitudes of diagonal and cross-peaks as a function of mixing time for various values of $\omega\tau_c$ is shown in Fig. 4.57. The figure caption explains the various results. The important thing to note is that when $\omega\tau_c \ll 1$ the cross-peaks appear 180° out of phase with respect to diagonal peaks, whereas when $\omega\tau_c \gg 1$ all four peaks are in phase and are of equal intensity. When $\omega\tau_c = 1.12$, irrespective of the mixing time τ_m, there is no effective cross-relaxation and hence no cross-peaks.

The case of $\omega\tau_c \gg 1$ is the so-called spin diffusion limit or the "black" spectral limit and is of particular relevance in the case of large molecules (proteins and nucleic acids or molecules in highly viscous liquids) because the spectral densities in this case are:

$$\begin{aligned} J(\omega_{\mathrm{A}}) = J(\omega_{\mathrm{B}}) = J(\omega_{\mathrm{A}} + \omega_{\mathrm{B}}) \approx 0 \\ J(\omega_{\mathrm{A}} - \omega_{\mathrm{B}}) = q\tau_c \neq 0 \end{aligned} \tag{119}$$

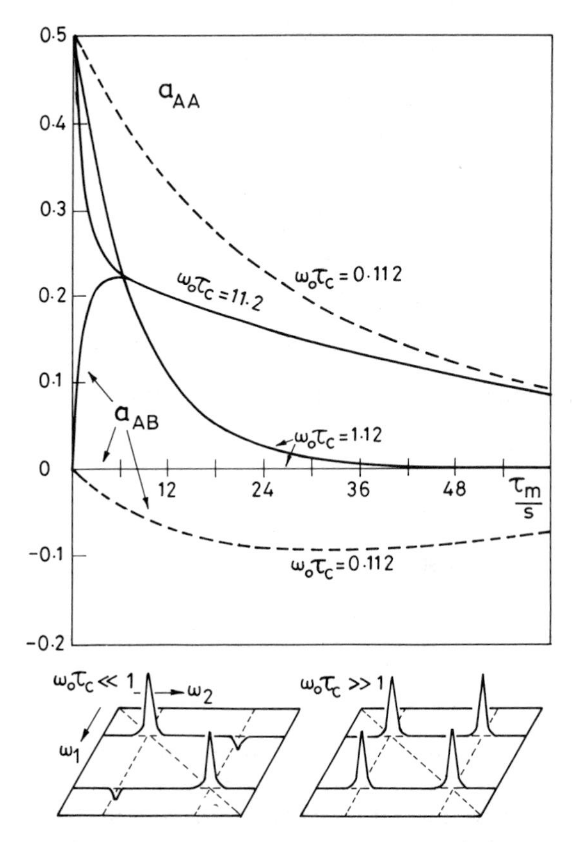

Figure 4.57. Dependence of mixing coefficients a_{AA} and a_{AB} on τ_m in a two-spin AB system for three typical correlation times corresponding to $\omega_0\tau_c \gg 1$, $\omega_0\tau_c \approx 1$, $\omega_0\tau_c \ll 1$. The 2D exchange lineshapes are also represented for the two extreme situations. [Reproduced by permission. S. Macura and R.R. Ernst, *Mol. Phys.*, **41**, 95 (1980), Taylor & Francis Ltd., London]

Thus, the rate constants of magnetization transfers are:

$$R_C = 2q\tau_c$$
$$R_L = R_1 \tag{120}$$

so that if external relaxation can be damped then almost equal intensities can be achieved for diagonal and cross-peaks, since the magnetization transfer is entirely controlled by spin diffusion. This situation is very important, since in the black spectral limit the cross-relaxation is governed by $J(\omega_A - \omega_B)$ [$J(0)$ in the usual nomenclature, i.e., the spectral density at zero frequency]. This has an r^{-6} dependence, and it should be possible, in principle, to discriminate between strongly coupled or weakly coupled through-space connectivities. Thus, for short mixing times only cross-peaks between nearest neighbor spins (nevertheless not necessarily bonded) would appear, while a weakly coupled distant partner would require a larger mixing period to generate sufficient magnetization transfer and hence the corresponding cross-peaks. The relative phases of the signals are also apparent from the processes that dominate the white and black spectral regimes. Thus, when $\omega\tau_c \gg 1$, W_0 is the only effective relaxation mechanism corresponding to a flip–flop transition $\alpha\beta \rightleftharpoons \beta\alpha$ so that one spin loses magnetization while the other gains, leading to reduction in the diagonal peak intensity and a growth in the cross-peak, giving diagonal and cross-peaks both of which are positive in phase. On the other hand, in the white spectral limit ($\omega\tau_c \ll 1$) W_2 dominates, corresponding to $\alpha\alpha \rightleftharpoons \beta\beta$. Therefore, when one spin loses a quantum of energy a second spin also loses energy concomitantly, a process that does not conserve energy, leading to cross-peaks negative in phase relative to the diagonal peaks.

The apparent discrepancy between Overhauser saturation experiments, wherein long correlation times result in negative intensity for the transferred partner and short correlation times increase the intensity of the transferred partner, and the present 2D experiments is no discrepancy at all. In the former saturation is being transferred corresponding to negative magnetization, whereas positive magnetization is being exchanged in 2D experiments.

Similar arguments when applied to system A_nB_n lead to:

$$\begin{aligned} R_C &= 10nq_{AB}\tau_c^{AB} \\ R_L &= [5(3n-1)\lambda + n]q_{AB}\tau_c^{AB} \end{aligned} \tag{121}$$

for $\omega\tau_c \ll 1$, where λ is the ratio of intragroup relaxation to intergroup relaxation given by:

$$\lambda = \frac{q_{AA}\tau_c^{AA}}{q_{AB}\tau_c^{AB}} = \frac{q_{BB}\tau_c^{BB}}{q_{AB}\tau_c^{AB}} \tag{122}$$

Cross-peaks are negative as in the AB case and intergroup cross-peak intensity is suppressed by stronger intragroup interaction by a factor f

$$f = \frac{3(n-1)}{n}\lambda \tag{123}$$

For $\lambda = 10$ and for two methyl groups separated by 2.5 Å, this gives a cross-peak of intensity 0.0012 under ideal mixing time conditions. This is exactly analogous to very low Overhauser effects observed between methyl groups. On the other hand, where $\omega\tau_c \gg 1$,

$$\begin{aligned} R_C &= 2nq_{AB}\tau_c^{AB} \\ R_L &= 0 \end{aligned} \tag{124}$$

so that intragroup relaxation causes only a distribution among equivalent spins and strong cross-peaks are expected. It is therefore obvious that 2D NOESY spectroscopy is ideal in the black spectral regime (1) to obtain

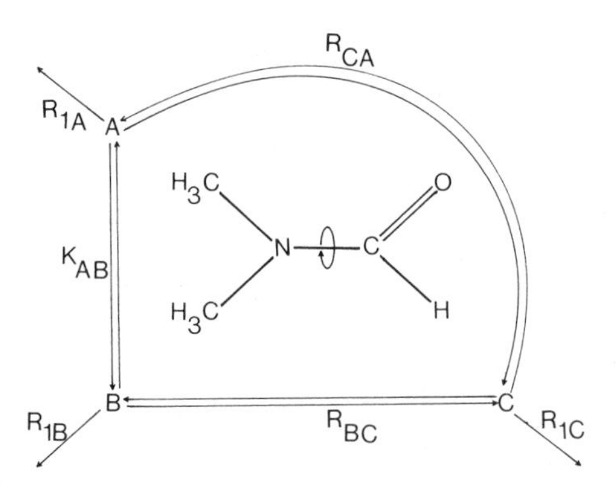

Figure 4.58. Cross-relaxation and exchange network in *N,N*-dimethylformamide (DMF) based on Macura and Ernst. [Reproduced by permission. S. Macura and R.R. Ernst, *Mol. Phys.*, **41**, 95 (1980), Taylor & Francis Ltd., London]

through-space connectivities, which help unravel molecular conformation, and (2) to distinguish proximal and distant coupled partners by way of mixing time variation.

A similar approach to examining two-spin intermolecular cross-relaxation requires translational self-diffusion constants, relative mole fractions of the two spins, as well as an estimate of their distance of closest approach. Assuming equimolar concentrations of the two molecules and identical self-diffusion coefficients, it can be shown that the cross-relaxation, for fast translational diffusion, is only half as efficient as the loss of magnetization to the surroundings, that too in the absence of other external relaxation processes. This leads to very weak cross-peaks. However, the situation is modified when the concentrations are very different.

We give below the excellent example of a cross-relaxation and chemical exchange study by 2D spectroscopy on *N,N*-dimethylformamide (see Fig. 4.58).

Here K_{AB} is the chemical exchange rate constant, R_{AC} and R_{BC} are cross-relaxation rates, and R_{1A}, R_{1B}, and R_{1C} are the so-called external relaxation rates. To understand the cross-relaxation and exchange effects in the 2D spectrum it is enough to start with the following basic statements. The R_{BC} is likely to be predominantly intramolecular with a correlation time τ_c. The relaxation R_{AC} will be mainly intermolecular, controlled by diffusion constant D and the radius of closest approach of B and C, r_{BC}. The magnetization transfer between A and B will be almost entirely governed by exchange, since cross-relaxation in A_3B_3 is rather inefficient. We need also consider "external" relaxation rates R_{1A}, R_{1B}, R_{1C}, which take into account inter- and intramolecular AA, BB, and CC interactions. We shall neglect dipolar AB cross-relaxation (*vide supra*), intermolecular BC relaxation, and intramolecular AC relaxation. With these assumptions the differential equation governing the dynamic processes is:

$$\begin{pmatrix} dM_A/dt \\ dM_B/dt \\ dM_C/dt \end{pmatrix} = \boldsymbol{R} \begin{pmatrix} M_A \\ M_B \\ M_C \end{pmatrix} \tag{125}$$

where $\boldsymbol{R}$ is governed by the above-mentioned relaxation and exchange rates:

$$\begin{aligned}
&R_{AA} = R_{1A} + K_{AB} + 2R_{CA} \\
&R_{AB} = -K_{AB}; \quad R_{AC} = 3R_{CA} \\
&R_{BA} = -K_{AB}; \quad R_{BB} = R_{1B} + K_{AB} + 2R_{CB} \\
&R_{BC} = 3R_{CB}; \quad R_{CA} = 5q_{AC}\tau_c \\
&R_{CB} = \tfrac{1}{2}n_C\rho_{CB}; \quad R_{CC} = R_{1C} + 6R_{CA} + 6R_{CB}
\end{aligned}$$

with:

$$q_{AC} = \frac{1}{10}\gamma^4 \frac{h^2}{4\pi^2} r_{AC}^{-6} \left(\frac{\mu_0}{4\pi}\right)^2$$

and:

$$\rho_{CB} = \frac{2\pi}{15}\gamma^4 \frac{h^2}{4\pi^2} \frac{1}{Dr_{BC}} \left(\frac{\mu_0}{4\pi}\right)^2 \tag{126}$$

where n_C is the number of molecules per unit volume. Correct to second order in mixing time, the six cross-peaks are given by the following expressions for short τ_m;

$$\begin{aligned}
a_{AB}(\tau_m) &= a_{BA}(\tau_m) \\
&= \frac{3}{7}M_0\left[K_{AB}\tau_m\left(1 - \frac{R_{AA} + R_{BB}}{2}\tau_m\right) + \frac{15}{4}q_{AC}\tau_c n_C\rho_{CB}\tau_m^2\right] \\
a_{BC}(\tau_m) &= a_{CB}(\tau_m) = -\frac{1}{7}M_0\left[\frac{3}{2}n_C\rho_{CB}\tau_m\left(1 - \frac{R_{BB} + R_{CC}}{2}\tau_m\right)\right. \\
&\qquad \left. + 15K_{AB}q_{AC}\tau_c\tau_m^2\right] \\
a_{AC}(\tau_m) &= a_{CA}(\tau_m) = -\frac{1}{7}M_0\left[15q_{AC}\tau_c\tau_m\left(1 - \frac{R_{CC} + R_{AA}}{2}\tau_m\right)\right. \\
&\qquad \left. + \frac{3}{2}K_{AB}n_C\rho_{CB}\tau_m^2\right]
\end{aligned} \tag{127}$$

The cross-peaks (ω_A, ω_B) and (ω_B, ω_A) will be possible for all mixing times, whereas the other cross-peaks, (ω_A, ω_C), (ω_C, ω_A), and (ω_B, ω_C), (ω_C, ω_B), will start with negative intensity for short τ_m. In these expressions the terms linear in τ_m correspond to direct transfer either via exchange or cross-relaxation, while those quadratic in τ_m are two-step successive dipolar transfer (cf. the section on Relayed COSY). Thus magnetization transfer can take place even

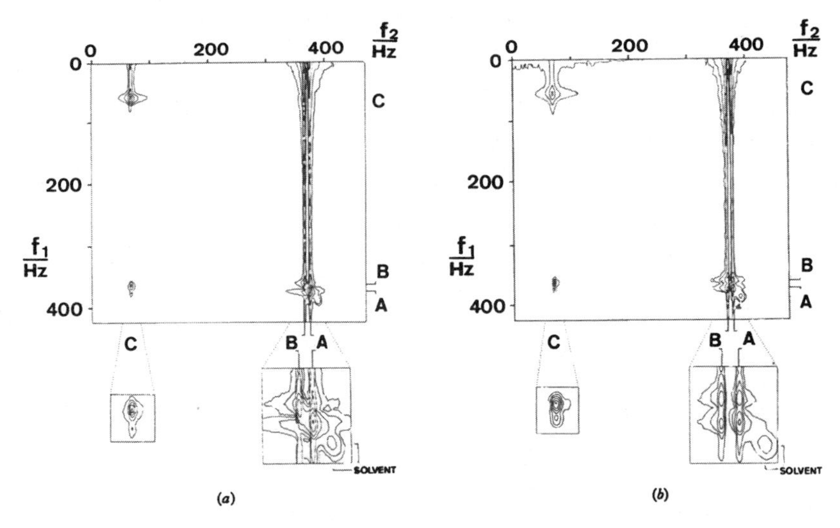

Figure 4.59. Contour plots of 2D exchange in DMF at 30°C and 60°C in DMSO-d_6. At 30°C, because of slow exchange, AB cross-peaks do not occur; at 60°C they become apparent. [Reproduced by permission. S. Macura and R.R. Ernst, *Mol. Phys.*, **41**, 95 (1980), Taylor & Francis Ltd., London]

in the absence of any direct dipolar link. However, the efficiency, involving only cross-relaxation, will rapidly decrease as the number of mediations increase.

At room temperature in *N*,*N*-dimethyl formamide, no apparent methyl (A), methyl (B) cross-peaks are visible. However, cross-peaks between the formyl (C) proton and the two inequivalent methyls are present. The BC cross-peak (see Fig. 4.59) must arise via intramolecular dipolar relaxation and hence fairly intense, while the weak AC cross-peak is due to intermolecular interaction.

At 60°, internal rotation about the N–C bond sets in and then exchange cross-peaks between A and B are visible. Because the exchange is rapid the AB, BA cross peaks are as intense as the AA and BB autopeaks. While the intermolecular dipolar induced AC cross-peak is not modified, there is a further gain in the BC cross-peak through intramolecular cross-relaxation and magnetization transfer via $A \rightarrow B$ by exchange and $B \rightarrow C$ cross-relaxation. Further, exchange broadening also reduces the resolution of AC and BC cross-peaks. The phase-sensitive spectra also exhibit the predicted positive intensity for exchange cross-peaks and negative intensity for cross-relaxation cross-peaks.

Elimination of Undesirable Artifacts in the NOESY Scheme. While a field gradient pulse can get rid of all unwanted transverse coherences that may interfere with the 2D NOESY spectrum, they can also be eliminated by phase cycling the three $\pi/2$ pulses (Fig. 4.54) as follows: In the first experiments the phases of $\pi/2$ pulses can be (x, x, x) and in the second experiment $(-x, -x, x)$.

In this way all transverse coherence of the type $I_{1z}I_{2y}$, $I_{1y}I_{2x}$, etc., alternate their signs in the two experiments; the longitudinal magnetizations are unaffected because they go through a combined $\pm 180°$ rotation by the first two $\pi/2$ pulses. Thus, coaddition of signals from two such experiments should cancel all COSY-type interferences.

In addition, during the mixing period some magnetization would have recovered along the z axis and so would have no "frequency labeling" prior to the detection pulse. The third "read" pulse would then create signals at $S(0, \omega_2)$ corresponding to axial peaks. For extended mixing periods required for investigating large molecules in the slow motional regime such axial peaks could be very intense and have to be suppressed. This can be done by coadding the spectra obtained from two experiments in which the second and third pulses are phase alternated; i.e., the two experiments have the sequence:

$$\left[\left(\frac{\pi}{2}\right)_x - t_1 - \left(\frac{\pi}{2}\right)_x - \tau_m - \left(\frac{\pi}{2}\right)_x - t_2\right]$$

and

$$\left[\left(\frac{\pi}{2}\right)_x - t_1 - \left(\frac{\pi}{2}\right)_{-x} - \tau_m - \left(\frac{\pi}{2}\right)_{-x} - t_2\right] \tag{128}$$

This cancels the z magnetization which has been recovered during the mixing period. It is also possible to eliminate axial peaks and COSY interference by coadding results from two experiments with:

$$[(\pi/2)_x - t_1 - (\pi/2)_x - \tau_m - (\pi/2)_x - t_2]$$

and

$$[(\pi/2)_{-x} - t_1 - (\pi/2)_x - \tau_m - (\pi/2)_{-x} - t_2] \tag{129}$$

Thus, 2D cross-relaxation and exchange spectroscopy opens up new vistas in the conformational analysis of complex molecules by NMR spectroscopy. Samples that exhibit negative Overhauser effects ($^{15}N-^{1}H$ systems) in the normal saturation transfer experiments give strong positive cross-peaks in 2D NOESY. In the black spectral limit the method promises a wealth of information regarding distances in dipolar-coupled networks by careful variation of the mixing period.

For the determination of rate parameters that govern cross-relaxation, it may be necessary to perform the experiment for several mixing periods τ_m including short values of τ_m. Under these circumstances coherence transfer peaks of the COSY type occur in the NOESY spectra and detailed analysis is possible only when we suppress these peaks completely. In addition, although transverse coherences of the single quantum and other higher quantum types can be eliminated by phase cycling, zero-quantum coherences are not affected by phase cycling. The following technique helps remove magnetization that emanates from ZQC in the final detection period. By progressively increasing the mixing period from experiment to experiment such that $\tau_m = \tau_{m0} + kt_1$, the incremented mixing time will lead to displacement of all J cross-peaks (of the COSY type), including those which arise from ZQC and MQC along the ω_1 dimension, and does not produce counterparts in the mirror symmetry position about the diagonal line. They can then be eliminated by symmetrization, a software routine which eliminates all peaks except those which occur in symmetrical pairs about the diagonal. For a concerted increment of τ_m with t_1, an array of cross-peaks will be generated at $\omega_1 = \delta_A + k\delta_B$ and $\omega_2 = \delta_B$. Thus, all coherent magnetization processes lead to displaced cross-peaks as a function of the mixing period, while incoherent transfer via chemical exchange/NOE are unaffected by τ_m except for intensities. Thus, a symmetrization in a 2D NOESY with τ_m incrementation should eliminate all J cross-peaks, making dipolar connectivity unambiguous, a method extremely useful in the three-dimensional analysis of large molecules by 2D NOESY (see Fig. 4.60). One danger could be accidental overlap of J cross-peaks with NOESY peaks giving rise to anomalous intensities. Differentiation can be obtained by using two different incrementation schemes, where displacement steps are different such that accidental superpositions can be ascertained.

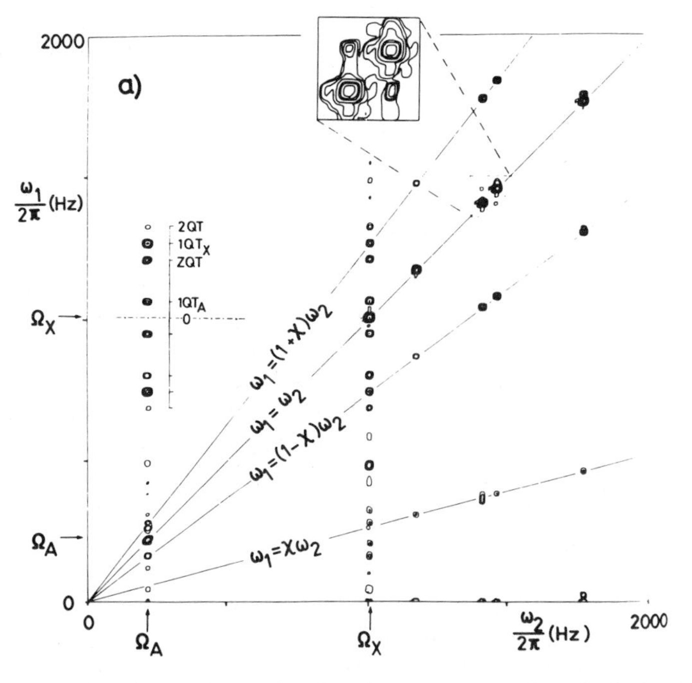

Figure 4.60. (a) Absolute value mode 2D exchange contour plot of a 1 : 1 mixture of DMA and 1,1,2-trichloroethane (TCE) recorded with incrementation of τ_m in the range 100–132 ms. (b) Same as in (a), but after symmetrization [Reproduced with permission. S. Macura, K. Wüthrich, and R.R. Ernst, *J. Magn. Reson.*, **46**, 269 (1982), copyright 1982, Academic Press, New York]

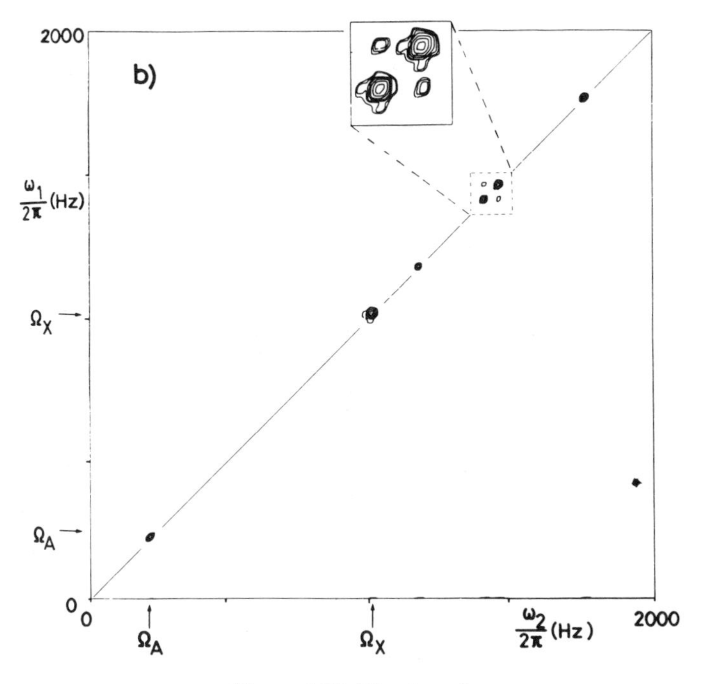

Figure 4.60. (Continued)

Combined COSY–NOESY 2D Spectroscopy (COCONOESY). We have seen that whereas the COSY scheme gives chemical shift correlation among coupled partners and is limited to only nuclei that have resolved scalar coupling, the NOESY scheme gives connectivity between dipolar coupled spins. For a complete understanding and unraveling of molecular structure and conformation one often needs to perform both experiments. For example, in reasonably sized proteins/nucleic acids COSY allows the identification of families of signals belonging to the same coupled networks, e.g., specific amino acid residues, whereas NOESY, which gives interresidue cross-peaks, helps identify possible sequencing. Thus, for a total conformational analysis of large molecules by 2D NMR one would like to perform both COSY and NOESY at identical experimental conditions. A new scheme, which allows COSY and NOESY information to be obtained simultaneously, has been reported and the pulse scheme is as shown in Fig. 4.61. The first two $\pi/2$ pulses transfer magnetization among scalar-coupled spins, so that acquisition after the second pulse generates $S(t_1, t_2^1)$ for COSY information. The second pulse, the COSY acquisition time together with the delay (if any), and the third pulse form the mixing period for NOESY. An optional field gradient pulse prior to the third $\pi/2$ pulse can get rid of all observable and other higher order coherences except ZQC and the t_2^2 period should give $S(t_1, t_2^2)$ that would be obtained in a normal NOESY scheme. In this way Fourier transformation of

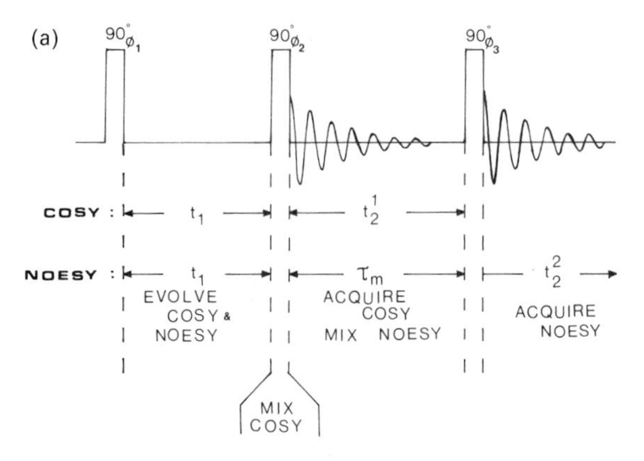

Figure 4.61. (a) Pulse scheme for COCONOESY. (b) the 500-MHz absolute value contour plot of the COSY spectrum of hybrid DNA–RNA oligonucleotide, *d*(CG) *r*(CG) *d*(CG), employing COCONOESY. (c) The NOESY spectrum from the experiment in (b). [Reproduced by permission. C.A.G. Haasnoot, F.J.M. van de Ven, and C.W. Hilbers, *J. Magn. Reson.*, **56**, 343 (1984), copyright 1984, Academic Press, New York]

$S(t_1, t_2^1)$ and $S(t_1, t_2^2)$ will give COSY and NOESY from the same experiment. This would save considerable amount of measuring time especially in large molecular weight systems under low concentrations, where each scheme individually can take a day or two. Here also, elimination of axial peaks in both spectra as well as J cross-peaks in NOESY is achieved when the phases are cycled as follows: Cycling the above four-step experiments by incrementing all phases by 90°, 16 experiments are performed for each t_1 and τ_m and the result coadded. This way undesirable artifacts are eliminated and also positive and negative frequencies can be discriminated. The t_2 period for COSY in COSY–NOESY experiment is limited by the mixing period τ_m for NOESY, so a short mixing period poses problems in proper acquisition of COSY signals. The lower limit for the mixing period cannot be prescribed a priori without knowledge of the resolution needed for COSY, the rotational correlation time of the molecules, etc.

It should be pointed out in conclusion that through NOESY spectra it is possible to establish dipolar correlation between homo- and heteronuclear systems. The NOESY spectra in the homonuclear case also can be presented in the SECSY form by advancing the mixing period to the middle of the t_1 period. The SECSY form of NOESY and the pulse scheme for pure 2D exchange spectroscopy for ^{13}C are shown in Fig. 4.62. The diagonal peaks now occur at $(0, \delta_i)$ and cross-peaks at $(\delta_i - \delta_j, \delta_i)$ and $(\delta_j - \delta_i, \delta_j)$.

zz-Peak Suppression in NOESY. After the first pair of 90° pulses on a two-spin-1/2 AX system, the density operator is composed not only of the frequency-labeled longitudinal magnetizations I_{1z} and I_{2z}, but also of zero-, single-, and double-quantum coherences. The signal contributions of the last two terms following the third $\pi/2$ pulse are suppressed by phase cycling, and

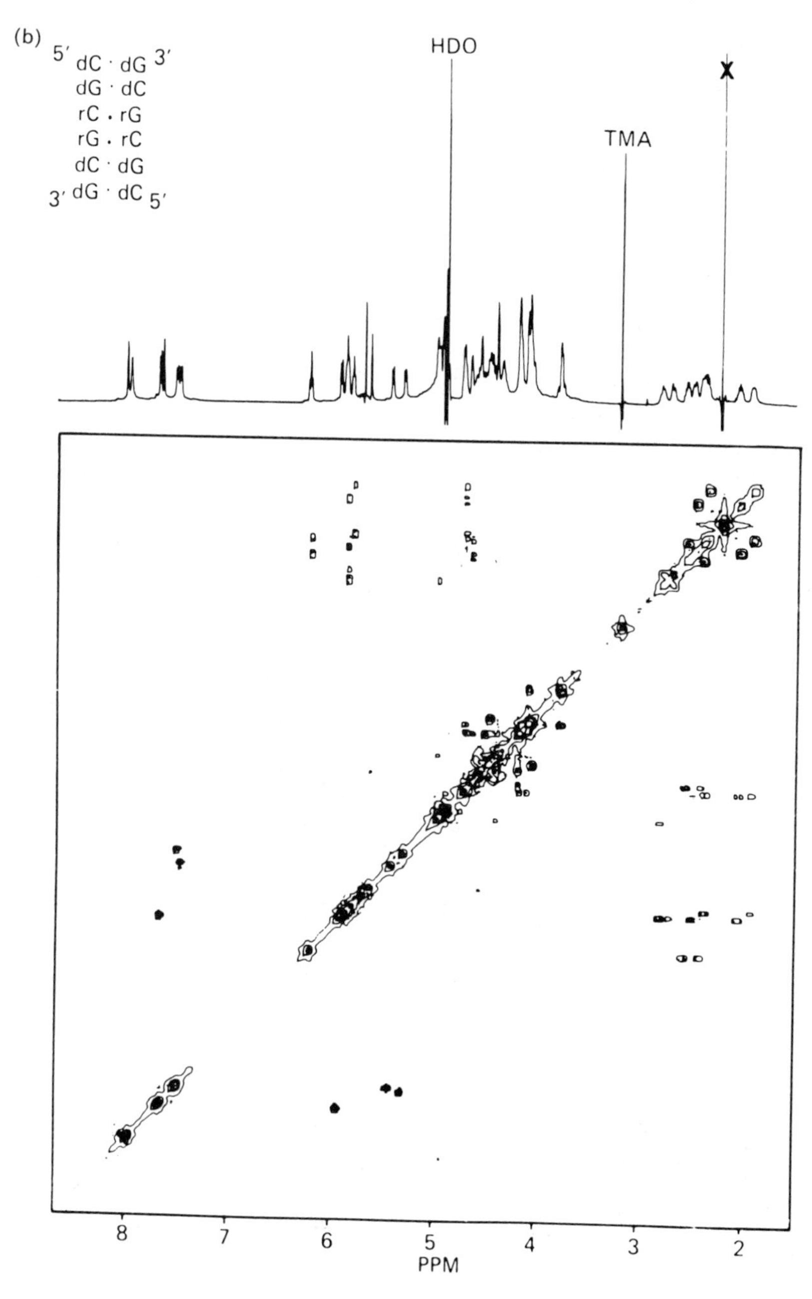

Figure 4.61. (Continued)

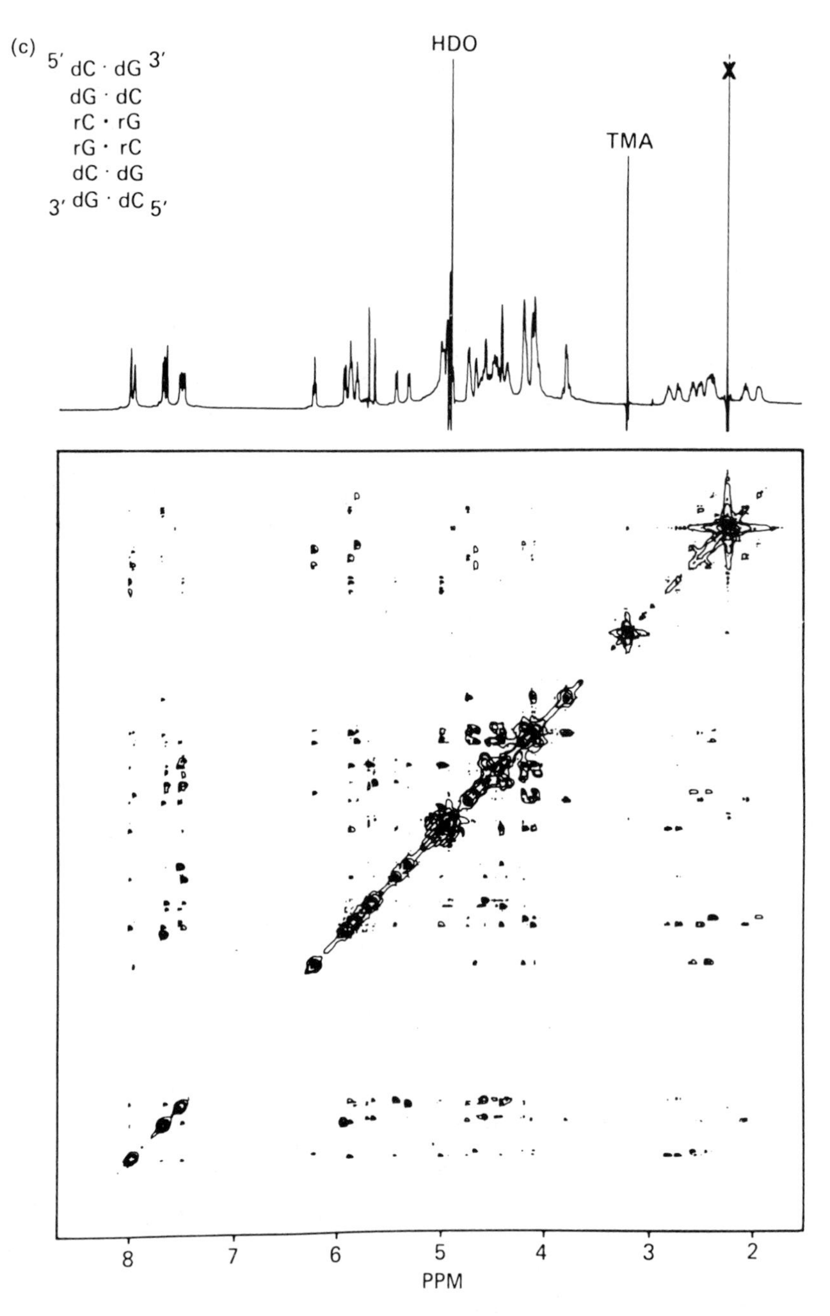

Figure 4.61. (Continued)

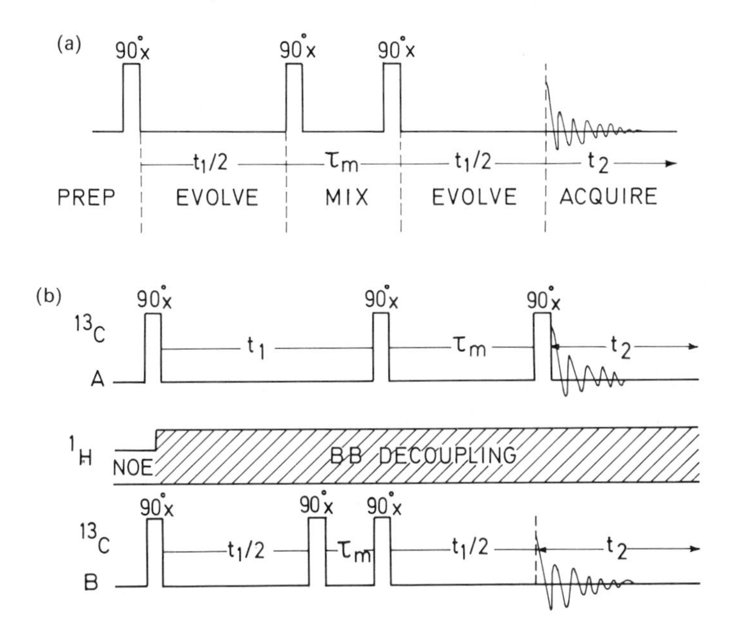

Figure 4.62. (a) Pulse scheme for SECSY form of NOESY. (b) Pulse scheme for ^{13}C 2D exchange spectroscopy. (A) COSY format; (B) SECSY format.

those of the first are suppressed by suitable "refocusing" strategies. However, there is a further artifact in NOESY spectra that arises from pulse imperfections, in systems with resolved couplings, and that is easily identified because it leads to multiplets with reduced symmetry. The rf inhomogeneity effects commonly result in non-$\pi/2$ flip angles, which result in the creation of a longitudinal two-spin order, otherwise known as *zz* order. Assuming the flip angle of the first mixing pulse to be actually θ, the anti-phase terms resulting after the evolution period are converted, for example, into:

$$I_{1z}I_{2y} \xrightarrow{\theta(I_{1x}+I_{2x})} c_\theta s_\theta I_{1z}I_{2z} - c_\theta s_\theta I_{1y}I_{2y} - s_\theta^2 I_{1y}I_{2z} + c_\theta^2 I_{1z}I_{2y} \quad (130)$$

while in-phase terms are converted into:

$$I_{1y} \xrightarrow{\theta(I_{1x}+I_{2x})} c_\theta I_{1y} + s_\theta I_{1z} \quad (131)$$

In the absence of relaxation of the two-spin order during τ_m, the second mixing pulse, also of actual flip angle θ, leads to reconversion:

$$I_{1z}I_{2z} \xrightarrow{\theta(I_{1x}+I_{2x})} -c_\theta s_\theta(I_{1z}I_{2y} + I_{1y}I_{2z}) + \text{unobservable terms} \quad (132)$$

and:

$$I_{1z} \xrightarrow{\theta(I_{1x}+I_{2x})} -s_\theta I_{1y} + c_\theta I_{1z} \quad (133)$$

The *zz* term gives rise to an anti-phase doublet structure of the 2D cross-peaks provided J_{12} is resolved, whereas the NOESY responses are in phase. If τ_m is suf-

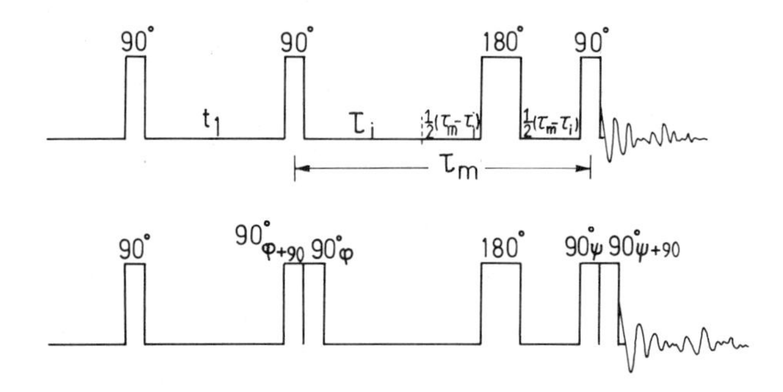

Figure 4.63. Pulse sequences for NOESY with ZQC suppression (top trace) and for *zz* peak suppression (bottom trace).

ficiently long, the two-spin order migrates under dipolar coupling, leading to:

$$I_{kz}I_{lz} \rightarrow I_{kz}I_{mz}$$

or:

$$I_{kz}I_{lz} \rightarrow I_{mz}I_{nz} \tag{134}$$

the latter situation involving pair migration, e.g., owing to chemical exchange. In such cases, anti-phase doublet structure is exhibited provided J_{kl} and J_{km} are resolved in the first case, and provided J_{kl} and J_{mn} are resolved in the second.

The *zz* signal artifacts in NOESY spectra may be suppressed by employing the idea of composite pulses during the mixing period. Figure 4.63 gives the normal NOESY pulse sequence and the corresponding compensated pulse version. In the uncompensated version, the *zz* signal intensity is proportional to $\sin^2\theta\cos^2\theta$, where θ is nominally 90°. In the above version of the mixing pulse compensated experiment, on the other hand, the intensity is proportional to $\sin^2\theta\cos^4\theta$. *zz* signals are attenuated therefore by $\cos^2\theta$ by this first-order scheme of compensating for rf inhomogeneity.

The *zz* signals and NOESY signals may be separated on noting that π pulses affect the *zz* and NOESY terms differently:

$$\begin{aligned} I_{kz} &\xrightarrow{\pi} -I_{kz} \\ I_{kz}I_{lz} &\xrightarrow{\pi} I_{kz}I_{lz} \end{aligned} \tag{135}$$

Insertion of a π pulse at the commencement of τ_m on alternate scans, as shown in Fig. 4.64, and subtracting the corresponding FID's results therefore in pure

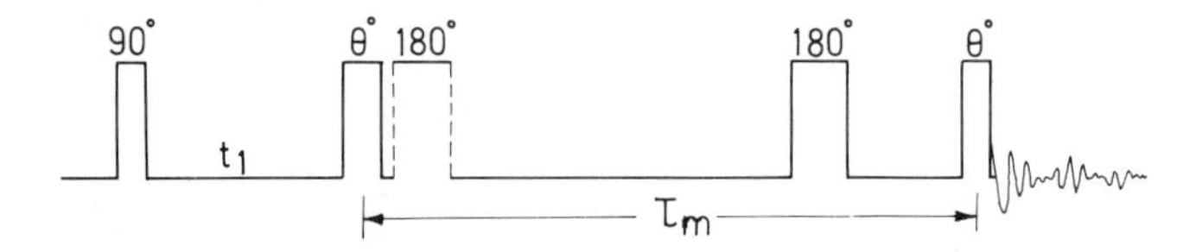

Figure 4.64. Pulse sequence for separation of *zz* and NOESY spectra.

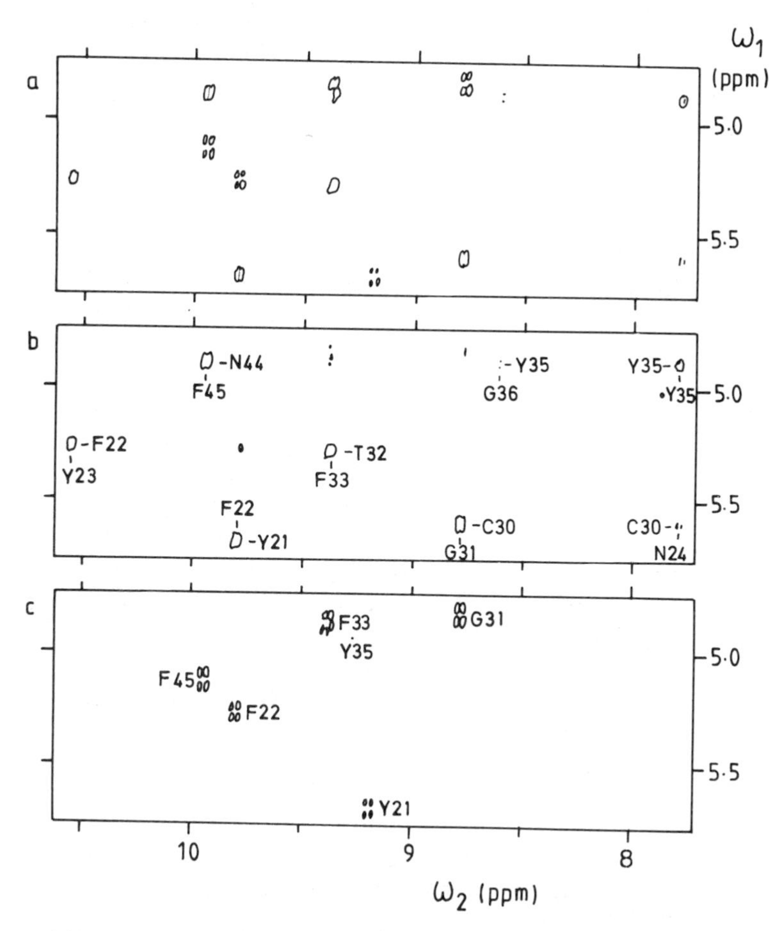

Figure 4.65. Portions of 2D spectra of BPTI: (a) Superposition of NOESY and zz cross-peaks; (b) pure NOESY spectrum; (c) pure zz spectrum. The cross-peaks have been identified with IUB-IUPAC one-letter symbols for amino acids and their location in the sequence. [Reprinted with permission. G. Bodenhausen, G. Wagner, M. Rance, O.W. Sørensen, K. Wüthrich and R.R. Ernst, *J. Magn. Reson.*, **59**, 542 (1984), copyright 1984, Academic Press, New York]

NOESY spectra from which the zz artifacts have been removed. On the other hand, addition of the resulting FID's generates pure zz spectra, which for short τ_m during which the zz order has not migrated, are essentially equivalent to DQF–COSY spectra. For longer τ_m, the zz spectra could provide new insight into cross-relaxation and chemical exchange processes. The separation of zz and NOESY spectra is exemplified in Fig. 4.65.

It may be noted that the π pulse strategy really only discriminates between odd- and even-spin longitudinal order terms; the separation into individual n-spin longitudinal order terms may be achieved as in DEPT editing. For example, in the simple situation treated above, separation of zz and NOESY spectra may alternatively be achieved by setting both the last pulses deliberately to $\pi/4$ or $3\pi/4$ and adding/subtracting the resulting FID's.

For short τ_m's such that migration of zz terms may be neglected, it is interesting to note that the acquisition of data with $\theta = \pi/4$, and a π pulse at the beginning of τ_m on alternate scans, for example, and subsequent processing to generate pure NOESY and zz spectra is an alternative method of obtaining NOESY and COSY spectra from a single set of measurements. This method has the advantage that the F_2 resolution in the COSY spectrum, which depends on the COSY acquisition time t_2, may be selected independently of the mixing time τ_m. It may be noted, however, that for this choice of θ, the intensity of the NOESY and zz cross-peaks is 50% of the normal situation with separate measurements of NOESY and DQF–COSY, respectively.

Accordion Spectroscopy

In such experiments as the chemical exchange or NOE 2D spectroscopy we have three periods after the preparation, namely the evolution period t_1, the mixing period τ_m, and the detection period t_2. In a given 2D experiment τ_m is fixed and only t_1 and t_2 are varied, t_1 systematically, while t_2 is the normal acquisition of the spectrum in the detection period. It is possible to conceive a 3D spectroscopy by systematically varying the mixing period τ_m with a view to studying dynamics of exchange, in addition to varying t_1 systematically. This would truly constitute a three-dimensional time response $S(t_1, \tau_m, t_2)$ and would warrant a triple Fourier transformation, giving rise to various sine and cosine components, such as $S^{ccc}(F_1, F_m, F_2)$, $S^{ccs}(F_1, F_m, F_2)$. Not only is the storage requirement of the computer increased, but even when one achieves such a triple Fourier transform there is no sensible way of presenting four variables F_1, F_m, F_2 and intensity in a three-dimensional perspective. To circumvent this problem and to present the results much the same as in 2D experiments is to increment τ_m in a concerted fashion with t_1 such that t_1 and τ_m are related as $\tau_m = kt_1$, in the so-called accordion spectroscopy. Thus the detected signals are modulated as function of t_1 in the evolution period, whereas in the mixing period the modulation is a function of kt_1; therefore, a double FT of $S(t_1, kt_1 = \tau_m, t_2)$ gives the 2D frequency domain where one can label the usual F_1 axis by F_1 on one side and $F_m = F_1/k$ on the parallel opposite side. While the diagonal and cross-peaks follow exactly the same principle of 2D NOESY/exchange spectroscopy, the effect of the mixing pulse can be seen in the lineshapes of the diagonal/cross-peaks. The spectra contain all the information for kinetic studies and lineshape analysis, and the mixing functions can be retrieved from $S(F_1, F_m, F_2)$ by an inverse Fourier transform with respect to F_m.

A simple two-site exchange with rate of exchange K under conditions of equal populations and relaxation rate R will give for the peaks:

$$\begin{aligned} S_{AA}^{(F_m)} = S_{BB}^{(F_m)} &= \tfrac{1}{2}[(R/(R^2 + \Delta F_m^2)) + (2K + R)/((2K + R)^2 + \Delta F_m^2)] \\ S_{AB}^{(F_m)} = S_{BA}^{(F_m)} &= \tfrac{1}{2}[(R/(R^2 + \Delta F_m^2)) - (2K + R)/((2K + R)^2 + \Delta F_m^2)] \end{aligned} \qquad (136)$$

The time evolution during τ_m can now be retrieved by an inverse fourier transform of the lineshape as a function of τ_m. This integral for a cross-peak S_{kl} is:

$$S_{kl}(\tau_m) = \frac{1}{2\pi}\int S(\omega_m)g(\omega)\exp(-i\omega_m\tau_m)\,d\omega_m \qquad (137)$$

where $g(\omega)$ is a window function that is 1 between $\Omega \pm \Delta\Omega$ and zero elsewhere and just takes into account of the two superposed Lorentzian lines in the F_m dimension (Eq. (136)). In case of magnetization transfer of the type $A \rightleftharpoons B \rightleftharpoons C$, the A to C transfer will be of second order so that inverse Fourier transformation of a cross-peak will show vanishing value at $\tau_m = 0$, while inverse Fourier transform of the diagonal peaks should lead to a nonvanishing value at $\tau_m = 0$. Thus, the accordion scheme is a simplified approach to 3D spectroscopy in which F_1 and F_m are represented along one axis and F_2 is along the other axis. A careful analysis of the lineshape function in the F_m dimension should hold clues as to the mechanism of magnetization transfer processes such as spin–lattice relaxation, spin diffusion, cross-polarization, etc. (see Fig. 4.66).

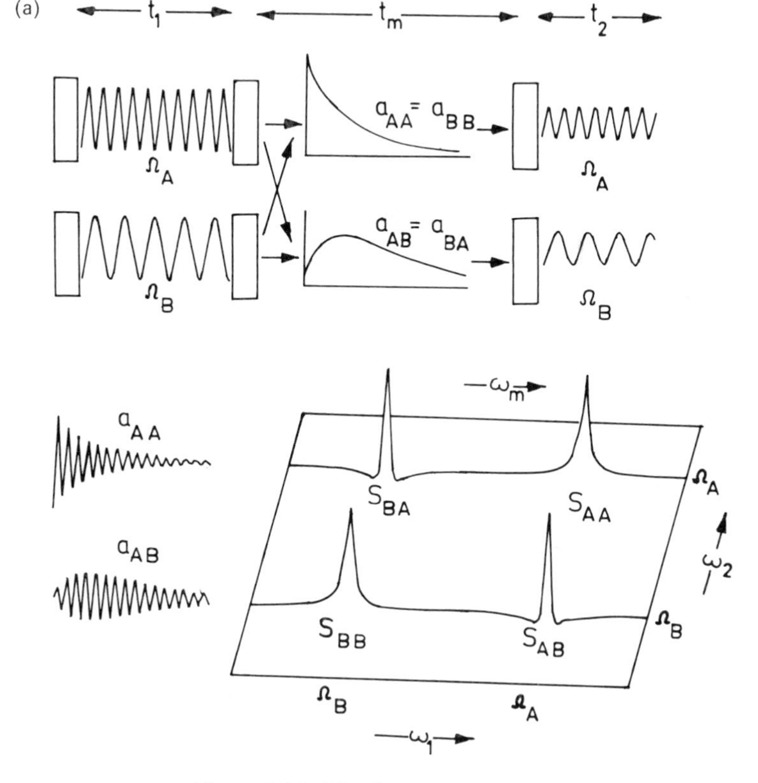

Figure 4.66. (Caption on page 218).

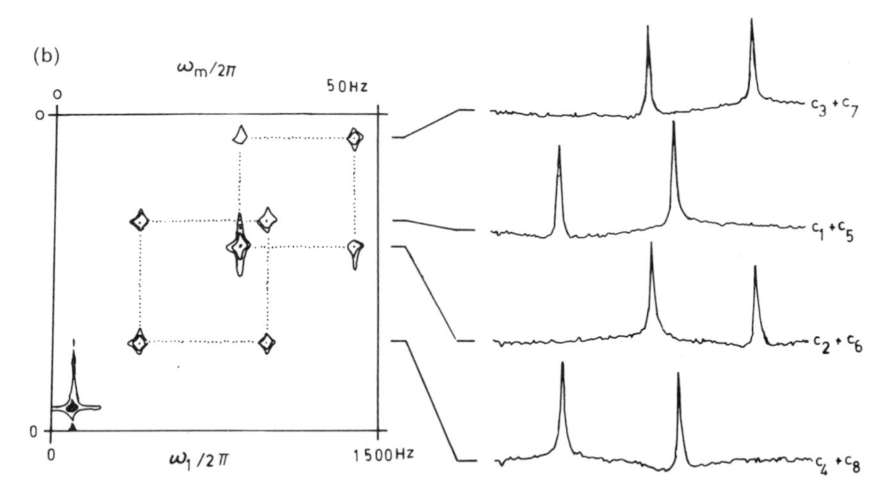

Figure 4.66. (a) Schematic of an accordion experiment, showing mixing functions and lineshapes for a system with two-site exchange. (b) The 75-MHz ^{13}C accordion spectrum of the ring puckering process in *cis*-decalin at 240K. The figure on the right shows phase-sensitive cross-sections exhibiting the characteristic accordion lineshapes. [Reproduced by permission. G. Bodenhausen and R.R. Ernst, *J. Magn. Reson.*, **45**, 367 (1981), copyright 1981, Academic Press, New York]

Coherence Transfer Pathways

Multiple-pulse experiments excite several coherences in a spin system, of which the experimenter is normally interested in but one. The selective detection of the coherence of interest is achieved by exploiting characteristic dependences of the desired coherence on pulse flip angles and phases. We discuss in the following the general principles of phase cycling, which employs rf pulse phases as the discriminant. This strategy is a generally valid one without regard to the details of the spin system, such as the strength of couplings. Note, however, that no phase cycling can distinguish between longitudinal magnetization and homonuclear ZQC's.

The description of phase cycling is most conveniently carried out employing the coherence transfer pathway formalism. As discussed in Chapter 1, the coherence order of a spin system is conserved during free evolution, but may be altered by rf pulses subject to the condition that a nonselective "hard" rf pulse can create only the coherence orders ± 1 when acting on a system in thermal equilibrium. The most general form of the density operator may be expressed as:

$$\sigma(t) = \sum_{p=-M}^{M} \sigma^p(t) \tag{138}$$

for an M-spin-1/2 system. The effect of a pulse at time t can then be expressed by the transformation:

$$U\sigma^{p}(t^{-})U^{-1} = \sum_{p'} \sigma^{p'}(t^{+}) \tag{139}$$

The effect of phase shifting the pulse by φ radians follows then on employing the characteristic response of multiple-quantum coherences to phase shifts, as discussed in Chapter 1.

$$\begin{aligned} U_{\varphi}\sigma^{p}(t^{-})U_{\varphi}^{-1} &= \exp(-i\varphi F_z)U\exp(+i\varphi F_z)\sigma^{p}(t^{-}) \\ &\quad \times \exp(-i\varphi F_z)U^{-1}\exp(i\varphi F_z) \\ &= \exp(-i\varphi F_z)U\sigma^{p}(t^{-})\exp(ip\varphi)U^{-1}\exp(i\varphi F_z) \\ &= \exp(-i\varphi F_z)\left[\sum_{p'}\sigma^{p'}(t^{+})\exp(ip\varphi)\right]\exp(i\varphi F_z) \\ &= \sum \sigma^{p'}(t^{+})\exp(-i\Delta p\varphi) \end{aligned} \tag{140}$$

where:

$$\Delta p = p'(t^{+}) - p(t^{-}).$$

The pathway of a coherence during a pulse sequence may be specified by the vector $\mathbf{\Delta p}$, which is the array of changes in coherence order wrought by the succession of pulses. The initial coherence level is 0, while the final coherence level detected by the receiver is $+1$ or -1 depending on the setting of the supposedly perfect quadrature receiver. After n successive pulses therefore, we have, assuming detection of the $p = -1$ coherence,

$$\begin{aligned} &\sigma^{p=-1}(\varphi_1, \varphi_2, \ldots, \varphi_n, t) \\ &\quad = \sigma^{p=-1}(\varphi_1 = \varphi_2 = \ldots = \varphi_n = 0, t) \\ &\qquad \times \exp(-i(\Delta p_1\varphi_1 + \Delta p_2\varphi_2 + \cdots + \Delta p_n\varphi_n)) \\ &\quad = \sigma^{p=-1}(\boldsymbol{\varphi} = 0)\exp(-i\mathbf{\Delta p}\cdot\boldsymbol{\varphi}) \end{aligned} \tag{141}$$

$\boldsymbol{\varphi}$ being the array of phase shifts on the n pulses. The detected signal, $s(t)$, similarly works out to be the sum of the contributions of the individual pathways:

$$s(t) = \sum_{\mathbf{\Delta p}} s^{\mathbf{\Delta p}}(t)$$

with:

$$s^{\mathbf{\Delta p}}(\boldsymbol{\varphi}, t) = s^{\mathbf{\Delta p}}(\mathbf{0}, t)\exp(-i\mathbf{\Delta p}\cdot\boldsymbol{\varphi}) \tag{142}$$

The characteristic phase shift associated with each pathway for a given pulse is employed to separate the different pathways by Fourier analyzing with respect to the phase of the rf pulse in question. Upon performing N_i experiments with the rf phase of the ith pulse incremented systematically as:

$$\varphi_i = k_i 2\pi/N_i, \qquad k_i = 0, 1, \ldots, N_i - 1 \tag{143}$$

and combining the N_i detected signals $s(\varphi_i, t)$ according to:

$$s^{\Delta p_i}(t) = \frac{1}{N_i} \sum_{k_i=0}^{N_i-1} s(\varphi_i, t) \exp(i\Delta p_i \varphi_i) \tag{144}$$

which is a discrete Fourier analysis with respect to φ_i with Δp_i as the conjugate variable, one selects all pathways which undergo a change in coherence order Δp_i under the ith pulse. Owing to aliasing, however (see Appendix 3), the selection is really $\Delta p_i \pm nN_i$ ($n = 0, 1, 2, \ldots$). The criterion for the minimum number of steps N_i in a phase cycle is that selection of a unique Δp_i from a range of r consecutive values requires that $N_i \geqslant r$.

In cases where pathway selection is required for m pulses, the phases of each are cycled separately:

$$\varphi_1 = k_1 2\pi/N_1, \ldots, \varphi_m = k_m 2\pi/N_m$$

for:

$$k_1 = 0, 1, \ldots, N_1 - 1; \ldots; k_m = 0, 1, \ldots, N_m - 1 \tag{145}$$

Here, k_1 is incremented through all N_1 steps before k_2 is incremented, and so on. The total number of experiments to be performed is therefore:

$$N = N_1 N_2 \cdots N_m$$

The signals are then combined in accordance with:

$$s^{\mathbf{\Delta p}}(t) = (1/N) \sum_{k_1=0}^{N_1-1} \sum_{k_2} \cdots \sum_{k_m} s(t) \exp(i \mathbf{\Delta p} \cdot \boldsymbol{\varphi})$$

where:

$$\mathbf{\Delta p} \cdot \boldsymbol{\varphi} = \Delta p_1 k_1 \frac{2\pi}{N_1} + \Delta p_2 k_2 \frac{2\pi}{N_2} + \cdots + \Delta p_m k_m \frac{2\pi}{N_m} \tag{146}$$

The manifold of pathways surviving this process of selection are:

$$\mathbf{\Delta p} = \{\Delta p_1 \pm n_1 N_1, \ldots, \Delta p_m \pm n_m N_m\} \tag{147}$$

It is usually possible to retain a unique pathway by relatively small increment numbers N_i since $|p_{\max}| \leqslant M$ and the amplitude of coherence transfer into very high orders is small.

The phase factors required for the discrete Fourier analysis may be obtained conveniently by shifting the phase of the receiver reference channel. The reference phase must be:

$$\varphi^{\text{ref}} = -\sum \Delta p_i \varphi_i \tag{148}$$

if the $p = -1$ coherence is detected. This completes the prescription to design a phase cycle in terms of the number of steps and the phase shifts, given the Δp_i values to be passed and those to be blocked.

We briefly describe below the application of the coherence transfer pathway formalism to some 2D experiments. Figure 4.67 shows the coherence transfer pathway associated with two- and three-pulse experiments.

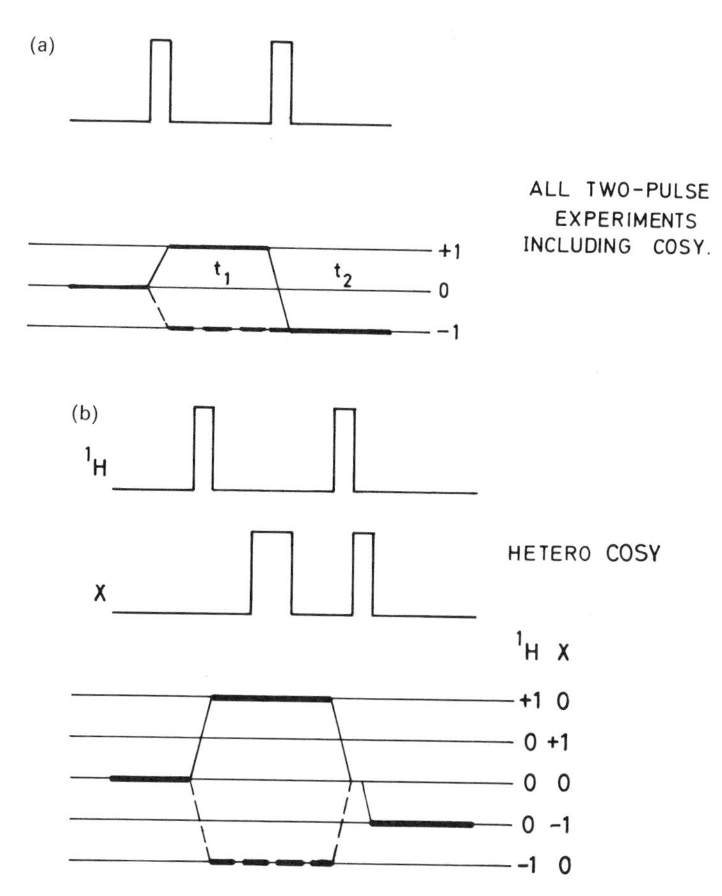

Figure 4.67. Coherence transfer pathways. (a) For all two-pulse experiments. (b) For hetero COSY.

During the evolution period t_1, two coherences with opposite order $\pm p$ and opposite frequencies are always associated with each pair of connected states r and s. Both coherences, associated with mirror image pathways, must be retained if pure phase 2D spectra are required. On the other hand, retention of a single pathway is sufficient if absolute value spectra will do and allows discrimination between the signs of the frequencies F_1. Experiments may also be designed that allow pure phase 2D spectra to be obtained with the sign information of F_1 preserved. To summarize, to calculate the effect of a phase cycling procedure, the phase factor for each pulse is calculated from the phase of the pulse and the change in the coherence level. All the phase factors are multiplied, along with the receiver phase, to get the overall phase factor for a given pathway for the whole pulse sequence, for each step in the phase cycling. If the sum over the full cycle vanishes, the pathway in question will not contribute to the signal; on the other hand, a nonvanishing sum is a necessary

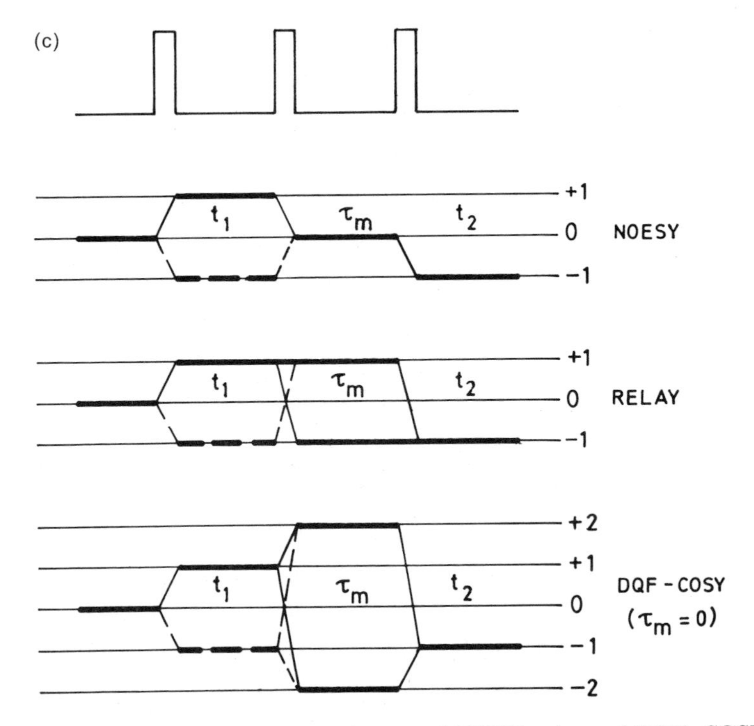

Figure 4.67. (c) For three pulse experiments—NOESY, relay and DQF–COSY.

but not sufficient condition for the pathway to contribute to the signal: in this case, the appearance of the signal depends on the flip angle.

It may also be noted that if the FID's that result from each step in a phase cycle are stored in *separate* memory locations, *one* set of experimental data may be combined in different ways—corresponding to the selection of different coherence transfer pathways—to yield different classes of 2D spectra (Relay, NOESY, MQS, etc.), provided the mixing time requirements are compatible.

Comparison of 1D and 2D Spectroscopy with Respect to Efficiency

***t_1* Noise.** In the course of a repetitive one-pulse experiment, instabilities of rf pulse amplitude and phase merely mean that signal averaging must be performed slightly longer than otherwise in order to attain the desired signal to noise (S/N) ratio. In any one-dimensional experiment, in fact, the magnetization excited appears with slight distortions in amplitude or phase, but not in frequency, as a result of such instabilities in the excitation.

In a 2D experiment, however, or one in which an FT is performed on an interferogram collected point by point on successive scans, such phase and amplitude variations are translated to the frequency domain. This manifests itself in a 2D spectrum as a ridge of noise running parallel to the F_1 axis, at each value of F_2 that corresponds to a transition, and is known as t_1 noise. This noise is white and is proportional to the signal—it is, in other words, multiplicative and not additive. The t_1 noise, therefore, is most dominant exactly where $\mathscr{S}/\mathscr{N}$ (in terms of the normal additive noise) would be expected to be excellent. Since t_1 noise arises essentially from instabilities in excitation and not in detection, the normal strategies of improving it by increasing the field strength or sample size or lowering the lattice or spin temperature are counterproductive. With such methods, the signal increases, but so does the multiplicative t_1 noise, which is really nuclear magnetization gone out of control. The method to reduce t_1 noise is to improve the stability of the spectrometer to minimize instabilities in excitation! Apart from this general prescription, wherever multiplicative noise dominates, it is useful to increase the repetition rates of successive scans beyond the optimum used in 1D experiments, where only additive noise is considered. Special strategies, such as coherence transfer echo filtering (see Chapter 5), may also be employed especially in MQS to reduce the t_1 noise by reducing the number of NMR signals reaching the receiver. It may also be remarked that a pseudo-filtration in F_1 may be achieved by imposing a t_1 jitter, if necessary with a suitable modulation of the receiver attenuation.

Sensitivity of 1D versus 2D experiments. One might wonder what is the sensitivity of a 2D, high-resolution experiment in comparison with a 1D spectrum under identical spectrometer conditions and time spent. The process of 2D involves acquiring N point free induction decays in the t_2 period (detection period), with M increments of the evolution period. The data matrix represented by $M \times N$ may be upto 1024×1024 for high-resolution applications. If signal averaging is involved to increase the $\mathscr{S}/\mathscr{N}$ ratio then each of the FID's for a given t_1 period may have to be time averaged. The signal to noise ratio of 2D spectrum (S_{kl}) can be defined in a manner analogous to the 1D experiment:

$$\mathscr{S}/\mathscr{N} = \frac{\text{Signal peak value}}{\text{Noise rms value}} = \max S_{kl}/\sigma_N \tag{149}$$

With a matched filter function one can improve the $\mathscr{S}/\mathscr{N}$ ratio and under ideal matching, for a reference peak, with $0 \leqslant k \leqslant M$ and $0 \leqslant l \leqslant N$:

$$(\mathscr{S}/\mathscr{N})_{\max} = \left[\sum_{k=0}^{M-1}\sum_{l=0}^{N-1} S_{kl}^2\right]^{1/2} \Big/ \sigma_N \tag{150}$$

This is proportional to the total signal energy of the 2D reference peak, and the weighting function of the 2D matched filter is equal to the inverted lineshape:

$$H_{kl}^{\text{matched}} = S_{-k-l} \tag{151}$$

so that weighted convolution sum would be:

$$S'_{kl} = \sum_{r=0}^{M-1} \sum_{s=0}^{N-1} S_{rs} S_{-r-s} \tag{152}$$

Let us consider a specific example of 2D resolved spectrum. Suppose a magnetization component $M^{(ab)}$ of the coherence ab processes in period t_1 under a Hamiltonian $\bar{\mathscr{H}}^{(1)}$ with frequency $\omega_{ab}^{(1)}$; at the end of t_1 it may be under another Hamiltonian $\bar{\mathscr{H}}^{(2)}$ and precesses with the modified frequency $\omega_{ab}^{(2)}$ during the detection period t_2,

$$\begin{aligned} M^{(ab)}(t_1, t_2) = M^{(ab)}(0)\,&[\cos(\omega_{ab}^{(1)} t_1 + \omega_{ab}^{(2)} t_2)] \\ &\times \exp(-t_1/T_{2ab}^{(1)}) \exp(-t_2/T_{2ab}^{(2)}) \end{aligned} \tag{153}$$

M points for $t_1 = 0, t_1/M, \ldots, (M-1)t_1/M$ and N sampling values taken at $t_2 = 0, t_2/N, \ldots, (N-1)t_2/N$ give rise to the data matrix $(M_{rs}^{(ab)})$. The real part of the complex Fourier transform is:

$$S_{kl}^{(ab)} = \sum_{r=0}^{M-1} \sum_{s=0}^{N-1} M_{rs}^{(ab)} \cos\left[\frac{2\pi kr}{M} + \frac{2\pi ls}{N}\right] \tag{154}$$

For the corresponding M identical 1D experiments, recorded under identical conditions and coadded and Fourier transformed,

$$S_l^{(ab)} = M \sum_{s=0}^{N-1} M_{0s}^{(ab)} \cos(2\pi ls/N) \tag{155}$$

Since the variances σ_{2D}^2 and σ_{1D}^2 for 2D and 1D spectra can be shown to be identical, we can compare the peak heights directly. Replacing the sums in the above equations by integrals from $0 - t_1$ and $0 - t_2$ for 2D,

$$\begin{aligned} S^{(ab)}(\omega_{ab}^{(1)}, \omega_{ab}^{(2)}) = M^{(ab)}(0) \frac{1}{2} \frac{T_{2ab}^{(1)}}{t_1} \frac{T_{2ab}^{(2)}}{t_2} &[1 - \exp(-t_1/T_{2ab}^{(1)})] \\ &\times [1 - \exp(-t_2/T_{2ab}^{(2)})] \end{aligned} \tag{156}$$

and for 1D:

$$S^{(ab)}(\omega_{ab}^{(2)}) = M^{(ab)}(0) \frac{1}{2} \frac{T_{2ab}^{(2)}}{t_2} [1 - \exp(-t_2/T_{2ab}^{(2)})] \tag{157}$$

so that:

$$\frac{(\mathscr{S}/\mathscr{N})_{2D}}{(\mathscr{S}/\mathscr{N})_{1D}} = \frac{T_{2ab}^{(1)}}{t_1} [1 - \exp(-t_1/T_{2ab}^{(1)})] \tag{158}$$

With matched filtering it can be shown that,

$$(\mathscr{S}/\mathscr{N})_{2D}/(\mathscr{S}/\mathscr{N})_{1D} = \left[\frac{1}{2} \frac{T_{2ab}^{(1)}}{t_1} (1 - \exp(-2t_1/T_{2ab}^{(1)}))\right]^{1/2} \tag{159}$$

While in matched filtering the signal to noise improves, especially for higher values of $t_1/T_{2ab}^{(1)}$, the sensitivity of 2D is only one-half that of the corresponding 1D experiment. In 2D-correlated experiments, the signal to noise ratio is somewhat poorer. The crux of the matter is that whether 1D or 2D, each experiment (or a scan) adds to the total information, and it is the total signal energy that matters for sensitivity, whether it originates from M-equal or M-different experiments.

CHAPTER 5

Multiple-Quantum Spectroscopy

Multiple-quantum transitions (MQT's), which are the subject of this chapter, involve a change in magnetic quantum number (M) of p units (i.e., $\Delta M = \pm p$) and occur by the simultaneous absorption of p equal quanta. The number of p-quantum transitions ($p = 0, 1, 2, \ldots, N$) in a system of N inequivalent spin-1/2 nuclei is given by:

$$\begin{aligned} p = 0&: \ \frac{1}{2}\left[\binom{2N}{N} - 2^N\right] \\ p \neq 0&: \ \binom{2N}{N-p} \end{aligned} \qquad (1)$$

These functions are given for systems of upto eight spins in Table 5.1. (When equivalences are involved, the number of transitions is reduced for all orders except the highest.) The counting procedure also includes the combination lines for each order, which are M-spin p-quantum transitions ($M > p$) that are of vanishing intensity even for $p = 1$, in a conventional one-dimensional (1D) spectrum. It is clear from the table that single-quantum transitions (SQT's) are the most numerous single group, but beyond $N = 3$, the total number of MQT's is higher than that of SQT's. At the same time, the number of MQT's drops very rapidly for higher values of p in each case. It is often the case that the single-quantum spectrum is too crowded for analysis and assignment; in such situations, it may be of value to study the high-order MQ spectra of the system, which exhibit fewer transitions and yet contain all the desired information on the spin system.

The study of multiple-quantum coherences (MQC's) is in fact not only a convenient way of getting information about spin systems in terms of the parameters of their time-independent Hamiltonians but actually adds new information about dynamic parameters. Classically, multiple-quantum co-

Table 5.1. Number of MQT's in Systems of N Spin-1/2 Nuclei

	N							
$p = \Delta M$	1	2	3	4	5	6	7	8
0	0	1	6	27	110	430	1,652	6,307
1	1	4	15	56	210	792	3,003	11,440
2	0	1	6	28	120	495	2,002	8,008
3	0	0	1	8	45	220	1,001	4,368
4	0	0	0	1	10	66	364	1,820
5	0	0	0	0	1	12	91	560
6	0	0	0	0	0	1	14	120
7	0	0	0	0	0	0	1	16
8	0	0	0	0	0	0	0	1

herences were excited and detected employing strong resonant rf, but the technical difficulties of excitation—and of detection in the presence of a complicating background of single-quantum coherences driven to saturation—insured that this continuous wave (CW) technique never really achieved wide popularity.

With the advent of pulsed techniques, the scenario has changed. For instance, one can create transverse magnetization and allow the internal Hamiltonian of the spin system to "prepare" it to exchange n photons coherently with resonant radiation. The MQC's created after a subsequent pulse are unobservable and need to be reconverted into single-quantum coherences by a third pulse. On systematically incrementing the evolution period of the MQC's, t_1, and sampling either one point or the whole of the free induction decay (FID), the frequencies of the MQC's result, in a 1D or 2D correlated format.

What kind of information may one hope to get by such studies? Considering the simplest spin system that can exhibit an MQC, i.e., a system of spin-1 nuclei, it may be recalled from Chapter 1 that the double-quantum coherence (DQC) in such a system is invariant to the quadrupole coupling. Although it is hopeless to attempt to measure chemical shifts in the presence of the strong quadrupole coupling in solids which splits and/or broadens single-quantum resonances, a study of the DQC enables such shifts to be discerned. It is for the same reason, i.e., the independence of the DQC frequency on the quadrupole coupling, that spin-1 nuclei in solids may be decoupled from observed spin-1/2 nuclei in spite of the immense single-quantum spectral width of the spin-1 nucleus with quadrupole coupling. This has in fact allowed proton shifts in solids to be measured, even where line-narrowing techniques have failed, by taking recourse to deuterium isotope substitution and then measuring the "dilute" proton NMR in the presence of deuterium decoupling.

The advent of pulsed techniques, furthermore, has enabled the characterization of zero-quantum coherences (ZQC's), which is truly a new result compared to the classical CW methods. The simplest spin system that can exhibit ZQC is an ensemble of two-spin-1/2 particles. Multiple-quantum spectroscopy (MQS) also allows uniquely the characterization of magnetic equivalence, without having to take recourse to multiplet structure determination. Also, observation of the MQS is a practical alternative to isotopic labeling and is, in fact, very akin to it. Synthesis is completely avoided, however, and isotopic distortions are out of the question in the NMR parameters measured.

In this chapter we review the selection rules for creation and detection of MQC's and discuss the study of spin connectivities by two-dimensional double-quantum spectroscopy, paying attention to reconversion pulse flip angle effects on the sensitivity of the experiment, on the possibility of F_1 sign determination, and on the nature of the transfer of DQC's to single-quantum coherences of "active" or "passive" nuclei. Spin connectivity studies by 2D zero-quantum spectroscopy are also mentioned and "flip angle filtering" is discussed to distinguish ZQC's from z magnetizations. The applications of multiple-quantum spectroscopy in identifying spin networks and magnetic equivalence are discussed for high-resolution studies in the liquid state. Multiple-quantum NMR studies of systems exhibiting partial ordering in liquid crystals and in the solid state are briefly treated, along with methods of selectively exciting and detecting MQC's of a given order. The chapter also includes discussion of multiple-quantum cross-polarization in solids and some recent results in multiple-quantum spin imaging.

Excitation of MQC's

Multiple-quantum coherences may be excited in a system of N coupled spin-1/2 nuclei by a pair of 90° pulses with a pulse interval on the order of the inverse of the bilinear couplings. The basic pulse sequence has been set out in Chapter 3. It can be shown quite easily, however, that this strategy fails to create MQC's in a system of equivalent spins.

Consider, for instance, a pair of equivalent spins, A_2. The NMR Hamiltonian is given by ($\mathscr{H}''$, J', Δ in radians per second):

$$\mathscr{H}'' = J'\mathbf{I}_1 \cdot \mathbf{I}_2 + \Delta(I_{1z} + I_{2z}) \tag{2}$$

whereas the equilibrium state is represented by:

$$\sigma_0 \sim (I_{1z} + I_{2z}) \tag{3}$$

Since the two terms of $\mathscr{H}''$ commute, we have:

$$\exp(\pm i\mathscr{H}''t) = \exp(\pm i\Delta(I_{1z} + I_{2z})t)\exp(\pm iJ'\mathbf{I}_1 \cdot \mathbf{I}_2 t). \tag{4}$$

At time t after the first $\pi/2$ pulse we have:

$$\begin{aligned}\sigma(t) &= \exp(-i\mathscr{H}''t)\sigma_0^+ \exp(i\mathscr{H}''t)\\ &= \exp(-i\Delta(I_{1z} + I_{2z})t)\exp(-iJ'\mathbf{I}_1 \cdot \mathbf{I}_2 t)(I_{1i} + I_{2i})\\ &\quad \exp(+iJ'\mathbf{I}_1 \cdot \mathbf{I}_2 t)\exp(+i\Delta(I_{1z} + I_{2z})t)\\ &= \exp(-i\Delta(I_{1z} + I_{2z})t)(I_{1i} + I_{2i})\exp(+i\Delta(I_{1z} + I_{2z})t)\end{aligned} \tag{5}$$

since

$$[\mathbf{I}_1 \cdot \mathbf{I}_2, (I_{1i} + I_{2i})] = 0 \ldots \qquad (i = \pm x, \pm y) \tag{6}$$

The effect of such an evolution is simply the precession of the transverse magnetization of both spins, leaving the system in a state where it can interact with but one photon upon being given a second pulse. Thus, MQC's are not created by the pulse pair.

However, it is possible to create and detect p-quantum coherence of A_N in an $A_N X$ system. Consider, for instance, a heteronuclear $I_N S$ spin system. Following a 90°_x pulse on the S spins,

$$\sigma_1 \sim -S_y \tag{7}$$

With chemical shift refocusing during the preparation period t by a 180°_x pulse pair, and with $\cos \pi J_{IS} t \to 0$, we find for N even:

$$\sigma_2 = (-)^{N/2} s_J^N 2^N \prod_{i=1}^{N} I_{iz} S_y \tag{8}$$

which is a state capable of multiphoton exchange, leading to:

$$\sigma_3 = (-)^{N/2} s_J^N 2^N \prod_{i=1}^{N} I_{iy} S_z \tag{9}$$

following a 90°_x pulse pair. Even-order coherences are thus created. For N odd, on the other hand, we find:

$$\sigma_3 = (-)^{(N+1)/2} s_J^N 2^N \prod_{i=1}^{N} I_{ix} S_z \tag{10}$$

with the evolution period terminated by a pair of 90°_y pulses, thus creating odd-order coherences. Here, $s_J = \sin \pi J_{IS} t$. Provided the coupling J_{IS} is resolvable, therefore, MQC of the eqivalent spins may be created.

Note also that heteronuclear coherences of odd order $[(N+1), (N-1), \ldots]$ are created in the first case when the second pulse on S is in quadrature with the first, while heteronuclear coherences of even order $[(N+1), (N-1), \ldots]$ are created in the second case when the second pulse on S has the same phase as the first. It is a general result that a pair of in-phase 90° pulses produce even-order coherences in a system of spin-1/2 nuclei when chemical shifts are refocused, while a pair of 90° pulses in quadrature to each other produce odd-order coherences in these conditions.

As for the reconversion process, MQC's involving a set of q spins may be transferred to single-quantum transitions of a spin A only if the scalar couplings between A and all q spins are resolved; if A belongs to the set of q spins, only the couplings to the remaining $(q - 1)$ spins need to be resolved. This implies for instance that an A_N MQC of an $A_N X$ system cannot be transferred to the A transitions. This selection rule is valid for single-pulse detection but is relaxed in situations with strong coupling.

Two-Dimensional INADEQUATE

The INADEQUATE experiment on ^{13}C gives rise to anti-phase AB or AX multiplets for every ^{13}C–^{13}C coupling. The identification of coupling partners, however, still remains ambiguous in all but the simplest molecules, since the coupling constants are very similar. Extremely accurate couplings measured from INADEQUATE spectra with high signal to noise ratios are required to make any headway at all along these lines.

The ambiguity in such spectra may be resolved, however, by performing a 2D version of the experiment where the delay Δ between the final 90° pulse of the excitation sandwich and the reconversion pulse is systematically incremented, allowing evolution of the DQC's. (Recall that this delay is fixed at about 10 μs in the one-dimensional experiment to allow rf phases to settle: Fig. 3.26.) The pulse sequence is given in Fig. 5.1. Proton decoupling is applied throughout the experiment; the phase ϕ is cycled through 0°, 90°, 180°, and 270°, and the resulting FID's are alternately added and subtracted. The evolution frequencies during t_1 are the sum of the chemical shifts of the two spins whose DQC has been excited, such evolution in a two-spin system being independent of couplings.

The F_2 dimension exhibits AB/AX patterns; different two-spin systems, however, are displaced along the F_1 axis differently, according to their common double-quantum frequency. The absolute value mode of display is generally employed. With such a two-dimensional presentation it is possible to trace the carbon skeleton of the molecule by inspection, breaks in the connectivity pattern occurring at the sites occupied by heteroatoms. The simple pulse sequence of Fig. 5.1, however, does not "enable" strong coupling connectivities with $\tau = (4J)^{-1}$.

It may be noted, moreover, that this method of generating 2D INADEQUATE spectra suffers from two shortcomings. Recall that double-quantum coherence of the y phase $(I_{1x}I_{2y} + I_{1y}I_{2x})$ is produced by the excitation sandwich. The 1D INADEQUATE strategy reconverts essentially all of this DQC into an observable signal because Δ is only about 10 μs; however, DQC precession during the evolution period t_1 in the 2D version causes appearance of the x component of the DQC $(I_{1x}I_{2x} - I_{1y}I_{2y})$, which is clearly not convertible into signal by a 90° pulse. The pulse sequence in Fig. 5.1 therefore leads to poor sensitivity in the 2D spectrum owing to the partial "loss" of DQC. At the same time, the sign of F_1 is left undetermined because the DQC of the x phase is not reconverted and detected; therefore the ^{13}C carrier frequency may

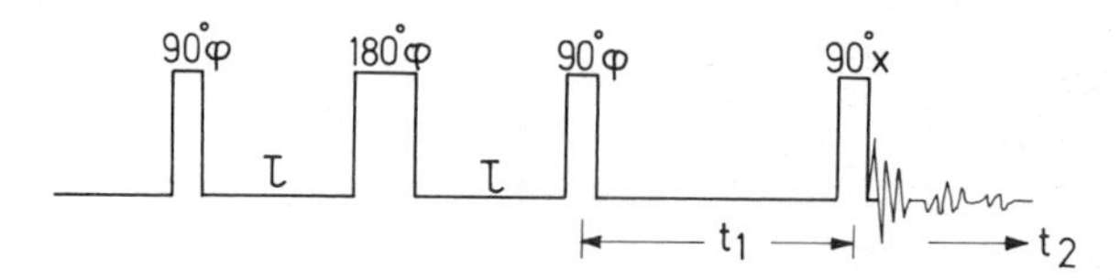

Figure 5.1. Pulse sequence for 2D INADEQUATE.

not be placed in the middle of the spectrum. To remedy these problems, flip angle effects are employed in this special case of a two-spin-1/2 system. Upon varying the flip angle θ of the reconversion pulse we find:

$$\begin{aligned} (I_{1x}I_{2y} + I_{1y}I_{2x}) &\xrightarrow{\theta_x} s_\theta(I_{1x}I_{2z} + I_{1z}I_{2x}) \\ (I_{1x}I_{2x} - I_{1y}I_{2y}) &\xrightarrow{\theta_x} -c_\theta s_\theta(I_{1y}I_{2z} + I_{1z}I_{2y}) \end{aligned} \tag{11}$$

For $\theta \neq (2n+1)\pi/2$ $(n = 0, 1, 2, \ldots)$ therefore, both the y and x components of the DQC are converted to observable single-quantum coherences.

By choosing $\theta = 135°$ for example, one achieves 20.7% sensitivity improvement over the $\theta = 90°$ case, besides retrieving the sign of F_1, with the N-type peaks selected: the N to P discrimination ratio achieved is 6 : 1 in this case. If sensitivity enhancement were the objective, on the other hand, maximal (29.9%) improvement would result for $\theta = 60°$ or 120°, compared to the situation with $\theta = 90°$.

An example of the results with such a pulse sequence ($\theta = 135°$) is shown in Fig. 5.2. An alternative, general method of extracting the sign of F_1 without taking recourse to flip angle effects is to introduce a 90° phase shift on the DQC together with a 90° phase shift of the receiver reference phase. The FID is then acquired $2N$ times for each value of t_1, N times each with and without the phase shift on the DQC. Note that a 90° phase shift of the DQC requires 45° rf phase shifting in the preparation sandwich. Such an effect may be generated by suitable rf hardware or alternatively by a composite z-pulse approach employing the result:

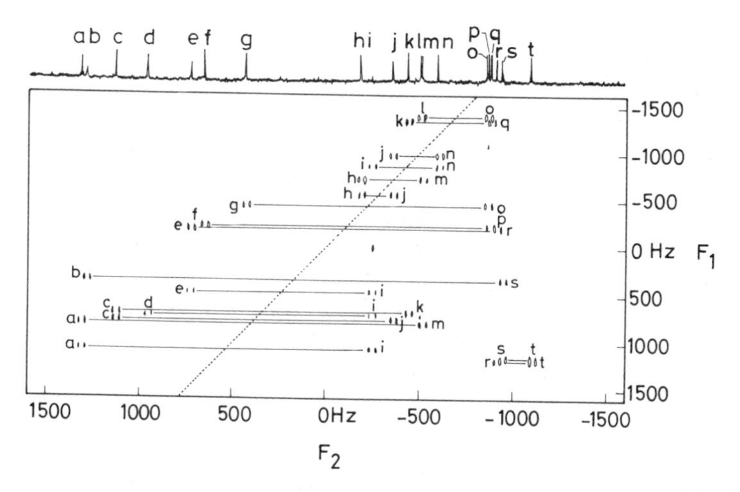

Figure 5.2. Carbon-13 2D-INADEQUATE spectrum of panamine at 50 MHz with quadrature detection in F_1, employing a 135° reconversion pulse. The carbon connectivity is apparent. [Reproduced by permission. T.H. Mareci and R. Freeman, *J. Magn. Reson.*, **48**, 158 (1982), copyright 1982, Academic Press, New York]

$$\begin{aligned} \exp(-i\varphi I_z) &= \exp\left(-i\frac{\pi}{2}I_x\right)\exp(-i\varphi I_y)\exp\left(+i\frac{\pi}{2}I_x\right) \\ &= \exp\left(+i\frac{\pi}{2}I_x\right)\exp(+i\varphi I_y)\exp\left(-i\frac{\pi}{2}I_x\right) \end{aligned} \tag{12}$$

The z-pulse may be introduced right at the commencement of the evolution period. The last pulse of the excitation sandwich may then be "merged" with the z-pulse, giving, for instance, the following sequence:

$$90^\circ_x - \tau - 180^\circ_y - \tau - 45^\circ_{-y} - 90^\circ_x - t_1 - 90^\circ_\phi - t_2 \tag{13}$$

Note that the composite z pulse really works efficiently only if radiofrequency inhomogeneity and resonance offset effects are negligible.

Two-dimensional INADEQUATE spectra of ^{13}C at natural abundance exhibit direct connectivity peaks, known as type I signals, which appear in pairs symmetrically disposed on either side of the skew diagonal $F_1 = 2F_2$. Signal intensities in the two quadrants with $F_2 > 0$ are asymmetrical in this case, as noted above.

In dealing with double-quantum spectra of protons, one encounters in addition to the direct connectivity (type I) signals, two other classes of signals, known as type II signals, arising from magnetic equivalence, and type III signals, associated with remote connectivity. Type II and type III signals do not exhibit the intensity asymmetry characteristic of type I signals.

Type II peaks are lone standing in the F_1, F_2 quadrant and intersect the skew diagonal at a chemical shift, say of A. This indicates the presence of at least two magnetically equivalent A nuclei. Such signals are symmetrical about $F_1 = 0$ regardless of the value of θ. Strong coupling effects give rise to additional signals coinciding with the skew diagonal.

Type III peaks are also lone standing in the F_1, F_2 quadrant but intersect the skew diagonal at an F_2 frequency that does not coincide with any of the chemical shifts. Such peaks arise from "remote" nuclei, not coupled together but having a common coupling partner. These peaks afford information similar to the relay correlation experiment.

One may derive information on direct and remote connectivities as well as magnetic equivalence from 2D MQ spectroscopy even without taking recourse to identification of the multiplet structure in the multiple-quantum spectrum. An instructive example is the 300-MHz ^1H double-quantum (2Q) spectrum of 3-aminopropanol, $DO{-}CH_2{-}CH_2{-}CH_2{-}ND_2$, shown in Fig. 5.3. This spectrum exhibits all three types of signals. On the basis of the discussion so far, it is clear at once from the spectrum that the spin system may be characterized as $A_aM_mX_x$, with a, m, $x \geqslant 2$ and $J_{AX} \to 0$. If any of a, m, or x were greater than 2, a triple-quantum (3Q) spectrum would show a signal at $F_1 = 3\delta_i$ and $F_2 = \delta_j$, where $i, j = A, M$, or X and $i \neq j$. In actual fact, none of the lines drawn through the centers of multiplets in the 3Q spectrum intersect the skew diagonal ($F_1 = 3F_2$) at any of the chemical shifts (of A, M, or X) in

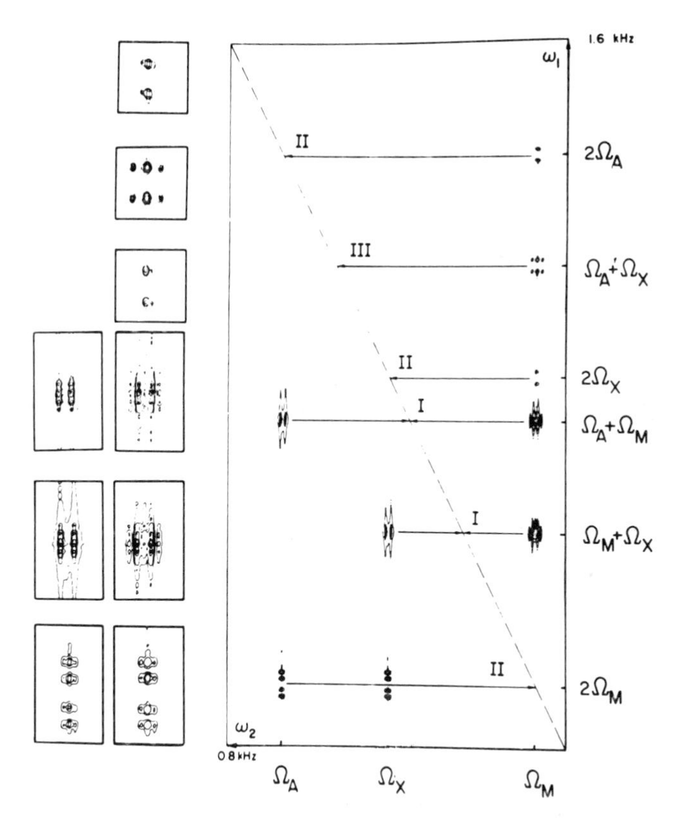

Figure 5.3. The 300-MHz proton absolute value DQS of 3-aminopropanol, $\tau = 19\ \mathrm{ms} \approx (8J)^{-1}$. [Reproduced by permission. L. Braunschweiler, G. Bodenhausen, and R.R. Ernst, *Mol. Phys.*, **48**, 535 (1983), Taylor & Francis, London]

this case. Assuming uniform excitation of MQC's therefore, this is conclusive evidence that $a, m, x < 3$. The 2Q and 3Q spectra together therefore establish, without using any multiplet structure and signal intensity information, that the spin system in question is of the $A_2M_2X_2$ type.

Transfer of MQC to passive spins giving rise to the type III remote connectivity peaks may be suppressed by suitably adjusting the flip angle of the reconversion pulse. Consider for example an AMX spin system in which AM DQC's have been excited. The AM DQC's evolve under coupling to the X spin, corresponding to a doublet structure in the AM DQ spectrum at the F_1 frequencies $(\delta_A + \delta_M) \pm (J_{AX} + J_{MX})/2$. These two AM DQC's may be represented as:

$$\mathrm{DQC}_x\text{: } (I_xS_x - I_yS_y)(1 \pm 2T_z)$$

and

$$\mathrm{DQC}_y\text{: } (I_xS_y + I_yS_x)(1 \pm 2T_z) \tag{14}$$

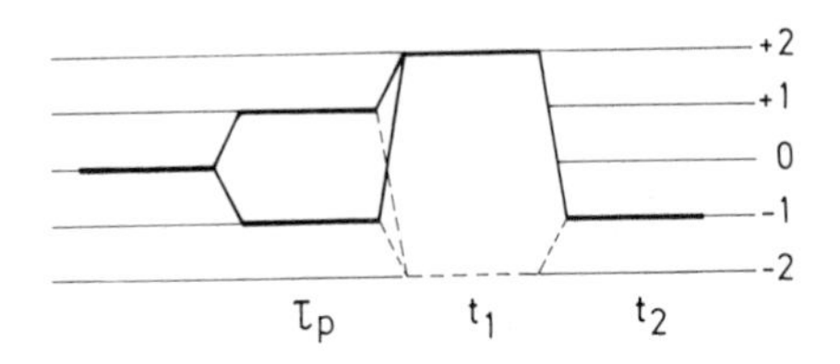

Figure 5.4. Coherence transfer pathway diagram for DQS with a three-pulse sequence.

where the spins of A, M, and X have been represented by **I**, **S**, and **T**, respectively. We may now compute the flip angle dependence for an experiment in which quadrature detection is performed in F_1 (phase shifting the DQC's by 90° employing 45° rf phase shift hardware or the composite z-pulse approach). We focus on the situation corresponding to selection of the $p = +2$ coherence level during t_1, detecting the $p = -1$ level during t_2, as shown in the coherence transfer (CT) pathway diagram given in Fig. 5.4 (the dotted pathway is suppressed by this procedure). We have:

$$\begin{aligned}
(I_xS_x - I_yS_y) &\xrightarrow{\theta_x} -c_\theta s_\theta(I_yS_z + I_zS_y) \\
2(I_xS_x - I_yS_y)T_z &\xrightarrow{\theta_x} 2s_\theta^3 I_zS_zT_y - 2c_\theta^2 s_\theta(I_yS_z + I_zS_y)T_z \\
(I_xS_y + I_yS_x) &\xrightarrow{\theta_x} s_\theta(I_xS_z + I_zS_x) \\
2(I_xS_y + I_yS_x)T_z &\xrightarrow{\theta_x} 2c_\theta s_\theta(I_xS_z + I_zS_x)T_z
\end{aligned} \tag{15}$$

where we have listed only the observable terms, with $c_\theta = \cos\theta$ and $s_\theta = \sin\theta$. From the energy level diagram of the AMX system shown in Fig. 5.5, it is clear that I_xS_z and $I_xS_zT_z$ reinforce each other if the A transitions are connected to the AM double-quantum transition in question, whereas they are opposed if nonconnected transitions are considered. We therefore have from the above expressions:

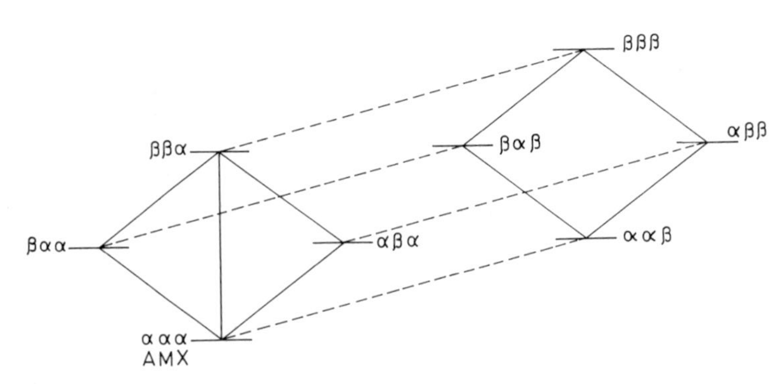

Figure 5.5. Energy level diagram for an AMX system, indicating an AM double-quantum transition.

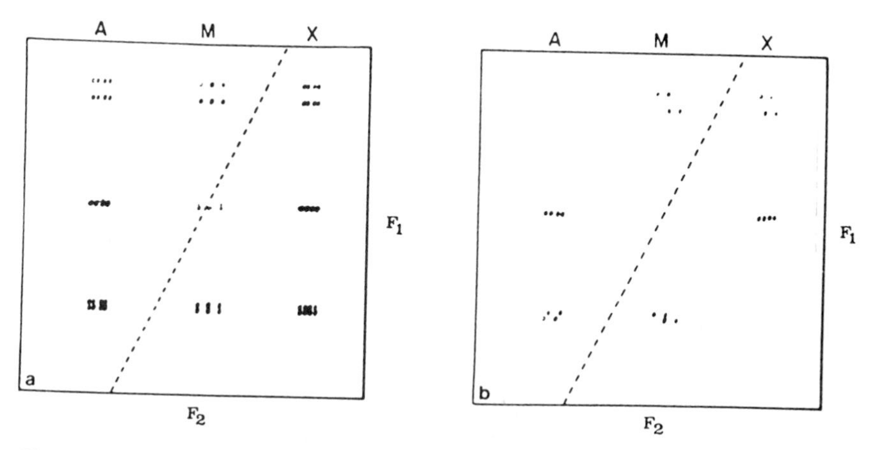

Figure 5.6. The 200-MHz proton DQS of 2,3-dibromopropionic acid, (a) with a 90° reconversion pulse and (b) with a 135° reconversion pulse. Note the significant reduction in the number of peaks in the latter case. [Reproduced by permission. T.H. Mareci and R. Freeman, *J. Magn. Reson.*, **51**, 531 (1983), copyright 1983, Academic Press, New York]

$$\begin{aligned}\text{Active spin connected peak intensity} &= -c_\theta s_\theta(1 + c_\theta) + s_\theta(1 + c_\theta)\\ &= s_\theta^3 \sim \cos^3\frac{\theta}{2}\sin^3\frac{\theta}{2}\end{aligned} \tag{16}$$

$$\begin{aligned}\text{Active spin nonconnected peak intensity} &= -c_\theta s_\theta(1 - c_\theta) + s_\theta(1 - c_\theta)\\ &= s_\theta(1 - c_\theta)^2 \sim \cos\frac{\theta}{2}\sin^5\frac{\theta}{2}.\end{aligned} \tag{17}$$

$$\text{Passive spin peak intensity} = s_\theta^3 \sim \cos^3\frac{\theta}{2}\sin^3\frac{\theta}{2}. \tag{18}$$

It is clear from the above that a 90° reconversion pulse puts equal intensities in all these classes of signals, the active and passive spin signals being in quadrature in F_2. Figure 5.6 shows 2D 2Q ^{1}H spectra of 2, 3-dibromopropionic acid, which is a nonlinear AMX network (J_{AM}, J_{MX}, $J_{AX} \neq 0$). The spectrum with $\theta = 90°$ exhibits 72 peaks with significant intensity (four nonconnected, four connected, and four passive spin transitions observed for each of the six MQC's of the nonlinear AMX system), while the spectrum with $\theta = 135°$ exhibits only 24 peaks (four nonconnected transitions for each of the six possible MQC's in the system). The ratio of connected, nonconnected, and passive spin intensities is computed from the above to the 0.04 : 0.26 : 0.04 for this flip angle. Inevitably, the suppression of transfer to passive spins does not work as well with strongly coupled spin systems.

It should be emphasized, however, that while simplifying the spectrum by suppressing the remote connectivity peaks, we are also losing valuable infor-

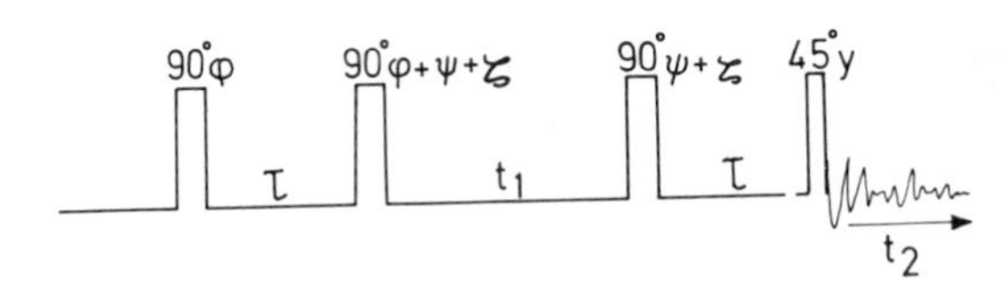

Figure 5.7. Pulse sequence for spectral editing of proton DQS.

mation contained in these peaks. A uniform excitation–detection strategy with a purge prior to acquisition, on the other hand, allows one to obtain absorptive in-phase multiplets in both F_1 and F_2. This gives positive peaks for direct connectivities and negative peaks for remote connectivities as well as magnetic equivalences, so that spectral editing into these classes may be achieved. The pulse sequence shown in Fig. 5.7 may be employed.

For odd-quantum excitation $\psi = \pi/2$, while for even-quantum excitation $\psi = 0$. Chemical shift effects during τ average out magnetizations starting and ending at different spins: this process is much more efficient than averaging dependent on couplings. For magnetizations starting and ending on the same spin, the chemical shift dependence of the multiquantum excitation efficiency is avoided by subtracting experiments with $\zeta = 0$ and $\pi/2$.

One context in which suppression of "remote" transfer is useful is the MQ relay experiment. This multiple-quantum version of the heteronuclear relay experiment is useful when the single-quantum relay spectrum is ambiguous owing to chemical shift degeneracies of remote and neighbor protons. In this experiment, ^{13}C shifts are correlated with ^{1}H DQ frequencies. Remote transfer of proton MQC's, which is not of interest here, is avoided by the flip angle effect discussed above, employing a reconversion pulse of 45°. The pulse sequence employed is shown in Fig. 5.8 (see also Table 5.2).

One may also design a DQ NOESY experiment where DQ frequencies evolve during t_1, followed by reconversion to *longitudinal order*, cross-relaxation during the mixing period, and reconversion to SQC's for acquisition during t_2. Reconversion of DQC's into longitudinal order may be achieved with a $90°-\tau-180°-\tau-90°$ pulse sandwich once again.

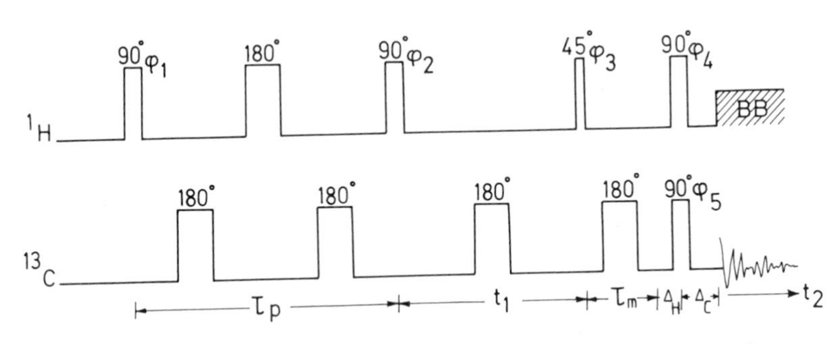

Figure 5.8. Pulse sequence for heteronuclear multiple-quantum relay.

Table 5.2. Phase Cycling for a Heteronuclear Multiple-Quantum Relay

ϕ_1	ϕ_2	ϕ_3	ϕ_4	ϕ_5
x	y	x	x	x
y	$-x$	$-x$	x	x
$-x$	$-y$	x	x	x
$-y$	x	$-x$	x	x
x	y	x	y	y
y	$-x$	$-x$	y	y
$-x$	$-y$	x	y	y
$-y$	x	$-x$	y	y
x	y	x	$-x$	$-x$
y	$-x$	$-x$	$-x$	$-x$
$-x$	$-y$	x	$-x$	$-x$
$-y$	x	$-x$	$-x$	$-x$
x	y	x	$-y$	$-y$
y	$-x$	$-x$	$-y$	$-y$
$-x$	$-y$	x	$-y$	$-y$
$-y$	x	$-x$	$-y$	$-y$

Heteronuclear Multiple-Quantum Spectroscopy (HMQS)

A pulse sequence suitable for the creation of heteronuclear multiple-quantum coherence is shown in Fig. 5.9. Selection of even- or odd-order MQC's is achieved by the 180° pulse during preparation, depending on the relative phase of the two 90° pulses that flank it: the order p now refers to $(p^I + p^S)$. A heteronuclear coherence transfer echo is also included to detect a particular MQC selectively depending on the choice of Δ_1 and Δ_2:

$$\Delta_2 = \Delta_1(\gamma_S p^S + \gamma_I p^I)/\gamma_I \tag{19}$$

if I magnetization is detected.

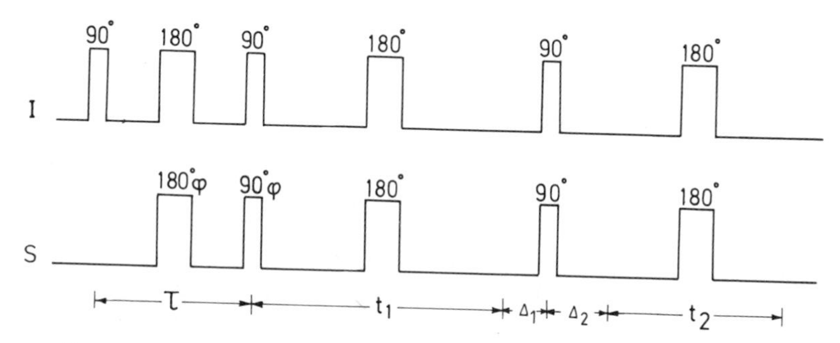

Figure 5.9. Pulse sequence for heteronuclear MQS, for example, for "indirect" excitation/detection of ^{14}N DQS in solution, via nonvaishing $^1J_{NH}$.

The S-spin preparation pulses are phase incremented by ϕ on successive scans (by time proportional phase incrementation, or TPPI) to label the MQC produced, $p = (p^I + p^S)$, with the order p^S. Separation of orders with respect to both p^I and p^S in general requires independent preparation phase incrementing on both the I- and S-spin channels.

This sequence has been employed to obtain the ^{14}N DQC ($p^I = 0, p^S = 2$) of the I_4S system, $[NH_4]^+$, in isotropic solution where irradiation of ^{14}N alone fails to create ^{14}N DQC owing to the fact that the quadrupole coupling vanishes in solution.

For the preparation sequence indicated in the figure, the time evolution operator (also called propagator or time displacement operator) may be written:

$$\begin{aligned} U(\tau) &= \exp\left(-i\frac{\pi}{2}(I_x + S_x)\right)\exp(-i\mathscr{H}\tau/2)\exp(-i\pi(I_x + S_x)) \\ &\quad \times \exp(-i\mathscr{H}\tau/2)\exp\left(-i\frac{\pi}{2}I_x\right) \\ &= \exp(-2\pi i\tau J I_y S_y)\exp\left(-3i\frac{\pi}{2}S_x\right) \end{aligned} \tag{20}$$

for an even number of equivalent spin-1/2 nuclei, I_i. The density operator at the end of the preparation therefore takes the form:

$$\begin{aligned} \sigma(\tau) &= U(\tau)\sigma(0)U^{-1}(\tau) \\ &= \exp(-2\pi i\tau J I_y S_y) I_z \exp(+2\pi i\tau J I_y S_y) \\ &= I_z(1 + S_y^2(\cos 2\pi J\tau - 1)) + I_x S_y \sin 2\pi J\tau \end{aligned} \tag{21}$$

for $S = 1$.

The preparation sequence has thus created double-quantum coherence, $I_z S_y^2$, of the spin-1 nucleus through the resolved heteronuclear coupling, although quadrupole coupling of the spin-1 nucleus vanishes in isotropic solution. In addition, the sequence has also created heteronuclear ZQC and DQC, $I_x S_y$. With the choice $\tau = (2J)^{-1}$, spin-1 DQC is maximal and heteronuclear MQC vanishes. Note the presence of longitudinal I-magnetization at the end of τ. The DQC evolves during t_1 and is converted to I-spin magnetization by the 90° pulse pair. The delays Δ_1 and Δ_2 are chosen such that:

$$2\gamma_S \Delta_1 = \gamma_I \Delta_2 \tag{22}$$

whereby the coherence transfer echo is observed on the proton channel at $t_2 = \tau$. This strategy is useful in filtering large extraneous signals that are not S-spin DQC in origin. The TPPI procedure allows field inhomogeneity effects to be suppressed while retaining separation of MQ orders. Figure 5.10 shows that proton-detected ^{14}N DQS of 8 M NH_4NO_3 in acidified aqueous solution,

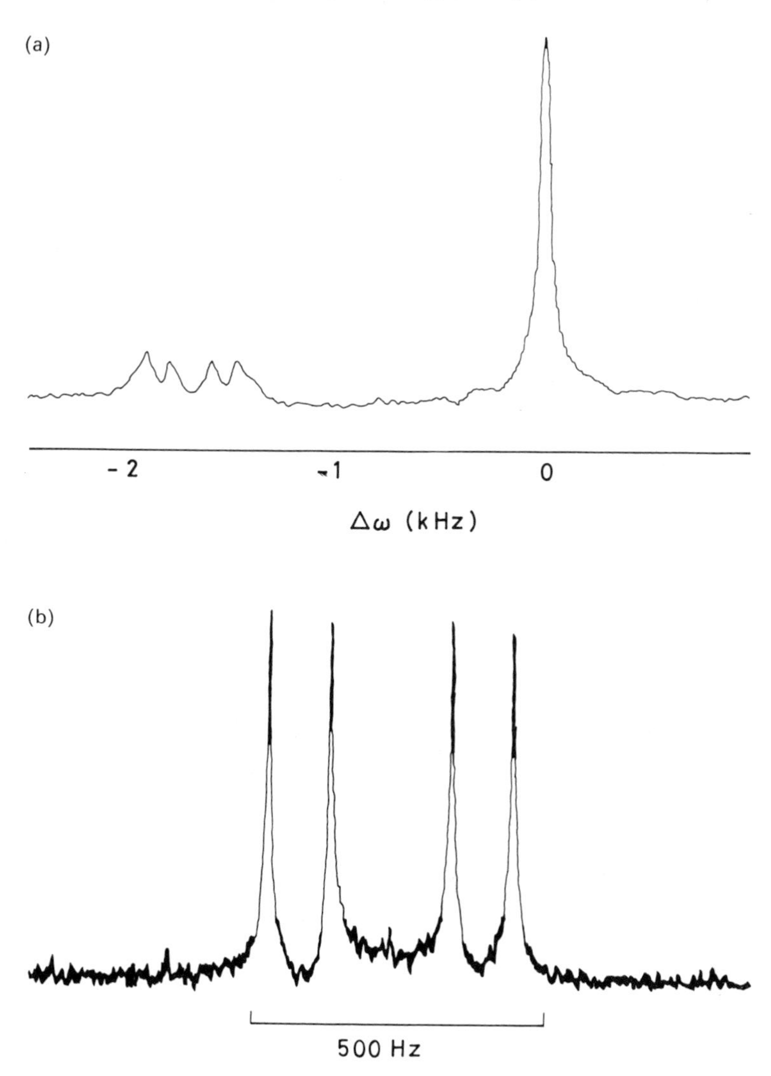

Figure 5.10. (a) The 185-MHz proton-detected ^{14}N DQS of 8 M NH_4NO_3 in acidified aqueous solution; $\tau = t_2 = 9.6$ ms; no TPPI/t_1 refocusing; ^{14}N carrier 0.85 kHz off resonance. $\Delta t_1 = 200$ μs; $\Delta_1 = 11.327$ ms; $\Delta_2 = 1.618$ ms. The signal at 0 Hz arises predominantly from longitudinal H_2O magnetization present during t_1. (b) Same as (a) but with TPPI/t_1 refocusing; $\tau = t_2 = 10$ ms; ^{14}N and ^1H carrier on resonance. [Reproduced by permission. Y.S. Yen and D.P. Weitekamp, *J. Magn. Reson.*, **47**, 476 (1982), copyright 1982, Academic Press, New York]

with and without TPPI/t_1 refocusing. The spectrum exhibits the (1, 1, 0, 1, 1) quintet expected, with splittings $2J$.

Zero-Quantum Spectroscopy. Among MQC's of various orders, ZQC's have certain special properties that put them in a class of their own. Homonuclear ZQC's may be created instantly in an incoherent spin system by suddenly changing from one high-field spin Hamiltonian to another, e.g., in CIDNP (chemically induced dynamic nuclear polarization) situations. Also, because homonuclear ZQC's are insensitive to field inhomogeneities they may be detected selectively in the presence of other MQC's with the aid of field gradient pulses followed by detection processes. In fact, it is even possible to do this in a 1D manner by employing small angle detection pulses ($1°$ to $2°$) at regular intervals preceded by gradient pulses: thus, continued evolution of the bulk of the ZQC is possible while the mixing and detection processes are occurring as well!

In weakly coupled homonuclear spin systems, the pulse sandwich for shift-independent excitation of even-order MQC's, i.e., 90°_ϕ–$\frac{\tau}{2}$–180°_ϕ–$\frac{\tau}{2}$–90°_ϕ, fails to excite ZQC's. It turns out that couplings as well as shifts have to be active in such systems to produce ZQC's by a $90°$ pulse pair. The $180°$ refocusing pulse must therefore be omitted, or at least displaced from the middle of the preparation period, to produce ZQC's by this method. This, however, makes the excitation efficiency dependent on the shifts. After evolution for the variable period t_1, detection may be effected by a third $90°$ pulse. The phase of the 90°_ϕ–τ–90°_ϕ sandwich is incremented by $180°$ and the FID's are added to get rid of the contributions from odd-order coherences in t_1; z magnetization, ZQC, and DQC remain during t_1 for a two-spin system.

In order to suppress the DQC contribution, ϕ is incremented by $\frac{\pi}{2}$ and $\frac{3\pi}{2}$ and the resulting FID's are also added. If the flip angle of the detection pulse is set to $45°$ and in a subsequent scan to $135°$, one may distinguish between z magnetization and ZQC's by subtracting the resultant FID's, resulting in suppression of the signals at $F_1 = 0$ originating from z magnetization. This procedure may be termed flip angle filtering, because it employs the characteristic flip angle dependence of ZQC conversion to SQC, which is a different functional dependence compared to that for longitudinal magnetization. Note that this is the only means to distinguish homonuclear ZQC's from z magnetizations, because no phase cycling will work in this case, both homonuclear ZQC's and z magnetizations being invariant to phase shifts.

In isotropic solution, zero-quantum transitions (ZQT's) occur only between levels belonging to the same irreducible representation. The frequencies of ZQT's involve differences of shifts and differences of couplings. For example, a three-spin-1/2 system, AMX, exhibits three doublets centered at the three shift differences, the splitting being $|J_{AX} - J_{BX}|$ for the doublet at $(\delta_A - \delta_B)$. The relative signs of couplings may be evaluated by a comparison of the normal single-quantum spectrum (SQS) which involves the sum of J's and the Zero-quantum spectrum (ZQS), which involves their differences. Note also

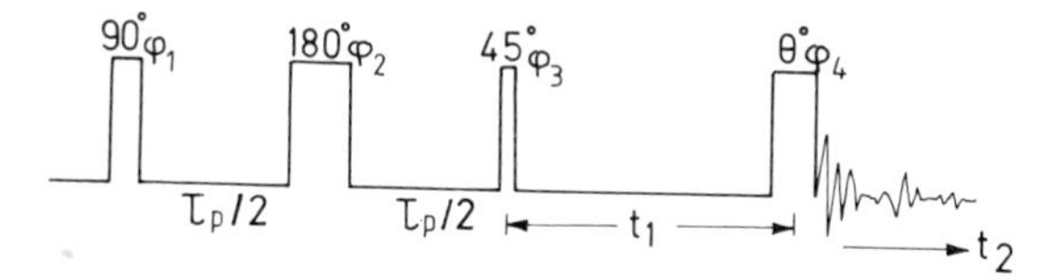

Figure 5.11. Pulse sequence for producing homonuclear zero-quantum spectra with uniform excitation.

Table 5.3. Phase Cycling for ZQS without Homospoil

ϕ_1	ϕ_2	ϕ_3	ϕ_4	ϕ_R
x	y	y	y	x
x	y	y	$-x$	y
x	y	y	$-y$	$-x$
x	y	y	x	$-y$
x	$-y$	y	y	x
x	$-y$	y	$-x$	y
x	$-y$	y	$-y$	$-x$
x	$-y$	y	x	$-y$
x	y	$-y$	y	$-x$
x	y	$-y$	$-x$	$-y$
x	y	$-y$	$-y$	x
x	y	$-y$	x	y
x	$-y$	$-y$	y	$-x$
x	$-y$	$-y$	$-x$	$-y$
x	$-y$	$-y$	$-y$	x
x	$-y$	$-y$	x	y

that the ZQS exhibits, in general, fewer lines than the SQS. For example, a four-spin system with no equivalences exhibits 56 transitions in the SQS but only 27 in the ZQS. This feature often helps simplify spectral analysis.

A modified method for producing ZQ spectra employs the pulse sequence shown in Fig. 5.11. By this means, a more uniform excitation of ZQC's is possible since the shift dependence is removed; density operator terms of the form I_xS_z at the end of τ_p are converted partly by the final pulse into I_xS_x, which contains the desired ZQC. The magnetizations of uncoupled spins are unaffected by this pulse. The phase cycling scheme of Table 5.3 suppresses unwanted signals with the exception of residual Zeeman polarization arising from pulse imperfections and terms corresponding to longitudinal two-spin order. If a homogeneity spoil pulse is applied at $t_1 = 0$, on the other hand, the following phase cycling is adequate (Table 5.4). If the reconversion pulse at the end of t_1 has a flip angle $\theta \neq (2n + 1)\frac{\pi}{2}$, we again encounter familiar effects from DQS: transfer to passive spins is suppressed and transfer to one quadrant of the 2D spectrum results preferentially, allowing quadrature detection in F_1.

Table 5.4. Phase Cycling for ZQS with Homospoil

ϕ_1	ϕ_2	ϕ_3	ϕ_4	ϕ_R
x	y	y	y	x
x	$-y$	y	y	x
x	y	$-y$	y	$-x$
x	$-y$	$-y$	y	$-x$

This similarity may be understood at once upon recalling the form of the ZQC's:

$$\begin{aligned} ZQC_x &= (I_x S_x + I_y S_y) \\ ZQC_y &= (I_y S_x - I_x S_y) \end{aligned} \tag{23}$$

which implies that, like DQC_x, ZQC_x is also not converted into signal by a 90° pulse, in homonuclear spin systems. The intensity ratio of a peak to its mirror image in F_1 is once again given by $\tan^2 \theta/2$, which is about 6 : 1 for $\theta = 45°$ or 135°. Transfer to passive spins is again proportional to $\sin^3 \frac{\theta}{2} \cos^3 \frac{\theta}{2}$; that to active spins is about six times stronger for $\theta = \frac{\pi}{4}$ and there is once again a 20% improvement in the intensity of transfer to active spins, compared with the $\theta = 90°$ situation. An example of a 2D ZQS with $\theta = 45°$ is given in Fig. 5.12. Note the appearance of the spectrum, which is reminiscent of SECSY. The peaks at $F_1 = 0$ originate in longitudinal multispin order and pulse imperfections. Note that the choice of τ_p may be made to emphasize only direct connectivities, corresponding to large couplings, so that for ^{13}C studies, information of the carbon skeleton is once again available, as with INADEQUATE. Recalling that magnet inhomogeneities do not affect linewidths in F_1 and that frequency *differences* are active during t_1, allowing a narrower spectral width in F_1, this ZQS technique may yet turn out to be an ADEQUATE alternative to INADEQUATE!

Heteronuclear zero-quantum coherences may be excited by the pulse sequence shown in Fig. 5.13. The phase ϕ is cycled through 0°, 90°, 180°, and 270° and the resulting FID's added. This phase cycling insures that the acquired FID relates only to the $(ZQC)_y$ created at the end of the preparation period.

Consider, for example, the case $I = {}^1H$, $S = {}^{31}P$. The frequency of the heteronuclear zero-quantum signal in the absence of ^1H–^1H couplings is the difference between the ^1H and ^{31}P offsets. Linewidths in the ZQS method will be 60% of those in the ^1H–^{31}P heteronuclear correlation experiment, since the ZQC operative in t_1 corresponds to $\gamma_{eff} \approx 0.6\gamma_{1_H}$. The resulting ZQ spectrum corresponds to the ^{31}P-decoupled proton spectrum if only one proton is attached to ^{31}P. The technique is thus of value in getting simplified proton spectra indirectly.

MQS in CIDNP Experiments. In CIDNP situations, the nuclear spin system

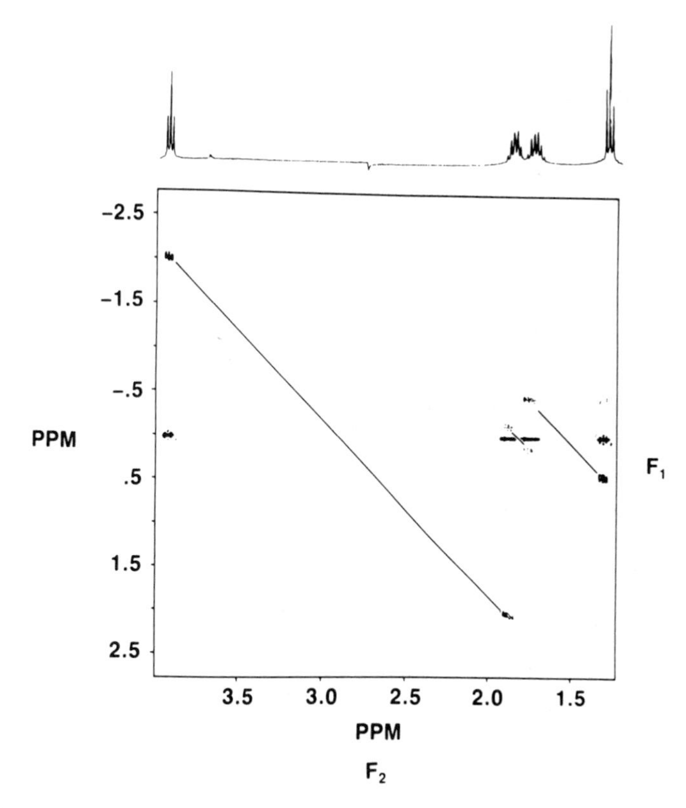

Figure 5.12. The 360-MHz proton ZQS of *n*-butanol; $\theta = 45°$. [Reproduced by permission. L. Müller, *J. Magn. Reson.*, **59**, 326 (1984), copyright 1984, Academic Press, New York]

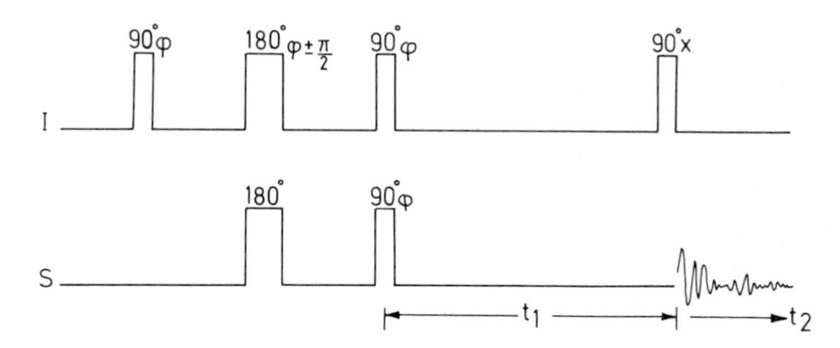

Figure 5.13. Pulse sequence for heteronuclear ZQS.

comes prepared in a nonequilibrium state of the first kind; i.e., the density operator is diagonal, but the diagonal elements (i.e., populations P) do not correspond to Boltzmann equilibrium. A two-spin-1/2 (IS) system in this state may be represented by:

$$\begin{aligned}\sigma_0 = \tfrac{1}{2}[&(P_{\alpha\alpha} + P_{\alpha\beta} - P_{\beta\alpha} - P_{\beta\beta})I_z + (P_{\alpha\alpha} - P_{\alpha\beta} + P_{\beta\alpha} - P_{\beta\beta})S_z \\ &+ 2(P_{\alpha\alpha} - P_{\alpha\beta} - P_{\beta\alpha} + P_{\beta\beta})I_zS_z + \tfrac{1}{2}(P_{\alpha\alpha} + P_{\alpha\beta} + P_{\beta\alpha} + P_{\beta\beta})\mathbf{1}]\end{aligned} \tag{24}$$

If a θ_x pulse is applied, we have:

$$\begin{aligned}\sigma_1 = \tfrac{1}{2}[&(P_{\alpha\alpha} + P_{\alpha\beta} - P_{\beta\alpha} - P_{\beta\beta})(c_\theta I_z - s_\theta I_y) \\ &+ (P_{\alpha\alpha} - P_{\alpha\beta} + P_{\beta\alpha} - P_{\beta\beta})(c_\theta S_z - s_\theta S_y) \\ &+ 2(P_{\alpha\alpha} - P_{\alpha\beta} - P_{\beta\alpha} + P_{\beta\beta})(c_\theta^2 I_zS_z + s_\theta^2 I_yS_y \\ &- c_\theta s_\theta(I_zS_y + I_yS_z))]\end{aligned} \tag{25}$$

Clearly, if $\theta = 90^\circ$, natural multiplets alone are obtained, regardless of the initial state. On the other hand, if θ is small ($\cos\theta \to 1$), the Fourier transform of the resulting FID produces a spectrum equivalent to the CW response, because all product operators including the anti-phase terms now contribute to the signal. However, the loss in sensitivity should be noted.

At the same time, however, the θ pulse produces ZQC and DQC which is maximal for $\theta = \pi/2$. Provided t_1 is kept short enough to avoid relaxation, therefore, a 2D experiment may be performed, with a reconversion pulse at the end of t_1. The signals transferred from MQC by the second $\pi/2$ pulse are of the form:

$$(I_zS_x - I_xS_z)\sin\Delta_0 t_1 - (I_zS_x + I_xS_z)\sin\Delta_2 t_1 \tag{26}$$

where Δ_i's represent the precessional frequencies of iQC's. Thus, 90° pulses may be employed in the sequence to achieve maximum sensitivity, while retaining the CIDNP information through MQC's in t_1.

Selective Detection of Various Orders of pQC's

It is clear from Table 5.1 that a plethora of transitions show up in MQS, making it necessary to separate them according to their order. Note that in the rotating frame the various orders all have a similar spectral range, so the experimenter must employ special tricks to achieve this separation. One simple strategy would be to shift the phase-sensitive detector reference frequency off resonance by $\Delta\omega$. The pQC's are shifted, then, by $p\Delta\omega$. To avoid overlaps of adjacent orders $\Delta\omega$ should be of the order of the SQS spectral width. This would require resonant pulses for effective excitation at a frequency *different* from the phase sensitive detector (PSD) reference and also demands that digitization be at least twice as fast as otherwise; most important, however, MQ refocusing during the evolution period t_1 would cause failure of the strategy.

As described in Chapter 4, the rf phase of the preparation period labels the

Table 5.5. Order Selection of MQC's by Phase Fourier Transformation

Preparation phase for co-added FID's	Observed orders, p									
0°	0	1	2	3	4	5	6	7	8	9
0° 180°	0		2		4		6		8	
0° $\overline{180^\circ}$		1		3		5		7		9
0° 90° 180° 270°	0				4				8	
0° $\overline{90^\circ}$ 180° $\overline{270^\circ}$			2				6			
0° $\overline{60^\circ}$ 120° $\overline{180^\circ}$ 240° $\overline{300^\circ}$				3						9

resulting coherences with distinct, order-dependent phase factors. When the preparation is phase cycled, the resulting phase-dependent modulation of the signals may be sorted out orderwise by a phase Fourier transform (PFT), as pointed out in the preceding chapter. Table 5.5 summarizes the possibilities.

The FID's resulting from preparations with a phase indicated with a bar, $\bar{\phi}$, should be subtracted. The phase ϕ should be implemented only on the nonselective 90° pulse when dealing with nonequilibrium states of the first kind prepared for example by CIDNP or by selective π pulses. When employing methods involving nonequilibrium states of the second kind, on the other hand, the phase shift should be applied to both nonselective 90° pulses as well as the refocusing 180° pulse if shift-insensitive excitation is employed.

A powerful alternative method of such phase sorting of orders is the time proportional phase incrementation method (TPPI). In this scheme, the preparation phase is set to $\phi = \Delta\omega t_1$; in other words, it is incremented in proportion to the increment in the evolution period: $\Delta\phi = (\Delta\omega)(\Delta t_1)$. A phase factor of $\exp(+ip\Delta\omega t_1)$ is thus introduced and separation by order p follows straightway upon the usual FT with respect to t_1. It may be noted that refocusing during t_1 does not affect this separation of orders, since the $\Delta\omega$-dependent phase shift of the MQC's is an effect of the *preparation* period and does not involve the *evolution* period t_1. The required phase shifts, π/p_{max} or smaller, may be generated: (1) by digital synthesis, (2) by switching delay lines, (3) by the less accurate composite z pulse method, or (4) by phase-coherent frequency switching of the spectrometer reference by $\Delta\omega$ for a duration Δt during t_1, such that $\Delta\phi = (\Delta\omega)(\Delta t)$. Figure 5.14 shows an MQ spectrum of oriented benzene with the orders separated by TPPI.

Coherence Transfer Echoes. Consider a pQC dephasing under magnet inhomogeneities during t_1; at the end of t_1, a mixing pulse or mixing sequence effects coherence transfer from the pQC to other coherences. The frequency of precession before and after mixing is therefore different in general, but there are components whose *sense* of precession in the rotating frame is *reversed* as well. Such situations lead to echoes, which may be termed coherence transfer echoes. Notice that this is really a generalized spin echo phenomenon: the spin echoes that were described in Chapter 2 involved no change in the precession

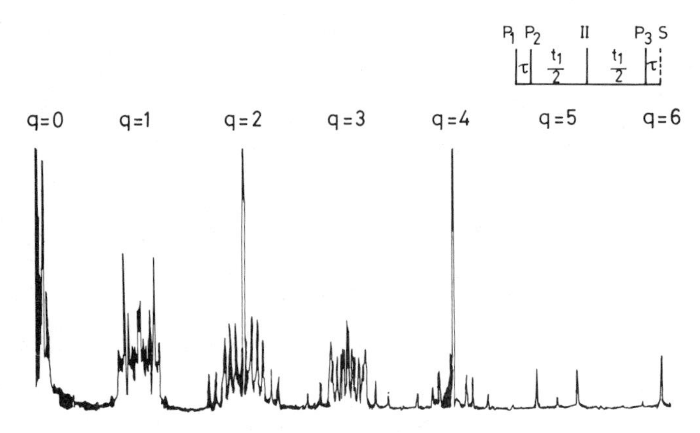

Figure 5.14. The 185-MHz proton MQS of benzene dissolved in a liquid crystal, measured at 22°C, with the TPPI procedure for order separation. $P_1 = 90^\circ_\varphi$, $P_2 = 90^\circ_{\bar{\varphi}}$, $P_3 = 90^\circ_x$, $\varphi = \Delta\omega t_1$; $\Delta\varphi = 29.5^\circ$, $\Delta t_1 = 10\ \mu s$. 1024 values of t_1 were used and the magnitude spectra were added in each case for eight values of τ between 9 and 12.5 μs. The magnetization was sampled at $t = \tau$. [Reproduced by permission. G. Drobny, A. Pines, S. Sinton, D.P. Weitekamp, and D. Wemmer, *Symp. Faraday. Soc.*, **No. 13**, 49 (1979), copyright 1979, the Royal Society of Chemistry, London]

frequency before and after the refocusing pulse (barring diffusion effects), but merely involved reversal of the sense of precession in the rotating frame. When the order of the coherence is not changed refocusing occurs, leading to an echo maximum after a time equal to the dephasing time; a coherence transfer echo, on the other hand, is retarded or advanced in time depending on the change in coherence order, $\Delta|p|$. When a pQC that has dephased for t_1 seconds refocuses as an SQC, for example, the CT echo occurs pt_1 seconds after mixing, since pQC is p times as sensitive to $\mathbf{B}_0$ inhomogeneities as SQC and so dephases p times more rapidly. Recall that in many 2D experiments described in this chapter, as well as the last, phase cycling was performed to select those coherence transfer components that reverse their sense of precession in going from the evolution to the detection period in order to take advantage of this refocusing effect. This procedure was termed selection of N-type peaks.

It may be noted that 180° pulses have the special property of leaving the magnitude of the coherence order unchanged, changing only its sign. In heteronuclear IS spin systems, therefore, a π pulse on one of the spins interchanges heteronuclear ZQC and DQC.

Coherence transfer echoes may be employed as a means of filtering to separate MQC orders. Consider the pulse sequence shown in Fig. 5.15a. Multiple-quantum coherences created at the end of τ evolve during t_1; de phasing and offset effects are refocused at the end of the evolution period; a further constant dephasing period T follows, before mixing to SQC's; the signal is acquired at a point of time $t_2 = pT$, and the experiment is repeated for incremented evolution periods t_1, to get the pQS selectively, suppressing

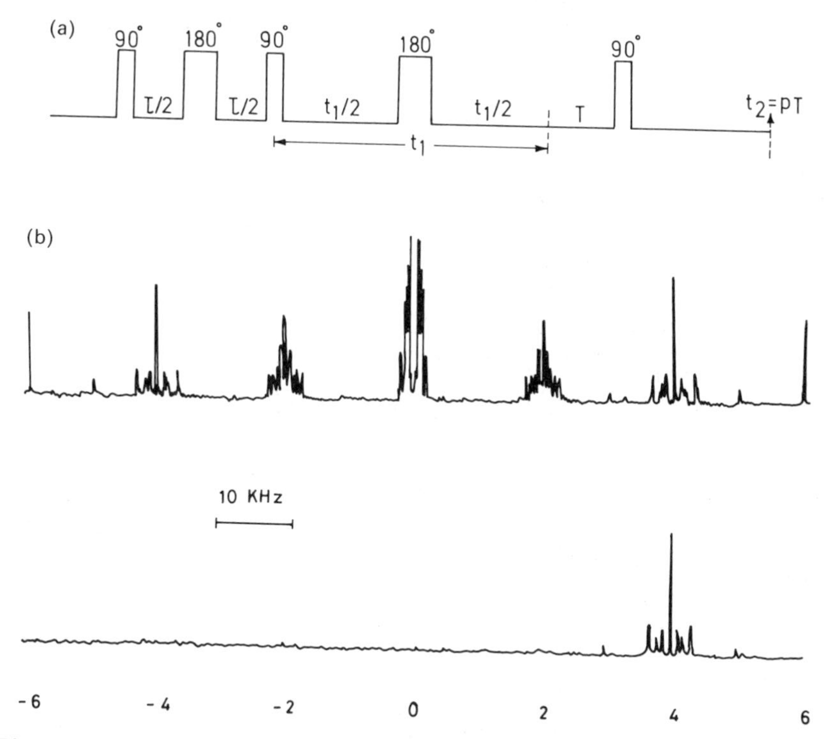

Figure 5.15. (a) Pulse sequence for coherence transfer echo filtering to separate MQC orders. (b) The CTEF spectrum of benzene in a liquid crystal, for four-quantum selection, following even-quantum excitation. [Reproduced by permission. D.P. Weitekamp, *Adv. Magn. Reson.*, **11**, 111 (1983), copyright 1983, Academic Press, New York]

other MQ spectra. An example is given in Fig. 5.15b. This experiment effectively samples only the CT echo maximum arising from pQC. There is therefore no need for separation of orders by PFT or TPPI: this reduces the minimum experiment time and data storage required. Such CT echo filtering (CTEF) methods always result in loss of signal, however, because signal is suppressed for most values of p and t_2. In homogeneous fields, a field gradient may be switched on during T and pT to achieve the same order selectivity, and the π pulse at $t_1/2$ may be omitted when working on resonance with systems with a single shift. The full FID may be acquired, beginning at $t_2 = pT$.

It may be mentioned, however, that whereas such CTEF methods result in loss of signal, multiplicative t_1 noise is also minimized, because extraneous transitions are suppressed during detection.

A special coherence transfer echo experiment is the so-called TSCTES, total spin coherence transfer echo spectroscopy, which may be used to record HR spectra in inhomogeneous fields. In this experiment, total spin coherence (NQC) of the N-spin system is prepared and allowed to evolve for a time pt_1/N, at the end of which a mixing sequence transfers the NQC to pQC; the CT echo occurs at t_1 after this mixing, whereupon further mixing creates observable signals if $p \neq 1$. The method retains full sensitivity to chemical shifts among coupled spins and generates any pQS with properly phased lines and predictable intensities. At the same time, strong coupling effects do not complicate the spectrum by producing additional transitions. The method relies on the fact that in the resonant rotating frame (i.e., sum of chemical shifts is zero) the total spin coherence evolves solely under magnet inhomogeneity effects and is independent of shifts and couplings.

A pulse sequence suitable for single-quantum TSCTES spectra is shown in Fig. 5.16. The echo maximum is sampled at time $t_2 = 0$; FT with respect to t_1 generates the homogeneous SQS. The TSC evolution for time t_1/N is refocused with the echo maximum at the end of t_1, SQC's having been created at the end of TSC evolution. The resulting spectrum has transitions at the same frequencies and with the same resolution obtainable in absolutely homogeneous fields. Figure 5.17 shows the 200-MHz ^{1}H SQS of a nonspinning

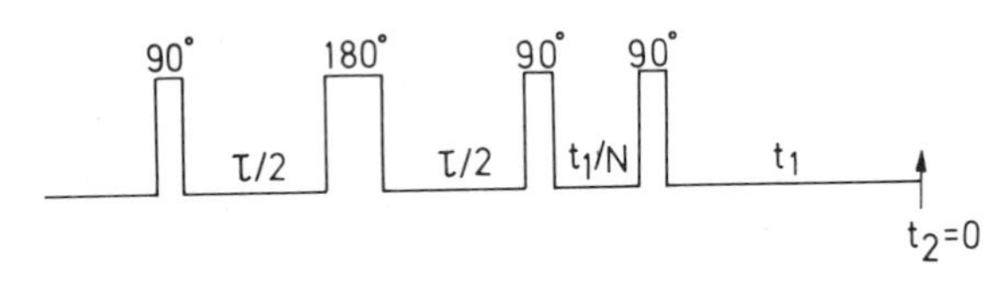

Figure 5.16. Pulse sequence for single-quantum TSCTES.

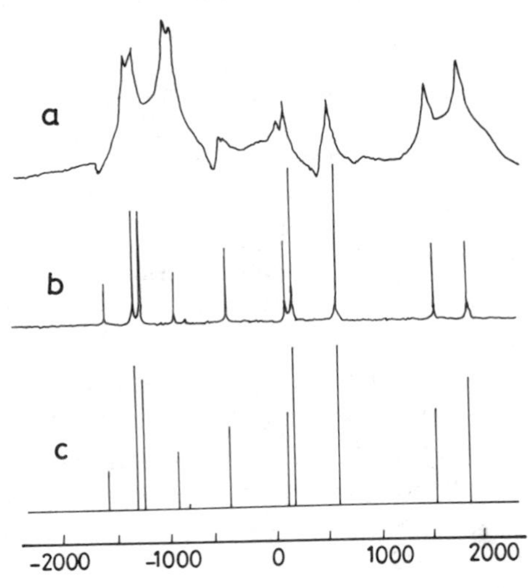

Figure 5.17. The 200-MHz proton NMR spectrum of a nonspinning 10-mm sample of partially oriented acetaldehyde. Note the vast improvement in resolution afforded by the TSCTES spectrum in comparison to the normal spectrum. [Reprinted with permission. D.P. Weitekamp, J.R. Garbow, J.B. Murdoch and A. Pines, *J. Amer. Chem. Soc.*, **103**, 3578 (1981). Copyright (1981) American Chemical Society]

10-mm sample of partially oriented acetaldehyde. Even though the magnet inhomogeneity is worse than 1 ppm, 0.02-ppm linewidths result with TSCTES. In this case, $N = 4$.

It should be remarked that although TSCTES retains the chemical shift differences within a group of coupled spins, shift differences between isolated systems cannot be measured. Therefore, chemical shift referencing to a standard cannot be performed with TSCTES. Also, the TSCTES lines always belong to the totally symmetric (A_1) representaton, since the observed coherence originates in total spin coherence, which is totally symmetric! Transitions of uncoupled subsystems, as well as those belonging to representations other than the totally symmetric one, are missed in the TSCTES experiment. In fact, CT from TSC and $(N - 1)$QC in an N-spin-1/2 system is symmetry selective, giving rise to SQT's of A_1 symmetry alone.

Although the method is in principle applicable to spin systems in liquids, liquid crystals, or solids, the efficiency of excitation of the total spin coherence for more than just a few spins is passable only with oriented molecules.

Order-Selective Excitation of MQC's

It is generally impractical to attempt to calculate the phases and intensities in MQS under various preparation and mixing sequences. Consider, for example, a system of strongly coupled spins (such as in oriented molecules), subject to the three-90°-pulse sequence. Practically all single-quantum frequencies contribute to the excitation dynamics during the preparation period τ. Consequently, any given line in the MQS is found in practice to occur with wide variations in phase and amplitude as a functon of τ. This suggests that a statistical model may be adopted, treating $\sigma_{ij}(\tau)$, the elements of the density operator, as random variables, with τ as the parameter specifying different events. On this model, equal fractions of the equilibrium magnetization detected in t_2 on the average are found oscillating in t_1 at any frequency ω_{ij}. This simple system-independent model serves to emphasize that most of the magnetization appears in the numerous low-order transitions when such nonselective means of excitation are used. The interesting and interpretable spectra of higher order are therefore excited very inefficiently. Selective *detection*, as discussed earlier, is clearly to be preceded or even substituted by selective *excitation*, in order to remedy the situation. We discuss below in brief some approaches to selective excitation.

Consider a sequence of rf irradiation subcycles, each of period $\Delta\tau_p$, n successive subcycles being phase shifted by $k2\pi/n$ ($k = 0, 1, 2, 3, \ldots$) to form a cycle. The average Hamiltonian of this overall cycle has in the high-field eigenbasis, matrix elements of the form:

$$(i|\bar{\mathscr{H}}^{(0)}|j) = (i|\bar{\mathscr{H}}_{\phi=0}|j)\delta(nk - n_{ij}) \tag{27}$$

where $n_{ij} = M_i - M_j$, $k = 0, 1, 2, \ldots$, and ϕ is the phase. This average Hamiltonian, in other words, allows the system to exchange photons with the radiation field only in groups of n. Such a cycle is characterized as being

Table 5.6. Dependence on τ of MQC's produced by the sequence $90^\circ_\phi - \tau - 90^\circ_y$

	ΔM	$\phi = -90^\circ$	$\phi = 0^\circ$	$\phi = -45^\circ$
Anisotropic systems	0	2	1	1
	n (even)	$n - 1$	n	$n - 1$
	n (odd)	n	$n - 1$	$n - 1$
Isotropic systems	0	2	1	1
	n (even)	$2n - 1$	$2n$	$2n - 1$
	n (odd)	$2n - 1$	$2n$	$2n - 1$

zero-order nk-quantum selective. It is to be borne in mind that selectivity requires that $\Delta\tau_p$, the period of the basic cycle, be kept short; higher order transitions, however, require more time to be pumped: too short a cycle can excite only low-order coherences efficiently. With a two 90° pulse sequence, for example, for vanishing interpulse spacing τ, only SQC's are created; MQC's appear as τ increases (see Table 5.6). This unfavorable dependence on the cycle time may be avoided if the Hamiltonian responsible for creation of MQC's is made to act nonlinearly.

The basic subcycle in an nk-quantum selective cycle could be a line-narrowing sequence (see Chapter 6) with interpulse spacing too large to suppress the dipolar Hamiltonian $\mathscr{H}_{II}^{\mathrm{D}}$ effectively. This causes the $\mathscr{H}_{II}^{\mathrm{D}}$ to act in a highly nonlinear fashion, resulting in selective and effective excitation of the MQC. Alternatively, a "time-reversal sandwich" may be employed for the basic subcycle. This is composed, for example, of a $[\frac{\tau}{2} - 90^\circ_y - 2\tau - 90^\circ_y - \tau - 90^\circ_y - 2\tau - 90^\circ_y - \tau - 90^\circ_{\bar{y}} - 2\tau - 90^\circ_{\bar{y}} - \tau - 90^\circ_{\bar{y}} - 2\tau - 90^\circ_{\bar{y}} - \frac{\tau}{2}]$ sequence followed by free evolution under $\mathscr{H}_{zz}$ for a period $\Delta\tau_p$, followed by a $[\frac{\tau}{2} - 90^\circ_x - 2\tau - 90^\circ_x - \tau - 90^\circ_x - 2\tau - 90^\circ_x - \tau - 90^\circ_{\bar{x}} - 2\tau - 90^\circ_{\bar{x}} - \tau - 90^\circ_{\bar{x}} - 2\tau - 90^\circ_{\bar{x}} - \frac{\tau}{2}]$ sequence. It may be noted that the average Hamiltonians of these two sequences are, respectively, $\pm\frac{1}{3}(\mathscr{H}^{\mathrm{D},xx} - \mathscr{H}^{\mathrm{D},yy})$, where:

$$(\mathscr{H}^{\mathrm{D},xx} - \mathscr{H}^{\mathrm{D},yy}) = 3 \sum_{i<j} \mathrm{D}_{ij}(I_{iy}I_{jy} - I_{ix}I_{jx}) \tag{28a}$$

In this case, although the time-reversal sandwich is of duration $2T + \Delta\tau_p$ (where $T \gtrsim T_2$) to produce higher order MQC's efficiently, the effective subcycle time $\Delta\tau_p \ll T_2$, so that $\|\bar{\mathscr{H}}_0 \Delta\tau_p\| \ll 1$, insuring selectivity.

Note that nk-quantum selective cycles always produce ZQC's ($k = 0$). If $\mathscr{H}_{zz}$ is reversed in every alternate subcycle, the cycle becomes $\frac{1}{2}(2k + 1)\,n$-quantum selective ($k = 0, 1, 2, \ldots$). Excitation of ZQC's may thus be avoided. For the full cycle then, the average Hamiltonian corresponds once again to nonlinear action of the dipolar terms responsible for creating MQC's ultimately.

Such nonlinear action insures that $\Delta\tau_p$ is kept short enough to result in selective excitation, while at the same time this time factor is prevented from restricting excitation to low-order MQC's alone.

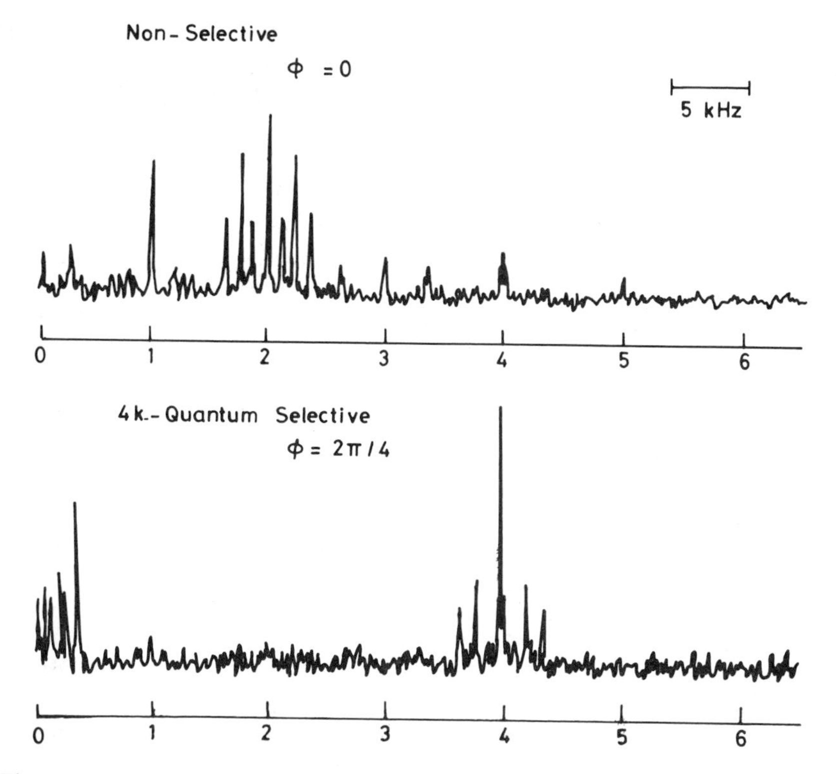

Figure 5.18. Spectrum of benzene in a liquid crystalline solvent following selective excitation of $4k$-quantum coherence, compared with the result of nonselective excitation. [Reproduced by permission. W.S. Warren, S. Sinton, D.P. Weitekamp and A. Pines, *Phys. Rev. Lett.*, **43**, 1791 (1979) copyright (1979) American Physical Society]

A selective four-quantum spectrum of benzene is shown in Fig. 5.18.

It may be mentioned that, owing to the form and the small magnitude of the couplings involved in an isotropic solution, selective excitation strategies of this type are not ideally suited to the isotropic phase. The selectivity is generally a very complicated function of τ. For instance, $4k$-quantum selection in isotropic systems may be achieved by the pulse sequence: $90^\circ_x-\tau-180^\circ_{-x}-\tau-90^\circ_x-90^\circ_y-\tau-180^\circ_{-y}-\tau-90^\circ_y-90^\circ_{-x}-\tau-180^\circ_x-\tau-90^\circ_{-x}-90^\circ_{-y}-\tau-180^\circ_y-\tau-90^\circ_{-y}$. For methanol at 270 MHz, a large 4Q signal is expected with three cycles of the first-order $4k$-selective version of this sequence. Here, $\tau = 12$ ms, the total sequence duration being 576 ms.

Multiple-Quantum Imaging

Nuclear magnetic resonance imaging is a technique developed in the last decade that attempts to produce a map of the distribution of NMR parameters (density of spins, T_1, T_2, chemical shift, etc.) as a function of location in a sample. In essence, different regions of a sample object are frequency labeled by imposing on the sample a calibrated magnetic field gradient in an NMR experiment. In a linear field gradient g, the frequency spread across a slice of thickness ΔZ is $g\Delta Z$. If features of the sample are to be observed with a spatial resolution on the order of ΔZ, then the externally imposed field $g\Delta Z$ must of course be resolved with respect to any internal field. In solids, the local dipolar field B_L can be quite strong, amounting for instance to 5 G in protonated solids with restricted molecular motions. A gradient greater than 50 G cm^{-1} will then be required to attain a spatial resolution of 1 mm. Line-narrowing sequences may be employed to reduce B_L. Alternatively, the effective g may be increased by working with MQC's. Note that the effective B_L's roughly remain comparable for high- and low-order MQC's for large spin systems. The pulse sequence shown in Fig. 5.19 may be employed. The effective Hamiltonians indicated during the preparation and mixing periods may be made operative employing the eight-pulse sequences discussed in the previous section on order selective excitation of MQC's. Separation of orders is achieved by incrementing in proportion to t_1, the phase ϕ of the preparation period with respect to that of the mixing period. The time-domain data are Fourier transformed with respect to t_1. Figure 5.20 shows the n-quantum ^{1}H spectra at 360 MHz of a phantom object composed of three parallel glass melting point tubes (1.3 mm i.d, 1.65 mm o.d.), arranged linearly. The outer tubes are loaded with a 4-mm length of compressed adamantane, while the middle one is empty. The sample is aligned with its cylindrical axes perpendicular to the z-field gradient. The spectra have been recorded with and without a z gradient of 20 kHz cm^{-1}. Clearly, the single-quantum spectrum is merely broadened by this modest gradient and does not resolve the two adamantane plugs; resolution sets in from the 4QS onwards; the two peaks are well resolved at $n = 10$. The gradient is thus effectively amplified 10-fold. The preparation time chosen ($\tau = 396$ μs) in this case is sufficient to excite up to 20QC's with reasonable intensity.

Similar ideas may also be employed in diffusion measurements.

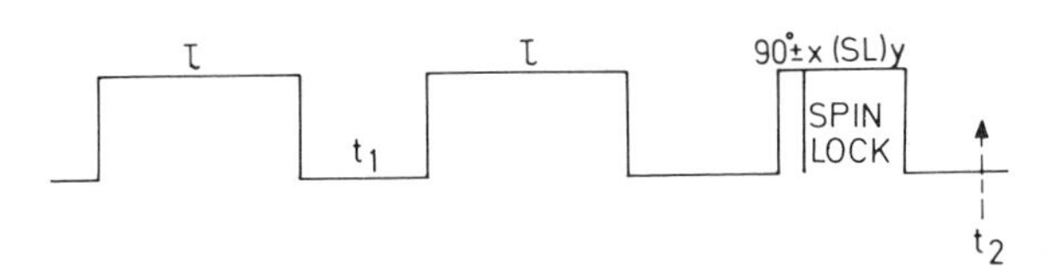

Figure 5.19. Pulse sequence for multiple quantum imaging. During the first τ period, $\mathscr{H} \sim \sum_{i<j}(I_{iy}I_{jy} - I_{ix}I_{jx})$, and during the second τ period $\mathscr{H} \sim -\sum_{i<j}(I_{iy}I_{jy} - I_{ix}I_{jx})$. Separation of orders is achieved by incrementing the relative phase between preparation and mixing in proportion to t_1. The z magnetization is monitored about 1 ms after mixing by a $90^\circ_{\pm x}$ pulse followed by spin locking for about 100 μs. One point is sampled for each value of t_1.

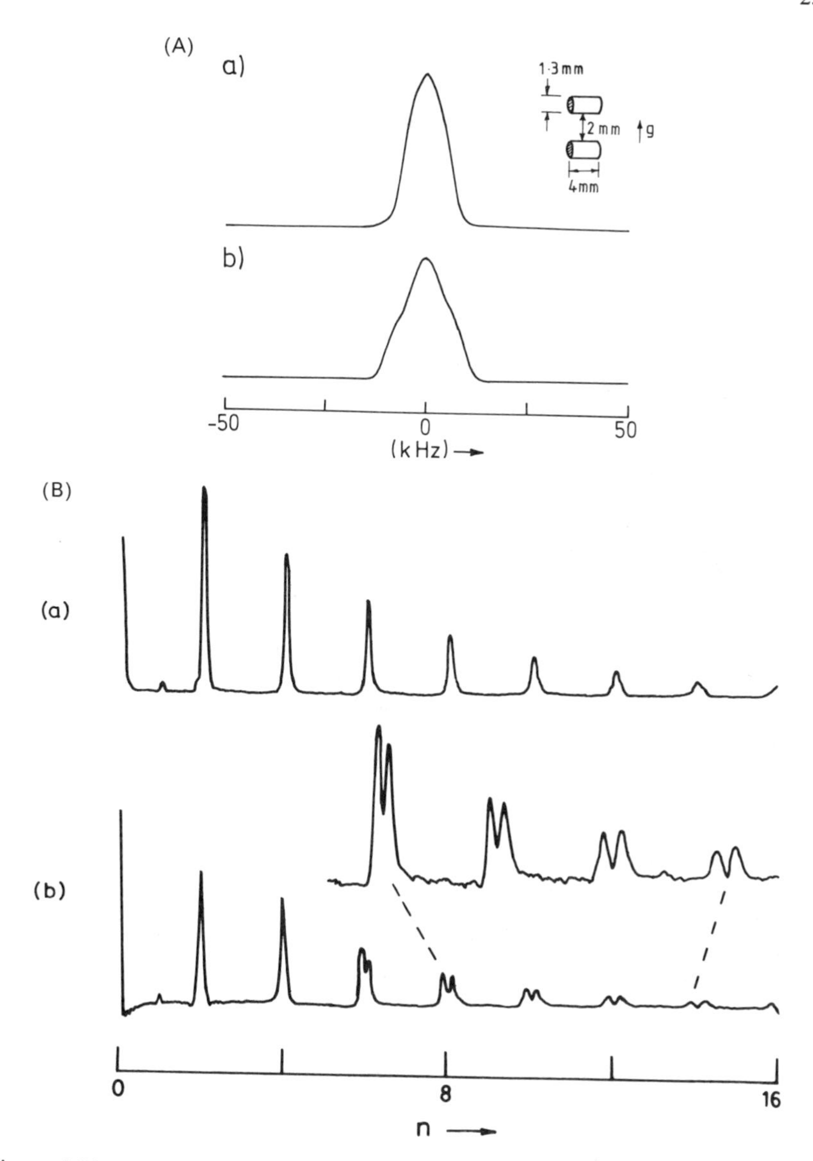

Figure 5.20. (A) Single-quantum spectra of adamantane for the sample geometry shown, without and with a field gradient. (B) The MQS of the same sample with and without a field gradient. [Reproduced by permission. A.N. Garroway, J. Baum, M.G. Munowitz and A. Pines, *J. Magn. Reson.*, **60**, 337 (1984), copyright 1984, Academic Press, New York]

Multiple-Quantum NMR of Solids

Chemical Shift Resolution of Spin-1 Nuclei. As noted in Chapter 1, the double-quantum coherence in spin-1 systems does not evolve under quadrupolar coupling. This situation reflects the fact that the first-order quadrupolar

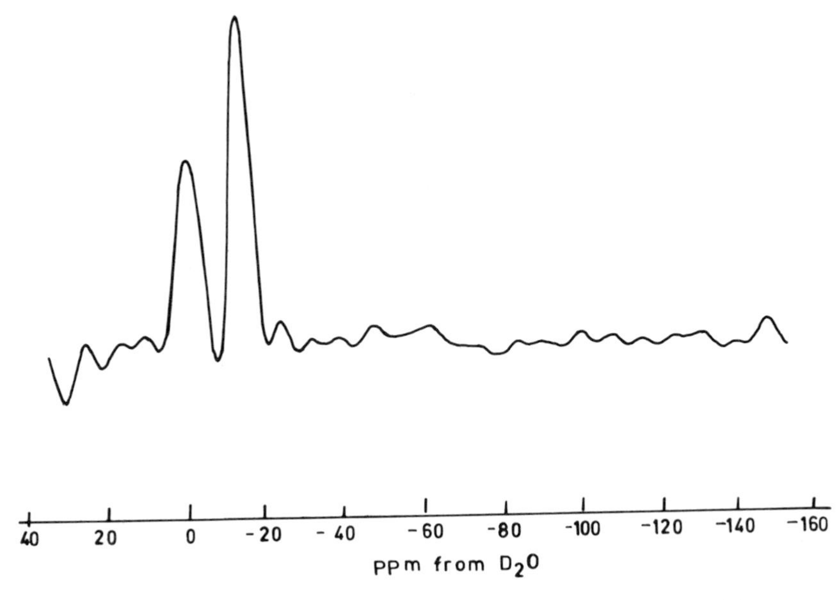

Figure 5.21. The 28.4-MHz deuterium DQS of a 10% deuterated oxalic acid dihydrate crystal. The peak on the right corresponds to the carboxyl group. [Reproduced by permission. S. Vega, T.W. Shattuck and A. Pines, *Phys. Rev. Lett.* **37**, 43 (1976), copyright 1976, American Physical Society]

Hamiltonian

$$\mathscr{H}_Q'' = \tfrac{1}{3}\omega_Q(3I_z^2 - I(I+1)) \tag{28b}$$

shifts both the uppermost and lowest energy levels equally, leaving their separation unchanged. Chemical shifts of spin-1 nuclei are normally masked by the quadrupole interaction. When the evolution of the DQC is measured, however, the chemical shift information may be retrieved. Figure 5.21 shows an example of deuterium shift resolution in a single crystal of 10% deuterated oxalic acid dihydrate, where the carboxyl and water deuterons are well resolved, the protons being decoupled. It is to be noted, on the other hand, that resolution was impossible in either proton or single-quantum deuterium spectra in this case. The isotropic shift in powder samples may be measured by the DQ method with magic angle spinning (see Chapter 6); rotor alignments need only be accurate enough to diminish dipolar and shift anisotropy broadening, because the DQC is invarient to first-order quadrupolar interaction anyway. Although the quadrupole coupling is averaged to zero over a rotor cycle, DQC's may still be studied since it takes but a fraction of the period of rotation to create DQC. In contrast, it is to be noted that single-quantum studies on spin-1 nuclei, which depend on the magic angle spinning efficiency to achieve line-narrowing, are extremely sensitive to small errors in the rotor alignment.

The DQC in spin-1 systems is excited by a selective pulse at the Larmor frequency, which is the center of the quadrupolar doublet. Such a weak rf pulse causes the DQC to nutate with the frequency:

$$\omega_{\text{eff}} = \omega_1^2/\omega_Q \tag{29}$$

The rf field therefore appears attenuated by the factor ω_1/ω_Q; the excitation of the DQC varies across a powder pattern because the flip angle is inversely related to the quadrupolar splitting. Triple-quantum coherences may also be created in $I = 3/2$ nuclei such as ^{23}Na by a 3Q selective resonant pulse. For $I = 3/2$, one finds for the nutation frequencies:

$$\omega_{\text{eff, DQC}} = \frac{7\omega_1^2}{4\omega_Q}$$

whereas: (30)

$$\omega_{\text{eff, 3QC}} = \frac{3}{8}(\omega_1^3/\omega_Q^2)$$

Double-Quantum Cross-Polarization. The DQ spectra of spin-1 nuclei discussed in the last section suffer from poor sensitivity because only partial deuteration is done, in order to keep the sample dilute with respect to deuterium spins and get narrow deuterium resonances. The sensitivity may be improved by cross-polarization techniques using the abundant protons as polarization source.

The allowed single-quantum transitions of a spin-1 nucleus may be cross-polarized with the following modified Hartmann–Hahn matching condition (see Chapters 1, 3, and 6) to insure maximal rate of transfer:

$$\sqrt{2}\omega_{1I} = \omega_{1S} \tag{31}$$

where $I = 1$ and $S = 1/2$. The irradiation is close to one of the transitions of the quadrupolar doublet. The relevant rotating frame energy level diagram is given in Fig. 5.22.

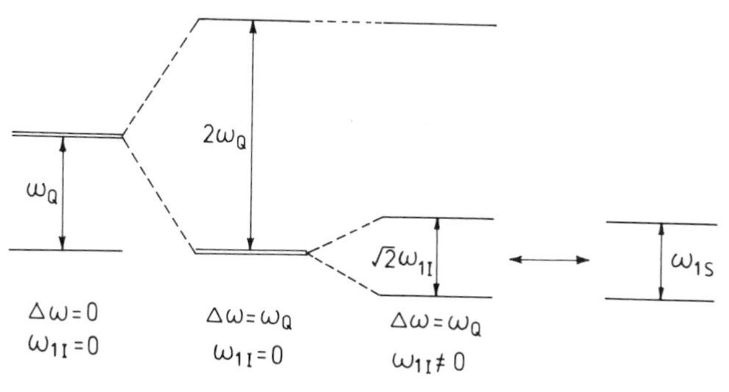

Figure 5.22. Rotating-frame energy level matching for single-quantum cross-polarization of a spin-1 nucleus from a spin-$\frac{1}{2}$ source.

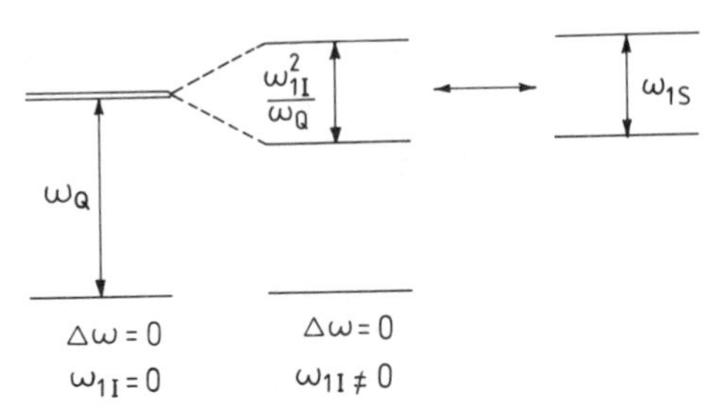

Figure 5.23. Rotating-frame energy level matching for double-quantum cross-polarization of a spin-1 nucleus from a spin-$\frac{1}{2}$ source.

However, the double-quantum transition may be *directly* cross-polarized from the protons as well, the Hartmann–Hahn matching condition now being:

$$(\omega_{1I}^2/\omega_Q) = \omega_{1S} \tag{32}$$

Irradiation is now at the center of the quadrupolar doublet. The corresponding rotating-frame energy level diagram is shown in Fig. 5.23.

The efficiency of cross-polarization (CP), as measured by the destruction of the S-spin order upon mixing, is greater for 2Q CP when compared to 1Q CP. Also, Hartmann–Hahn matching is less critical for 2Q CP. As with CP involving only spin-1/2 nuclei (see Chapter 6), adiabatic demagnetization in the rotating frame (ADRF) leads to a higher efficiency of CP compared to the spin-lock strategy discussed above, for both 1Q CP and 2Q CP between spin-1/2 and spin-1 nuclei. These characteristics are shown in Fig. 5.24. Two-quantum CP may also be applied to advantage in ^{14}N studies. Figure 5.25 shows the 2Q spectrum of ^{14}N in $(NH_4)_2SO_4$, following 2Q CP. The presence of two nonequivalent ^{14}N sites in the crystal is borne out by the occurrence of two DQ frequencies. After taking into account the effect of second-order quadrupolar shifts, the relative chemical shift of the two sites turns out to be about 8 ppm for this crystal orientation.

Multiple-quantum CP may be applied to $I > 1$ nuclei as well. For example, 2Q CP and 3Q CP have been demonstrated and analyzed for ^{23}Na ($I = 3/2$) in sodium ammonium tartarate.

Two-Dimensional INADEQUATE on Solids. Observation of ^{13}C–^{13}C dipolar couplings in solids may be achieved by employing the INADEQUATE strategy to suppress the signals at natural abundance from lone carbons not dipole coupled to other carbons. The sequence of Fig. 5.26 may be employed to cross-polarize the carbons and then pick out the dipole-coupled carbon pairs. Phase cycling is performed as usual. If t_1 is set to about 10 μs and the experiment performed for various values of τ, F_2 contains carbon chemical shifts and dipolar couplings, while F_1 has ^{13}C dipolar splittings. Coupled ^{13}C pairs share a common frequency in F_1. The F_1 frequency dispersion is limited to about 6 kHz. On the other hand, with an estimate of the operative ^{13}C–^{13}C dipolar couplings, τ may be set for efficient excitation of DQC's and t_1

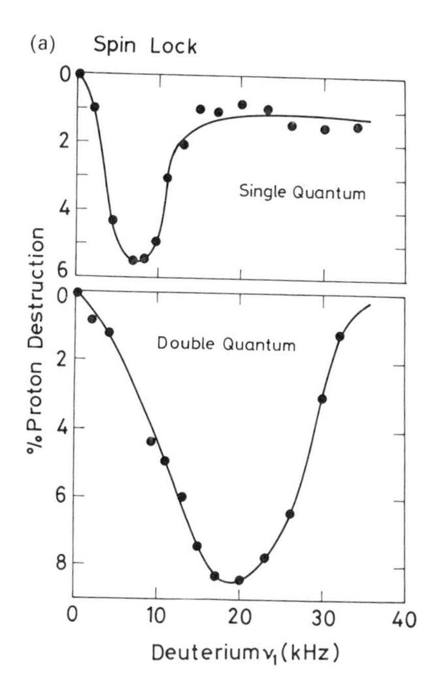

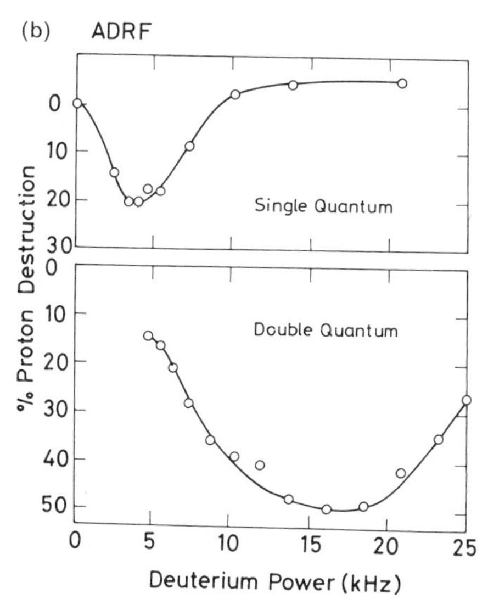

Figure 5.24. (a) Efficiencies of single- and double-quantum cross-polarization by spin locking, for a spin-1 nucleus. (b) Efficiencies of single- and double-quantum cross-polarization by ADRF for a spin-1 nucleus. [Reproduced by permission. S. Vega, T.W. Shattuck and A. Pines, *Phys. Rev.* **A22**, 638 (1980), copyright 1980, American Physical Society]

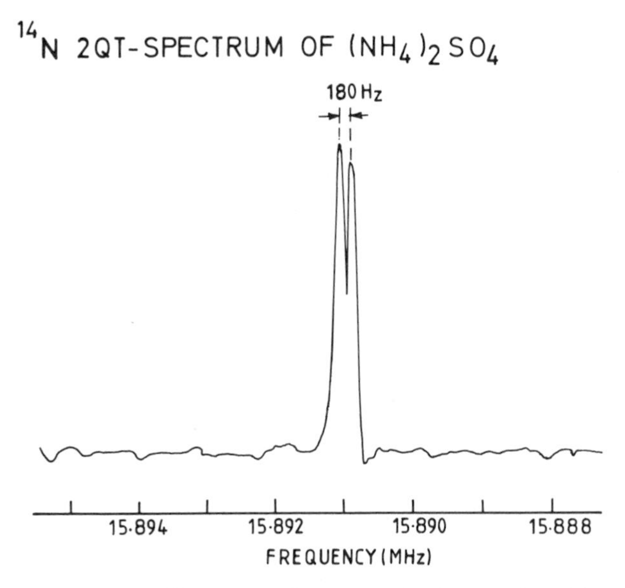

Figure 5.25. The 16-MHz ^{14}N DQS of $(NH_4)_2SO_4$. [Reproduced by permission. P. Brunner, B.H. Meier, P. Bachmann and R.R. Ernst, *J. Chem. Phys.*, **73**, 1086 (1980), copyright 1980, American Institute of Physics]

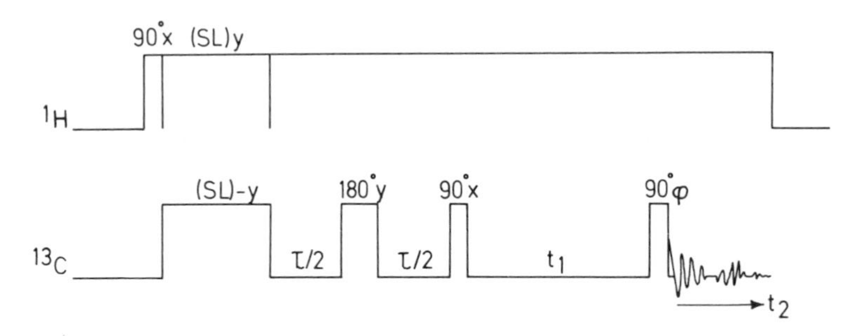

Figure 5.26. Pulse sequence for observation of ^{13}C–^{13}C dipolar couplings at natural abundance, following cross-polarization.

incremented; in this case, F_1 corresponds to the sums of the chemical shifts, once again allowing connectivities to be established as usual. Note that the F_1 frequency spread is now much higher than with the previous strategy of τ variation.

Heteronuclear MQS on Solids. For systems with a small number of coupled spins, such that at least high-order MQ spectra exhibit high resolution, pulse-interrupted free precession methods may be employed to effect polarization transfer from spins I to the spin S, and more interestingly, to measure the IS couplings with high resolution. A homonuclear (^{13}C) example of CT by such a pulse-interrupted free precession strategy was discussed in the last section, involving creation of homonuclear DQC's. Resolved couplings are possible in this case because at natural abundance, and with ^{1}H decoupling, the number of coupled spins is very well defined: even three-^{13}C-spin systems are a hundred times rarer than two-^{13}C-spin systems. Heteronuclear MQS may be employed to advantage in determining molecular geometry in anisotropic phases, through the measurement of direct heteronuclear dipolar couplings. It is of advantage to measure heteronuclear couplings because for a spin system of given size, the number of couplings is smaller in this case, compared to the homonuclear situation. However, the abundant spins must in general be subject to homonuclear dipolar decoupling when $\mathscr{H}_{IS}$ is sought to be measured. It is possible to assign the observed dipolar couplings to specific pairs of nuclei by adopting a 2D strategy with heteronuclear decoupling during the detection period t_2 giving rise to chemical shifts in F_2, while the dipolar couplings are in F_1. Such local field spectroscopy studies involving ^{1}H and ^{13}C, for example, have been shown to give valuable information on proton positions and molecular geometry.

If an HMQ strategy is adopted together with homonuclear decoupling, one gains the advantage of a substantial reduction of the number of lines as compared to the conventional single-quantum local field spectra; also the line positions in HMQ spectroscopy are sensitive to the signs of heteronuclear couplings.

Relaxation of MQC's

Relaxation studies of MQC's frequently afford additional insights into a spin system. Transverse relaxation of multiple-quantum coherence, for instance, contains information that is often complementary to single-quantum T_2 results. Experimentally, 180° refocusing pulses may be applied to refocus the evolution of MQC's under magnetic field inhomogeneities; a 180° train may also be employed to minimize diffusion effects, as in the classical Carr–Purcell method. The Meiboom–Gill modification to avoid phase reversal of successive echoes may be implemented for pQC's by phase shifting the π pulses with respect to the preparation pulses by $\pi/2p$. Similarly, spin locking of DQC's may be achieved with a spin-lock pulse of phase $\pi/4$ with respect to the selective DQ excitation pulse. Relaxation studies on MQC's can provide information on the *correlation* of fluctuating random local fields, which is not available from T_1 studies or T_2 measurements on SQC's in weakly coupled spin systems.

CHAPTER 6

High-Resolution Pulse NMR in Solids

Introduction

In liquids the scalar chemical shifts and the indirect spin–spin coupling constants are the two important NMR parameters that hold clues as to molecular structure and conformation. In solids, however, because of the relatively high rigidity of the environment and the frozen state of the molecules, these two interactions become minor and the homo and heteronuclear dipolar Hamiltonians become the most dominant interactions. One can easily calculate the dipolar interaction from a knowledge of the positional parameters of the dipoles in a rigid lattice and can also calculate what would be the energy levels in a randomly oriented powder assuming statistical distribution of the magnetic dipoles. The dipolar interaction in solids is on the order of several kilohertz and the resulting broad structureless resonance absorption will mask completely the chemical shift information ($\sim$kHz) and wipe out spin–spin coupling ($\sim$0.01–0.2 kHz). One cannot get much information from such broad lines except in favorable cases, where dipolar splitting between neighbor interactions far exceeds the lattice contribution so that distances and direction cosines of dipolar-coupled pairs can be derived. In fact, this method has been used successfully to locate proton positions in hydrated crystals, as a supplement to x-ray crystallography. One may also analyze the moments of spectral distribution in solids and the effect of molecular motions on these spectral moments. Temperature dependence of lineshapes has been successfully analyzed in terms of specific models of molecular libration, rotation, etc.

Apart from the dipolar coupling dominating the scene in solids, it must be remembered that the chemical shift, which is often referred to by a single number in high-resolution NMR, is no longer its liquid state average but a symmetric second-rank tensor. As such these dipolar broadened lines are also subject to chemical shift anisotropies; but then to evaluate them without doing anything to get rid of the dipolar interaction from solids is next to impossible.

Not only the dipolar interaction but other interactions that do not show up in liquids because of the random motion of the molecules are now eminently present and generally mask finer information. In this chapter we shall briefly touch upon some of the techniques to achieve high-resolution NMR in solids. In this connection it is better to start with all the various interactions in terms of their Hamiltonians; these are:

$$
\begin{aligned}
\mathscr{H}_Z &= \sum_i \omega_0 \mathbf{I}_z^i \\
\mathscr{H}_{CS} &= \sum_i \gamma^i \mathbf{I}^i \cdot \boldsymbol{\sigma}^i \cdot \mathbf{B}_0 \\
\mathscr{H}_{SR} &= \sum_m \sum_i^{\text{all nuclei}} \mathbf{I}^i \cdot \mathbf{C}^{im} \cdot \mathbf{J}^m \\
\mathscr{H}_Q &= \sum_i (e^2 Q^i q / 6I^i(2I^i - 1)) \mathbf{I}^i \cdot \mathbf{P}^i \cdot \mathbf{I}^i \\
\mathscr{H}_D &= \sum_{i<k} \frac{\gamma^i \gamma^k}{r_{ik}^3} \frac{h^2}{4\pi^2} \left[\mathbf{I}^i \cdot \mathbf{I}^k - \frac{3(\mathbf{I}^i \cdot \mathbf{r}_{ik})(\mathbf{I}^k \cdot \mathbf{r}_{ik})}{r_{ik}^2} \right] \\
&= \sum_{i<k} (-2\gamma^i \gamma^k h^2 / 4\pi^2) \sum_{\alpha,\beta=1}^{3} I_\alpha^i \cdot \mathbf{D}_{\alpha\beta} \cdot I_\beta^k \\
\mathscr{H}_J &= \sum_{i<k} \mathbf{I}^i \cdot \mathbf{J}^{ik} \cdot \mathbf{I}^k \\
\mathscr{H}_{rf} &= B_1(t) \cos(\omega t + \phi(t)) \sum_i \gamma^i I_x^i
\end{aligned}
\tag{1}
$$

Here $\mathscr{H}_Z$ is the Zeeman Hamiltonian. The anisotropic chemical shift Hamiltonian, $\mathscr{H}_{CS}$, describes the magnetic field induced by the electronic charge distribution, which is not necessarily spherical and in general depends on the direction of the external magnetic field relative to the molecular frame. Here we have written this in dyadic notation, where $\mathbf{I}^i$ and $\mathbf{B}_0$ are the spin and static field vectors and $\boldsymbol{\sigma}^i$ is the anisotropic chemical shift tensor at the site of the ith nucleus.

The spin rotation interaction Hamiltonian, $\mathscr{H}_{SR}$, is a measure of the coupling of rotational angular momentum J of the molecule with the nuclear spin; and $\mathbf{C}^{im}$ is the spin–rotational coupling tensor.

The quadrupole Hamiltonian, $\mathscr{H}_Q$, is nonzero only when $I \geqslant 1$ and arises as a result of the interaction of the nuclear electric quadrupole moment eQ with the electric field gradient eq at the site of the nucleus and the corresponding interaction energy, called the quadrupole coupling constant, is a tensor usually cylindrically symmetric and in some systems orthorhombic. The tensor is traceless.

The well-known (and in fact the dominant term in solids for spin-1/2 nuclei) dipolar Hamiltonian, $\mathscr{H}_D$, is strictly cylindrically symmetric and traceless.

The indirect spin–spin coupling, $\mathscr{H}_J$, is anisotropic in solids and has a finite trace and is responsible for the fine structure in high-resolution NMR.

The perturbing rf Hamiltonian, $\mathscr{H}_{rf}$, applied along the x axis of the laboratory frame and in the form given, corresponds to general phase- and amplitude-modulated rf at the carrier frequency.

$\mathscr{H}_Z$, $\mathscr{H}_{rf}$, $\mathscr{H}_{CS}$, $\mathscr{H}_Q$, and $\mathscr{H}_{SR}$ are single-spin Hamiltonians. However, the dipolar and the spin–spin coupling Hamiltonians couple all spins in the system: although the latter is intramolecular, the former is both intra- and intermolecular. The Zeeman and rf Hamiltonians are called "external" Hamiltonians (they are under the direct control of the experimeter), while the others, which are the inherent property of a given nuclear system, are called "internal" Hamiltonians.

The Effect of Rotation on the Internal Hamiltonians

The fact that liquids give rise to high-resolution NMR because of fast random tumbling at once suggests that we must impart some sort of motions to the solid sample to "average out" certain interactions. In order to examine the effect of any rotation it is best to express the various internal Hamiltonians in terms of components of irreducible spherical tensor operators (see Appendix 6). Thus all internal Hamiltonians can be expanded as:

$$\mathscr{H}_\lambda = C^\lambda \sum_{\alpha,\beta=1}^{3} I_\alpha R^\lambda_{\alpha\beta} A^\lambda_\beta = C^\lambda \sum_{\alpha,\beta=1}^{3} R^\lambda_{\alpha\beta} T^\lambda_{\beta\alpha} \tag{2}$$

where C^λ is a constant depending on the nuclear property, while $R^\lambda_{\alpha\beta}$ depend on the electronic state of the molecule or crystal and are all tensors of rank two and as shown in Appendix 6, they can be decomposed into their irreducible components of the rotation group. The $T^\lambda_{\alpha\beta}$ are dyadic products of two vectors, one of which is always the nuclear spin vector I_α, while the other, A^λ_β, can be the same nuclear spin (when $\mathscr{H}_\lambda$ is the quadrupole Hamiltonian), another nuclear spin (when $\mathscr{H}_\lambda$ is the dipolar or indirect spin–spin coupling Hamiltonian), the external field vector (when $\mathscr{H}_\lambda$ is chemical shift), or the molecular angular momentum vector (when $\mathscr{H}_\lambda$ is the spin–rotational Hamiltonian). Of these, dipolar and quadrupolar tensors are traceless and symmetric in nature, while the chemical shift, spin–rotational interaction and indirect spin–spin coupling contain, in principle, isotropic, traceless symmetric and traceless antisymmetric parts. The traceless antisymmetric part is not resolvable and can be ignored. We shall treat all these tensors as symmetric tensors. For all these tensors, depending on the molecular electronic structure and/or the symmetry of crystalline environment, there exists a principal axis system in which they are diagonal. These diagonal values are then the so-called principal values of the tensor, which we shall label R_{xx}, R_{yy}, and R_{zz} and by convention:

$$|R_{zz} - R| \geqslant |R_{xx} - R| \geqslant |R_{yy} - R| \tag{3}$$

where $R = 1/3\ \mathrm{Tr}\,\mathbf{R}$. In the principal axis system we can define the tensor by these three parameters or, alternatively, knowing $\mathrm{Tr}\,\mathbf{R}$, by three parameters

R, δ', and η such that:

$$\delta' = R_{zz} - \tfrac{1}{3}\mathrm{Tr}\,\mathbf{R}; \qquad \eta = (R_{yy} - R_{xx})/\delta' \tag{4}$$

so that:

$$\mathbf{R} = \begin{bmatrix} R_{xx} & 0 & 0 \\ 0 & R_{yy} & 0 \\ 0 & 0 & R_{zz} \end{bmatrix} = R\mathbf{1} + \delta' \begin{bmatrix} -(1+\eta)/2 & 0 & 0 \\ 0 & -(1-\eta)/2 & 0 \\ 0 & 0 & 1 \end{bmatrix} \tag{5}$$

$R = 0$ for dipolar and quadrupolar interaction and $1/3\ \mathrm{Tr}\,\mathbf{R}$ of chemical shift and indirect coupling is what we measure in high-resolution NMR of liquids. The spin–rotational interaction tensor, which in general is traceless, causes mainly line broadening and never causes line shifts or line splitting in NMR and we need not consider this any further.

Let us now consider techniques that are used in averaging internal Hamiltonians by rotating about an axis. To do this, we express our internal Hamiltonians in terms of irreducible spherical tensor operators T_{lm} and R_{lm} Wigner rotation matrices. The form of the Hamiltonian now is:

$$\mathscr{H}_\lambda = C^\lambda \sum_l \sum_{m=-l}^{+l} (-1)^m R^\lambda_{l,-m} T^\lambda_{lm} \tag{6}$$

where R^λ_{lm} and T^λ_{lm} are derived from the $R^\lambda_{\alpha\beta}$ and $T^\lambda_{\alpha\beta}$ mentioned above. For symmetric second-rank tensors the nonzero R^λ_{lm}'s exist only for $l = 0$ and $l = 2$ and in the principal axis system only $m = 0, \pm 2$ survive. Thus, in the principal axis system the spherical tensor (ST) components are given by:

$$\begin{aligned} \mathrm{ST}_{00} &= \frac{1}{3}\mathrm{Tr}\,\mathbf{R} = R \\ \mathrm{ST}_{20} &= \sqrt{\frac{3}{2}}\left(R_{zz} - \frac{1}{3}\mathrm{Tr}\,\mathbf{R}\right) = \sqrt{\frac{3}{2}}\delta' \\ \mathrm{ST}_{2\pm 2} &= \frac{1}{2}(R_{yy} - R_{xx}) = \frac{1}{2}\eta\delta' \end{aligned} \tag{7}$$

The $T^\lambda_{\beta\alpha}(I_\alpha \cdot A^\lambda_\beta)$ are basically represented in the laboratory axis system. We can transform the R^λ_{lm}'s into the laboratory frame by using the Wigner rotation matrices (see Appendix 6). For example:

$$R^\lambda_{lm} = \sum_{m'} \mathscr{D}^l_{m'm}(\alpha^\lambda, \beta^\lambda, \gamma^\lambda)\mathrm{ST}^\lambda_{lm'} \tag{8}$$

where $\mathscr{D}^l_{m'm}$ are the elements of the Wigner rotation matrices. Similarly the products T^λ_{lm}'s can also be reduced using group theoretic methods to the irreducible representations. This can be done once for all for every dyadic tensor product. The important ones are listed in Table A6.1 of Appendix 6.

Thus, the dipolar Hamiltonian in irreducible spherical tensor operators is given by:

$$\mathscr{H}_{\mathrm{D}}^{ik} = -\sqrt{6}\gamma^i\gamma^k \frac{h}{2\pi} r_{ik}^{-3} \sum_m (-1)^m \mathscr{D}_{0,-m}^2(\Omega^{ik}) T_{2m}^{\mathrm{D},ik} \tag{9}$$

where Ω^{ik} is the set of Euler angles that make the principal axes of the dipolar tensor coincide with the laboratory axes. In terms of polar angles θ^{ik} and ϕ^{ik}, of the internuclear vector r_{ik} in the laboratory frame, the dipolar Hamiltonian can be rewritten as:

$$\mathscr{H}_{\mathrm{D}} = -2\sqrt{\frac{6\pi}{5}} \frac{h}{2\pi} \sum_{i<k} \gamma^i\gamma^k r_{ik}^{-3} \sum_m (-1)^m Y_{2,-m}(\theta^{ik}, \phi^{ik}) T_{2m}^{\mathrm{D},ik} \tag{10}$$

where we have expressed the matrix elements of the rotation matrices in terms of spherical harmonics Y_{2m}'s.

In a similar way the shift Hamiltonian can be expressed as:

$$\mathscr{H}_{\mathrm{CS}} = \sum_{l=0,2} \sum_{m=-l}^{+l} (-1)^m T_{lm} \gamma \sum_{m'} \mathscr{D}_{m',-m}^l S T_{lm'} \tag{11}$$

Of the various terms, the secular terms ($m = 0$) are of the greatest importance and taking the secular terms alone we can rewrite $\mathscr{H}_{\mathrm{CS}}$ as:

$$\begin{aligned} \mathscr{H}_{\mathrm{CS}} &= \omega_0 I_0 \left\{ R^\delta + \sqrt{\frac{8\pi}{15}} \delta'^\delta \left[\sqrt{\frac{3}{2}} \cdot Y_{20} + \frac{1}{2}\eta^\delta (Y_{22} + Y_{2,-2}) \right] \right\} \\ &= \omega_0 I_0 \delta_{zz} \end{aligned} \tag{12}$$

where δ_{zz} is the shielding component in the laboratory frame. From the transformation properties of the shift tensor, we see that it does not depend on the Euler angle α and hence is invariant to rotation about an axis parallel to the applied dc field.

Qualitative Aspects of Lineshapes of Systems Showing Chemical Shift Anisotropy

The resonance lineshape of systems that show resolved large dipolar anisotropy is usually the so-called Pake diagram, from which it is possible to read off the "parallel" and "perpendicular" dipolar coupling constants. Dipolar coupling is strictly cylindrically symmetric and hence the lineshape can be easily interpreted. If we can suppress the dipolar interaction by a suitable manipulation of the Hamiltonian (*vide infra*) and bring the chemical shift anisotropy to surface, then the shape of the spectrum will be governed by the nature of the anisotropy, whether axially symmetric or orthorhombic. In powder samples there will be a statistical distribution of molecular axes and hence the NMR spectrum will be the sum total of each individual orientation with equal weightage. If we represent the various possible orientations as areas on the surface of a sphere, then the intensity $I(\omega)$ of the NMR spectrum between frequencies ω_i and ω_f is proportional to the area between the curves $\omega = \omega_i$ and $\omega = \omega_f$ on the surface of the sphere, i.e.,

$$\int_{\omega_i}^{\omega_f} I(\omega)\,d\omega = N \iint \sin\theta\,d\theta\,d\phi \tag{13}$$

where N is the normalization constant such that

$$\int_{-\infty}^{+\infty} I(\omega)\,d\omega = 1 \tag{14}$$

For the axially symmetric chemical shift tensor ($\eta = 0$), it can be shown that the $\delta(\theta)$ at any angle depends only on the polar angle θ:

$$\delta(\theta) = (\delta_{\parallel}^2 \cos^2\theta + \delta_{\perp}^2 \sin^2\theta)^{1/2} \tag{15}$$

and the intensity in a range $\omega_a - \omega_b$ is given by:

$$\int_{\omega_a}^{\omega_b} I(\omega)\,d\omega = \frac{1}{\sqrt{3\omega_0\delta'}} \int_{\omega_a}^{\omega_b} \frac{d\omega}{\left(1 + \dfrac{2\omega}{\omega_0\delta'}\right)^{1/2}}$$

so that:

$$I(\omega) = \frac{1}{\sqrt{3\omega_0\delta'}} \frac{1}{\left(1 + \dfrac{2\omega}{\omega_0\delta'}\right)^{1/2}} \qquad (-\omega_0\delta'/2 \leqslant \omega_0 \leqslant \omega_0\delta') \tag{16}$$

For the orthorhombic case, where we define:

$$\begin{aligned} \omega_3 &= \omega_0\delta' \\ \omega_2 &= -\tfrac{1}{2}(\omega_0\delta')(1 - \eta) \\ \omega_1 &= -\tfrac{1}{2}(\omega_0\delta')(1 + \eta) \end{aligned} \tag{17}$$

the intensity as a function of ω is given by:

$$\begin{aligned} I(\omega) &= \pi^{-1}[(\omega_3 - \omega_2)(\omega - \omega_1)]^{-1/2} F\left(\sqrt{\frac{(\omega_3 - \omega)(\omega_2 - \omega_1)}{(\omega_3 - \omega_2)(\omega - \omega_1)}}, \frac{\pi}{2}\right) \\ &\qquad\qquad (\omega_2 \leqslant \omega \leqslant \omega_3) \\ &= \pi^{-1}[(\omega_3 - \omega)(\omega_2 - \omega_1)]^{-1/2} F\left(\sqrt{\frac{(\omega_3 - \omega_2)(\omega - \omega_1)}{(\omega_3 - \omega)(\omega_2 - \omega_1)}}, \frac{\pi}{2}\right) \\ &\qquad\qquad (\omega_1 \leqslant \omega \leqslant \omega_2) \end{aligned} \tag{18}$$

where $F(k, \pi/2)$ is the incomplete elliptic integral of the first kind. The corresponding lineshapes are also indicated in Fig. 6.1, where it can be seen that, in favorable cases, the principal values of the δ tensors can be read off from the lineshapes. However, if one wants to evaluate the direction cosines of the principal values of the δ tensor with respect to any crystal-fixed axes, then one must rotate the crystal about three mutually orthogonal axes and follow the angular variations of the chemical shift. The number of resonances will now depend on the number of spatially distinct sites in the unit cell, and their

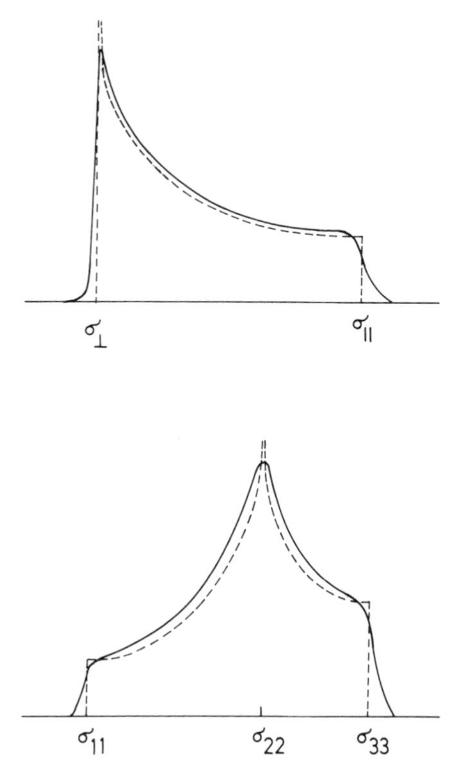

Figure 6.1. Powder lineshapes for axially symmetric and orthorhombic chemical shift tensors. Dotted lines correspond to δ-function lineshapes.

orientation dependence will have a direct relation to the crystal space group. Unless one knows the relationship between the shift tensor and molecular framework one may not be able to derive much useful information from a measurement of these tensors. Except in cases where symmetry arguments and the simplicity of the molecule can lead to assignment of powder values to various symmetry axes of the molecule, in most cases the δ-tensor has to be evaluated using single crystals just as in the evaluation of electron paramagnetic resonance (EPR) **g** and hyperfine tensors.

Averaging in Coordinate Space: Magic Angle Spinning (MAS)

Nature itself endows the molecular system with characteristic motions corresponding to complete and incomplete averaging of some of the tensors in coordinate space.

Thus, random isotropic translational and reorientational motions that are present in liquids of low viscosity average out all R_{lm}^{λ}'s except R_{00}. Therefore only the scalar spin–spin coupling and the average of the chemical shift survive, which lead to the high-resolution NMR in liquids. The indirect spin–spin couplings between different molecules are averaged by translational diffusion, while intermolecular shielding gives rise to the so-called solvent effects.

In liquid crystals where there is no complete freedom for random tumbling, although motion along one or two dimensions is restricted by the orientational order, all intermolecular interactions are wiped out. Intramolecular isotropic and reduced anisotropic interactions survive. This reduced dipolar interaction inextricably mixed with indirect J, reduced quadrupole interaction, and reduced chemical shift anisotropies lead to highly complicated spectra with an increased number of transitions compared to isotropic spectra.

If the order parameter and the orientation matrix is known, then ratios of internuclear distances of a molecule can be derived from spectra taken in an anisotropic medium. Thus the anisotropic spectrum obtained from a liquid crystalline medium simulates a single orientation of a molecule but with reduced anisotropic components, usually reduced to cylindrical symmetry about the director axis.

Random molecular jumps resulting from free or restricted rotation about bonds or interchange of nuclei among equivalent or inequivalent positions by inter- or intramolecular exchange processes correspond to exchange of Hamiltonian parameters such that the results depend on the frequency of the dynamic processes.

In a rigid lattice, however, one can impart a motion by spinning the sample about an axis; since the scalar parameters are unaffected by any motion, we focus our attention on the second-rank tensors. It is well known that to increase the resolution in high-resolution NMR spectrometers, the sample is spun about an axis either parallel or perpendicular to the field. The field strength $B_0(\mathbf{r})$ at a point $\mathbf{r}(r, \theta, \phi)$ can be given in terms of the field $B_0(\mathbf{0})$ at the center and a field gradient, by expanding $\mathbf{r}$ in spherical harmonics, as:

$$B_0(\mathbf{r}) = B_0(\mathbf{0}) + \sum_{l=1}^{\infty} \left(\frac{4\pi}{2l+1}\right)^{1/2} \sum_{m=-l}^{+l} a_{lm} r^l Y_{lm}(\theta, \phi)$$

with:

$$a_{l,-m} = (-1)^m a_{lm}^* \tag{19}$$

since $B_0(\mathbf{r})$ describes the real field. Thus,

$$\begin{aligned} B_0(\mathbf{r}) - B_0(\mathbf{0}) = {} & a_{10}z - Re\, a_{11}\sqrt{2}x + Im\, a_{11}\sqrt{2}y \\ & + a_{20}\frac{3z^2 - r^2}{2} + Re\, a_{22}\sqrt{\frac{3}{2}}(x^2 - y^2) \\ & + \sqrt{6}(Im\, a_{21}yz - Re\, a_{21}xz \\ & - Im\, a_{22}xy) + \cdots \end{aligned} \tag{20}$$

The various polynomials in x, y, z are the familiar labels on the shim control units, using which we try to adjust the coefficients a_{lm} to zero, for example, in a high-resolution NMR spectrometer. When the sample is rotated fast with a frequency ω_r about the y axis, for example, the gradients acquire time dependence given by:

$$B_0(\mathbf{r}(t)) = B_0(\mathbf{0}) + \sum_{l=1}^{\infty} \left(\frac{4\pi}{2l+1}\right)^{1/2} r^l \sum_{m=-l}^{+l} a_{lm} \times \sum_{m'=-l}^{+l} \mathscr{D}^l_{m'm}(0, \omega_r t, 0)\, Y_{lm}(\theta, \phi) \tag{21}$$

The elements of $\mathscr{D}^l_{m'm}$ are those of the Wigner rotation matrices and all we need to calculate is the time average of $\overline{\mathscr{D}^l_{m'm}(0, \omega_r t, 0)}$. Including up to $l = 2$, spinning leads to:

$$\overline{B_0(\mathbf{r}(t))} = B_0(\mathbf{0}) + Im\, a_{11}\sqrt{2}y + a_{20}\frac{1}{8}[3z^2 - r^2 + 3(x^2 - y^2)] + Re\, a_{22}\frac{1}{4}\sqrt{\frac{3}{2}}[3z^2 - r^2 + x^2 - y^2] \tag{22}$$

which substantially reduces the field variation at point $\mathbf{r}$, although a number of terms still survive.

Let us now consider the spinning of the sample about an angle 54°44′ to the quantizing field, the so-called magic angle spinning (MAS). With this concept we will examine, in particular, the result of such a spinning on solids. To appreciate the effect let us consider the sample spinning at angle β to the external field B_0 and consider the effect on the dipolar Hamiltonian. In spherical tensor operator notation:

$$\mathscr{H}_{\mathrm{D}} = -2\frac{h}{2\pi}\sum_{i<k}\gamma^i\gamma^k \sum_{m=-2}^{+2}(-1)^m R^{\mathrm{D},ik}_{2,-m} T^{\mathrm{D},ik}_{2m} \tag{23}$$

The transformation of the dipolar tensor from the principal axis system to the laboratory frame is given by:

$$\begin{aligned} R^{\mathrm{D},ik}_{2m}(\mathrm{lab}) &= \sum_{m'} \mathscr{D}^2_{m'm}(\Omega'') R^{\mathrm{D},ik}_{2m'}(CR) \\ &= \sum_{m'} \mathscr{D}^2_{m',-m}(\Omega'') \sum_{m''} \mathscr{D}^2_{m''m'}(\Omega_{ik}) S T^{\mathrm{D},ik}_{2m''} \end{aligned} \tag{24}$$

where $\mathscr{D}(\Omega)$ relates the principal axis system to the crystal reference (CR) frame and $\mathscr{D}(\Omega'')$ relates the crystal reference frame to the laboratory frame. Restricting ourselves to secular terms ($m = 0$) we need to consider, of the $\mathscr{D}^2_{m'm}(\Omega'')$, only $\mathscr{D}^2_{00}(0, \beta'', \omega_r t)$ where the rotation axis is making an angle β'' with the magnetic field and the frequency of rotation is ω_r. If the rotation is fast then we talk of a time average of $\overline{\mathscr{D}^2_{m'0}}$ and for dipolar Hamiltonian:

$$\overline{\mathscr{D}^2_{00}(0, \beta'', \omega_r t)} = \frac{1}{2}(3\cos^2\beta'' - 1) \tag{25}$$

If the rotation is perpendicular to the magnetic field, then:

$$\overline{\mathscr{D}_{00}^{2}\left(0, \frac{\pi}{2}, \cdots\right)} = \frac{1}{2}\left(3\cos^2\frac{\pi}{2} - 1\right) = -\frac{1}{2} \tag{26}$$

We can now talk of effective secular dipolar Hamiltonian as:

$$\begin{aligned} \mathscr{H}_{\text{sec,eff}}^{\text{D}} &= \frac{h}{2\pi}\sum_{i<k}\gamma^i\gamma^k\overline{\mathscr{D}_{00}^{2}(\Omega'_{ik})}\sqrt{\frac{3}{2}}\,r_{ik}^{-3}\,T_{20}^{ik} \\ &= -\frac{1}{2}\frac{h}{2\pi}\sum_{i<k}\gamma^i\gamma^k\frac{(3\cos^2\beta''_{ik} - 1)}{2}r_{ik}^{-3}(3I_0^iI_0^k - \mathbf{I}^i\cdot\mathbf{I}^k) \end{aligned} \tag{27}$$

whereas in the static case:

$$\mathscr{H}_{\text{sec}}^{\text{D}} = \frac{h}{2\pi}\sum_{i<k}\gamma^i\gamma^k\frac{(3\cos^2\beta_{ik} - 1)}{2}r_{ik}^{-3}(3I_0^iI_0^k - \mathbf{I}^i\cdot\mathbf{I}^k) \tag{28}$$

Thus, the effective secular dipolar Hamiltonian under sample spinning conditions is different from its static counterpart in two respects. First, the angle β''_{ik} is the angle between the dipole–dipole vector and crystal reference frame, in the effective Hamiltonian, while it relates the dipole–dipole vector to the laboratory axis in the static Hamiltonian. Such a difference is not important when we deal with powder samples. Second, in the sample spinning case there is an additional factor of $-1/2$. While this sign does not affect the line-broadening mechanism of $\mathscr{H}_{\text{sec,eff}}^{\text{D}}$, it can have interesting consequences when one exploits the change in sign in some "time reversal" experiments. When the spinning axis makes an angle 54°44′ (MAS) the time average of $\overline{\mathscr{D}_{00}^{2}(0, 54°44', \omega_r t)}$ vanishes and so does $\mathscr{H}_{\text{sec,eff}}^{\text{D}}$.

We shall now give a physical picture of magic angle spinning. Any tensor quantity, whether it be cylindrically symmetric or orthorhombic, when rotated about any axis sufficiently fast "reduces" to effective tensors which are all cylindrically symmetric. Thus all symmetric second-rank tensors become reduced to cylindrically symmetric tensors such that the unique axis of the resultant tensor lies along the rotation axis. Now if the quantizing external field makes an angle 54°44′ to the unique axis of a cylindrically symmetric tensor, then the interaction effectively vanishes. This is pictorially represented in Fig. 6.2. Thus, all anisotropies, whether chemical shift, dipolar interaction, or quadrupole interaction, are made to vanish. Therefore, magic angle spinning leads to indiscriminate averaging to zero of all anisotropic interactions, provided the spinning speed is sufficiently high. The average chemical shift one obtains in such a MAS experiment is usually equal to the isotropic chemical shift one gets in a solution of the same system; however, if specific packing effects in the solid lead to considerable changes in the chemical shift anisotropy, then the isotropic average and MAS results need not coincide. Figure 6.3 shows some examples of MAS results in some solid systems, where one can see a dramatic reduction in the linewidths and in favorable circumstances get excellent ideas of chemical shift differences in solids, otherwise masked by dipolar broadened Gaussian lines in the static case.

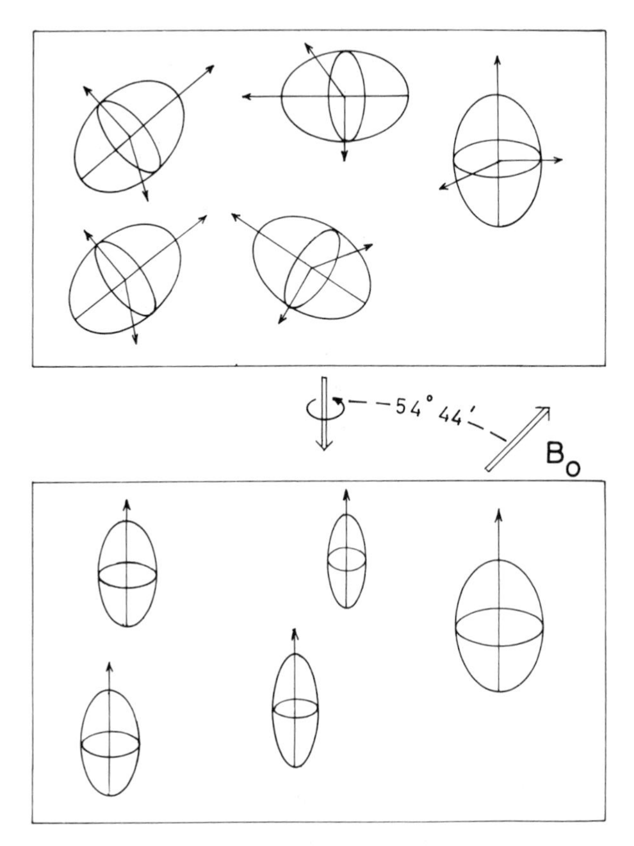

Figure 6.2. Schematic representation of MAS. Top picture shows random orientations of second-rank symmetric tensor ellipsoids. For rotation about the magic angle, all are time averaged to axially symmetrical tensors of different magnitude but with the unique axis along the rotation axis.

To effectively average all dipolar interactions one needs MAS speeds of the order of the interaction itself, so that a 2.5 kHz dipolar coupling would need speeds greater than 2500 rps to get rid of the same effectively. In this connection it is interesting to look into cases where the spinning of the sample leads to rotary echoes. Transforming the principal axis to the spinner frame we can express the spherical tensor components:

Figure 6.3. (a) Static and MAS ^{19}F spectra of polycrystalline $KAsF_6$. The bottom figure shows the dramatic reduction in dipolar broadening. The quartet arises from coupling to ^{75}As ($I = 3/2$). [Adapted from E.R. Andrew in *Biennial Rev. Magn. Reson.*, Ed. C.A. McDowell, Vancouver, 1974.] (b) The 22.6-MHz ^{13}C MAS spectrum of polycarbonate, spinning at 3 kHz. [Reprinted with permission from J. Schaefer, E.O. Stejskal and R. Buchdahl, *Macromolecules*, **10**, 384 (1977); copyright (1977), American Chemical Society.] ▶

(a)
19F in KAsF6

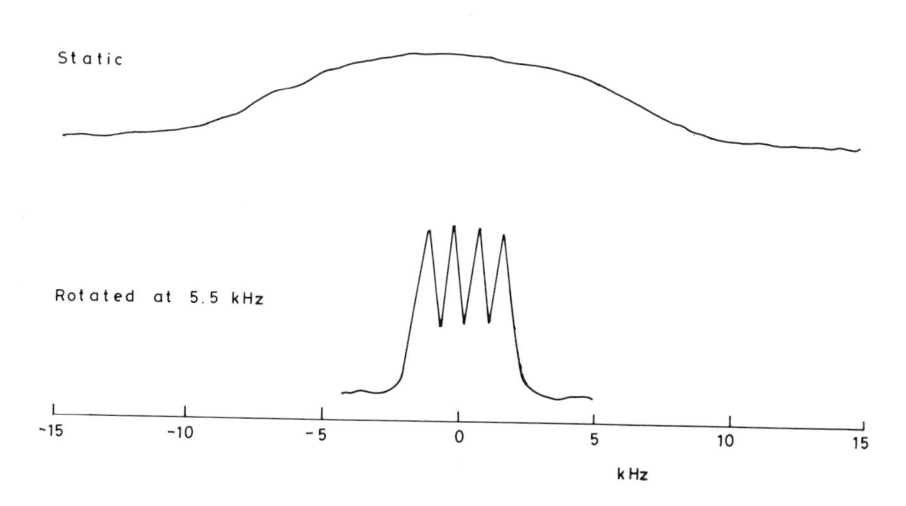
Static
Rotated at 5.5 kHz
-15
-10
-5
0
5
10
15
kHz

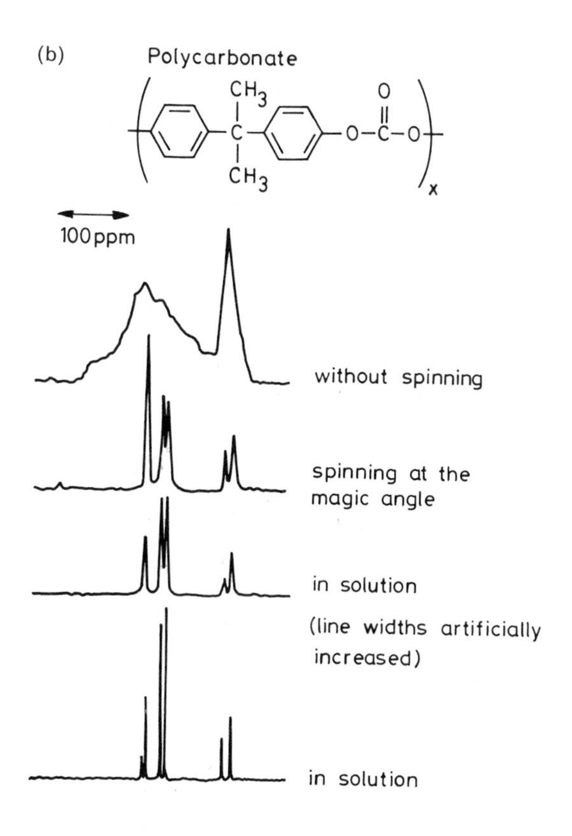
(b)
Polycarbonate
CH3
CH3
O
O–C–O
x
100ppm
without spinning
spinning at the magic angle
in solution
(line widths artificially increased)
in solution

$$R_{2m}^{\mathrm{D}} = \sum_{m'=-2}^{+2} ST_{2m'}^{\mathrm{D}} \exp(-i\alpha m')\mathscr{D}_{m'm}^{2}(\beta)\exp(i\gamma m) \tag{29}$$

where (α, β, γ) are the Euler angles relating the principal axis to the frame coinciding with the rotor axis we have already defined:

$$\begin{aligned} ST_{20} &= \sqrt{\frac{3}{2}}(\delta_{33} - \delta_{\mathrm{av}}) \\ ST_{2\pm 1} &= 0 \\ ST_{2\pm 2} &= \frac{1}{2}(\delta_{11} - \delta_{22}) \end{aligned} \tag{30}$$

where δ_{11}, δ_{22}, δ_{33} are principal values of the $\boldsymbol{\delta}$ tensor. Imparting a rotation at an angle θ about the static field leads to time-dependent frequencies $\omega(t)$ at:

$$\omega(t) = \omega_0 \left[\delta_{\mathrm{av}} + \sqrt{\frac{3}{2}} \sum_{m,m'=-2}^{+2} ST_{2m'} e^{i\alpha m'} \mathscr{D}_{m'm}^{2}(\beta)\mathscr{D}_{m0}^{2}(\theta) e^{-im(\gamma+\omega_r t)} \right] \tag{31}$$

where $\mathscr{D}_{m'm}^{2}$ are elements of the Wigner rotation matrix. The $\omega(t)$ can be separated into two parts: a static part, which depends on only the principal values of the tensor and Euler angles α and β, and a time-dependent term, which depends on the Euler angle γ and the spinning frequency ω_r. The time-dependent part is cyclic with a period $2\pi/\omega_r$ and hence generates side-bands at frequencies $\pm n\omega_r$ whose intensities depend critically on ω_r and n and the range of anisotropy of the tensor. A Fourier analysis of the time-dependent part of the free induction decay leads to the following conclusion (a full derivation of the results is beyond the scope of the book).

Defining:

$$\begin{aligned} \eta'' &= -\left(\frac{\gamma B_0}{\omega_r}\right)(\delta_{xx} - \delta_{yy}) \\ \delta'' &= -\left(\frac{\gamma B_0}{\omega_r}\right)3(\delta_{\mathrm{av}} - \delta_{zz}) \end{aligned} \tag{32}$$

with the convention $\delta_{zz} > \delta_{xx} > \delta_{yy}$, and further,

$$\begin{aligned} \mathrm{A}_1 &= \frac{\sqrt{2}}{6}\{[\eta'' \cos 2\alpha + \delta''] \sin 2\beta\} \\ \mathrm{A}_2 &= \frac{1}{24}\{\eta'' \cos 2\alpha[3 + \cos 2\beta] + \delta''[\cos 2\beta - 1]\} \\ \mathrm{B}_1 &= \frac{\sqrt{2}}{3}(-\eta'' \sin 2\alpha \sin \beta) \\ \mathrm{B}_2 &= \frac{1}{6}(-\eta'' \sin 2\alpha \cos \beta) \end{aligned} \tag{33}$$

the intensity of the nth sideband is given by:

$$I_n = \frac{1}{4\pi}\int_0^{\pi}\int_0^{2\pi} |F|^2 \, d\alpha \sin\beta \, d\beta$$

where:

$$F = \sum_{j=-\infty}^{+\infty}\sum_{k=-\infty}^{+\infty}\sum_{m=-\infty}^{+\infty} J_j(\mathrm{A}_2)J_k(\mathrm{B}_2)J_{n-2j-2k-m}(\mathrm{A}_1)J_m(\mathrm{B}_1) \tag{34}$$

Here J's are the spherical Bessel functions of the first kind. The infinite sum can be avoided by writing it in terms of the following integral:

$$F = \frac{1}{2\pi}\int_0^{2\pi} \exp\left[i(n\theta + \mathrm{A}_2 \sin 2\theta + \mathrm{B}_2 \cos 2\theta + \mathrm{A}_1 \sin\theta + \mathrm{B}_1 \cos\theta)\right] d\theta \tag{35}$$

It can be shown that the sum of the intensities of the sidebands and the central resonance is a constant and the spectral pattern and relative intensities are independent of the sign of ω_r. At $\omega_r \gg \delta_{\max} - \delta_{\min}$ (range of anisotropy) the sidebands vanish. Thus looking at the relative intensities of a few sidebands one can get a fairly reasonable estimate of the principal components of the chemical shift tensor. Figure 6.4 shows an example of these sideband patterns as a function of rotor speed and it can be seen that at low speeds the sideband progressions do peak at approximate positions of the principal values of the chemical shift tensor.

Thus magic angle spinning techniques bring into the realm of NMR a large number of specimens that cannot be studied in solution. Additionally, the study of high-density polymers and other materials used in application devices, such as ceramics and refractories, can be pursued at higher precision using the MAS technique. Chemical shift differences can give very important clues as to the structural and conformational status, distinguish the crystalline and amorphous nature of polymers, quantify aromatic versus aliphatic carbon content in coal, etc.

Whereas the presence of sidebands can lead to a determination of chemical shift anisotropy, this is useful and simple only when a few chemically different nuclei are present in the molecule. When a large number of chemically shifted nuclei are present, then unless sidebands are suppressed it is almost impossible to distinguish the actual centrebands of interest from the sidebands. The methods of sideband suppression are described briefly toward the end of this chapter.

It is of interest to note that even for HR NMR of liquids, MAS has been shown to be of advantage because it gets rid of susceptibility broadening effects.

Averaging in Spin Operator Space: Multiple-Pulse Line Narrowing in Solids

We have seen that MAS indiscriminately averages all anisotropic interactions in solids, causing in favorable instances, only scalar chemical shifts and spin–spin coupling to be retained. In typical homonuclear cases, such as the

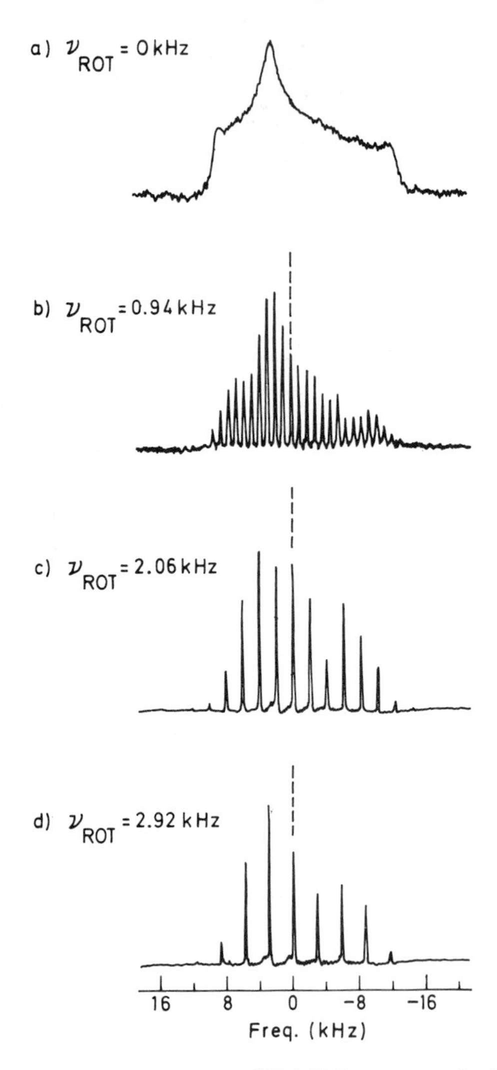

Figure 6.4. Proton-decoupled 119.05-MHz ^{31}P MAS spectrum of solid Barium diethylphosphate at the indicated rotor speeds. Note that the intensity profiles of the sideband patterns at moderate speeds reflect the shift anisotropy. [Reproduced by permission. J. Herzfeld and A.E. Berger, *J. Chem. Phy.*, **73**, 6024 (1980), copyright 1980, American Institute of Physics]

^1H and ^{19}F systems, the dipolar coupling may be tens of kilohertz, requiring very high spinning speeds. The MAS technique, which is averaging in coordinate space, induces a time dependence on the second-rank tensors $\mathbf{R}^{\lambda}$ leading to a time average that nulls the traceless part of the tensors. This can be viewed as a dipole vector **I**–**S** being spun around an axis making angle 54°44′ with it. In a laboratory Cartesian axis system, with the z axis coinciding with the magnetic field, the dipolar coupling during one cycle of rotation averages exactly to zero (being given by the classical expression of a field arising from a bar magnet of pole strength M on a point dipole at a distance d in the z, x, and y directions as $2M/d^3$, $-M/d^3$, $-M/d^3$ which vanishes on an average for equal occupancy of the three directions). There is yet another way of achieving an averaging process, this time by imparting a time dependence to the I_{lm}^{λ} operators. Here we do not physically move the nuclei in space, but subject the spin system to a series of short-duration intense pulses which lead to imparting time dependence to the internal Hamiltonians. Such an averaging, where we scramble the spin operators leading to spin-operator-space averaging, can be done much more efficiently and at very high frequencies compared to MAS. Not only can we achieve line narrowing (*vide infra*), we have an additional important leeway. The radiofrequency pulses that can cause nutations of the spins are selective to only one species of spins, so that it is possible to discriminate, for example, between homo- and heteronuclear interactions. When we apply rf fields on or close to the resonance of one type of spins only the corresponding spin operators are subject to toggling. The T_{lm}^{λ} operators therefore need not transform under rotations in I spin subspace as irreducible tensor operators of rank l and order m. Thus the ranks l of T_{00}^{cs}, T_{20}^{cs}, $T_{20}^{\text{D}}(IS)$, $T_{00}^{J}(IS)$, $T_{20}^{J}(IS)$ are different in ordinary space and I-spin subspace. Therefore, they can be discriminated from $T_{00}^{\text{D}}(I^iI^k)$ and $T_{20}^{\text{D}}(I^iI^k)$, which transform identically under rotations in both ordinary space and I-spin subspace (see Appendix 6).

Let us write the spin Hamiltonian of an NMR system in the absence of rf perturbation as:

$$\mathscr{H} = \mathscr{H}_{\text{Zee}} + \mathscr{H}_{\text{int}} \tag{36}$$

where the $\mathscr{H}_{\text{int}}$ corresponds to all interactions except the nuclear Zeeman term. The time development of the density matrix is given by:

$$\dot{\sigma}(t) = -i[\mathscr{H}, \sigma(t)] \tag{37}$$

and the point here is that $\mathscr{H}$ is time independent as far as the spin operators are concerned. We now represent the density matrix in the rotating frame in the so-called interaction representation by defining:

$$\sigma(t) = T_z \sigma_R(t) T_z^{-1}$$

where

$$T_z = \exp(-i\mathscr{H}_{\text{Zee}}t) \tag{38}$$

Differentiating the equation:

$$\begin{aligned}
\dot{\sigma} &= \dot{T}_z \sigma_R T_z^{-1} + T_z \dot{\sigma}_R T_z^{-1} + T_z \sigma_R \dot{T}_z^{-1} \\
&= -i[\mathscr{H}_{\text{Zee}}, T_z \sigma_R T_z^{-1}] + T_z \dot{\sigma}_R T_z^{-1} \\
&= -i[(\mathscr{H}_{\text{Zee}} + \mathscr{H}_{\text{int}}), T_z \sigma_R T_z^{-1}]
\end{aligned} \tag{39}$$

Evaluation of the commutators leads to:

$$T_z \dot{\sigma}_R T_z^{-1} = -i[\mathscr{H}_{\text{int}}, T_z \sigma_R T_z^{-1}] \tag{40}$$

Left multiplying with T_z^{-1} and right multiplying with T_z we get,

$$\dot{\sigma}_R = -i[\mathscr{H}_{\text{int}}(R), \sigma_R]$$

where

$$\mathscr{H}_{\text{int}}(R) = T_z^{-1} \mathscr{H}_{\text{int}} T_z \tag{41}$$

The factors on which the T_z operators operate are the spin operators contained in the T_{lm}^{λ}, so that:

$$T_{lm}^{\lambda} \to T_{lm}^{\lambda}(R) = T_z^{-1} T_{lm}^{\lambda} T_z \tag{42}$$

Similarly, by subjecting the spin system to a series of pulses the spin Hamiltonian can be made to appear time dependent in a controlled manner, while in laboratory frame the Hamiltonian is static in a rigid system.

Consider a spin system subject to a series of delta pulses (so narrow that we can ignore evolutions of spin system during the pulse). The spin system, initially in a quantum state $|\psi_0)$, is taken to a state $|\psi_1)$ by the first pulse. During the interval between the first and second pulses, the state $|\psi_1)$ evolves freely under the internal Hamiltonian $\mathscr{H}_{\text{int}}$, say for the duration τ_1. The resulting state, $|\psi_2)$, can be subject to the second pulse, followed by free evolution for a period τ_2, and so on. Arbitrarily narrow and intense pulses induce rotation of the spin operators corresponding to the resonant spins by an instantaneous transformation given by:

$$P_k = \exp(-i\theta_k \mathbf{n}_k \cdot \mathbf{I}) \tag{43}$$

where θ_k is the nutation angle ($\gamma B_1 t_p$ radians) and $\mathbf{n}_k$ is the unit vector along which the kth pulse has been applied. The spin state of the system just before the $(n + 1)$th pulse is:

$$\left|\psi_0 + \sum_{k=1}^{n} \tau_k\right) = \prod_{k=1}^{n} [\exp(-i\mathscr{H}\tau_k) P_k] |\psi_0) \tag{44}$$

with the k's arranged in increasing order from right to left. By repeatedly using the identity $P_k P_k^{-1} = 1$, we can rewrite the above equation as:

$$\left|\psi_0 + \sum_{k=1}^{n} \tau_k\right) = \left\{\prod_{m=1}^{n} P_m\right\} \prod_{k=1}^{n} \exp(-i\mathscr{H}_k \tau_k) |\psi_0)$$

with:

$$\mathscr{H}_k = \left(\prod_{l=1}^{k} P_l\right)^{-1} \mathscr{H} \left(\prod_{l=1}^{k} P_l\right) \tag{45}$$

When the system is thus subjected to pulse-interrupted-free evolutions we can define the final stage $|\psi(t_n))$ in terms of a transformation U_t^n such that:

$$|\psi(t_n)) = U_t^n|\psi(t_0)) \tag{46}$$

where for three pulses and three evolution times τ_1, τ_2, and τ_3:

$$\begin{aligned} U_t^n &= U_{\text{int}}(\tau_3)P_3 U_{\text{int}}(\tau_2)P_2 U_{\text{int}}(\tau_1)P_1 \\ &= P_3P_2P_1[P_1^{-1}P_2^{-1}P_3^{-1}U_{\text{int}}(\tau_3)P_3P_2P_1] \\ &\quad \times [P_1^{-1}P_2^{-1}U_{\text{int}}(\tau_2)P_2P_1][P_1^{-1}U_{\text{int}}(\tau_1)P_1] \end{aligned} \tag{47}$$

If we so adjust the sequence of pulses such that:

$$P_3P_2P_1 = 1 \tag{48}$$

we can talk of a cyclic pulse sequence by which the dressed internal Hamiltonians return to their original state. For a cyclic pulse sequence P_1, P_2, P_3,

$$U_t = U_{\text{int}}(\tau_3)[P_1^{-1}P_2^{-1}U_{\text{int}}(\tau_2)P_2P_1][P_1^{-1}U_{\text{int}}(\tau_1)P_1] \tag{49}$$

For a unitary operator P (pulse rotation, for example), a Hamiltonian $\mathscr{H}$ and the corresponding transformation $U(t - t_0)$ we have:

$$\begin{aligned} P^{-1}U(t-t_0)P &= P^{-1}[\exp(-i\mathscr{H}(t-t_0)2\pi/h)]P \\ &= P^{-1}\left[1 - i\mathscr{H}(t-t_0)\frac{2\pi}{h} - \frac{\mathscr{H}^2(t-t_0)^2}{2!}\left(\frac{2\pi}{h}\right)^2 + \cdots\right]P \end{aligned} \tag{50}$$

By inserting $P^{-1}P = 1$ between factors the above expression can be rewritten:

$$P^{-1}U(t-t_0)P = \exp\left(-\frac{2\pi}{h}i(P^{-1}\mathscr{H}P)(t-t_0)\right) \tag{51}$$

The effect of $P^{-1}UP$ is to cause U to evolve in time under a transformed Hamiltonian. The exponent in the expression for $U_{\text{int}}(\tau_2)$ is given by $P_1^{-1}P_2^{-1}\mathscr{H}_{\text{int}}(\tau_2)P_2P_1$; i.e., we first do the transformation $P_2^{-1}\mathscr{H}_{\text{int}}P_2$ and then sandwich the results between P_1^{-1} and P_1 and evaluate the same. The order of application of rotation operators to $\mathscr{H}_{\text{int}}(t)$ is the reverse of the order of applying the same to spin operators. Phenomenologically this is not surprising, because we are applying a sequence of operations corresponding to pulse rotation of spin operators, which obviously make the internal Hamiltonians (which were otherwise undisturbed) to be operated upon in the reverse sense.

Let us now define a cycle of pulses and the so-called cycle time. Suppose for some value n the pulses have the property such that:

$$\prod_{m=1}^{n} P_m = 1 \tag{52}$$

Such a sequence is called a cycle, during which the spin operators go through a "toggling" excursion and end up at the original status. The duration of the cycle is given by:

$$\sum_{m=1}^{n} \tau_m = t_c \tag{53}$$

called the cycle time. We then have:

$$|\psi_0 + t_c) = \left\{ \prod_{k=1}^{n} \exp(-i\mathscr{H}_k \tau_k) \right\} |\psi_0) = U_{t_c} |\psi_0) \tag{54}$$

If there are N such cycles then the periodicity with frequency t_c^{-1} gives:

$$|\psi_0 + Nt_c) = (U_{t_c})^N |\psi_0) \tag{55}$$

It is possible to show (see Appendix 5) that as t_c/T_2 approaches zero the system evolves over long times Nt_c according to the time-independent average Hamiltonian $\bar{\mathscr{H}}$ such that:

$$\lim_{\substack{t_c/T_2 \to 0 \\ N \to \infty \\ Nt_c = t}} [U_{t_c}]^N = \exp(-i\bar{\mathscr{H}} Nt_c)$$

with:

$$\bar{\mathscr{H}} = \sum_{k=1}^{N} \mathscr{H}_k(\tau_k/t_c) \tag{56}$$

Here, $\exp(-i\bar{\mathscr{H}} Nt_c)$ is the time development operator under the influence of the pulse sequence which would be used to calculate the Bloch decay. Fourier transformation of this last should lead to the unsaturated slow passage spectrum of the system evolving under the average (fictitious) Hamiltonian. For the averaging process to be efficient all units of "action" $\mathscr{H}_i \tau_i$ during a given cycle should be infinitesimal quantities. Experimentally the τ_i's, which can be termed the state residence times, are the convenient parameters that can be manipulated to satisfy this condition. Because of the periodicity introduced by the pulse trains the magnetization also will experience strong periodicities, leading to distinct sidebands in the frequency spectrum. By restricting our observations at instances separated by t_c, i.e., by observing the magnetization through windows during evolutions (by so-called stroboscopic observations), we eliminate these periodicities and focus attention only on the central components. Such a rosy picture, however, will be marred by not being able to satisfy the condition $t_c/T_2 \to 0$ as well as possible finite width for the nutating pulses. These effects can be taken into account in a suitable way.

Thus time evolution of the density matrix is given by:

$$\sigma(t) = U_t \sigma(0) U_t^\dagger$$

where:

$$U_t = T \exp\left(-i \int_0^t [\mathscr{H}_{\text{int}} + \mathscr{H}_1(t')]\, dt' \right) \tag{57}$$

where $\mathscr{H}_1(t)$ is the rf Hamiltonian and T is the Dyson time ordering operator, which makes sure that "actions" that occur later in time are arranged in the order right to left. To separate out part of the motion due to the rf Hamiltonian alone we write:

$$U_t = L_1(t)\mathscr{L}(t) \tag{58}$$

where:

$$L_1(t) = T\exp\left(-i\int_0^t \mathscr{H}_1(t')\,dt'\right) \tag{59}$$

so that:

$$\mathscr{L}(t) = T\exp\left(-i\int_0^t \mathscr{H}_R(t')\,dt'\right) \tag{60}$$

where:

$$\mathscr{H}_R(t) = L_1^{-1}(t)\mathscr{H}_{\text{int}}L_1(t) \tag{61}$$

Subjecting the system to a periodic progression of cyclic pulses over a cycle time t_c:

$$\mathscr{H}_1(t + Nt_c) = \mathscr{H}_1(t)$$

$$L_1(Nt_c) = T\exp\left(-i\int_0^{Nt_c} \mathscr{H}_1(t)\,dt\right) = 1$$

Therefore:

$$\tilde{\mathscr{H}}(t + Nt_c) = \tilde{\mathscr{H}}(t) \tag{62}$$

That is, the periodicity of $\mathscr{H}_1(t)$ is transferred to the now time-dependent internal Hamiltonian $\tilde{\mathscr{H}}(t)$. Also,

$$\sigma(Nt_c) = \mathscr{L}(Nt_c)\sigma(0)\mathscr{L}^\dagger(Nt_c)$$

and:

$$\mathscr{L}(Nt_c) = [\mathscr{L}(t_c)]^N \tag{63}$$

To calculate the time evolution of the density matrix of the system at any integer multiple of the cycle time t_c, it is enough to calculate over the short time t_c and then this one-cycle propagator is raised to the Nth power.

To obtain the time evolution in the form of a single exponential we resort to Magnus expansion (see Appendix 5) since the $\tilde{\mathscr{H}}_{\text{int}}(t)$ in general does not commute with itself at all times:

$$\begin{aligned}\mathscr{L}(t_c) &= T\exp\left[-i\int_0^{t_c} \tilde{\mathscr{H}}_{\text{int}}(t')\,dt'\right] \\ &= \exp\left[-i(\bar{\mathscr{H}} + \bar{\mathscr{H}}^{(1)} + \bar{\mathscr{H}}^{(2)} + \cdots)t_c\right]\end{aligned} \tag{64}$$

where:

$$\begin{aligned}
\bar{\mathscr{H}} &= \frac{1}{t_c}\int_0^{t_c} \tilde{\mathscr{H}}_{\text{int}}(t)\,dt \\
\bar{\mathscr{H}}^{(1)} &= -\frac{i}{2t_c}\int_0^{t_c} dt_2 \int_0^{t_2} dt_1\,[\tilde{\mathscr{H}}_{\text{int}}(t_2), \tilde{\mathscr{H}}_{\text{int}}(t_1)] \\
\bar{\mathscr{H}}^{(2)} &= -\frac{1}{6t_c}\int_0^{t_c} dt_3 \int_0^{t_3} dt_2 \int_0^{t_2} dt_1\,\{[\tilde{\mathscr{H}}_{\text{int}}(t_3), [\tilde{\mathscr{H}}_{\text{int}}(t_2), \tilde{\mathscr{H}}_{\text{int}}(t_1)]] \\
&\quad + [\tilde{\mathscr{H}}_{\text{int}}(t_1), [\tilde{\mathscr{H}}_{\text{int}}(t_2), \tilde{\mathscr{H}}_{\text{int}}(t_3)]]\}
\end{aligned} \tag{65}$$

If the internal Hamiltonian $\tilde{\mathscr{H}}_{\text{int}}(t)$ commuted with itself at all times, then $\bar{\mathscr{H}}$ (zeroth-order average Hamiltonian) alone describes the time development of the system. If the observation time is restricted to integer multiples of t_c then,

$$\begin{aligned}
\sigma(t) &= \exp[-i(\bar{\mathscr{H}} + \bar{\mathscr{H}}^{(1)} + \bar{\mathscr{H}}^{(2)} + \cdots)t]\sigma(0) \\
&\quad \times \exp[i(\bar{\mathscr{H}} + \bar{\mathscr{H}}^{(1)} + \bar{\mathscr{H}}^{(2)} + \cdots)t]
\end{aligned} \tag{66}$$

where $\bar{\mathscr{H}}^{(i)}$ is the ith order correction to the average Hamiltonian.

We can also talk of symmetric cycles, where:

$$\tilde{\mathscr{H}}_{\text{int}}(t) = \tilde{\mathscr{H}}_{\text{int}}(t_c - t) \tag{67}$$

in which case Magnus expansion shows that all odd-order corrections to the average Hamiltonian vanish; also, for antisymmetrical cycles defined by:

$$\tilde{\mathscr{H}}_{\text{int}}(t) = -\tilde{\mathscr{H}}_{\text{int}}(t_c - t) \tag{68}$$

the average Hamiltonian of all orders vanish.

Now we are reasonably equipped with the basics required to analyze multiple-pulse experiments.

We have included in Appendix 5 the results for a Carr–Purcell echo train. This sequence leads to positive and negative echoes and the static field inhomogeneities are eliminated. However, an error in the setting of the π pulses accumulates proportional to the length of the echo train and destroys the echo envelope. The Meiboom–Gill modification corrects this by a 90° phase shift of the initializing pulse. Also discussed in that Appendix is the phase-alternated sequence, which is merely a series of equally spaced δ pulses alternating in phase. It leads to scaling of the chemical shifts by a factor of $\cos(\beta/2)$, where β is the flip angle corresponding to the pulses. It also scales the dipolar interactions. We shall now briefly describe the multiple-pulse line-narrowing sequences used in studying solids.

The WAHUHA-4 and MREV-8 Pulse Line-Narrowing Sequences

The WAHUHA-4 sequence. The WAHUHA-4 pulse sequence can be written as (τ–P_{-x}–τ–P_y–2τ–P_{-y}–τ–P_x–τ) and is shown in Fig. 6.5, which consists of a cycle of four 90° pulses with subcycles of two pulses each. The cycle time is 6τ. For δ pulses we can look at the transformations of the operators corresponding to chemical shift and dipolar Hamiltonian under the influence of the pulses (Table 6.1). Thus homonuclear dipolar coupling is averaged to zero,

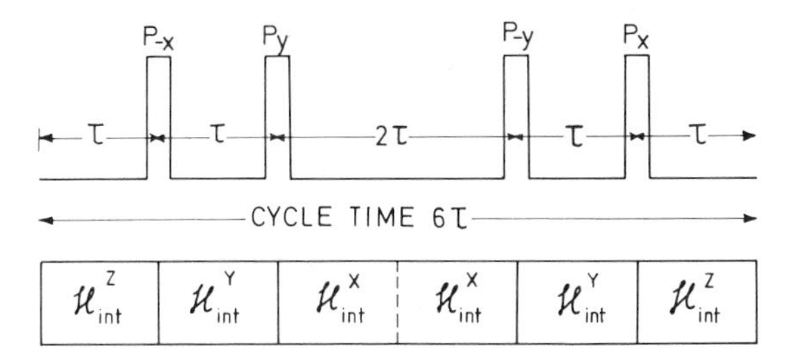

Figure 6.5. The WAHUHA-4 pulse sequence, inclusive of the states of the internal Hamiltonian in the toggling frame.

Table 6.1. Transformation of Spin Operators in the Toggling Frame During Windows of WAHUHA-4 Pulse Sequence

Time	U_{rf}	$U_{rf}^{-1}I_zU_{rf}$	$U_{rf}^{-1}(3I_z^iI_z^k - \mathbf{I}^i\cdot\mathbf{I}^k)U_{rf}$	$\mathscr{H}_T^{\text{int}}(t)$
$0\ldots\tau$	1	I_z	$3I_z^iI_z^k - \mathbf{I}^i\cdot\mathbf{I}^k$	$\mathscr{H}_z^{\text{int}}$
$5\tau\ldots6\tau$				
$\tau\ldots2\tau$	$e^{-i(\pi/2)I_x}$	I_y	$3I_y^iI_y^k - \mathbf{I}^i\cdot\mathbf{I}^k$	$\mathscr{H}_y^{\text{int}}$
$4\tau\ldots5\tau$				
$2\tau\ldots3\tau$	$e^{i(\pi/2)I_y}e^{-i(\pi/2)I_x}$	I_x	$3I_x^iI_x^k - \mathbf{I}^i\cdot\mathbf{I}^k$	$\mathscr{H}_x^{\text{int}}$
$3\tau\ldots4\tau$				
Average		$(1/3)(I_x + I_y + I_z)$	0	

while the shift Hamiltonian (and likewise the heteronuclear dipolar Hamiltonian) are scaled down. The shift Hamiltonian according to average Hamiltonian theory in the zeroth order is:

$$\bar{\mathscr{H}}_{\text{CS}} = \frac{1}{3}\sum_k(\delta_k + \Delta)(I_{kx} + I_{ky} + I_{kz}) \tag{69}$$

where δ_k is the chemical shift of the kth nucleus and Δ is the offset. Infact chemical shift and offset cannot be separated. Rewriting the above expression as:

$$\bar{\mathscr{H}}_{\text{CS}} = \frac{1}{\sqrt{3}}\sum_k(\delta_k + \Delta)\mathbf{I}_{k\langle 111\rangle}$$

where:

$$\mathbf{I}_{k\langle 111\rangle} = \frac{1}{\sqrt{3}}(I_{kx} + I_{ky} + I_{kz}) \tag{70}$$

shows that chemical shifts are scaled by a factor of $1/\sqrt{3}$. Alternatively, in the toggling frame the effective static field is $(1/\sqrt{3})B_0$. This scale factor of $1/\sqrt{3}$ is true only for δ pulses. In actual practice the scale factor is to be found from a reference liquid standard by measuring the chemical shift under the influence of the sequence and comparing with its normal chemical shift. Since the cycle is symmetric, $\bar{\mathscr{H}}_{\mathrm{D}}^{(n)} = 0$ for n odd. Thus the Magnus expansion contains no term quadratic in $\mathscr{H}_{\mathrm{D}}$ nor any cross-terms of the type $\mathscr{H}_{\mathrm{D}}\mathscr{H}_{\mathrm{CS}}$.

Second-order correction to dipolar interaction is given by:

$$\bar{\mathscr{H}}_{\mathrm{D}}^{(2)} = \frac{1}{648} t_c^2 [(\mathscr{H}_{\mathrm{D}}^x - \mathscr{H}_{\mathrm{D}}^z), [\mathscr{H}_{\mathrm{D}}^x, \mathscr{H}_{\mathrm{D}}^y]]$$

where:

$$\mathscr{H}_{\mathrm{D}}^i = \gamma^2 \frac{h}{2\pi} \sum_{j<k} \frac{1}{2}(3\cos^2\theta_{jk} - 1)\frac{1}{r_{jk}^3}(\mathbf{I}^j \cdot \mathbf{I}^k - 3I_i^j I_i^k); \quad i = x, y, z \qquad (71)$$

A few examples of the application of the WAHUHA-4 pulse sequence are given in Fig. 6.6. It is also possible to eliminate any remaining heteronuclear coupling by applying 180° pulses at the middle of every cycle. That is, observation windows and decoupling windows alternate through the subcycles (Fig. 6.7). The second-order correction to chemical shift interaction from Magnus expansion gives:

$$\bar{\mathscr{H}}_{\mathrm{CS}}^{(2)} = -\frac{\tau^2}{18}\sum_k (\delta_k + \Delta)^3 (I_{ky} - 2I_{kx} + 4I_{kz}). \qquad (72)$$

The MREV-8 pulse sequence. There is a family of 8 pulse–12 τ complementary pulse cycles of which MREV-8 cycle was the first to be proposed. The sequence and the corresponding $\tilde{\mathscr{H}}_{\mathrm{int}}(t)$ are indicated in Fig. 6.8.

Since $\tilde{\mathscr{H}}_{\mathrm{int}}(t)$ is toggling equally between the x, y, and z axes of the rotating

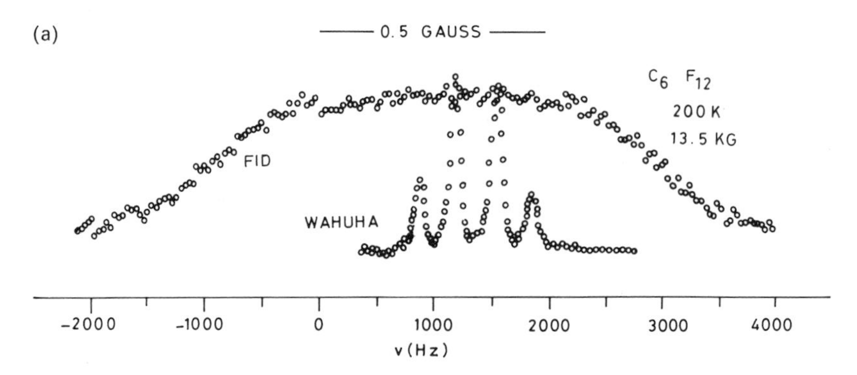

Figure 6.6. (a) The WAHUHA-4 multipulse line-narrowed ^{19}F spectrum of C_6F_{12} at 200K. The upper trace shows the dipolar-broadened spectrum, while the lower curve shows the AB pattern from axial and equatorial fluorines. [Reprinted with permission from D. Ellett, U. Haeberlen and J.S. Waugh, *J. Amer. Chem. Soc.*, **92**, 411 (1970), copyright (1970), American Chemical Society.] (b) The WAHUHA-4 ^1H spectrum of a single crystal of malonic acid at two different orientations. [Reproduced by permission. U. Haeberlen, *High Resolution NMR in Solids*, Academic Press, New York, 1976).]

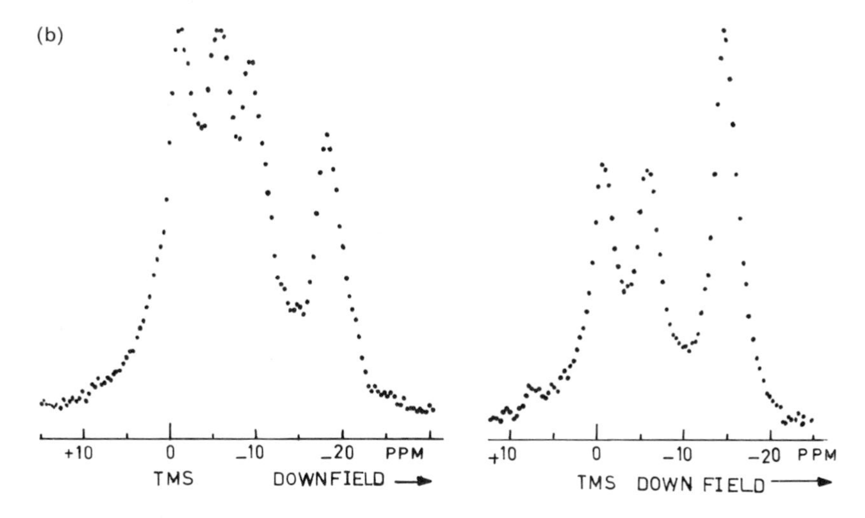

Figure 6.6. (Continued)

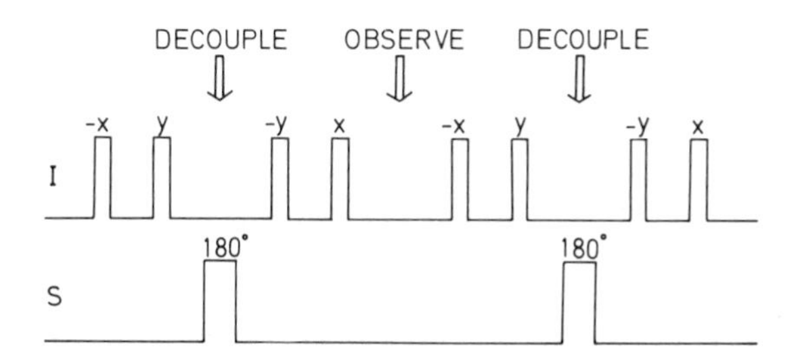

Figure 6.7. Pulse scheme for heterodecoupled acquisition of WAHUHA-4 line-narrowed spectrum.

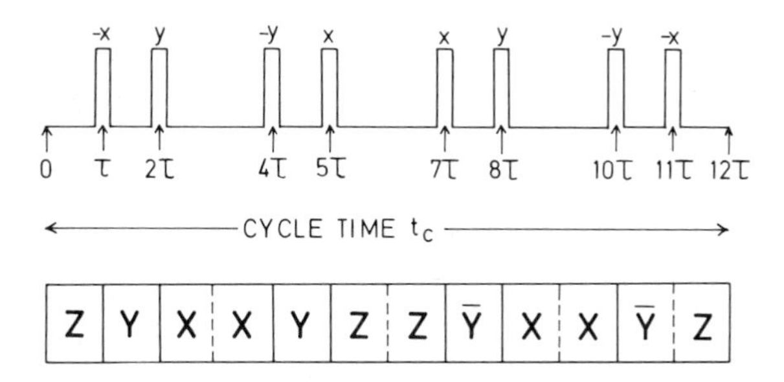

Figure 6.8. MREV-8 pulse sequence and the corresponding states of the internal Hamiltonian in the toggling frame.

frame, the zeroth-order average dipolar coupling $\bar{\mathscr{H}}_{\mathrm{D}}^{(0)}$ is zero. The chemical shift tensor is given by:

$$\begin{aligned} \bar{\mathscr{H}}_{\mathrm{CS}} &= \frac{1}{3}\sum_k (\delta_k + \Delta)(I_{kx} + I_{kz}) \\ &= \frac{\sqrt{2}}{3}\sum_k (\delta_k + \Delta)\mathbf{I}_{k\langle 101\rangle} \end{aligned} \tag{73}$$

where $\mathbf{I}_{k\langle 101\rangle} = (1/\sqrt{2})(I_{kx} + I_{kz})$ so that the scale factor for an idealized MREV-8 sequence is $\sqrt{2}/3$. As for the first-order $\bar{\mathscr{H}}_{\mathrm{D}}^{(1)}$, since the cycle is symmetrical, $\bar{\mathscr{H}}_{\mathrm{D}}^{(n)}$ (n = odd) vanishes. However, with respect to the shift Hamiltonian the cycle symmetry is broken and leads to:

$$\bar{\mathscr{H}}_{\mathrm{CS}}^{(1)} = \frac{\tau}{3}\sum_k (\delta_k + \Delta)^2 (I_{kz} - I_{kx}) \tag{74}$$

The second-order correction to the dipolar term is identical to that of the WAHUHA-4 pulse scheme since the subcycle structure is the same in both MREV and WAHUHA as far as the dipolar Hamiltonian is concerned. The MREV sequence scales the chemical shift further compared to WAHUHA sequence. However, with respect to the effect of pulse imperfections the MREV cycle is superior to WAHUHA.

Chemical shift scale factors. While we do not want to go elaborately into the effect of flip angle errors and finite pulse widths, etc., on the line-narrowing efficiency, we would like to touch briefly upon the nature of the chemical shift scale factors, and the aportioning of the magnetization into observable signal and pedestal. Each multiple-pulse line-narrowing sequence has its characteristic chemical shift scale factor besides reducing to zero frequency a certain fraction of magnetization in a frame rotating with the receiver frequency, the so-called pedestal mentioned above. Without paying attention to the line-narrowing efficiency *per se* of the WAHUHA-4 and MREV-8 sequences, we shall look into the scale factors, signal, and pedestal intensities as a function of offset (chemical shift); the spectral invariance or lack of it to the sign of offset; as well as the effect on these of the various preparation and detection phases.

The periods of evolution in an N pulse sequence are $(N + 1)$ when we deal with δ pulses or $(2N + 1)$ when we take into account of evolution during the finite pulse width. Action during each of these periods can be set up by (3×3) rotation matrices; the product of these matrices with the initial magnetization vector as determined by the preparation pulse gives the magnetization at the end of one cycle. Repeated application of this will generate the signal after n cycles. Thus, we could generate the time-domain function in a computer for any preparation, detection phase, pulse phase, and evolution duration for any cyclic sequence. The resulting FID can be exponentially weighted and suitably zero filled and Fourier transformed. The state of the system $|\psi(t_c)\rangle$ after each cycle of N pulses is given by:

$$|\psi(t_c)\rangle = \left[\prod_{i=1}^{N} \exp(-i\mathscr{H}_i\tau_i)\right] \exp(-i\mathscr{H}_0\tau_0)|\psi_0\rangle \tag{75}$$

where $|\psi_0\rangle$ is the initial state and τ_0 is the preevolution period before the first pulse. The transverse magnetization after the cycle is given by:

$$M_{x,y} = \gamma \frac{h}{2\pi} \langle\psi(t_c)| I_{x,y} |\psi(t_c)\rangle \tag{76}$$

which in the Heisenberg notation can be written:

$$\begin{aligned} M_x &= (\gamma h/2\pi)\langle 0| AI_x + BI_y + CI_z |0\rangle \\ M_y &= (\gamma h/2\pi)\langle 0| DI_x + EI_y + FI_z |0\rangle \end{aligned} \tag{77}$$

and:

$$M_z = (\gamma h/2\pi)\langle 0| PI_x + QI_y + RI_z |0\rangle$$

where $|0\rangle \equiv |\psi_0\rangle$ and, for example,

$$\begin{aligned} (AI_x + BI_y + CI_z) &= \exp(i\mathscr{H}_0\tau_0)\left[\prod_{i=1}^{N} \exp(i\mathscr{H}_i\tau_i)\right] I_x \\ &\quad \times \left[\prod_{i=1}^{N} \exp(i\mathscr{H}_i\tau_i)\right]^{-1} \exp(-i\mathscr{H}_0\tau_0) \end{aligned} \tag{78}$$

The nine coefficients, A, B, C, D, E, F, P, Q, and R are characteristic of the particular pulse sequence used. Magnetization after n cycles can be calculated by repeated application of the above equation. The mode of preparation/detection determines which of the three terms in I_x, I_y, or I_z finally contributes to the signal. In what follows we compare the results of average Hamiltonian theory and the exact calculations with respect to single spin properties, such as chemical shift scale factor dependence on the offset sign, magnitude, as well as preparation and detection phases. The coefficients A to R for both the sequences are given in Table 6.2 and Table 6.3.

For WAHUHA-4, average Hamiltonian theory predicts that the initial amplitude of the oscillating part of the NMR signal is two thirds of the initial amplitude of the FID, and the pedestals are half as large as the oscillation amplitudes. A 45° additional preparation pulse leads to 50% enhancement in signals and gets rid of the pedestal. First-rank I-spin interactions (chemical shift and heteronuclear coupling) are scaled by a factor $1/\sqrt{3}$, while the second-order correction term to scale factor is $(4\pi^2/\sqrt{3})\,(\Delta^2\tau^2/6)$, where τ is one-sixth the cycle time and Δ is the offset.

The results for the 24-μs (t_c) WAHUHA-4 in the δ pulse limit are shown in Fig. 6.9. The offset corresponding to the alias of the scaled frequency is about 60 kHz in this case. The figure includes a set of plots of the pedestal and absolute signal intensity as a function of offset for various modes of preparation and detection. The plots are all symmetric with respect to the sign of the

Table 6.2. Operator Evolution Matrix Elements[a] for the WAHUHA-4 Sequence

$$A = \frac{5}{8} - \frac{1}{16}c_2 + \frac{3}{8}c_4 + \frac{1}{16}c_6$$

$$B = \frac{1}{2}c - \frac{1}{2}c_3 - \frac{5}{16}s_2 - \frac{1}{4}s_4 - \frac{1}{16}s_6 = b_g + b_u$$

$$C = \frac{1}{4} - \frac{1}{4}c_4 + \frac{1}{4}s + \frac{3}{8}s_3 + \frac{1}{8}s_5 = c_g + c_u$$

$$D = b_g - b_u$$

$$E = -\frac{1}{8} + \frac{15}{16}c_2 + \frac{1}{8}c_4 + \frac{1}{16}c_6$$

$$F = \frac{1}{4}c - \frac{1}{8}c_3 - \frac{1}{8}c_5 - \frac{1}{2}s_2 - \frac{1}{4}s_4 = f_g + f_u$$

$$P = c_g - c_u$$

$$Q = f_g - f_u$$

$$R = -\frac{1}{4} + c_2 + \frac{1}{4}c_4$$

[a] $c_k = \cos 2\pi k\Delta\tau$, $s_k = \sin 2\pi k\Delta\tau$, Δ = offset, $\tau = \frac{1}{6}t_c$.

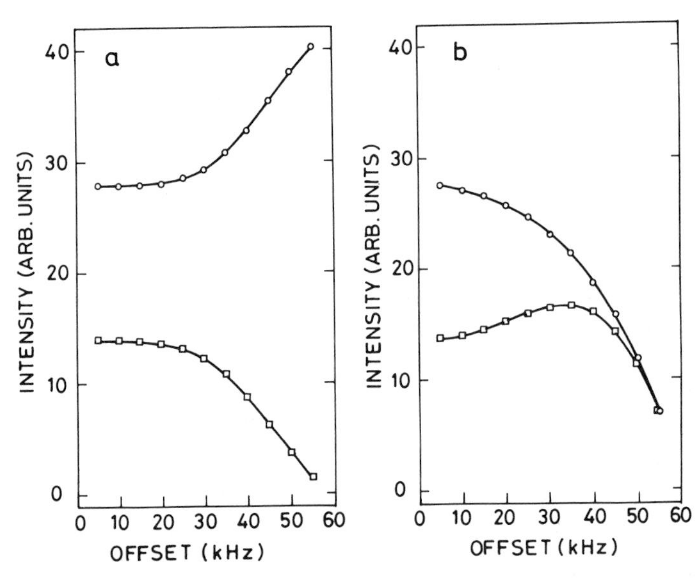

Figure 6.9. Plots of absolute intensity of pedestal and signal versus offset for a 24-μs WAHUHA-4 cycle in the delta pulse limit. ○, signal; □, Pedestal. (a) $90^\circ_x/\langle M_y \rangle$; (b) $90^\circ_x/\langle M_x \rangle$, $90^\circ_{-y}/\langle M_y \rangle$; (c) $90^\circ_{-y}/\langle M_x \rangle$; (d) $0/\langle M_y \rangle$; (e) $0/\langle M_x \rangle$ preparation/detection schemes. [Reprinted with permission. N. Chandrakumar, D. Ramaswamy and S. Subramanian, *J. Magn. Reson.*, **54**, 345 (1983), copyright 1983, Academic Press, New York]

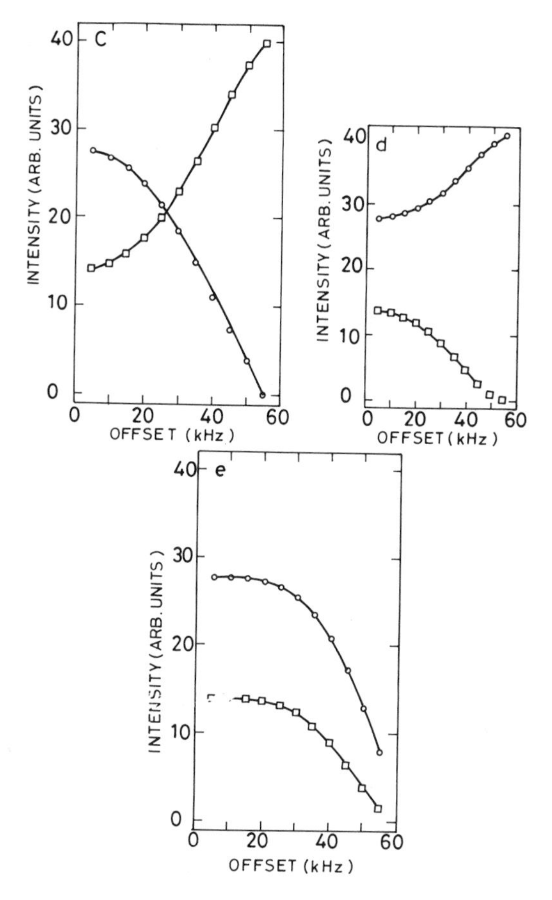

Figure 6.9. (Continued)

offset. The $90_x/\langle M_x \rangle$ and $90_y/\langle M_y \rangle$ preparation/detection methods give rise to mixed mode signals, as do the $\langle M_x \rangle$ and $\langle M_y \rangle$ detection schemes with no explicit preparation pulse at all. Although the scale factor is independent of the preparation/detection scheme, the $90_x/\langle M_y \rangle$ and $0/\langle M_y \rangle$ methods are clearly superior to the other possibilities in terms of signal and pedestal intensity behavior, especially at higher offsets, beyond 20 kHz. A composite 45°_y–90°_x preparation corresponds to an equally weighted linear combination of the $90^\circ_x/\langle M_y \rangle$ and $90^\circ_y/\langle M_y \rangle$ situations and leads, in fact, to signal enhancement and pedestal cancelation to within 10% up to about ± 20 kHz only, which is already the typical fluorine chemical shift anisotropy at 100 MHz. For the case of no explicit preparation pulse, quadrature detection corresponding to $(\langle M_y \rangle - \langle M_x \rangle)$ detection behaves similarly but gives net signals that have *opposite* signs for positive and negative offsets. The scale factor deviation at the alias frequency is also, in fact, very sizeable, being almost 40%. Application

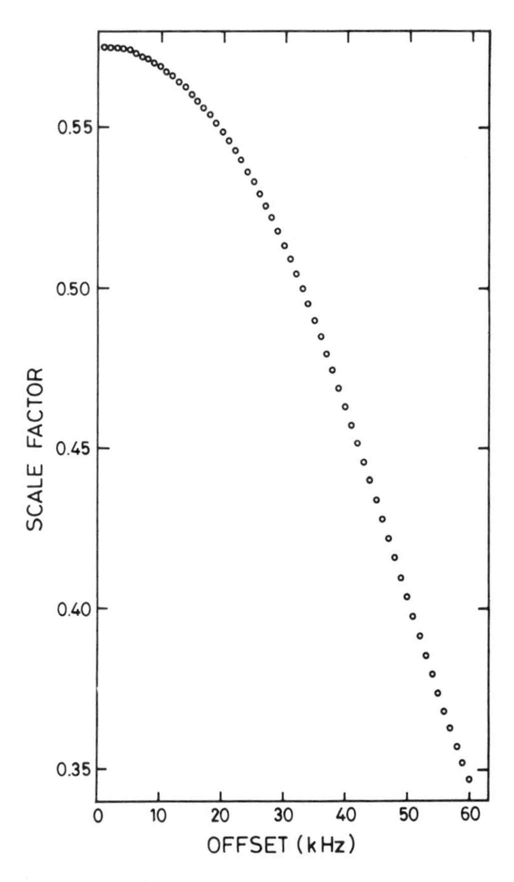

Figure 6.10. Scale factor versus offset for a 24-μs WAHUHA-4 pulse sequence, with pulsewidths of 1μs.

of average Hamiltonian expression, substituting the actual scale factor at alias and not $1/\sqrt{3}$ gives rise to an estimated deviation of 38%. When finite pulsewidths (t_w) are taken into account, average Hamiltonian theory predicts that the scale factor varies monotonically between 0.57735 and 0.56606 as (t_w/τ) increases from 0 to 1. Figure 6.10 is a plot of the scale factor versus offset for the WAHUHA-4 sequence with a cycle time of 24 μs and pulsewidths of 1 μs. The frequency-dependent scale factor deviates by as much as 40% at the Nyquist frequency of 20.833 kHz, at which the scale factor is 0.34, as against the average Hamiltonian value of 0.57532. The behavior of signal and pedestal intensities as a function of offset magnitude and sign is entirely similar to the δ-pulse limit.

The results of pedestal and absolute signal intensity as function of offset for the 27-μs MREV-8 sequence in the δ-pulse limit for various modes of preparation and detection are summarized in Fig. 6.11. The offset corresponding to

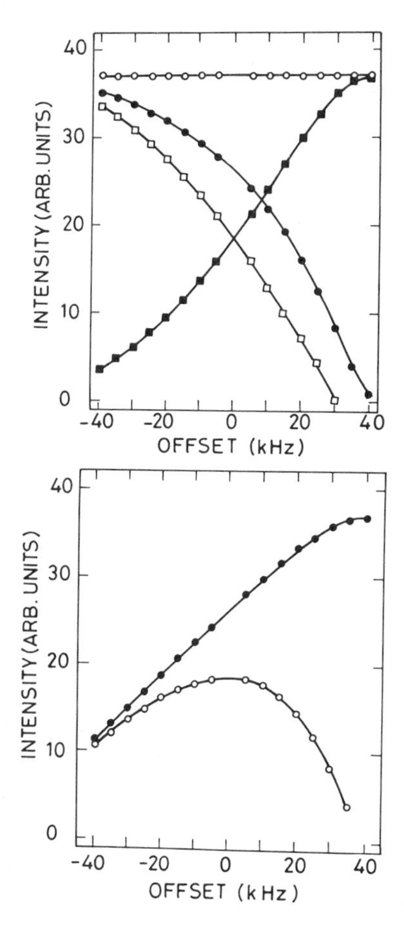

Figure 6.11. Plots of absolute intensity of pedestal and signal versus offset for a 27-μs MREV-8 cycle in the delta pulse limit. Top: ○, Signal for $90^\circ_x/\langle M_y \rangle$; pedestal vanishes. ●, Signal for $90^\circ_x/\langle M_x \rangle$ and $90^\circ_{-y}/\langle M_y \rangle$; pedestal vanishes. □, Signal for $90^\circ_{-y}/\langle M_x \rangle$. ■, Pedestal for $90^\circ_{-y}/\langle M_x \rangle$. Bottom: ●, Signal for $0/\langle M_y \rangle$; pedestal vanishes. ○, Signal and pedestal for $0/\langle M_x \rangle$. [Reprinted with permission. N. Chandrakumar, D. Ramaswamy and S. Subramanian, *J. Magn. Reson.*, **54**, 345 (1983), copyright 1983, Academic Press, New York]

alias of the scaled frequency is about 45 kHz in this case. In particular, except for the $90^\circ_x/\langle M_y \rangle$ situation, the plots are *not* symmetric with respect to zero offset, unlike the WAHUHA-4 sequence. This difference is to be related to the parity characteristics and the relationships between the elements of the operator evolution matrix characteristic of the sequence (Tables 6.2 and 6.3). While all the diagonal elements A, E, and R are even functions of the offset for the WAHUHA-4 sequence, only E is an even function of the offset for the MREV-8 sequence. Also, the signals, which are pure absorption or dispersion

Table 6.3. Operator Evolution Matrix Elements[a] for the MREV-8 Sequence

$$A = \frac{43}{128} + \frac{1}{16}c_2 + \frac{143}{256}c_4 - \frac{3}{32}c_6 + \frac{13}{128}c_8 + \frac{1}{32}c_{10} + \frac{1}{256}c_{12} - \frac{3}{8}s - \frac{1}{2}s_3 + \frac{3}{16}s_7 + \frac{1}{16}s_9$$

$$B = \frac{3}{16}c - \frac{5}{8}c_3 + \frac{3}{8}c_5 + \frac{1}{32}c_7 + \frac{1}{32}c_9 + \frac{7}{64}s_2 - \frac{69}{256}s_4 - \frac{33}{128}s_6 - \frac{7}{64}s_8 - \frac{3}{128}s_{10} - \frac{1}{256}s_{12}$$

$$C = \frac{1}{32} + \frac{7}{32}c_2 + \frac{1}{8}c_4 - \frac{13}{64}c_6 - \frac{5}{32}c_8 - \frac{1}{64}c_{10} + \frac{17}{64}s + \frac{13}{64}s_3 - \frac{21}{128}s_5 - \frac{9}{128}s_7 + \frac{5}{128}s_9 + \frac{1}{128}s_{11}$$

$$D = -B$$

$$E = -\frac{9}{128} - \frac{7}{32}c_2 + \frac{239}{256}c_4 + \frac{13}{64}c_6 + \frac{17}{128}c_8 + \frac{1}{64}c_{10} + \frac{1}{256}c_{12}$$

$$F = \frac{1}{64}c + \frac{11}{64}c_3 - \frac{5}{128}c_5 - \frac{15}{128}c_7 - \frac{3}{128}c_9 - \frac{1}{128}c_{11} - \frac{1}{32}s_2 - \frac{3}{8}s_4 - \frac{19}{64}s_6 - \frac{1}{16}s_8 - \frac{1}{64}s_{10}$$

$$P = C$$

$$Q = -F$$

$$R = \frac{19}{32} - \frac{9}{32}c_2 + \frac{3}{8}c_4 + \frac{19}{64}c_6 + \frac{1}{32}c_8 - \frac{1}{64}c_{10} + \frac{3}{8}s + \frac{1}{2}s_3 - \frac{3}{16}s_7 - \frac{1}{16}s_9$$

[a] $c_k = \cos 2\pi k\Delta\tau$, $s_k = \sin 2\pi k\Delta\tau$, Δ = offset, $\tau = \frac{1}{12}t_c$.

for all types of preparation and detection, do not lose any intensity to the pedestal except in the $90^\circ_y/\langle M_x\rangle$ and $0/\langle M_x\rangle$ cases. The $90^\circ_x/\langle M_y\rangle$ method is evidently optimal in all respects. The above trends are reflected for any practical range of cycle times. The scale factor, which is, of course, independent of the preparation and detection method, deviates by about 13% at the Nyquist frequency from the average Hamiltonian value. The behavior of signal, pedestal, and net transverse magnetization in the finite pulsewidth case closely follows the behavior in the δ-pulse limit. Figure 6.12 is a plot of scale factor versus offset. It is to be noted that the scale factor curve is similar for different pulsewidths for a given cycle time and is steeper for longer cycle times. Table 6.4 sums up the characteristics of the WAHUHA-4 and MREV-8 sequence as discussed above in a "truth table" format for ready reference.

It is also noted that the discussion above treats of a preparation pulse followed immediately by an FID sample, a time τ before the first pulse of the cycle is applied. This is equivalent to a preparation pulse followed by sampling after an interval τ, followed by the first pulse of the cycle after a further interval τ and subsequent frequency-dependent phase correction of the spectrum.

Scale factor curves, such as in Figs. 6.10 and 6.12, can be employed to deconvolute observed line-narrowed spectra to come up with more accurate chemical shift anisotropy values and information about the symmetry of chemical shift tensors. This should be of practical importance especially in studies of ^{19}F at high fields. Since the transverse magnetization and its aportioning into signal and pedestal depend critically on the preparation and

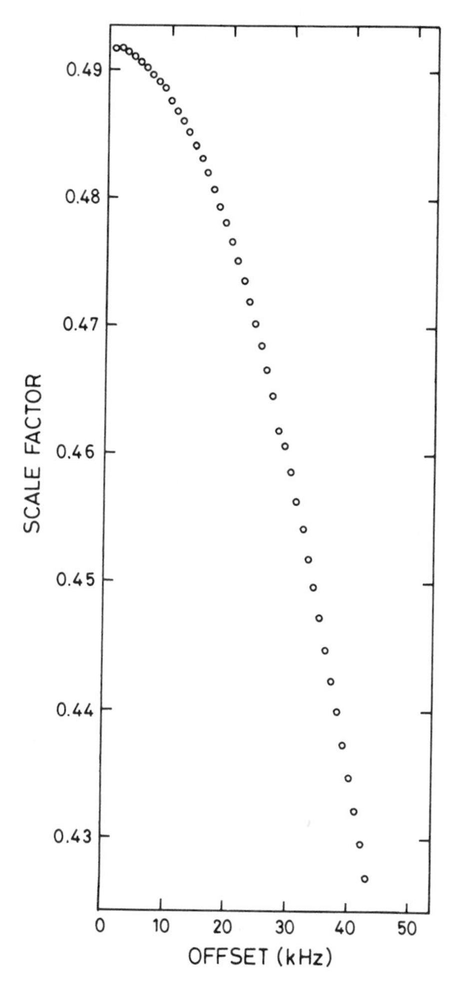

Figure 6.12. Scale factor versus offset for the 27-μs MREV-8 sequence with pulse width of 1 μs.

detection modes, offset sign, and magnitude, it is necessary to deconvolute the spectra (in powder samples) not only with respect to the scale factor profile, but also in general with the intensity profile which can be derived from the experiment on a liquid sample.

Other Multiple-Pulse Line-Narrowing Sequences. Let us represent the total Hamiltonian of the system as:

$$\mathscr{H} = \mathscr{H}_{\mathrm{rf}} + \mathscr{H}_{\mathrm{CS/O}} + \mathscr{H}_{\mathrm{D}} + \mathscr{H}_{\mathrm{PE}} \tag{79}$$

where $\mathscr{H}_{\mathrm{rf}}$ represents ideal rf pulses, $\mathscr{H}_{\mathrm{CS/O}}$ represents chemical shift and offset, $\mathscr{H}_{\mathrm{D}}$ is the dipolar Hamiltonian, and $\mathscr{H}_{\mathrm{PE}}$ takes into account pulse imperfec-

Table 6.4. Some Properties of the WAHUHA-4 and MREV-8 Sequences

	Preparation	Detection	WAHUHA-4	MREV-8
Offset sign invariance of spectrum	90°_x	$\langle M_x \rangle$	Yes	No
		$\langle M_y \rangle$	Yes	Yes
	90°_{-y}	$\langle M_x \rangle$	Yes	No
		$\langle M_y \rangle$	Yes	No
	0	$\langle M_x \rangle$	Yes	No
		$\langle M_y \rangle$	Yes	No
Signal intensity invariance to offset	90°_x	$\langle M_x \rangle$	No	No
		$\langle M_y \rangle$	No	Yes
	90°_{-y}	$\langle M_x \rangle$	No	No
		$\langle M_y \rangle$	No	No
	0	$\langle M_x \rangle$	No	No
		$\langle M_y \rangle$	No	No
Pedestal intensity invariance to offset	90°_x	$\langle M_x \rangle$	No	Yes (zero)
		$\langle M_y \rangle$	No	Yes (zero)
	90°_{-y}	$\langle M_x \rangle$	No	No
		$\langle M_y \rangle$	No	Yes (zero)
	0	$\langle M_x \rangle$	No	No
		$\langle M_y \rangle$	No	Yes (zero)

tions. The average Hamiltonian for a cyclic sequence is then:

$$\begin{aligned} \bar{\mathscr{H}} = {} & \bar{\mathscr{H}}_{\text{D}} + \bar{\mathscr{H}}_{\text{CS/O}} + \bar{\mathscr{H}}_{\text{PE}} \\ & + \bar{\mathscr{H}}^{(1)}_{\text{D}} + \bar{\mathscr{H}}^{(1)}_{\text{CS/O}} + \bar{\mathscr{H}}^{(1)}_{\text{PE}} \\ & + \bar{\mathscr{H}}^{(1)}_{\text{CS/O-D}} + \bar{\mathscr{H}}^{(1)}_{\text{PE-D}} + \bar{\mathscr{H}}^{(1)}_{\text{PE-CS/O}} \\ & + \bar{\mathscr{H}}^{(2)}_{\text{D}} + \bar{\mathscr{H}}^{(2)}_{\text{CS/O}} + \bar{\mathscr{H}}^{(2)}_{\text{PE}} + \cdots \end{aligned} \tag{80}$$

where the superscript corresponds to the order of the term in the Magnus expansion and the subscript to that of the interactions. The terms responsible for the efficiency of line narrowing are, e.g., $\bar{\mathscr{H}}_{\text{D}}$, $\bar{\mathscr{H}}^{(1)}_{\text{D}}$, $\bar{\mathscr{H}}^{(1)}_{\text{CS/O-D}}$, $\bar{\mathscr{H}}^{(1)}_{\text{PE-D}}$, $\bar{\mathscr{H}}^{(2)}_{\text{D}}$, $\bar{\mathscr{H}}^{(2)}_{\text{CS/O-D}}$, in order of significance. The term $\bar{\mathscr{H}}^{(1)}_{\text{CS/O}}$ corresponds to the chemical shift scale factor. When the pulse cycle has a reflection symmetry, then $\bar{\mathscr{H}}^{(1)}_{\text{D}}$ and odd-order terms $\bar{\mathscr{H}}^{(n)}_{\text{D}}$ as well as $\bar{\mathscr{H}}^{(1)}_{\text{CS/O-D}}$ vanish. It can also be shown that if in a subcycle of a full cycle $\bar{\mathscr{H}}_{\text{D}}$ vanishes, then $\bar{\mathscr{H}}_{\text{D}}$ vanishes for the full cycle. In addition, the contribution to average Hamiltonians from such terms as $\bar{\mathscr{H}}^{(1)}_{\text{D}}$, $\bar{\mathscr{H}}^{(1)}_{\text{CS/O-D}}$, $\bar{\mathscr{H}}^{(1)}_{\text{PE-D}}$, and $\bar{\mathscr{H}}^{(2)}_{\text{D}}$ from each unit of the total cycle is additive.

In designing new multipulse sequences we look for a cycle in which $\bar{\mathscr{H}}_{\text{D}}$ vanishes and combine with it its reflection, which makes $\bar{\mathscr{H}}^{(1)}_{\text{D}}$ and $\bar{\mathscr{H}}^{(1)}_{\text{CS/O-D}}$ also additionally vanish. For all enlarged cycles made from this basic cycle the above three terms will vanish in the average Hamiltonian, whereas the terms $\bar{\mathscr{H}}^{(1)}_{\text{PE-D}}$ and $\bar{\mathscr{H}}^{(2)}_{\text{D}}$, as mentioned before, will be additive. What we have to look for is additional combinations of reflected subcycles that can make $\bar{\mathscr{H}}^{(1)}_{\text{PE-D}}$ and $\bar{\mathscr{H}}^{(2)}_{\text{D}}$ also vanish. The BR-24 pulse sequence was introduced to improve the resolution. Defining,

$$\begin{aligned} P_1 &= \tau - 90^\circ_x - \tau - 90^\circ_y - 2\tau - 90^\circ_{-y} - \tau - 90^\circ_x - \tau \\ P_2 &= \tau - 90^\circ_{-x} - \tau - 90^\circ_y - 2\tau - 90^\circ_{-y} - \tau - 90^\circ_x - \tau \\ P_3 &= \tau - 90^\circ_y - \tau - 90^\circ_x - 2\tau - 90^\circ_{-x} - \tau - 90^\circ_y - \tau \\ P_4 &= \tau - 90^\circ_{-y} - \tau - 90^\circ_x - 2\tau - 90^\circ_{-x} - \tau - 90^\circ_y - \tau \end{aligned} \tag{81}$$

Then the BR-24 pulse sequence is:

$$P_1 - P_2 - P_3 - \tau - 90^\circ_{-y} - \tau - 90^\circ_x - \tau - P_3 - P_4 - \tau - 90^\circ_{-x} - \tau - 90^\circ_y - \tau \tag{82}$$

with a cycle time of 36τ. $\tilde{\mathscr{H}}_{\text{int}}(\text{t})$ toggles along $ZYXXYZZ\bar{Y}XX\bar{Y}ZZXY$-$YXZZXYY\bar{Z}XX\bar{Z}YYZXXZYYXZ$ axes, leading to:

$$\bar{\mathscr{H}}_{\text{CS/O}} = \frac{2}{3\sqrt{3}}(\delta + \Delta)\mathbf{I}_{\langle 111 \rangle} \tag{83}$$

corresponding to a scale factor 0.385.

$$\begin{aligned} \bar{\mathscr{H}}_{\text{D}} &= \bar{\mathscr{H}}^{(1)}_{\text{D}} = \bar{\mathscr{H}}^{(1)}_{\text{CS/O-D}} = 0; \\ \bar{\mathscr{H}}^{(1)}_{\text{PE-D}} &= -\frac{i}{t_c}\int_0^{t_c} dt_2 \int_0^{t_2} dt_1\, [\tilde{\mathscr{H}}_{\text{D}}(t_2), \tilde{\mathscr{H}}_{\text{PE}}(t_1)] \end{aligned} \tag{84}$$

where t_c is 36τ and $\tilde{\mathscr{H}}$ are the toggling frame Hamiltonians. For δ-pulses $\bar{\mathscr{H}}^{(1)}_{\text{PE-D}}$ vanishes. However, under finite pulsewidth conditions we can separate the contribution due to pulse shape and phase (PSP) transient effects, on the one hand, and flip angle, phase setting and rf inhomogeneity (FPI), on the other. This can be written:

$$\bar{\mathscr{H}}^{(1)}_{\text{PE-D}} = \bar{\mathscr{H}}^{(1)}_{\text{PSP-D}} + \bar{\mathscr{H}}^{(1)}_{\text{FPI-D}}$$

where:

$$\bar{\mathscr{H}}^{(1)}_{\text{PSP-D}} \propto \frac{1}{t_c} \tag{85}$$

being affected only at the rise and fall of the pulse, whereas:

$$\bar{\mathscr{H}}^{(1)}_{\text{FPI-D}} \propto \frac{t_w}{t_c} \tag{86}$$

which is constant throughout the cycle. Thus, $\bar{\mathscr{H}}^{(1)}_{\text{PE-D}}$ will become stronger when cycle time is reduced for a given pulse width t_w, while most other interactions tend to be coherently averaged out. That is, the errors contribute more substantially for an increased duty cycle. For MREV-8, $\bar{\mathscr{H}}^{(1)}_{\text{PE-D}}$ does not become dominant even for short cycle times, and longer cycles lead to dominance of $\bar{\mathscr{H}}^{(2)}_{\text{D}}$, limiting the resolution achievable. For BR-24, $\bar{\mathscr{H}}^{(2)}_{\text{D}}$ vanishes under δ-pulse approximation, but then rapid pulsing leads to an increase in $\bar{\mathscr{H}}^{(1)}_{\text{PE-D}}$ and hence to a loss of resolution.

Yet another extended multipulse scheme is the BR-52, which is a 52-

pulse sequence and can be derived from the basic four-pulse sequence of WAHUHA-4 by exploiting the symmetrization of the switched dipolar Hamiltonian substates.

The BR-24 and BR-52 pulse sequences, under δ-pulse approximation, lead to $\bar{\mathcal{H}}_D^{(2)} = 0$ and hence are extremely efficient in homonuclear dipolar decoupling besides being less susceptible to pulsewidth/cycle time changes. In fact, increasing the pulse width and keeping the cycle time constant leads to semiwindowless sequences in which the smaller space between the pulses disappear. To be able to perform multipulse experiments with the minimum requirement of rf power one should use the full duty factor and windowless (continuous) phase-switched rf irradiation is the answer. However, since there are no windows for stroboscopically observing the magnetization, such windowless sequences are primarily useful in a 2D experiment (Chapter 4) where one can use them to reduce dipolar interaction in the evolution period, t_1. Thus, windowless sequences applied to abundant spin enable study of rare spin interactions without the abundant spin–spin diffusion interfering with the observation. These windowless sequences are important in 2D solids work. An example is the BLEW-12 sequence consisting of the following windowless sequence:

$$X\ Y\ \bar{X}\ Y\ X\ Y\ \bar{Y}\ \bar{X}\ \bar{Y}\ X\ \bar{Y}\ \bar{X} \tag{87}$$

which consists of windowless $\pi/2$ pulses applied at the indicated phases. The chemical shift scale factor for BLEW-12 is 0.475. Whereas $\bar{\mathcal{H}}_D$, $\bar{\mathcal{H}}_D^{(1)}$, and $\bar{\mathcal{H}}_{CS/O\text{-}D}^{(1)}$ vanish, $\bar{\mathcal{H}}_D^{(2)}$ and $\bar{\mathcal{H}}_{PE\text{-}D}^{(1)}$ survive. BLEW-12 gives resolution almost independent of cycle time, but it is very sensitive to pulse errors. However, it can give sufficient resolution at very low rf power, which is difficult with MREV-8 or WAHUHA-4. An improved windowless sequence, by name BLEW-48, consists of 48 $\pi/2$ pulses; defining $P_1 = XY\overline{XYYX}$, BLEW-48 = P_1, $\bar{P}_1$, $\bar{P}_1$, P_1, P_1', $\bar{P}_1'$, $\bar{P}_1'$, P_1' where $\bar{P}_1$ is the complemented and reversed sequence of P_1 and P_1' is obtained from P_1 by complementing exclusively Y pulses. The chemical shift scale factor for BLEW-48 is 0.424, which is correct to second order independent of the finite pulsewidth. For the windowed sequences the scale factors (SF) correct to second order, as per average Hamiltonians, are:

$$\begin{aligned} \mathrm{SF}_{\mathrm{MREV}}^{\mathrm{CS}} &= \frac{\sqrt{2}}{3}(1 + 2a) \\ \mathrm{SF}_{\mathrm{BR\text{-}24}}^{\mathrm{CS}} &= \frac{2}{3\sqrt{3}}(1 + 2a) \\ \mathrm{SF}_{\mathrm{BLEW\text{-}48}}^{\mathrm{CS}} &= 0.424 \end{aligned} \tag{88}$$

where:

$$a = \frac{t_w}{4\tau}\left(\frac{4}{\pi} - 1\right)$$

Cross-Polarization: Proton-Enhanced Nuclear Induction Spectroscopy

While magic angle spinning and multiple pulse spectroscopy in solids can give rise to line-narrowing, such experiments cannot completely cover the whole range of situations. Under ideal conditions, MAS alone gets rid of all anisotropic interactions, but the measurement of rare spins is still difficult because of sensitivity problems. Likewise, the multiple-pulse sequences that have been detailed in the previous section get rid of homonuclear dipolar coupling but heteronuclear interactions are only scaled, not removed. Sensitivity with respect to rare spins is still a problem. These two techniques are quite useful in studying abundant spin systems with a fairly high sensitivity in the absence of other interfering abundant heteronuclei.

If, on the other hand, we want to attempt line narrowing and study finer interactions of rare nuclei with inherent lower sensitivity we have to look for alternative strategies. The nuclei of interest, especially in biological systems, are ^{13}C and ^{15}N. (Nitrogen-14 is a quadrupolar nucleus, prone to very short spin–spin and spin–lattice relaxation times due to the fluctuating quadrupole interaction; as such it gives broad lines in liquids and almost undetectable resonances especially when single-quantum coherences are concerned.) Because of their low natural abundance (^{13}C, 1.108% and ^{15}N, 0.37%) and very low gyromagnetic ratios ($\gamma_H/\gamma_{^{13}C} \approx 4$, $\gamma_H/\gamma_{^{15}N} \approx 10$), they have very poor NMR sensitivity (sensitivity being proportional to $\gamma^{5/2}$; Chapter 1). However, one may decouple the abundant spin in such systems and observe rare spin resonances because rare spin–rare spin dipolar interaction is negligible. We have seen in liquids that decoupling of the abundant spins leads to Overhauser enhancement (η) via the dipolar interaction with abundant spins (usually ^1H) as an added advantage of decoupling. We have also seen in Chapter 3 that it is possible to transfer polarization from abundant spin to rare spin (in fact from any spin to any other spin as long as there is a dipolar connectivity or spin–spin coupling) by a variety of methods. In particular, there is a close similarity between polarization transfer in the rotating frame for liquids and solids. In both cases transfer of polarization takes place when the Hartmann–Hahn condition ($\gamma_I B_{1I} = \gamma_S B_{1S}$) is satisfied; the mechanism is via direct dipolar interaction in solids. The mixing Hamiltonians are of identical form for liquids and solids. The rare spins are enhanced in their sensitivity by the ratio $\gamma_{source}/\gamma_{sink}$ for each Hartmann–Hahn contact. Although it is possible and more efficient to detect the rare spin resonance indirectly by monitoring the loss in abundant spin magnetization, the direct method of observing the enhanced rare spin magnetization under decoupled condition is the technique most used.

It is possible to describe such double-resonance experiments in solids using the concept of spin temperature. In liquids where one obtains high-resolution NMR spectra the dipolar coupling is averaged out and there are no overlaps between spin packets of different frequencies. Thus, when the various resonances are excited, say, by a nonselective pulse, each spin isochromat decays independently with its characteristic relaxation time; as such there is no common spin temperature to speak of in liquids. The presence of static dipolar interaction of several kilohertz range in solids, on the other hand, brings all nuclei, at least ones of a given type (homonuclear), to a common spin temperature by rapid spin diffusion. Hence the establishment of Boltzmann distribution is through a single relaxation time T_1. Even when the ensemble of spins is in thermal equilibrium with the lattice because of its link with the lattice via spin–lattice relaxation, it is possible selectively to "cool" or "heat" intensely the spins at least for times less than the relaxation times using radio frequency pulses. Any differential spin temperature (see below) between two spins will tend to be equalized via the dipolar link.

We shall briefly describe in this section the direct detection methods of obtaining high resolution and improved sensitivity for rare spins leading to measurement of chemical shift anisotropies in such systems.

We start with a system of an abundant spin I (^1H, ^{19}F, etc.) and rare spin S (^{13}C, ^{15}N); both I and S are coupled to the lattice and also among themselves. When placed in a magnetic field they reach the lattice temperature with characteristic relaxation times T_{1I} and T_{1S}. Any transfer of energy between I and S spins is governed by the cross-relaxation time T_{IS}.

The density matrix at thermal equilibrium (which is diagonal) is given by:

$$\sigma = Z^{-1} \exp(-\mathscr{H}/kT) \tag{89}$$

where Z is the partition function, given by:

$$Z = \sum_{\substack{m \\ \text{all states}}} \exp(-\mathscr{H}_m/kT) \tag{90}$$

For $N_I I$ spins and $N_S S$ spins the number of states would be $(2I + 1)^{N_I}(2S + 1)^{N_S}$. Defining an inverse spin temperature:

$$\beta = h/2\pi kT \tag{91}$$

we rewrite, expanding the exponential:

$$\sigma = Z^{-1}(1 - \beta\mathscr{H}'' + \cdots) \tag{92}$$

If we define a reduced Curie constant, say for I spins,

$$C_I = \frac{1}{3} N_I I(I + 1)\gamma^2 \frac{h}{2\pi} \tag{93}$$

then the magnetization at thermal equilibrium in a field of B_0 is given by:

$$M_{0I} = \beta_L C_I B_0$$

and

$$M_{0S} = \beta_L C_S B_0 \tag{94}$$

Here β_L is the inverse lattice temperature, which is the same as inverse spin temperature β_I and β_S at thermal equilibrium.

It is now possible to cool the I spins by spin locking these using 90°_x pulse followed immediately by a weak long pulse in the $\pm y$ direction. Under the circumstances the spins I are cooled in the rotating frame to a new spin temperature β_I such that:

$$\beta_I/\beta_L = B_0/B_1 \gg 1 \tag{95}$$

Under these conditions we can define rotating-frame magnetizations and energies as:

$$M_I^R = \beta_I C_I B_{1I}$$

and:

$$E_I^R = -\beta_I C_I B_{1I}^2 \tag{96}$$

The cooled I spins approach the inverse lattice temperature β_L with a characteristic time constant governed by spin–lattice relaxation in the rotating frame $T_{1\rho}$ ($T_2 \leqslant T_{1\rho} \leqslant T_1$). This is not the only way of cooling the abundant spin. It is possible to turn off the spin-locking rf field adiabatically and to leave the I spins in a state of dipolar order. We can describe this adiabatic demagnetization in the rotating frame, ADRF, as follows. Thus, under these conditions:

$$M_x = M_0 B_1/(\sqrt{(B_e^2 + B_L^2)})$$

and:

$$M_z = M_0\left(B_0 - \frac{\omega}{\gamma}\right)\Big/\sqrt{(B_e^2 + B_L^2)} \tag{97}$$

where B_e is the effective rf field in the rotating frame and B_L is the dipolar field. Typically,

$$B_L = \tfrac{1}{3}\langle \Delta H^2 \rangle \tag{98}$$

where $\langle \Delta H^2 \rangle$ is the second moment of the dipolar-broadened line. When ADRF is done exactly on resonance then $B_1 = B_e$, and

$$M_x = M_0 B_1/\sqrt{(B_1^2 + B_L^2)} \tag{99}$$

When B_1^2 is of the order of B_L^2, M_x will be less than M_0. When B_1 is adiabatically reduced, the magnetization could shrink to zero and the Zeeman order has been put into the dipolar order, bringing the spins to alignment along their dipolar local fields. One can slowly increase the rf field from zero and recover the remaining stored order by the so-called adiabatic remagnetization in the rotating frame (ARRF). In the state of dipolar order the magnetization would decay by $T_{1\mathrm{D}}$, the dipolar relaxation time. The inverse spin temperature in this case would be:

$$\beta_I = \frac{B_0}{B_{LI}}\beta_L \tag{100}$$

where B_{LI} is the strength of the I–I dipolar field. It is also possible to create dipolar order with somewhat less efficiency by a sequence of 90°_y–τ–45°_x pulse pair.

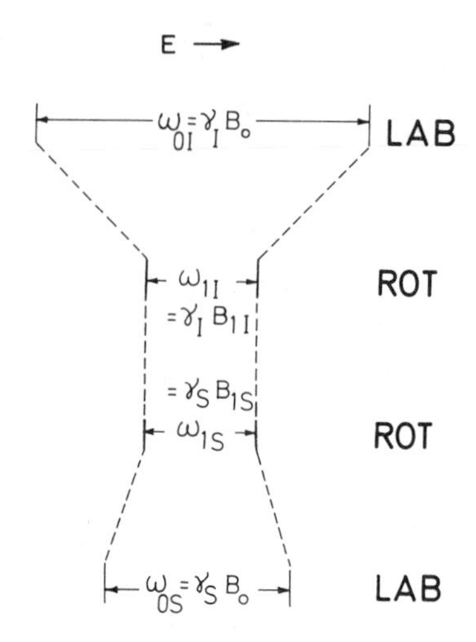

Figure 6.13. Schematic representation of Hartmann–Hahn condition.

Having cooled the I spins, the next step is to bring the relatively "hot" rare S spins into contact with the I spins, so that energy exchange can take place with the characteristic time constant T_{IS}. It is obvious that for efficient transfer, T_{IS} should be shorter than $T_{1\rho}^{I}$ (or $T_{1\mathrm{D}}^{I}$); then transfer can take place when the total energy is conserved. Such a transfer is not possible for solids in the laboratory frame because each spin species has its own Larmor frequency and cannot come to "speaking terms" with others. However, in the rotating frame energy conservation and rapid transfer can be achieved by the Hartmann–Hahn condition. A schematic representation of Hartmann–Hahn condition and such polarization transfer is shown in Fig. 6.13. Assuming a high inverse temperature for I spin and a zero inverse temperature for rare S spins, spin temperature equilibrium establishes itself as:

$$\beta_I(t) = (\beta_I - \beta_f)\exp(-t/T_{IS}) + \beta_f$$

and:

$$\beta_S(t) = \beta_f(1 - \exp(-t/T_{IS})) \tag{101}$$

where β_f is the final common temperature for both I and S. Conservation of total magnetic energy leads to:

$$\beta_I C_I B_{1I}^2 + \beta_S C_S B_{1S}^2 = \beta_f(C_I B_{1I}^2 + C_S B_{1S}^2)$$

giving:

$$\beta_f/\beta_I = 1/(1 + \varepsilon)$$

where:

$$\varepsilon = C_S B_{1S}^2/C_I B_{1I}^2 \tag{102}$$

corresponding to ratio of heat capacities of S and I spins. Under Hartmann–Hahn conditions:

$$\varepsilon = \frac{N_S S(S+1)}{N_I I(I+1)} \ll 1 \tag{103}$$

which for rare spins is a very small quantity. When steady state is reached,

$$M_I^f = M_I^i/(1+\varepsilon)$$

and:

$$M_S^f = \frac{\gamma_I}{\gamma_S}\frac{1}{(1+\varepsilon)} M_{0S} \tag{104}$$

where the superscripts i and f stand for initial and final stages. Considering that ε is small, for abundant to rare spin transfer we can see that while the abundant spin scarcely loses its magnetic energy the rare spin gains magnetization in the ratio of γ_I/γ_S. In order to take the maximum advantage of I-spin polarization, multiple contacts can be made. The spin-locking field, which is kept on during acquisition of the signal, provides the necessary heteronuclear decoupling for S spin resonances to get both sensitivity and resolution enhancement. This pulse sequence for achieving narrow rare spin resonances is shown in Fig. 6.14. Since $\varepsilon \ll 1$, we can approximate $1/(1+\varepsilon) \approx 1 - \varepsilon$ so that the I-spin and S-spin magnetization after N contacts, assuming zero S-spin inverse spin temperature to start with, are given by

$$M_I(N) = (1-\varepsilon)^N M_{0I} \approx \exp(-N\varepsilon) M_{0I}$$

and:

$$M_S(N) = \frac{\gamma_I}{\gamma_S}\exp(-N\varepsilon) M_{0S} \tag{105}$$

Summing up all S-spin signals by coadding successive FID's leads to:

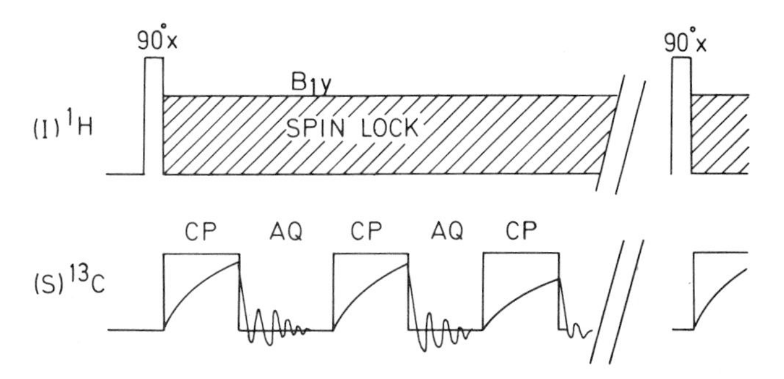

Figure 6.14. Multiple-contact cross-polarization scheme for direct observation of rare spins.

$$
\begin{aligned}
M_S(\text{total}) &= \sum_{k=1}^{N} M_S(k) \\
&= \frac{\gamma_I}{\gamma_S} M_{0S} \sum_{k=1}^{N} (1-\varepsilon)^k \\
&\approx \frac{\gamma_I}{\gamma_S} M_{0S} \sum_{k=1}^{N} \exp(-k\varepsilon)
\end{aligned} \tag{106}
$$

In the indirect method the I-spin magnetization is monitored after N contacts, while in the direct method N S-spin FID's are accumulated one after each Hartmann–Hahn contact.

We can now estimate the enhancement factor. For given spectrometer conditions (rf coil Q factor, filling factor, detector bandwidth, etc.) the signal to noise ratio for an induced magnetization M is given by:

$$\mathscr{S}/\mathscr{N} = K\omega_0^{1/2} M \tag{107}$$

where K depends on the spectrometer conditions mentioned above. If the rare S spins were measured under normal single resonance conditions, then:

$$(\mathscr{S}/\mathscr{N})_{\text{FID}} = K\omega_0^{1/2} M_{0S} \tag{108}$$

For N accumulated multiple contacts in a CP experiment:

$$(\mathscr{S}/\mathscr{N})_{\text{CP}} = \frac{1}{\sqrt{N}} K\omega_0^{1/2} M_{NS} \tag{109}$$

The enhancement over single FID under normal conditions is then:

$$\frac{(\mathscr{S}/\mathscr{N})_{\text{CP}}}{(\mathscr{S}/\mathscr{N})_{\text{FID}}} = \frac{\gamma_I}{\gamma_S} \frac{1}{\sqrt{N}} \sum_{k=1}^{N} \exp(-k\varepsilon) \tag{110}$$

This function has a maximum at $N\varepsilon \approx 1.3$, from which it can be shown that the gain in a multiple-contact CP experiment is given by:

$$\text{Gain}_{\text{CP}} = 0.64 \frac{\gamma_I}{\gamma_S} \left[\frac{N_I I(I+1)}{N_S S(S+1)} \right]^{1/2} \tag{111}$$

The value of N_I/N_S in natural abundance for ^{1}H and ^{13}C in organic solids is from 100 to 200 so that the gain factor can be in the range 25–35. This factor will be considerably higher for ^{1}H–^{15}N cross-polarization by more than an order of magnitude. Usually only a finite number, say 30–50 contacts, would be possible. The actual number of contacts possible will depend on $T_{1\rho}$. See Fig. 6.15 for examples of CP spectra.

If the proton transmitter is switched off for a brief period of 40–100 μs at the end of the contact time and acquisition is then commenced under proton decoupling, all resonances other than those of quaternary carbons are effectively suppressed because they dephase rapidly under conditions of proton-coupled evolution. This has proved a convenient method of measuring

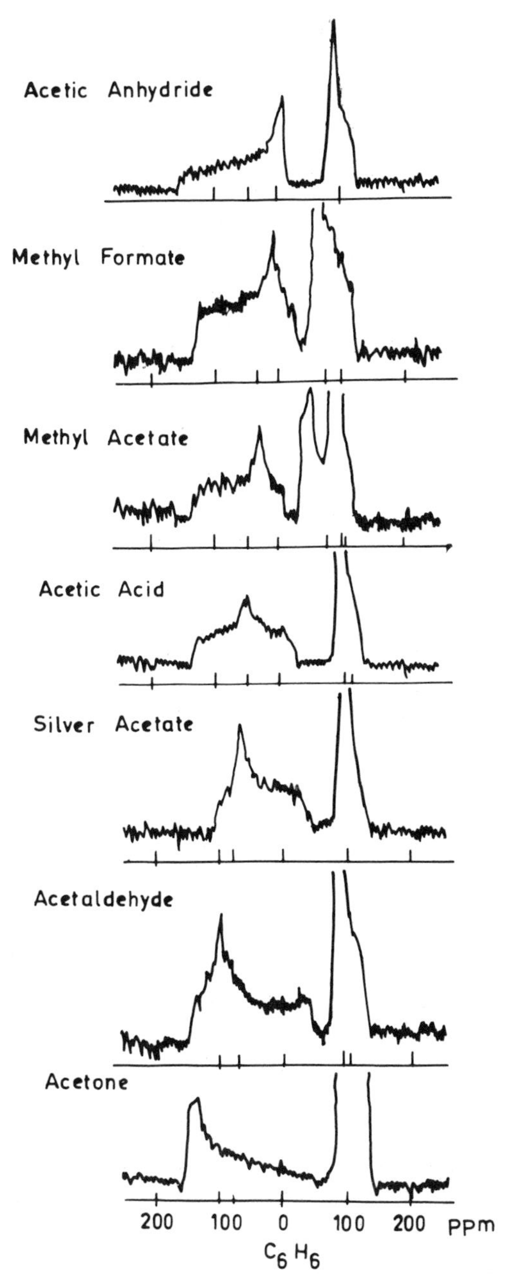

Figure 6.15. Carbon-13 cross-polarization spectra for a series of organic solids. The carbonyl groups, which resonate at low field, exhibit large shift anisotropies, whereas the methyl groups, resonating at high field, have a small anisotropy. [Reproduced by permission. A. Pines, M.G. Gibby and J.S. Waugh, *J. Chem. Phys.*, **59**, 569 (1973), copyright (1973), American Institute of Physics]

selectively the quaternary carbon spectrum in solids. Some crosstalk does occur into such a spectrum, from methyl groups in particular, because these exhibit effective C–H distances that are larger than the normal bond lengths, owing to rapid rotation.

In the indirect method of detecting loss in abundant spin magnetization each point in the ^{13}C spectrum is to be measured in a sequential manner. For the same quality factor of the spectrometer we can show that the indirect method, in the ^{1}H–^{13}C CP experiments, is approximately three times more efficient, say for $N_I/N_S = 150$ and the number of sampling points equal to 1024. For n sampling points,

$$\frac{(\mathscr{S}/\mathscr{N})_{\text{Indirect}}}{(\mathscr{S}/\mathscr{N})_{\text{Direct}}} = \frac{1}{\sqrt{n}}\left(\frac{\gamma_I}{\gamma_S}\right)^{3/2}\left[\frac{N_I I(I+1)}{N_S S(S+1)}\right]^{1/2} \tag{112}$$

Whereas the indirect method is sensitive, most work has been done only by monitoring directly the rare spin, because this way one is able to produce the spectrum in a short span of time compared to the indirect method. We can schematically represent CP strategies by four steps (Fig. 6.16). Some of the strategies used in the four steps are described very briefly.

During the "preparation" abundant I spins are polarized in a DC field. It is also possible to produce the I spin in a highly polarized state by optical polarization.

During the "cool" period the I spins can be spin locked (and thereby cooled in the rotating frame) along the rf axis. It is also possible to preserve the magnetization by converting the Zeeman order to dipolar order by ADRF.

In the "mixing" period, which is the most important step, the source spins and sink spins are made to have the same rotating-frame precessional frequencies by the Hartmann–Hahn condition. This could be perfectly matched or mismatched.

During the "observe" period one can have the spin-lock field on or off or one can also use pulsed I-spin decoupling. In the ADRF case, following the same lines as before in the spin-lock procedure, one can assume the following initial inverse spin temperatures before mixing.

$$\beta_{DI} = \beta_L B_0/B_{LI}; \qquad \beta_S = 0 \tag{113}$$

where β_{DI} is the inverse spin temperature of the dipolar ordered state and B_{LI}

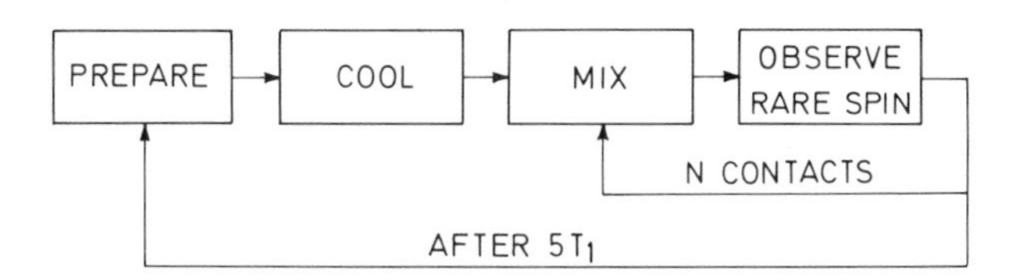

Figure 6.16. Schematic representation of the four-step strategy in the CP of dilute spins.

is the secular part of the dipolar Hamiltonian. The dipolar local field is given by:

$$\mathrm{Tr}\,(\mathscr{H}_{\mathrm{D}}^2) = \gamma_I^2 B_{LI}^2\,\mathrm{Tr}\,(I_z^2) \tag{114}$$

Assuming again the same final inverse spin temperature for I and S,

$$\beta_{\mathrm{D}I} C_I B_{LI}^2 = \beta_f [C_I B_{LI}^2 + C_S B_{1S}^2]$$

leading to:

$$\frac{\beta_f}{\beta_{\mathrm{D}I}} = \frac{1}{(1 + \varepsilon')} \tag{115}$$

where ε' is a measure of the ratio of the IS heteronuclear dipolar second moment to the I–I homonuclear dipolar second moment; i.e., ε' is given by:

$$\varepsilon' = \frac{N_S S(S + 1)}{N_I I(I + 1)} \alpha^2 = \varepsilon \alpha^2$$

with:

$$\alpha = \gamma_S B_{1S} / \gamma_I B_{LI} \tag{116}$$

When $\alpha = 1$ there will be perfect Hartmann–Hahn matching so that:

$$\frac{M_{\infty S}}{M_{0S}} = \frac{\beta_f}{\beta_L} \frac{B_{1S}}{B_0} = \frac{\gamma_I}{\gamma_S} \frac{\alpha}{1 + \varepsilon\alpha^2} = \frac{\gamma_I}{\gamma_S} \frac{1}{(1 + \varepsilon)} \tag{117}$$

and the gain in magnetization is nearly the ratio of the magnetogyric ratios.

When the Hartmann–Hahn condition is not satisfied, corresponding to $\alpha \gg 1$, the above expression reaches a maximum when $\alpha\sqrt{\varepsilon} = 1$ so that:

$$\frac{M_{\max S}}{M_{0S}} = \frac{1}{2} \frac{\gamma_I}{\gamma_S} \left[\frac{N_I I(I + 1)}{N_S S(S + 1)} \right]^{1/2} \tag{115}$$

With the ADRF and mismatched Hartmann–Hahn condition, therefore, the gain achieved is almost the same that one can obtain in a multiple contact in the spin-lock case. Figure 6.17 shows the schematic of this technique.

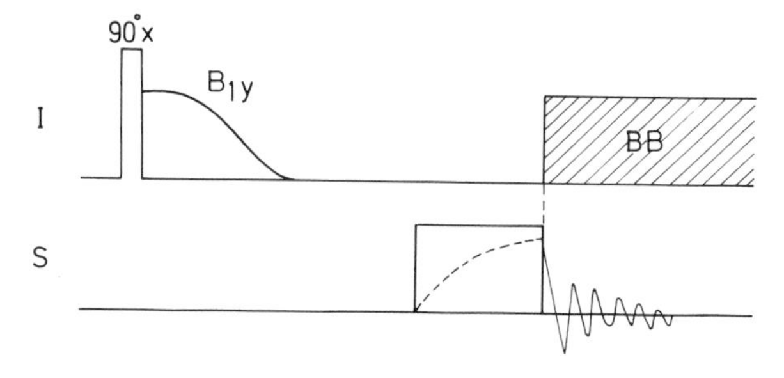

Figure 6.17. Pulse sequence for ADRF CP.

The efficiency of the experiment will be limited by T_{1D}. When $\alpha \gg 1$ this means poor matching and the transfer of polarization takes much longer time being an exponential function of α. Thus by the time the optimum value of α is reached T_{IS} may exceed T_{1D} and the maximum possible transfer would never happen. In systems where T_{1D} is much shorter than T_1 (e.g., in high polymers) it is better to use the spin-lock version with short contact times. It is sometimes advantageous to flip back any remaining I-spin magnetization back to the z axis before commencing the next series of contacts to improve the sensitivity.

The cross-polarization dynamics is fairly difficult to evaluate quantitatively; suffice it to say that it depends on the relative magnitudes of $T_{1\rho}$ (T_{1D} in the ADRF method), T_{IS}, and the inverse temperatures β_I, which is proportional to B_0/B_{1I} in the spin-lock case and B_0/B_{LI} in the ADRF case. In addition, these two mechanisms depend differently on the degree of matching of the Hartmann–Hahn condition.

The buildup of rare spin magnetization shows interesting oscillatory behavior in the initial stages. In this connection we would like to contrast this behavior to that in Hartmann–Hahn J cross-polarization in liquids. In liquids cross-polarization proceeds via a discrete number of scalar couplings and as such the transfer is oscillatory with definite frequencies (Chapter 3). In solids, however, since the dipolar mechanism is invoked it will depend on the dipolar coupling constants, and these generally have a Gaussian spread characteristic of the second moment of the spectral distribution and orientation of the dipolar vectors; in general an infinite number of frequencies can be found for the polarization transfer. Nevertheless, if there are two or three discrete neighbors with large dipolar coupling then one can see oscillatory behavior of the initial buildup which, however, soon gets damped by very rapid homonuclear I–I dipolar interactions. One can reduce these dampings by spin locking the I-spin magnetization by a magic angle pulse ($\gamma B_1 t_p = 54°44'$) followed by off-resonance locking, thereby bottlenecking the I-spin spin diffusion. An examination of the I-spin oscillations in a 2D sense should lead to quantification of heteronuclear dipolar coupling and hence distance information.

A detailed investigation of the cross-polarization and also the associated spin-decoupling dynamics is considered beyond the scope of the present edition and relevant literature quoted at the end of this book should be consulted for a detailed understanding of the same.

In solids containing a third nucleus X, having a quadrupole moment ($I \geqslant 1$) and a sizeable nuclear quadrupolar coupling constant and hence endowed with short spin–lattice relaxation time T_1, one can use the so-called cascaded enhancement of nuclear induction spectroscopy (CENIS). The technique uses the principle of "level crossing." The quadrupolar spins are first polarized in the strong magnetic field. Then the magnetic field is lowered or the sample is lifted so that the quadrupole split Zeeman levels of X match with the Zeeman energy of the I spins. This causes a rapid polarization of I spins, which in turn can be transferred to rare spin S via Hartmann–Hahn cross-polarization. This is especially suited to situations where the abundant spin 1/2 I nucleus has relaxation times of the order of several hours. An enormous gain in sensitivity is obtained compared to the normal cross-polarization, although the experiment is more difficult to perform.

Combined Rotation and Multiple-Pulse Spectra (CRAMPS)

We have seen that multiple-pulse sequences, such as WAHUHA-4, MREV-8, and BR-24, get rid of homonuclear dipolar coupling correct to second order in the average Hamiltonian, besides scaling the heteronuclear dipolar and chemical shift by a factor characteristic of the pulse sequence. In the proton NMR spectra of solids the range of chemical shift anisotropy is on the order of the average chemical shift differences between different protons. Therefore, the multiple-pulse line-narrowed spectra in powder or amorphous samples give rise to such an overlapping of chemical shift anisotropy patterns from different protons that an unequivocal analysis becomes formidably difficult. Only when large-sized single crystals are available can one study the angular variation of chemical shift as a function of crystal orientation and extract the magnitude and orientation of the shift tensors.

Because the average chemical shifts $[(1/3)\,\mathrm{Tr}\,\mathbf{R}^{CS}]$ should resemble the value obtained in solution, it will be good idea to average out chemical shift anisotropies further by magic angle spinning. Such a procedure, called the combined rotation and multiple pulse spectroscopy (CRAMPS), has become an important tool in the hands of solid-state materials scientists. We have already seen that the multiple-pulse line-narrowing sequences induce time dependences in the spin operators T_{lm}^{λ} and lead to averaging of the homonuclear dipolar interaction. The magic angle spinning strategy introduces time dependence on the R_{lm}^{λ}, i.e., the second-rank tensors. Since the transformation properties of all R_{lm}^{λ}'s are identical in coordinate space all anisotropies will average out during MAS of sufficient speed. One must digress for a moment, however, and ask the question what is the optimum MAS frequency versus the multipulse cycle time that is capable of achieving both homonuclear dipolar narrowing and averaging out the residual shift anisotropy. Usually the dipolar couplings in ^{1}H and ^{19}F systems are seldom less than 20 kHz. In order to remove homonuclear dipolar coupling effectively, multipulse sequences should have cycle times (or at least subcycle times) less than or of the order of the time equivalent of the dipolar coupling. A 20-kHz dipolar coupling will suggest a multiple-pulse cycle time of much less than 50 μs for efficient dipolar decoupling. Proton anisotropies are typically of the order of 3–6 kHz (at a spectrometer frequency of 100–200 MHz) so that for efficient averaging one needs a spinning speed corresponding to a MAS cycle time of the order of 200 μs. Therefore, nature itself dictates that the sequence of averaging is that we average first the homonuclear dipolar tensor in a MAS frame, which is very slow compared to the multiple-pulse cycle time. The resulting shift anisotropies are averaged out by MAS.

In fact it can be shown that if the MAS period and the multiple-pulse cycle time are of equal order then the pulsed dipolar narrowing will never work and the CRAMPS will fail. This is because while the T_{lm}^{λ}'s are being averaged to zero for $\bar{\mathscr{H}}_D$, the MAS induces further transformation in the $(3\cos^2\theta - 1)$ dependence of the anisotropic part of R_{lm}^{λ} so that effective nulling of the dipolar Hamiltonian is impossible. Typically, at least 10–12 subcycles of multiple-pulse cycle must be applied before MAS completes one revolution. Figure 6.18 shows the dramatic enhancement in resolution as one goes from simple multiple pulse to CRAMPS. The CRAMPS method is extremely useful

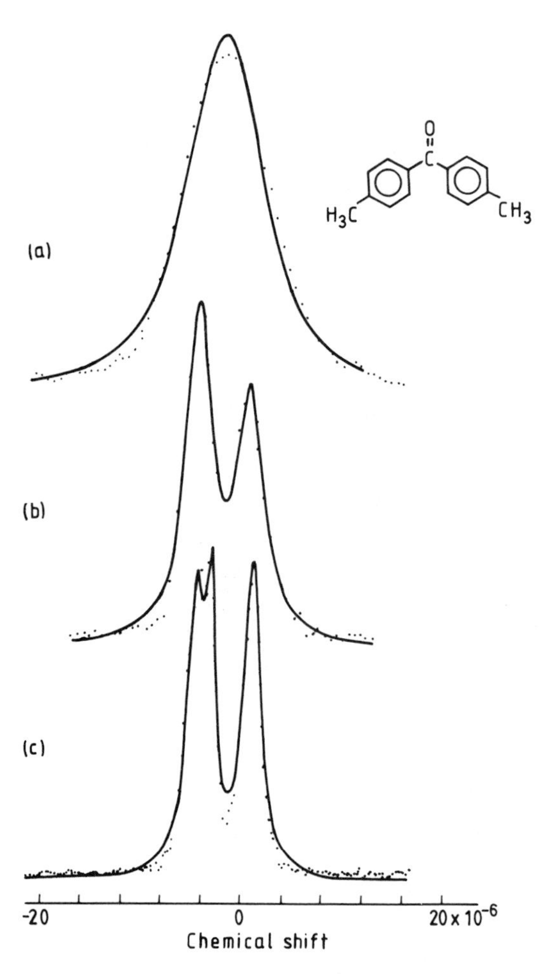

Figure 6.18. Proton spectra of 4,4′-dimethylbenzophenone. (a) MREV-8 spectrum; (b) MREV-8 + MAS at 2.5 kHz; (c) BR-24 + MAS at 25 kHz. [Reprinted with permission. B.C. Gerstean, *Phil. Trans. Royal Soc. London,* **A299** (1981). Copyright 1981 The Royal Society London 1981]

in getting highly resolved isotropic spectra in systems with large homonuclear dipolar coupling. As far as ^1H CRAMPS is concerned, since the shift anisotropy is only on the order of 3–6 kHz, sufficient rotor speed is possible to arrive at isotropic spectra with little interference from sidebands.

Cross-Polarization and Magic Angle Spinning (CPMAS)

Just as multiple-pulse sequences and MAS can be combined to produce isotropic spectra in solids that have nuclei prone to large homonuclear dipolar broadening, cross-polarization and MAS can be combined in systems containing rare spin (^{13}C) that is prone mainly to heteronuclear dipolar interaction. Thus the Hartmann–Hahn cross-polarization from abundant spin to rare spin is carried out under MAS conditions to produce rare spin isotropic spectra. Just as the MAS speed can interfere with the multiple-pulse dipolar averaging efficiency, MAS in CPMAS experiments can interfere with the cross-polarization efficiency, since the latter depends on the static heteronuclear dipolar coupling, which is scaled down considerably by MAS. In most situations the homonuclear dipolar coupling is much larger than the rare spin chemical shift anisotropy. Therefore, the modulation of heteronuclear dipolar coupling by the abundant spin homonuclear dipolar fluctuations will be much larger than the coherent modulation by the slow MAS. Only in cases where the rare spin shift anisotropy is on the order of homonuclear dipolar interaction among abundant spins do we get into problems of the influence of spinning rate on the cross-polarization efficiencies. Experimentally, it has been seen that for a static sample the transfer efficiency in a spin-lock CP experiment reaches a maximum when the Hartmann–Hahn condition is exactly matched. In fact the polarization transfer and the inverse cross-relaxation time reach a maximum at exact match. However, when the sample is subjected to MAS the cross-relaxation rate shows distinct oscillatory behavior as a function of Hartmann–Hahn mismatch, with the transfer becoming less efficient for Hartmann–Hahn match (see Fig. 6.19). If we tune the Hartmann–Hahn mismatch frequency $(\omega_{1I} - \omega_{1S})$, then the cross-relaxation time T_{IS} (which controls the polarization transfer rate), using a simplified model of relaxation in the weak collision limit (where the heteronuclear dipolar Hamiltonian is considered as a small perturbation on the large homonuclear abundant spin dipolar interaction $\mathscr{H}_{II}$), is given by:

$$T_{IS}^{-1} = f(\omega_{1I} - \omega_{1S}) = \langle \Delta\omega^2 \rangle_{IS} \int_0^\infty [\cos(\omega_{1I} - \omega_{1S})\tau] g(\tau)\, d\tau \quad (119)$$

Here, $\langle \Delta\omega^2 \rangle_{IS}$ is the second moment of the S resonance caused by coupling to I spins. Also, $g(\tau)$, which is an autocorrelation function of the fluctuating heteronuclear dipolar interaction as modulated by the abundant spin homonuclear interaction, is rather difficult to calculate and becomes formidably so in the presence of the additional motion from MAS. We can give only a brief

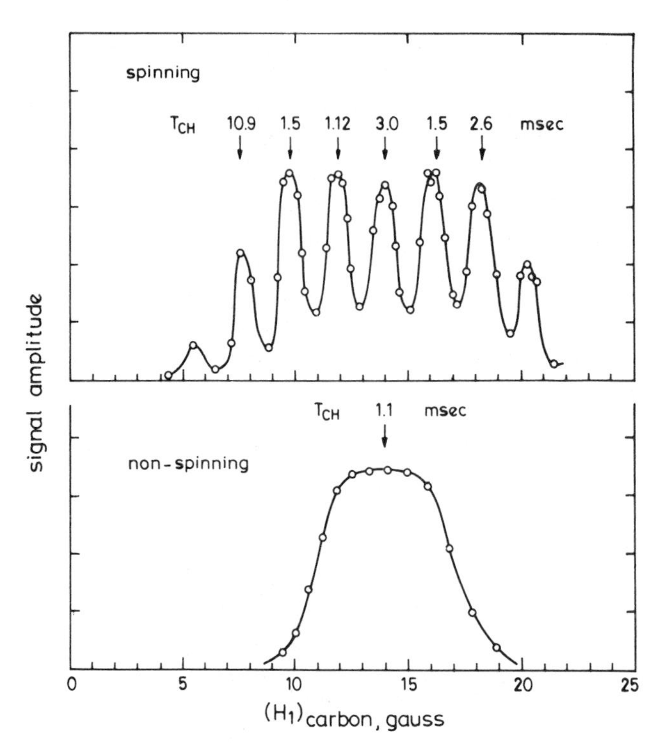

Figure 6.19. Cross-polarization efficiency versus Hartmann–Hahn (H-H) mismatch. *Lower trace*: Nonspinning sample shows maximum transfer efficiency at H-H match (corresponding to B_1 ca. 15 kHz in the figure). *Upper trace*: Spinning sample shows distinct oscillations in transfer efficiency. [Reproduced by permission. E.O. Stejskal, J. Schaefer and J.S. Waugh, *J. Magn. Reson.*, **28**, 106 (1977), copyright 1977, Academic Press, New York]

physical picture. The heteronuclear dipolar interaction that is modified by $\mathscr{H}_{II}$ will have a time dependence. For polarization transfer to take place with a definite mismatch, some of the Fourier components arising from the fluctuating $\mathscr{H}_{IS}$ (brought about by near-neighbor I–I flipflop) should exactly cancel the mismatch frequency. Suppose that there are many neighbors. This results in a power spectrum centered at $(\omega_{1I} - \omega_{1S}) = 0$ and a near Gaussian distribution with a characteristic bandwidth. When MAS is imparted, additional time dependences at the spinner frequency ω_r are introduced so that both $\mathscr{H}_{II}$ and $\mathscr{H}_{IS}$ are modulated at the additional frequency. The transfer of polarization now takes place under apparent mismatched conditions, which are internally compensated at frequencies $(\omega_{1I} - \omega_{1S} \pm \omega_r)$ and $(\omega_{1I} - \omega_{1S} \pm 2\omega_r)$ irrespective of the magnitude of $\mathscr{H}_{IS}$. Additionally, the rotationally modulated $\mathscr{H}_{II}$ further splits the above frequencies into sidebands at multiples of ω_r. Thus the presence of either molecular motion or magic

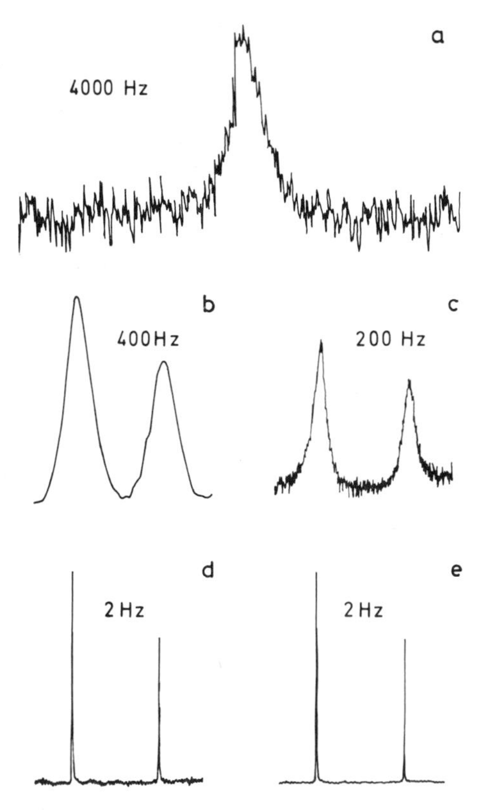

Figure 6.20. Carbon-13 CPMAS spectrum of adamantane. (a) Spectrum of static sample. (b) Proton-decoupled spectrum. (c) MAS spectrum. (d) Spectrum with MAS and proton decoupling. (e) CPMAS spectrum with proton decoupling. The figure indicates the respective linewidths. [Adapted with permission from CXP Application Note: "High Resolution NMR in Solids by Magic Angle Spinning" Fig. 3, page 5, Spectrospin AG, Zürich]

angle spinning would reduce the cross-polarization rate for exact Hartmann–Hahn matching. However, we can get sufficient transfer if we can sit at a mismatched situation when $\mathscr{H}_{IS}$ becomes effective again. In fact this poses no problem at all; since, in any case, CP work is done by adjusting the rare spin or abundant spin rf power on either side of the Hartmann–Hahn matching condition by trial and error, it is always possible to get the optimal "mismatch" for maximum polarization transfer. An example of CPMAS spectra is shown in Fig. 6.20.

Both CPMAS and CRAMPS are potentially useful in studying materials in their native form, whether crystalline or amorphous, and to characterize a number of solids that hitherto have not been amenable for study by NMR.

Suppression of Sidebands in CPMAS

Magic angle spinning experiments are intended to simplify spectra by averaging the anisotropic interactions in coordinate space. However, unless the spinning speed is on the order of the range of the anisotropic interaction the averaging will be incomplete and hence lead to a large number of spinning sidebands. As such MAS is generally combined with either multiple pulse (CRAMPS) to get rid of homonuclear dipole interactions in solids or cross-polarization (CPMAS) to eliminate heteronuclear dipolar interaction. It is then easier to average out the residual chemical shift anisotropy. However, even here large spinning speeds are required if one wants to eliminate ^{13}C or ^{19}F chemical shift anisotropy. If we want to obtain information on the anisotropic part of the chemical shift it is possible to rotate the sample about an axis away from $54°44'$ to the static field. For example, spinning the solid sample by an axis perpendicular to the field leads to a scale factor of -0.5 for the tensors. In general, it can be shown that for a rotation about an axis making an angle θ, if the Euler angles relating the transformation of the tensor principal axis system to that which rotates with the MAS axis are given by α, β, and a time-dependent γ, then the lineshape of the chemical shift tensor, for example, is given by:

$$\delta_{zz}(\text{lab}) = \frac{1}{3}\text{Tr}(\delta) + (3\cos^2\theta - 1)\frac{\delta'}{2}[(3\cos^2\alpha - 1) - (\eta\sin^2\alpha\cos^2\beta)] \tag{120}$$

where η and δ' depend on the principal values of the second-rank tensor concerned and have been defined previously. In any case, whether in CPMAS, CRAMPS, or just MAS alone, the presence of sidebands—sometimes a progression of them on either side of the average frequency, especially when a large number of different chemically shifted nuclei are present—makes spectral analysis rather difficult. In fact, when the spinning speed manages only to average the tensors partially the sidebands can be more intense than the isotropic line characteristic of $1/3\,\text{Tr}(\delta)$. The routine use of superconducting magnets operating at high spectrometer frequencies further aggravates the situation by producing much larger dispersion in chemical shifts.

One simple, but cumbersome, way of circumventing the sideband problem is to perform MAS experiments at two different frequencies and concentrate on those resonances which are not affected by the change in rotor speed. This can be successful only when a few chemically different species are present. Another way is to go down to lower field and faster spinning speeds and eliminate all sidebands. This is attractive, but then the inherent resolution is lost besides reducing the sensitivity of measurement.

A PASS (phase-altered suppression of sidebands) technique is now available that subjects the magnetization to a sequence of Hahn echo pulses at predetermined delays giving rise to an FID whose FT leads to selectively altered phases for sidebands of different order. The principle utilizes the fact that the sidebands of different order m, during a MAS experiment in a frame of reference rotating with the average chemical shift frequency, precesses with frequencies $m\omega_r (m = \pm 1, \pm 2, \ldots)$, where ω_r is the rotor frequency and the centerband remains unmodulated. Therefore, while a single experimental decay will result in the FID of a static sample on resonance, under coherent modulation by MAS the spinner will produce rotary echoes for every period of MAS, during which the magnetization vectors corresponding to different orders m would have excursed through $\pm 2\pi m$ radians. Therefore, perfect echoes are expected for a Hahn echo pulse of $90^\circ - \tau - \pi - \tau$ provided the rotors have completed integer multiples of revolutions. However, for π pulses that are applied at intervals much less than a cycle of rotation, very poor echoes result, although a number of closely spaced π pulses can recover a sizeable echo. By using a set of four π pulses with predetermined time delays it is possible to phase shift the various components in the FID such that the central line corresponding to the isotropic chemical shift is unaffected, while the sidebands of different orders are suitably phase shifted, compared to the ordinary unmanipulated FID. In fact, by such manipulations one can make a particular sideband to differ by 180° in phase in two different spectra and generate sum and difference spectra leading to sideband-eliminated or perhaps even centerband-eliminated spectra. The PASS technique imparts phase differences to sidebands in a process that does not produce any phase accumulation on the various isotropic central resonances of different chemically shifted nuclei. The PASS pulse scheme is shown in Fig. 6.21. For a given pulse delay set involving four π pulses the phases of the sidebands show a regular positive or negative twist depending on the order. Thus, in a CPMAS experiment as soon as the contact time is over for cross-polarization, the ^{13}C magnetization is allowed to evolve under the PASS sequence. If the rotor period is taken as the unit of time τ, then a PASS sequence (Prepare–0.1082τ–π–0.1082τ–π–0.4605τ–π–0.9706τ–π–0.5104τ–Acq) will lead to a phase twist

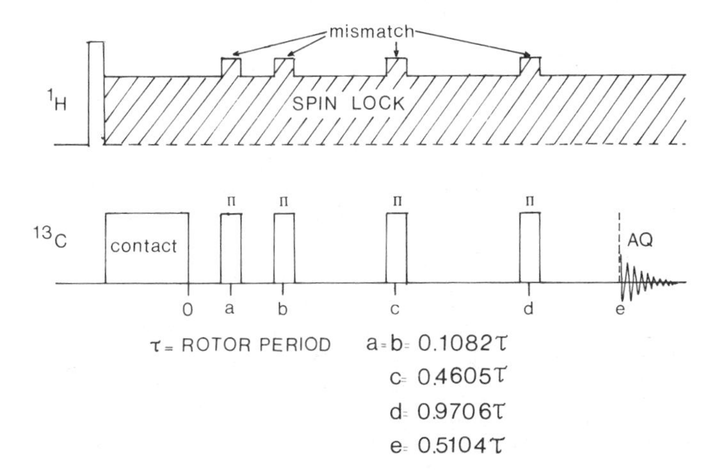

Figure 6.21. Pulse scheme for PASS.

of 60° per order of the sideband so that the phases of the side bands on either side vary as $\pm 60°$, $\pm 120°$, $\pm 180°$, and so on. During the π focusing pulses a Hartmann–Hahn mismatch may be introduced to avoid creating extra coherences by fresh polarization transfer by changing either the ^{13}C or the ^{1}H power. Suitable linear combination of spectra with different phases for sidebands can always be made to eliminate the sidebands.

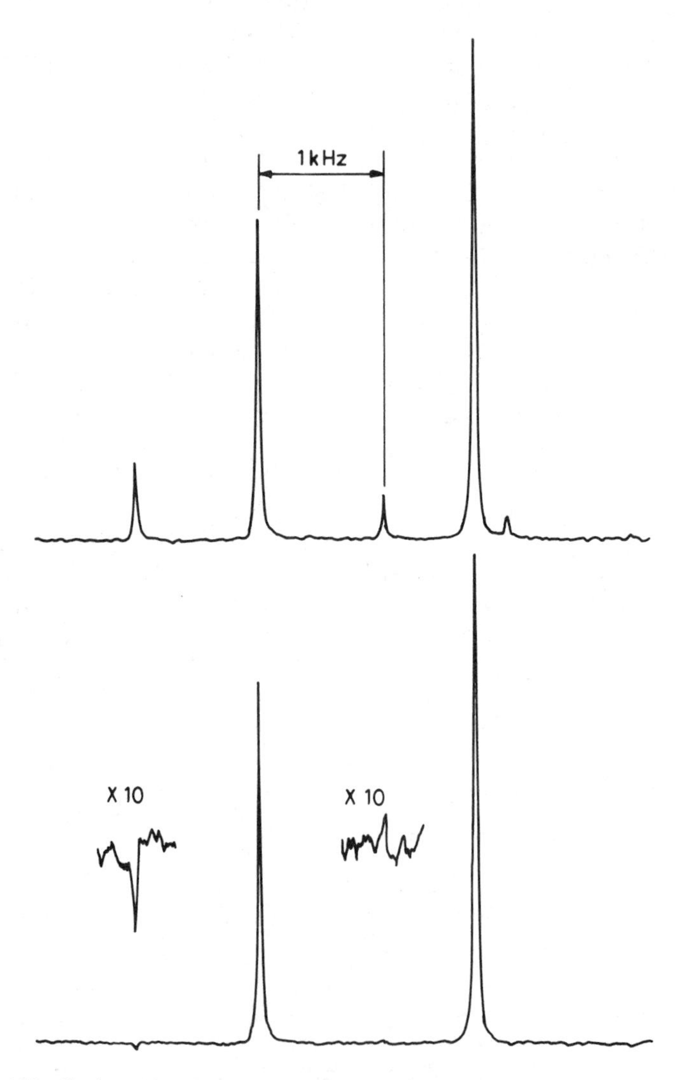

Figure 6.22. Carbon-13 TOSS spectrum of hexamethylbenzene spinning at 1 kHz. Upper trace: CPMAS spectrum showing distinct spinning sidebands. Lower trace: CPMAS–TOSS spectrum showing excellent sideband suppression. [Reproduced by permission. W.T. Dixon, *J. Chem. Phys.*, **77**, 1806 (1982), copyright (1982), American Institute of Physics]

It is also possible to so choose the pulse delays that the magnetization vectors corresponding to different orders of sidebands are distributed so as to maintain a constant phase distribution with respect to the receiver coil throughout the acquisition time. Consequently, there will be no rotary echoes, and hence no sidebands in the resulting spectrum. This technique, known as the total suppression of sidebands, (TOSS) must be extremely useful in simplifying MAS spectra and as such increase its potential use in analyzing a variety of materials that cannot be brought into solution form. A typical TOSS sequence in terms of rotor period τ is prepare–0.1885τ–π–0.0412τ–π–0.5818τ–π–0.9588τ–π–0.2297τ–Acq. One can use always delays that are further incremented from the values given above by integer multiples of τ without affecting the results. Example for application of TOSS is shown in Fig. 6.22. It may be mentioned in passing that TOSS should be useful in suppressing spinning sidebands in HR NMR of liquids also.

High-Resolution Dipolar NMR in Solids

We very briefly describe a few systems where one can get well-resolved heteronuclear dipolar coupling and hence precise distance information in solids. The systems should generally consist of a rare spin, or a low specific ratio of hetero spins, and should be measured (in either single crystal or powder form) under high-power decoupling of the most abundant spin. For example, in CHN systems one can measure ^{14}N and ^{13}C spectra and observe, respectively, ^{13}C and ^{14}N dipolar splittings under ^{1}H decoupling. This can be done with or without polarization transfer. Such spectra have linewidths less than heteronuclear dipolar splitting, whose $(3\cos^2\theta - 1)/r^3$ dependence on the dipole–dipole vector should lead to precise determination of positions of atoms and distances, and hence some clues as to molecular conformation in solids. In single crystals, one also gets additional information about the chemical shift anisotropy tensors, which is important in understanding the electronic structure. Figure 6.23 is the ^{13}C single-crystal spectrum of glycine, showing ^{13}C–^{14}N dipolar coupling and chemical shift differences. Measurement of ^{14}N under ^{1}H decoupling conditions can lead to a dispersion of ^{14}N resonances over several megahertz because of the quadrupole interaction. Precise determination of the site symmetry and quadrupole coupling constants is possible. Carbon-13 has a chemical shift dispersion of nearly tens of kilohertz, with a linewidth under ^{1}H decoupling conditions of 100 Hz, and yields a resolution of nearly 100, although the presence of too many different carbons in a molecule requires signal averaging for long times. Nitrogen-14, on the other hand, has 99.6% natural abundance and, because of its electric quadrupole moment, has resonances dispersed over the megahertz range depending on the symmetry of the environment; as such it holds good promise as long as the linewidth is not prohibitively large. One could get dipolar-resolved spectra under ^{1}H decoupling conditions within a matter of minutes with a decent $\mathcal{S}/\mathcal{N}$ ratio. Thus dipolar-resolved ^{14}N solid spectra should help as intrinsic probes for studying solid amino acids, peptides, proteins, etc. The

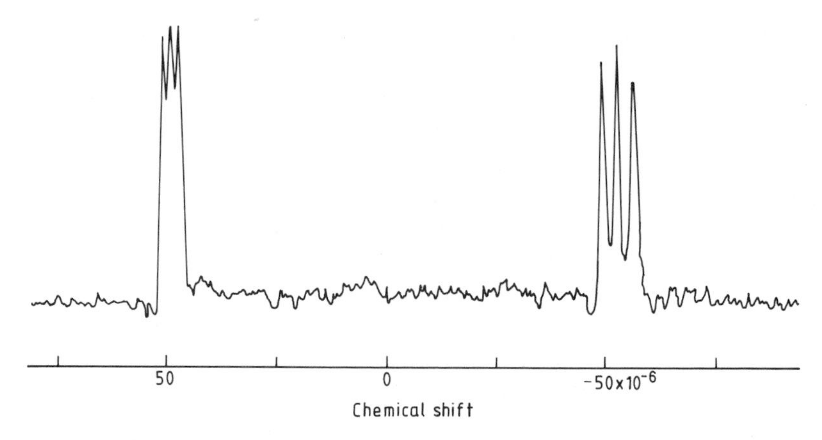

Figure 6.23. Carbon-13 single-crystal spectrum of glycine, showing ^{14}N dipolar splittings. The line on the right is from CH_2 and on the left is from $[COO]^-$. [Reproduced by permission. R.G. Griffin, G. Bodenhausen, R.H. Haberkorn, T.H. Huang, M. Munowitz, R. Osredkar, D.J. Ruben, R.E. Stark and H. van Willigen, *Phil. Trans. Roy. Soc. London*, **A 299**, 547 (1981), copyright (1981) Royal Society of Chemistry, London]

resolution that has been obtained in such ^{14}N spectra is nearly an order of magnitude better than in rare spin-1/2 nuclei in the solid state. One could also partly enrich ^{13}C and study $^{13}C-^{13}C$ dipolar-resolved spectra in solids to get structural information. These spectra behave either first order or second order depending on the relative magnitudes of chemical shift differences versus dipolar splitting, just as the liquid spectra depend on the relative magnitude of scalar chemical shift difference and indirect spin–spin coupling. In favorable circumstances $^{13}C-^{1}H$ and $^{14}N-^{1}H$ pairs have dipolar couplings much larger than the linewidth arising from lattice contribution to the second moment, so these could be resolved. Likewise, in $^{14}N-^{1}H$ pairs the quadrupole coupling constant can be much larger than the heteronuclear dipolar interactions, so one can observe these splittings in a simple one-dimentional cross-polarization experiment with or without ^{1}H decoupling. A large number of solid samples are being examined using such techniques as a useful supplement to x-ray studies.

Some Representative Examples of Improved Resolution in Solid-State NMR Spectroscopy by 2D Separation Techniques

We have seen in Chapter 4 that it is possible to use suitable pulse sequences and time delays to "prepare" the system and allow it to "evolve" in one time period under a suitably dressed Hamiltonian. The effects of the latter are transferred to another subsequent period with or without "mixing" where it is acquired under a different Hamiltonian and subsequently processed for 2D resolution or correlation. In liquids we have seen that we can use homo- and heteronuclear scalar coupling for *J*-resolved 2D spectroscopy and transfer of polarization either via scalar coupling or dipolar interaction and exchange to get shift correlation. We have seen in Chapter 5 that various multiple-quantum coherences can aid in getting structural information in a 2D experiment. In an exactly similar way solid systems are amenable for 2D spectroscopy, where we use the preparation–evolution–mix–acquire strategy to generate signals $S(t_1, t_2)$ and subject the same to 2D Fourier transformation. While a variety of 2D experiments have been carried out in solids we shall enumerate only a few examples.

Separated Local Field (SLF) 2D Spectroscopy. This 2D experiment is designed to separate the chemical shift along one axis and heteronuclear dipolar coupling along the other axis for a rare spin in solid state and as such is called separated local field (SLF) spectroscopy. In C–H systems the experiment is done by first preparing the rare spin using cross-polarization under Hartmann–Hahn condition. At the end of the contact period the ^1H field is switched off, and the polarized rare spin magnetization is allowed to evolve under conditions of pulsed homonuclear decoupling. The sequence used is WAHUHA-4, MREV-8, or BR-24 for a period t_1, during which the ^{13}C magnetizations evolve under chemical shifts and scaled heteronuclear (^{13}C–^1H) dipolar coupling. At the end of the t_1 period the ^1H decoupler is switched on again and the ^{13}C magnetization is acquired for a time t_2, where it evolves under the full shift Hamiltonian. The t_1 period is therefore an integer multiple of the multiple-pulse sequence cycle time. The t_1 is suitably incremented and the resulting signal $S(t_1, t_2)$, when subject to double Fourier transformation, should give chemical shift along the F_2 axis (normal CP spectrum) and shift and heteronuclear scaled dipolar coupling along the F_1 axis (Fig. 6.24).

The SLF spectra of single crystals are fairly easy to interpret. For each orientation of the crystal the relative orientation of the chemical shift tensor and the local heteronuclear dipolar tensor (symmetry axis parallel to the I–S vector) will dictate the nature of the spectrum (Fig. 6.25).

In powders, SLF spectra can in one shot give complete information of the chemical shift and heteronuclear dipolar coupling. One can easily reconstruct

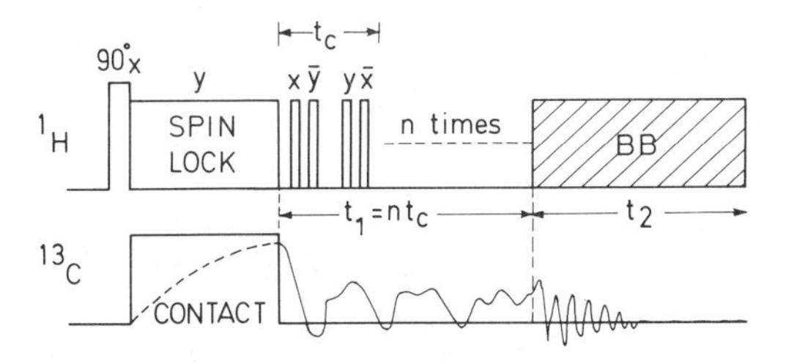

Figure 6.24. Pulse sequence for 2D separated local field (SLF) spectroscopy.

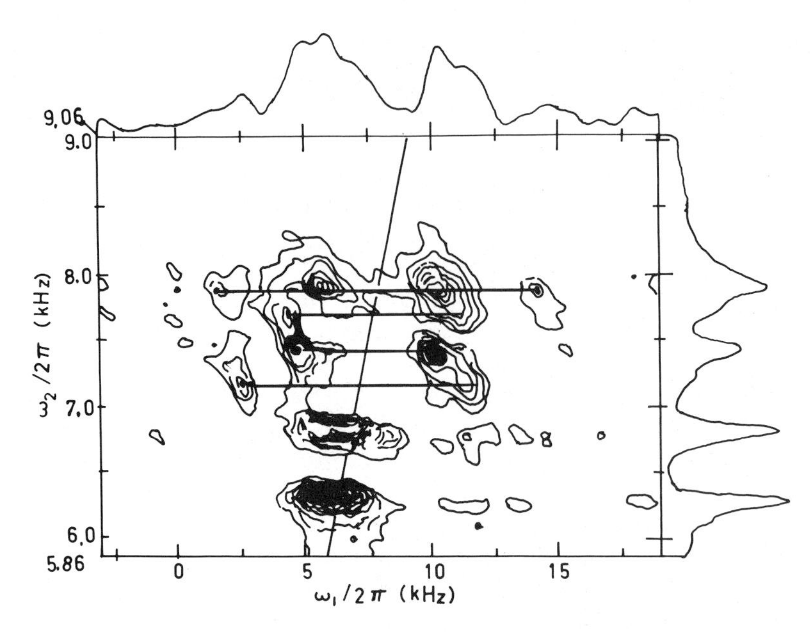

Figure 6.25. Carbon-13 SLF spectrum of a single crystal of calcium formate. Dipolar splittings for each of the observed seven lines are indicated by heavy tie lines. A projection onto the ω_2 axis gives the normal one-dimensional spectrum, and that onto the ω_1 axis yields the ^{13}C spectrum, including shifts and scaled heteronuclear dipolar couplings. [Reprinted with permission. R.K. Hester, J.L. Ackerman, B.L. Neff and J.S. Waugh, *Phys. Rev. Lett.*, **36**, 1081 (1976), copyright (1976) American Physical Society]

the 2D pattern by realizing that in a powder the molecules are randomly distributed. Typically, in an axially symmetrical case, the chemical shift profile can be simulated from a knowledge of θ (the angle between B_0 and $\delta_{\parallel}$) only. Now if the chemical shift tensor and the heteroatom dipolar tensors are coincident, then at B_0 parallel to $\delta_{\parallel}$, B_0 will also be parallel to the I–S vector, so the local field dipolar splitting will be a maximum. At the other extreme, when $B_0 \parallel \delta_{\perp}$, the dipolar splitting will be a minimum and of opposite sign because it follows a $(3\cos^2\theta - 1)$ dependence. Thus, in the powder profile, the extrema corresponding to $\delta_{\parallel}$ and $\delta_{\perp}$ will be split into doublets with general splitting $2A$ and $-A$ with the splitting reaching zero for those crystallities that are at the magic angle orientation. In fact, the observed 2D SLF spectra in powders can be computer simulated assuming the nature, whether cylindrically symmetric or not, of the chemical shift tensor along with the crucial information whether the local dipolar and shift tensors are coincident or not. In fact, it is possible in a single measurement to decide the noncoincidence of the tensors from the form of 2D spectra and to quantify these tensors by 2D SLF spectral simulation. The SLF spectra can also get quite complicated in the presence of a number of different chemically shifted nuclei with their local heteroatom dipolar coupling and the added complication of noncoincidence of the dipolar tensor principal axes with those of the chemical shift tensor, not to mention the possibility that the latter in general can also be orthorhombic. In addition, because the cross-polarization efficiency depends on the orientation of the heterodipolar vector with respect to B_0, intensity profiles cannot be very accurately predicted. Nevertheless, this is a unique method in that it is possible, in favorable cases, to determine the relative orientation of tensorial interactions using only powders (see Fig. 6.26).

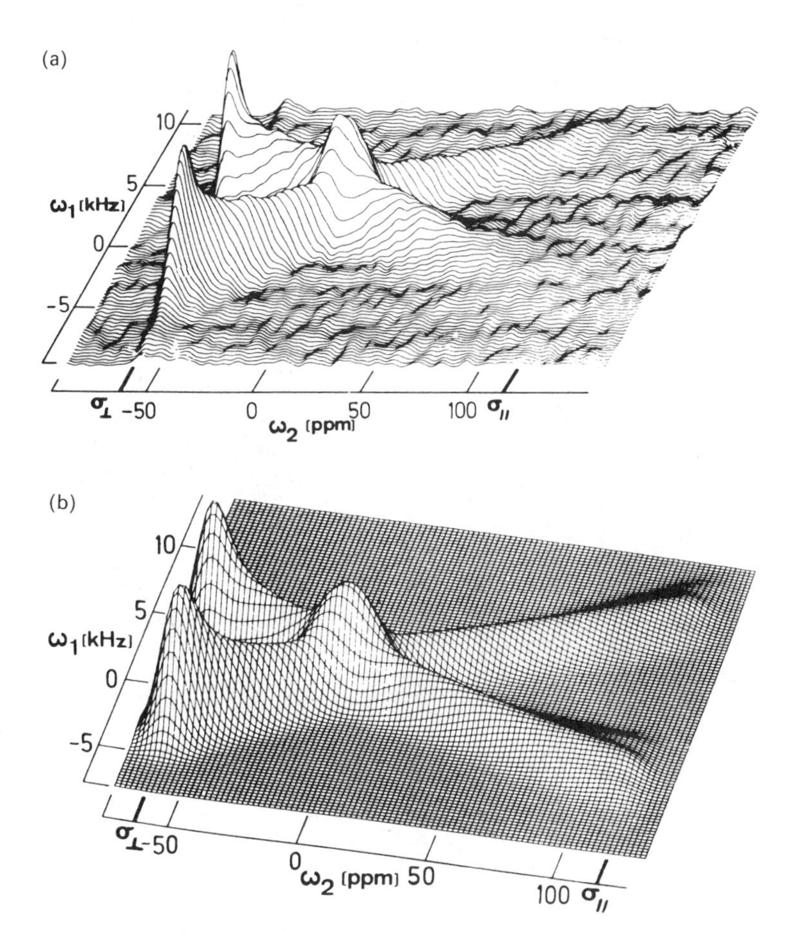

Figure 6.26. The 25-MHz ^{13}C shift-resolved dipolar 2D spectrum of benzene at 148K. (a) Experimental spectrum. (b) Computer simulated spectrum with the following parameters: $\sigma_{\parallel} = 117$ ppm, $\sigma_{\perp} = -63$ ppm, $D_{\parallel}$(C—H) = 11.66 kHz, and $D_{\parallel}$(C—C—H) = 1.52 kHz. (c) Computer-simulated shift-resolved dipolar spectrum of methyl formate with $\alpha = 0°$, $\beta = 90°$. (d) Same as above, with $\alpha = \beta = 90°$. (Here, α is the azimuthal angle of the dipolar vector in the chemical shift principal axis system, and β is the corresponding polar angle). [Reproduced by permission. M. Linder, A. Höhener and R.R. Ernst *J. Chem. Phys.*, **73**, 4959 (1980), copyright (1980) American Institute of Physics] (Parts (c) and (d) appear on page 318.)

(c)

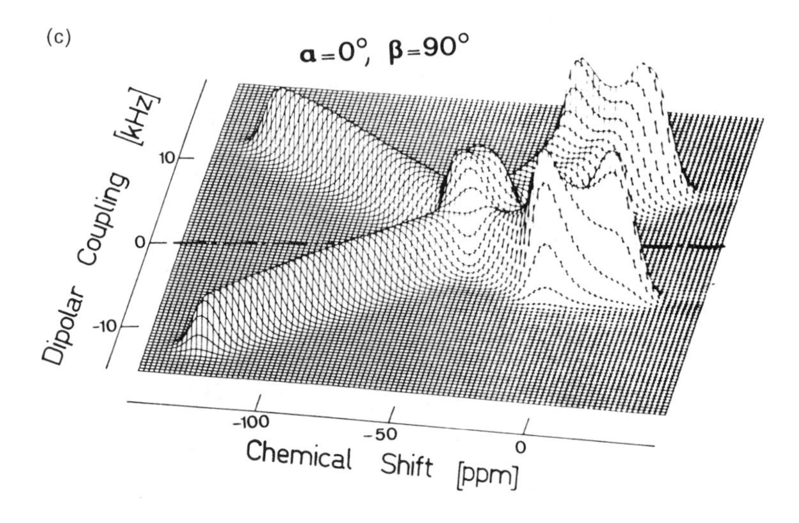

(d)

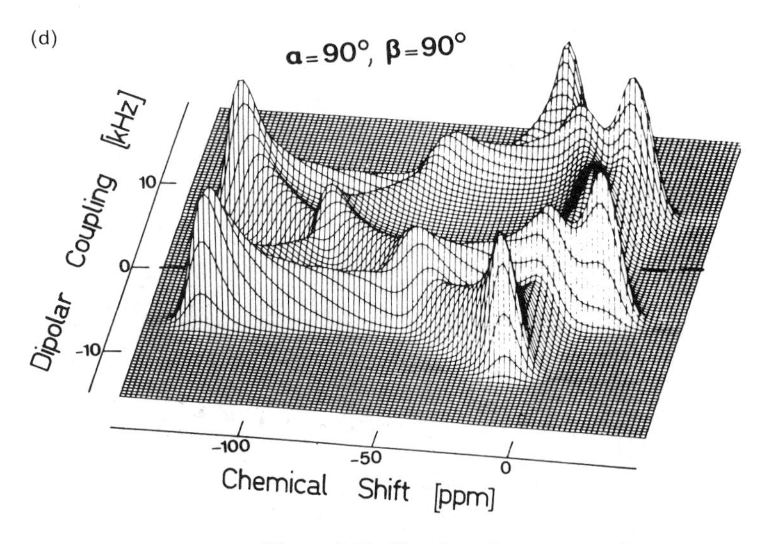

Figure 6.26. (Continued)

Two-Dimensional Separation of Isotropic and Anisotropic Parts of the Chemical Shift Tensor in CPMAS. We have seen that the CPMAS technique is used to examine crystals and powders with a view to studying chemical shift tensors of rare spin with increased sensitivity in the absence of heteronuclear coupling. We have also mentioned that at MAS speeds that are only half the shift anisotropy a number of sidebands develop and in the presence of a good number of different chemically shifted rare spin species the total overlap of center and sidebands leads to a pattern that is rather difficult to unravel. Besides, under moderate rotor speeds the sideband intensity profile still retains

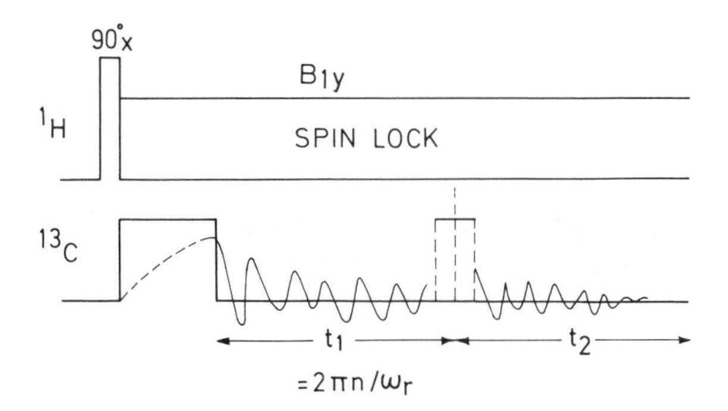

Figure 6.27. Pulse sequence for 2D separation of isotropic and anisotropic chemical shift in rotating solids.

the chemical shift anisotropy profile, so that one can in principle get an idea of the shift tensor by analyzing sideband intensities.

However, we can perform a 2D experiment in which it is possible to separate the isotropic chemical shift along one axis and the spinning sidebands, with their chemical shift anisotropy, along the other by the following strategy. The principle is that rotary echoes, which are produced at the frequency of the MAS rotor, correspond to a refocusing of magnetization vectors of all orders of sidebands on to that of the isotropic shift; as such this happens once in a rotation cycle time. If we have evolution periods that are integral multiples of the rotor period, then the accumulated phase will depend only on the isotropic shift frequency. In the final detection period furthermore, we can acquire the signal as in the normal CPMAS, starting from a point corresponding to an integral multiple of the rotor period. The sequence used is shown in Fig. 6.27. Here, t_1 is incremented by integer multiples of the rotor period $2\pi/\omega_r$ and hence carries accumulated phases corresponding only to isotropic chemical shifts. During t_2 the normal sideband spectra, which are centered around the various isotropic chemical shifts, are acquired. A double Fourier transformation should then lead to a separation of various isotropic chemical shifts parallel to F_1, each having its characteristic anisotropic sideband profile along F_2. The spectral lineshapes are superpositions of absorption and dispersion and the absolute value mode of presentation leads to broad lines. The dispersion contribution can be removed by reversing the accumulated phase during the t_1 period. This can be achieved by adding a π pulse every other scan at the end of the t_1 period and coadding the FID's (cf. Chapter 4). In addition, we can apply a quadrature phase cycle to the source and the sink spin polarization transfer pulses to get pure phase 2D with sign discrimination in F_1. A typical example of such 2D separation is indicated in Fig. 6.28.

Isotropic shifts and chemical shift anisotropies may be separated by two-dimensional experiments that involve special strategies for imparting motion to the sample. One possibility is the so-called magic angle "hopping" experi-

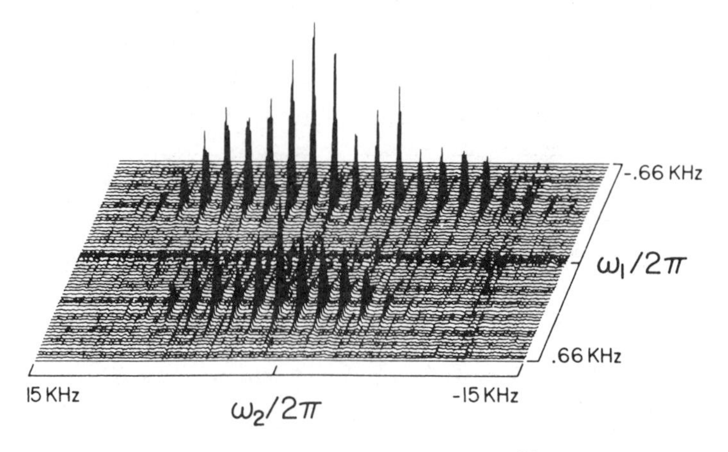

Figure 6.28. The 2D separated iso/aniso shift spectrum of ^{31}P in a mixture of barium diethyl phosphate (BDEP) and brushite ($CaHPO_4 \cdot 2\ HPO$). The BDEP signals occur in the upper half. [Reprinted with permission. W.P. Aue, D.J. Ruben and R.G. Griffin, *J. Magn. Reson.*, **43**, 472 (1981), copyright 1981, Academic Press, New York]

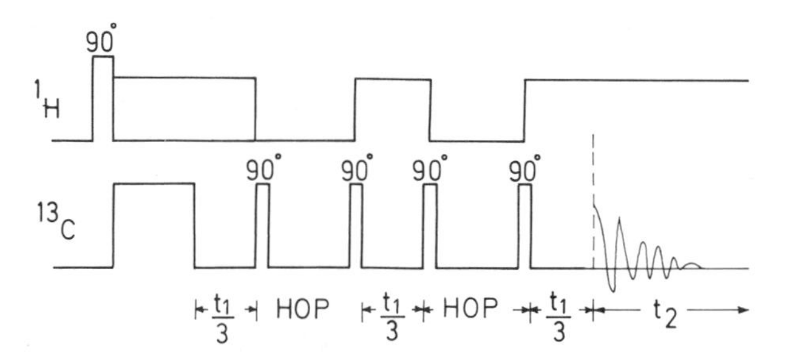

Figure 6.29. Pulse sequence for magic angle hopping spectroscopy to achieve 2D separation of isotropic and anisotropic shifts.

ment, where the rotor moves in 120° jumps during t_1 and the sample is static during signal detection. With phase modulation in t_1, the result of sampling the chemical shift of crystallites at orientations related by 120° rotations about the magic angle axis, corresponds to sampling the isotropic shift. This results in isotropic chemical shifts along F_1, and the shift anisotropy pattern along F_2. Figure 6.29 shows a suitable pulse sequence for the experiment. At the commencement of the "hop" period when the rotor jumps, one component of the transverse ^{13}C magnetization is stored longitudinally by a 90° pulse, while the other dephases during the hop period with the decoupler turned off. Phase modulation leading to quadrature detection in F_1 may be achieved by suitably phase cycling the receiver and the four carbon $\pi/2$ pulses.

An alternative experiment to generate such separations involves sample rotation about an axis perpendicular to $\mathbf{B}_0$ during t_1, followed by magic angle

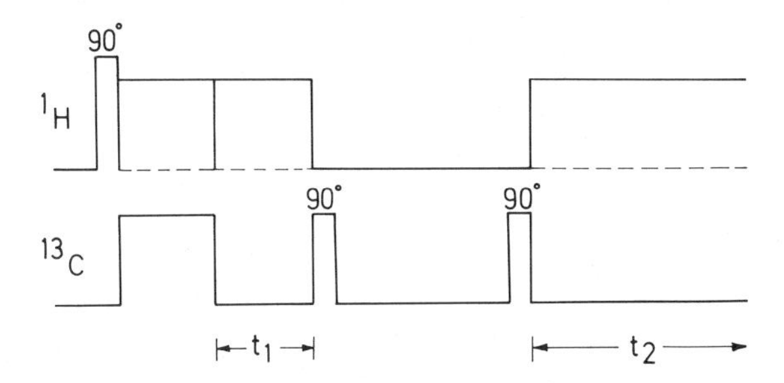

Figure 6.30. Pulse sequence for 2D separation of iso/aniso shifts by flipping of the spinning axis.

spinning during t_2. A suitable pluse sequence is given in Fig. 6.30. Cross-polarization is effected in this case with the sample spinning about an axis perpendicular to $\mathbf{B}_0$. Following evolution under decoupling one transverse component is stored longitudinally for the following period of about a second, during which the spinning axis is reoriented to make the magic angle with respect to $\mathbf{B}_0$; the stored magnetization is then brought back to the transverse plane. Note that the F_2 axis in this case corresponds to the isotropic shifts, whereas the F_1 axis corresponds to anisotropic shifts scaled by a factor of $-\frac{1}{2}$. It may be noted that in such experiments the rotor has to be under computer control. Such experiments have been termed SASS (switched angle sample spinning) or magic angle flipping and can lead to information on the absolute signs of couplings and the orientation of the dipolar tensors.

Heteronuclear Correlation in Solids

It is often not feasible to resolve proton chemical shifts in the spectrum of a solid that exhibits more than just a few signals, even with the best line-narrowing sequences. It would clearly be an advantage to use a directly bound heteronucleus, such as ^{13}C, to get well-dispersed chemical shift correlations. However, classical cross-polarization strategies are nonselective in nature because spin diffusion effectively leads to a thermodynamic interaction, where bond connectivities are lost sight of. Selective polarization transfer, however, effectively suppressing spin diffusion during the mixing period, can restore the connectivity information in solid-state heteronuclear correlation experiments. Following evolution under (scaled) chemical shifts and heteronuclear dipolar couplings, while suppressing homonuclear dipolar interactions of the abundant spins by a multipulse sequence, any one of several schemes may be employed for mixing to generate such 2D heteronuclear correlation spectra. A few important mixing strategies are briefly discussed here.

On-resonance Hartmann–Hahn cross-polarization (Fig. 6.31) may be employed with very short contact times ($\sim 20\ \mu s$) to gain reasonable selectivity; this, however, is achieved at the expense of sensitivity.

Pulsed polarization transfer, as in liquids, may be employed, following

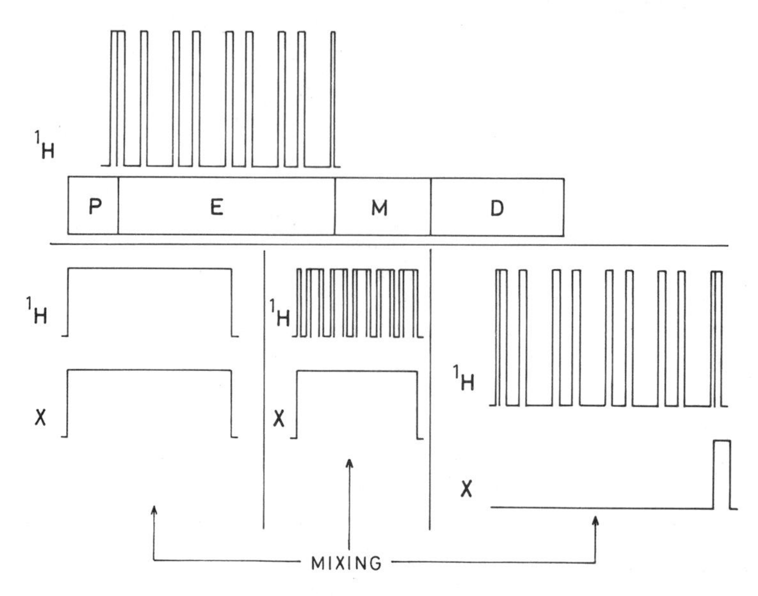

Figure 6.31. Pulse sequences for heteronuclear correlation in solids. Three different mixing strategies are shown in the figure, viz., on- or off-resonance Hartmann–Hahn cross-polarization, multiple-pulse cross-polarization, and pulsed polarization transfer.

suppression of abundant spin homonuclear dipolar interactions, to allow establishment of a state of heteronuclear dipolar order (Fig. 6.31).

Multiple-pulse cross polarization may also be employed to achieve cross-polarization with simultaneous homonuclear dipolar decoupling to suppress spin diffusion. A sequence of pulse sandwiches $(35.3^\circ_\theta, 120^\circ_\phi, 35.3^\circ_\Psi)_N$ may be used to this end, with $\theta = \phi + \frac{\pi}{2}$ and $\Psi = \phi - \frac{\pi}{2}$ causing a rotation about an axis at the magic angle from z in the $z\phi$ plane (Fig. 6.31).

When multipulse sequences are employed, large offsets must be avoided, as discussed earlier in this chapter. The abundant spin magnetization must eventually be tipped into the transverse plane by a pulse of suitable flip angle to compensate for the situation that the precession axis under multipulse excitation has a transverse as well as a longitudinal component.

Experimental results employing these mixing schemes are shown in Fig. 6.32 for a single crystal of ferrocene, which exhibits two ^{1}H and two ^{13}C resonances from the two inequivalent sites. In the 2D spectrum, correlation between C_B and H_A, for example, indicates nonselective transfer. The F_1 axis involves scaled proton shifts and scaled ^{1}H–^{13}C dipolar couplings, while the F_2 axis involves ^{13}C shifts. Quadrature detection in F_1 may be achieved by suitable phase cycling.

An alternative technique employs isotropic mixing, the evolution period exhibiting only scaled proton shifts; detection is under continuous wave (CW) proton decoupling. In this case, homonuclear dipolar interactions and

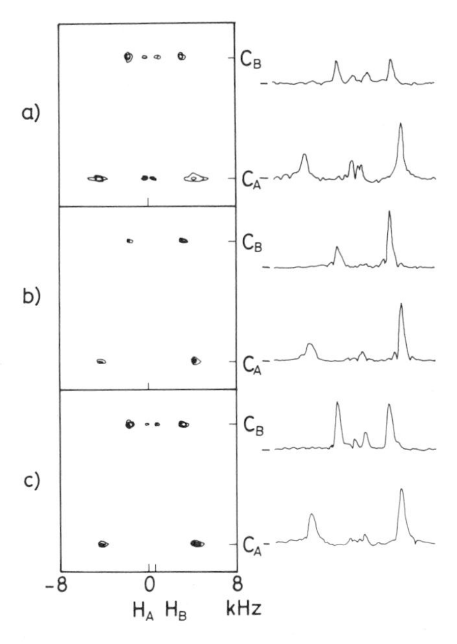

Figure 6.32. Heteronuclear correlation spectra of a single crystal of ferrocene in fixed orientation. Measurements were carried out on a 300-MHz spectrometer. Inequivalent sites are clearly resolved. The vertical F_2 axis shows ^{13}C shifts, while the F_1 axis shows scaled ^1H shifts and scaled heteronuclear couplings. The strength of peaks near $F_1 = 0$ is a measure of selectivity of transfer. (a) Transfer by 20-μs on-resonance CP; (b) multiple-pulse CP; (c) pulsed coherence transfer. [Reprinted with permission. P. Caravatti, G. Bodenhausen and R.R. Ernst, *Chem. Phys. Lett.*, **89**, 363 (1982), North-Holland Physics Publishing, Amsterdam]

chemical shifts are suppressed during mixing, while heteronuclear couplings are active: thus polarization transfer (PT) is highly selective. Since signals are not observed during the evolution and mixing periods, homonuclear dipolar decoupling may be performed by windowless sequences with low rf power (ca. 10–40 W). Composite pulse decoupling with the WALTZ sequence (see Chapter 7) is performed to suppress heteronuclear dipolar couplings during t_1. The experiment is performed with the sample spinning at the magic angle, the rotor period being at least six times the multipulse cycle time, so that coherent averaging may proceed efficiently without producing artifact signals. The pulse sequence for this experiment is shown in Fig. 6.33. Note that isotropic mixing automatically insures that the sign of F_1 is preserved without the need for any phase cycling, since both transverse components are exchanged, unlike the other techniques discussed earlier in this section. Experimental results on threonine are shown in Fig. 6.34 and clearly demonstrate efficient, selective transfer; this example also establishes that unresolved proton shifts may be "dispersed" via heteronuclear correlation.

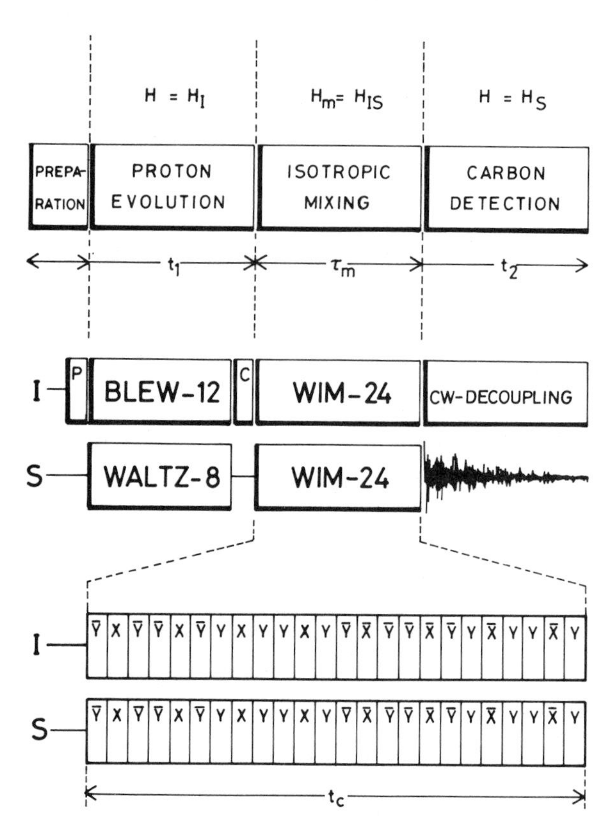

Figure 6.33. Pulse scheme for heteronuclear correlation in solids, using isotropic mixing, in this case by a windowless isotropic mixing sequence, WIM-24. [Reproduced by permission. P. Caravatti, L. Braunschweiler and R.R. Ernst, *Chem. Phys. Lett.*, **100**, 305 (1983), North-Holland Physics Publishing, Amsterdam]

Homonuclear Correlation in Solids

Two-dimensional homonuclear correlation may be achieved in solids by allowing the spin system to evolve in t_1 under the full internal Hamiltonian, which is dominated by dipolar interactions, followed by signal acquisition during t_2 under homonuclear dipolar decoupling, generating scaled chemical shifts. In order to distinguish positive and negative frequencies in F_1, one may combine the results of two experiments with the preparation pulse phase shifted by 90°. Concurrently a line-narrowing sequence may be employed in t_2 which has an average chemical shift Hamiltonian with no transverse spin operators. A suitable sequence is an extension of the MREV cycle, characterized as:

$$[ZYX][Z\bar{Y}X][ZY\bar{X}][Z\bar{Y}\bar{X}] \tag{121}$$

which has a nominal scale factor of 1/3.

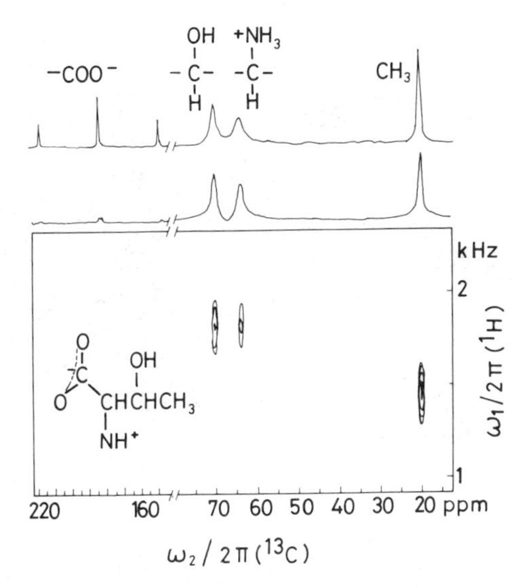

Figure 6.34. The 300-MHz contour spectrum of 2D hetero COSY of threonine. Top trace: Normal CP MAS spectrum. Second trace: Projection of the 2D spectrum. [Reproduced by permission from P. Caravatti, L. Braunschweiler and R.R. Ernst, *Chem. Phys. Lett.*, **100**, 305 (1983), North-Holland Physics Publishing, Amsterdam]

The spectral features in F_1 may be interpreted as follows: the cross-section along F_1 for the value of F_2 corresponding to the chemical shift of the ith inequivalent spin gives the ith single-spin dipolar spectrum. Such single-spin dipolar spectra correspond to selective excitation of the ith spin, keeping all dipolar couplings alive, or equivalently, to selective detection of the ith spin resonance following nonselective excitation.

When correlations between dipolar interactions and chemical shifts are generated in this manner, the assignment of chemical shift tensors to specific sites is greatly aided. In typical cases, computer simulations of the single-spin dipolar spectra of relatively small systems of coupled spins must be performed in order to analyze the F_1 cross-sections based on proton position data from NMR work itself or from other sources.

An example of a ^1H 2D correlation spectrum of a single crystal of malonic acid is shown in Fig. 6.35. In particular, F_1 cross-sections at the four chemical shifts in F_2 may be analyzed with the help of simulated single-spin dipolar spectra. This permits the assignment not only of the two carboxylic protons but also of the two methylene protons.

Zero-Field NMR

In the absence of a Zeeman field, no direction in space is imposed externally. Under these conditions, all orientations of a pair of dipoles are equivalent, yielding identical dipolar splittings. Polycrystalline powder specimens thus

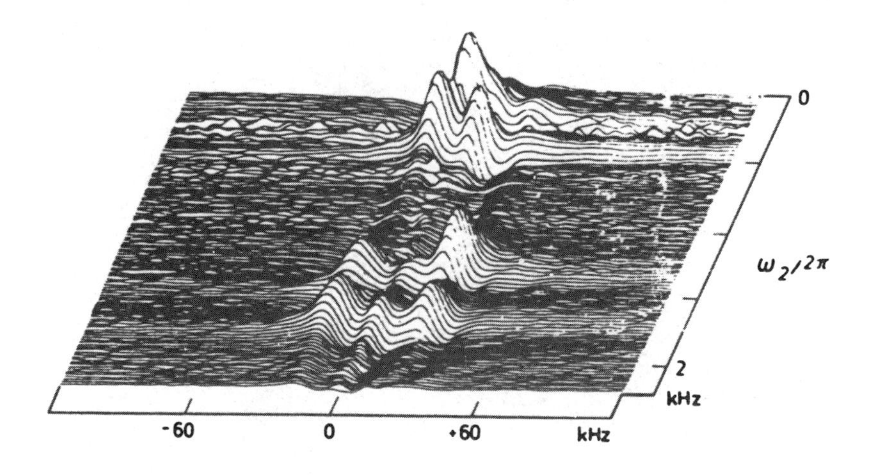

Figure 6.35. Homonuclear COSY of a single crystal of malonic acid. [Reprinted with permission. N. Schuff and U. Haeberlen, *J. Magna. Reson.* **52,** 267 (1983). Copyright 1983 Academic Press New York]

behave like a crystal in this experiment, which measures the spectrum of dipole–dipole interactions. For a pair of dipole-coupled spin-1/2 nuclei, for example, the observed splitting is a direct, orientation-independent measure of the internuclear distance. Line intensities, however, are orientation dependent. The sensitivity problems associated with zero-field NMR, however, may be avoided by preparing the spin system and detecting the NMR signals in a high Zeeman field. The polarized spin system is allowed to evolve for a variable period t_1 in zero field, and immediately after return of the sample to the high field at the end of t_1, the magnetization is sampled. An interferogram is built up point by point, incrementing the zero-field evolution period t_1. For a homonuclear pair of dipole coupled spins-$\frac{1}{2}$, we have:

$$S(t_1) = \tfrac{1}{3} + \tfrac{2}{3}\cos\tfrac{3}{2}\omega_{ij}^{\mathrm{D}}t_1 \tag{122}$$

where $\omega_{ij}^{\mathrm{D}} = \gamma^2 h/2\pi r_{ij}^3$. Figure 6.36 shows some typical results on the protons of $Ba(ClO_3)_2 \cdot H_2O$. The ^1H–^1H distance calculated from this spectrum of a powder is 1.6 Å, in agreement with the single-crystal NMR results.

For a heteronuclear pair of spins IS, on the other hand, we find:

$$\begin{aligned} S(t_1) = {} & \tfrac{1}{6}(a+b) + \tfrac{1}{3}(a+b)\cos(\tfrac{3}{2}\omega_{IS}^{\mathrm{D}}t_1) \pm \tfrac{1}{3}(a-b)\cos(\tfrac{1}{2}\omega_{IS}^{\mathrm{D}}t_1) \\ & \pm \tfrac{1}{6}(a-b)\cos(\omega_{IS}^{\mathrm{D}}t_1) \end{aligned} \tag{123}$$

where the density operator prepared prior to zero-field evolution corresponds to:

$$\begin{aligned} \sigma(0) &= aI_z + bS_z \\ &= \tfrac{1}{2}(a+b)(I_z + S_z) + \tfrac{1}{2}(a-b)(I_z - S_z) \end{aligned} \tag{124}$$

and the $\pm$ signs in $S(t_1)$ refer to I or S spin magnetization, respectively.

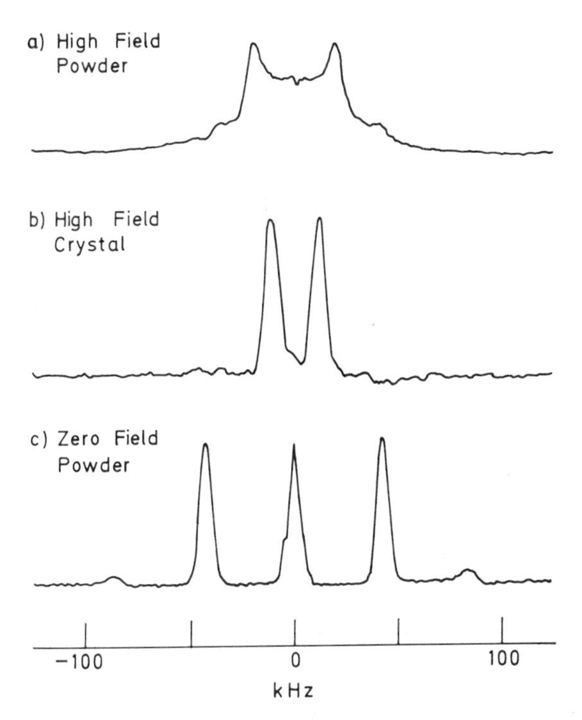

Figure 6.36. Proton NMR spectra of $Ba(ClO_3)_2 \cdot H_2O$. (a) High-field powder spectrum showing the Pake doublet. (b) High-field spectrum of a single crystal in an arbitrary orientation. (c) Zero-field powder spectrum. [Reprinted with permission. D.P. Weitekamp, A. Bielecki, D. Zax, K. Zilm and A. Pines, *Phys. Rev. Lett.*, **50**, 1807 (1983), copyright (1983), American Physical Society]

Additional frequencies of evolution in zero field are therefore available for the heteronuclear spin pair, with intensities depending on the way the system has been prepared. All these results follow from the equations of motion under dipolar coupling (see equation 1.79); although the dipolar Hamiltonian suffers no truncation in zero field, it may nevertheless be cast in the form of equation 1.78 for a two spin system, except there is no orientation dependence and **r** may without loss of generality be chosen to lie along the z axis in a molecular-frame.

Figure 6.37 shows the preparation-dependent proton spectrum of 90% ^{13}C-enriched polycrystalline β-$Ca(H^{13}COO)_2$. The C—H bond distances are 1.11 ± 0.02 Å on this basis, in agreement with other NMR and neutron diffraction data on this molecule. Various preparations may be achieved by the following sequence of operations, carried through before each cycle of the zero-field experiments:

1. Depolarization in zero field for 10 ms
2. Proton repolarization in high field

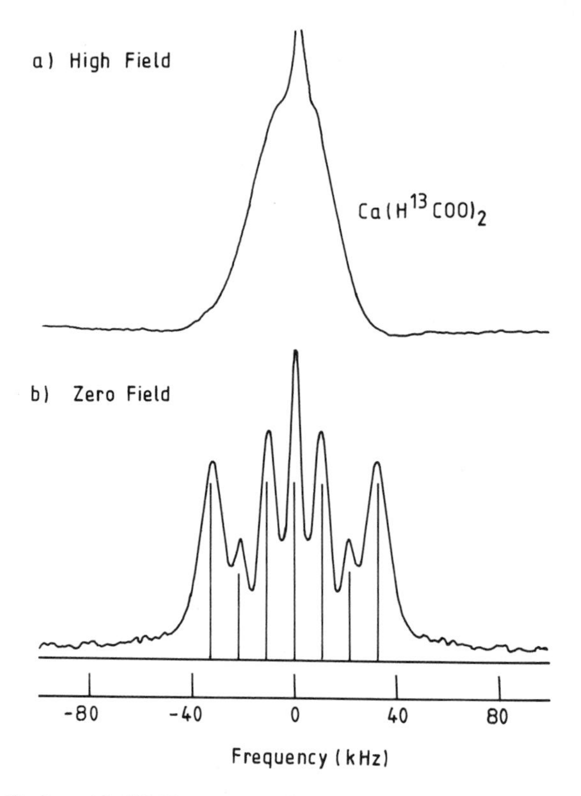

Figure 6.37. Carbon-13 NMR spectra of 90% enriched polycrystalline β-calcium formate. (a) Normal high-field spectrum, with a spike due to adsorbed water. (b) Zero-field spectrum and corresponding simulated stick diagram. (c) Zero-field spectra for various preparations of the system. Note that subsets of transitions (e.g., from different functional groups) may thus be selectively enhanced or eliminated. [Reproduced with permission. D. Zax, A. Bielecki, K.W. Zilm and A. Pines, *Chem. Phys. Lett.*, **106**, 550 (1984), North-Holland Physics Publishing, Amsterdam]

3. ^1H to ^{13}C magnetization transfer with a fixed zero-field interval of 32 μs
4. Proton repolarization in high field

Such a sequence creates equilibrium proton (I-spin) magnetization, and a persistent ^{13}C (S-spin) magnetization, that is 60% larger than that at equilibrium, i.e., $a/b = 2.5$. Finally, a resonant rf pulse may be applied to ^1H, destroying or inverting some of the initial ^1H magnetization.

Such zero-field evolution may clearly be employed for polarization transfer from protons to heteronuclei in solids and in liquids. An example of zero-field isotropic mixing in liquids has been discussed in Chapter 3, where it is noted that the antisymmetric components of the density operator evolve in zero field at the full coupling frequency.

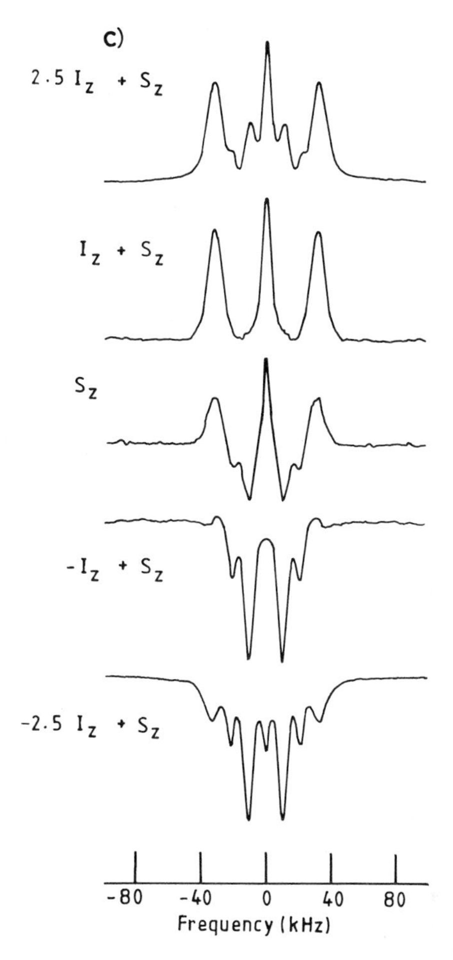

Figure 6.37. (Continued)

CHAPTER 7

Experimental Methods

Introduction

In this chapter, we briefly discuss a few basic experimental methods that have become standard in the arsenal of the NMR spectroscopist. First, we focus on the need to determine the signs of frequencies in the rotating frame and point out how this is achieved in practice with the suppression of the attendant errors. The effects of pulse errors are then mentioned, starting with rf inhomogeneity problems in line-narrowing and cross-polarization studies on solids. The composite pulse approach to surmounting the problem of pulse imperfections is dealt with in the context of specific applications, including coherence transfer experiments, broadband spin inversion in liquids and solids, broadband excitation of quadrupolar spin-1 nuclei in solids, and heteronuclear broadband decoupling in liquids. Finally, "retrograde" compensation schemes are mentioned in the context of depth resolution by accentuating B_1 inhomogeneity in surface coil imaging applications.

Quadrature Detection

Single-channel phase-sensitive detection does not determine the sign of frequencies in the rotating frame. This situation demands that the transmitter be placed at one end of the spectrum, resulting in power inefficiency and a poorer signal to noise ratio because the filter bandwidth is twice the spectral width. A procedure to determine the sign of frequencies in the rotationg frame would permit the transmitter frequency to be placed in the middle of the spectrum thus using transmitter power more efficiently. This would result also in a 40% improvement in the signal to noise ratio because the filter bandwidth would be halved in comparison to the single detection case. Signs of frequencies in the rotating frame may be retrieved by a quadrature detection process, which detects two components of the signal that are in quadrature with each other, i.e., have a 90° phase relationship. This requires two phase-sensitive detector channels that are associated with identical amplification factors, and reference phases with precisely a 90° phase difference. Imbalance in the amplification or phase leads to artifacts, which are known as image signals, that occur at $+\omega$ rad/s for a parent signal at $-\omega$ rad/s. Although careful balancing of the quadrature detectors reduces images to about 1%, further suppression by an order of magnitude or more may be achieved by phase cycling with a cyclically ordered phase sequence, CYCLOPS.

Quadrature detection may be considered as multiplication by, say, $(e^{+i\omega_0 t} + \eta e^{-i\omega_0 t})$, where the sense of the carrier circular polarization has been arbitrarily chosen. Here, η is a small complex number describing imperfect quadrature detection as a result of allowing a small contamination of the carrier by the opposite sense of polarization. Upon detection of the signals $Me^{-i\omega t}$ and $Me^{+i\omega t}$, the response $M(e^{-i(\omega-\omega_0)t} + \eta e^{+i(\omega-\omega_0)t})$ results. The image peak resulting on Fourier transformation is thus at $+(\omega - \omega_0)$ and has a relative intensity η compared to the parent signal at $-(\omega - \omega_0)$; it arises from the signal with the $+1$ coherence level, as shown in the coherence transfer pathway diagram (Fig. 7.1). When the phase of the pulse is shifted by $\pi/2$, the desired -1 level coherence is multiplied by $e^{-i\pi/2}$ so that the resulting signal is $-iMe^{-i\omega t}$. The $+1$ level coherence, on the other hand, leads to $iMe^{i\omega t}$. The nominal receiver phase change by $\pi/2$ (actually, interchange of real and imaginary parts with a sign reversal of one) results in the carrier $+i(e^{i\omega_0 t} + \eta e^{-i\omega_0 t})$. The solid pathway shown in (Fig. 7.1) thus contributes $+Me^{-i(\omega-\omega_0)t}$, while the dashed pathway leads to $-\eta Me^{i(\omega-\omega_0)t}$. When the two signals are added, the image signal at $(\omega - \omega_0)$ clearly vanishes. Phase shifting by 180° cancels any systematic ("coherent") noise from the receiver or computer. This leads then to the four-step phase cycle CYCLOPS:

Transmitter Phase	Increment to Computer Memory "A"	Increment to Computer Memory "B"
0°	I	II
90°	II	−I
180°	−I	−II
270°	−II	I

Here, the Roman numerals indicate the channels of the quadrature detector.

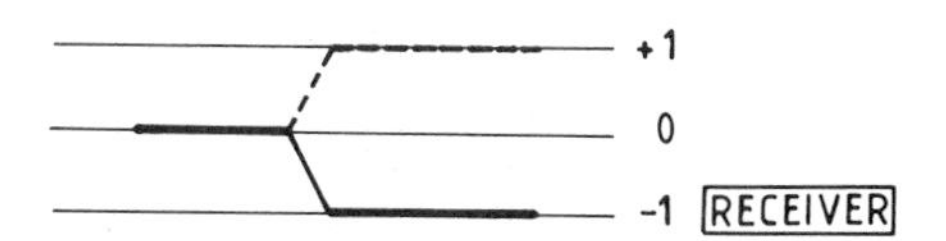

Figure 7.1. Coherence transfer pathway diagram for a one-pulse experiment, the receiver being set to detect the coherence of order -1.

Radiofrequency Homogeneity Effects in Multipulse and CP Experiments

Multipulse line-narrowing experiments on solids employ such sequences as WAHUHA-4, MREV-8, or BR-24. These sequences are composed of precisely adjusted 90° pulses (for WAHUHA-4, the optimal flip angle depends on the duty cycle, t_p/t_c) of precise phases, 0°, 90°, 180°, and 270°. The rf magnetic field created by a solenoid coil, however, is not uniform throughout the coil volume; in fact, B_1 fields are maximal near the center and are less intense at other locations. The rf inhomogeneity being a factor limiting the resolution attainable, line-narrowing experiments on solids are conducted with the sample, shaped perferably as a sphere, occupying only a limited volume of the coil around the center. Special tuning cycles are employed to set up the pulse flip angles and phases and to minimize phase transients, using a liquid sample of dimensions and shape similar to the solid sample to be studied. A coil with variable pitch leads to improved rf homogeneity.

Radiofrequency homogeneity is also vital to the success of cross-polarization (CP) experiments. Single coil probes that are doubly tuned to the two frequencies of interest are employed so that the rf inhomogeneity profiles at both frequencies are comparable. This allows the Hartmann–Hahn matching condition to be realized across the sample. It may be noted that, in contrast, pulsed polarization transfer experiments work well even with cross-coil probes, each coil being tuned to one of the frequencies of interest. Radiofrequency breakthrough during comparatively long spin-lock/mixing pulse in CP experiments leads to baseline artifacts that are best suppressed by "spin–temperature alternation."

Pulse Errors. Radiofrequency pulses may be characterized in terms of their carrier frequency, phase, pulse duration, amplitude, and shape. Modern experiments in NMR employ pulses with each one of these parameters taking on well-defined values for specific objectives. With increasing sophistication in tailoring spin dynamics under the action of rf pulses, tolerance to pulse errors is rapidly on the decline. We have encountered several situations involving coherence transfer, two-dimensional spectroscopy, and NMR of solids in which pulse errors have led to gross inefficiencies, to artifact signals, and even to failure of the experiment. In some cases, suitable phase cycling can suppress the unwanted responses arising from pulse errors. Improvement in the spectrometer hardware is another approach to minimizing pulse errors, but there are obvious practical limits on this direct approach, considering also the constant trend to NMR at ever higher field strengths.

Over the last few years, a trend has strongly emerged which seeks to achieve the required effect of a "perfect" pulse by employing a suitably tailored string of imperfect pulses. Some features of this strategy are discussed below.

Composite Pulses. It is commonly of interest in pulsed NMR to be able to design rf pulses that perform specific tasks regardless of the details of a particular experimental situation. Thus, it is extremely useful to be able to invert populations efficiently in a spin system, without regard to the spread in chemical shifts and/or with wide tolerance to misset pulse widths and rf inhomogeneity across the sample. Similarly, a pulse that can transfer *z* magnetization efficiently into the transverse plane in spite of misset flip angle or rf inhomogeneity is of considerable interest. In other contexts, it is of interest to excite a quadrupolar spin system efficiently so that true spectral lineshapes may be obtained without distortions resulting from the action of quadrupolar coupling during rf excitation. Similarly, it is of interest to have a pulse designed to invert spins over a large range of dipolar couplings in solids.

Composite pulses have proved to be suitable solutions to such problems. A composite pulse is really a finite, continuous sequence of pulses with various flip angles and phases, designed to do the job of a conventional single pulse but with drastically reduced dependence on certain selected spectral parameters.

In the context of transferring *z* magnetization into the transverse plane, e.g. a nominal 90°_x–90°_y sequence is equivalent to a 90°_x pulse for high-resolution applications in liquids but exhibits a markedly reduced dependence on rf inhomogeneity/flip angles. In fact, it essentially translates flip angle errors into phase errors. On the other hand, a 90° pulse has strong self-compensating properties for resonance offset errors essentially because the *tilt* of the effective field is largely compensated by its increased *magnitude*. Resulting phase errors are essentially a linear function of the offset.

In the context of inverting *z* magnetizations in HR applications in liquids, a 180°_x pulse may be replaced by a nominal 90°_x–240°_y–90°_x sequence, giving a reasonable measure of rf inhomogeneity compensation, as well as resonance offset compensation. Note that the first and last pulses must be identical to insure that the composite pulse has the same effect on *z* magnetization for offsets above and below resonance. This result can be proved by inspecting the explicit form of the relevant rotation matrices, as discussed in Chapter 6. It is of course, important to maintain this symmetry so that, in conjunction with quadrature detection, the effective range of allowed transmitter offsets may be doubled.

The sequence 90°_x–180°_y–90°_x may also be employed for spin inversion or refocusing. In spin inversion, i.e., taking *z* magnetization to the $-z$ axis, the sequence is more prone to offset errors than the 90°_x–240°_y–90°_x sequence. In refocusing transverse magnetization, the 90°_x–180°_y–90°_x composite pulse functions with second-order compensation in case there are no homonuclear couplings, whereas the compensation is first order if homonuclear couplings are present. In either case, the performance is better than the Meiboom-Gill 180°_y pulse strategy (following preparation with a 90°_x pulse). Flip angle errors with this sequence are converted into phase errors, which are canceled on every even echo.

While such composite pulses may be derived by arguments based on magnetization trajectories derivable from the Bloch equations of motion, or from a rotation operator approach to describe the effects of the pulses, it is possible to employ a recursive procedure to derive pulse sequences that are self-compensating to ever higher orders. From a composite $\pi/2$ pulse $P_0^{(m)}$, which transfers z magnetizations into the transverse plane, the inverse sequence, $(P_{\mp\beta}^{(m)})^{-1}$ may be concatenated in either order with $P_0^{(m)}$ to produce a new composite pulse. If $\beta = \pi/2$, this results in an improved composite $\pi/2$ pulse, $P_0^{(m+1)}$. If $\beta \neq \pi/2$, on the other hand, the resulting composite pulse is one of (arbitrary) flip angle β, with no additional compensation introduced by this final step unless $\beta \simeq \pi/2$. Good inverse sequences $(P_0^{(m)})^{-1}$ may be produced in the absence of significant resonance offsets simply by reversing the order of the individual pulses in $P_0^{(m)}$ and reversing each pulse phase. If resonance offset effects are significant, on the other hand, approximate inverse sequences may be found by employing the following prescription: a sequence C, which is *cyclic* to a high order (see Chapter 6) and which contains $P_0^{(m)}$ as a terminal element, is chosen; $P_0^{(m)}$ is removed from this sequence, which is then the desired inverse sequence $(P_0^{(m)})^{-1}$.

Phase-shifted composite pulses, $P_\beta^{(m)}$, are produced from $P_0^{(m)}$ by shifting the phase of each pulse in $P_0^{(m)}$ by β radians.

It is frequently the case that compensating to higher orders in one type of imperfection is at the expense of increased sensitivity to a different kind of imperfection. This cannot be ruled out even with composite sequences designed on the basis of recursive procedures, unless inverse sequences are derived with care, as described above.

On this basis, it is possible to design composite pulses with dual compensation for high-resolution applications on liquids. For example, an inversion efficiency of 99% may be maintained over an offset range that is $\pm 50\%$ of the rf amplitude and a pulsewidth/rf inhomogeneity range spanning some $\pm 20\%$ of the nominal value; this is done by employing the composite pulse $\overline{3X}\ 4X\ Y\ \overline{3Y}\ 4Y\ X$, where $\overline{3X}$, for example, indicates a 270°_{-x} pulse.

In heteronuclear 2D J-resolved spectroscopy, one frequently finds artifact signals that can be traced to the imperfection in the spin inversion pulse employed on the coupled nucleus, e.g., ^1H in studies on ^{13}C. Such imperfections produce several orders of satellite lines corresponding to the inversion of $(N-1), (N-2), \ldots, 0$ protons in a CH_N group. They may be suppressed by employing a composite inversion pulse on the protons, such as the 90°_x–180°_y–90°_x sequence.

In dealing with general pulse sequences in the HR NMR of liquids, one must distinguish different parts of the pulse sequence with respect to their specific function and apply composite pulses accordingly, after deciding whether offset errors or rf inhomogeneity/pulse misset errors dominate. The composite sequence must be constructed with the maximum possible overall symmetry to insure that phase error terms cancel out. Thus, in a multiple-quantum experiment, both excitation and detection pulses must be compensated. Rotation sandwiches, such as $90^\circ_x-\tau-180^\circ_x-\tau-90^\circ_x$, may be replaced by $(P^{-1}R)-\tau-R-\tau-P$ if offset errors dominate, and by $P_\pi^{-1}-\tau-P_\pi P^{-1}-\tau-P$ if pulse angle errors dominate. Here P denotes a composite $\pi/2$ pulse, while R

is a composite π pulse. If the rotation sandwich has no refocusing pulse one may employ $[P^{-1}C]$–τ–P for offset errors and P^{-1}–τ–P for flip angle errors. Single read pulses may be replaced with $[P^{-1}C]_{\pi/2}\,P$ for offset errors and with $P^{-1}_{\pi/2}P$ for flip angle errors. For the INADEQUATE experiment in the presence of large offsets, one may thus employ the following sequence:

$$[2X\overline{2X}X\text{–}\tfrac{\tau}{2}\text{–}3X\overline{2X}X\text{–}\tfrac{\tau}{2}\text{–}X]_{\chi}\text{–}\Delta\text{–}[2Y\overline{2Y}Y\overline{3Y}2Y\bar{Y}\,X]_{\xi}\text{–Acquire}_{\Psi},$$

where $\Psi = 2\chi - \xi$, both χ and ξ being multiples of $\frac{\pi}{2}$ ($k\pi/2$, $k = 0, 1, 2, 3$).

For offset errors, $P = (\frac{\pi}{2})_x$ is a good choice, while $R = (\frac{\pi}{2})_x\text{–}(\pi)_y\text{–}(\frac{\pi}{2})_x$ or $R = (\frac{3\pi}{2})_x\text{–}(\pi)_{-x}\text{–}(\frac{\pi}{2})_x$ may be employed. For flip angle errors, on the other hand, $P = (\frac{\pi}{2})_x\text{–}(\frac{\pi}{2})_y$ and $R = (\frac{\pi}{2})_x\text{–}(\pi)_y\text{–}(\frac{\pi}{2})_x$ are good choices. $C = (\frac{3\pi}{2})_x\text{–}(\pi)_{-x}\text{–}(\frac{\pi}{2})_x\text{–}(\frac{3\pi}{2})_{-x}\text{–}(\pi)_x\text{–}(\frac{\pi}{2})_{-x}$ is a useful cycle.

For anharmonic three-level systems, such as a spin-1 nucleus with quadrupole interaction, transverse coherence may be excited efficiently by the following composite pulses: $135^\circ_0\text{–}90^\circ_{180}\text{–}45^\circ_0$ or better still $90^\circ_{180}\text{–}180^\circ_0\text{–}90^\circ_{180}\text{–}135^\circ_0\text{–}45^\circ_{180}$.

It will have no doubt occurred to the reader that there is in general no *unique* recipe for the *basic* unit in a composite pulse sequence, although systematic procedures may be employed for *expansion* of a basic unit into a self-compensating sequence to ever higher orders. Hence, there is in the literature a large variety of prescriptions, and a number of competing claims. We have, in the above, attempted to highlight a few of the more promising methods.

In general, it may be noted that even with recursive procedures, imperfections in the individual pulses are translated into phase shifts; in some applications, the resulting phase distortions may be unacceptable. Composite pulses without phase distortion may be designed by employing a Magnus expansion approach to solve for the pulse flip angles and phases in the sequence. The resulting composite pulses generally require rather unusual phase shifts but tolerate errors of up to 10° in the phases, so that a phase shifter generating phases with a 15° increment (such as for multiple-quantum spectroscopy) would meet the needs. Compensation for rf inhomogeneity may be achieved with a composite π pulse of the form $180^\circ_0\text{–}180^\circ_{120}\text{–}180^\circ_0$ or, even better, with $180^\circ_0\text{–}180^\circ_{105}\text{–}180^\circ_{210}\text{–}360^\circ_{59}$, without serious phase distortions. A composite $\pi/2$ pulse with little phase distortion that compensates for rf inhomogeneity is $270^\circ_0\text{–}360^\circ_{169}\text{–}180^\circ_{33}\text{–}180^\circ_{178}$. A constant rotation $\pi/2$ pulse over a range of offsets is given by: $385^\circ_0\text{–}320^\circ_{180}\text{–}25^\circ_0$. Constant rotation π pulses may be achieved over a range of offsets with the sequence $90^\circ_0\text{–}270^\circ_{90}\text{–}90^\circ_0$ or, better still, with $336^\circ_0\text{–}246^\circ_{180}\text{–}10^\circ_{90}\text{–}74^\circ_{270}\text{–}10^\circ_{90}\text{–}246^\circ_{180}\text{–}336^\circ_0$. A composite π pulse for inversion in the presence of dipolar interactions is given by $45^\circ_0\text{–}180^\circ_{90}\text{–}90^\circ_{180}\text{–}180^\circ_{90}\text{–}45^\circ_0$.

Composite pulses of the form $90^\circ_x\text{–}\theta^\circ_y\text{–}90^\circ_{-x}$ may also be employed to generate θ° phase shifts, provided offset and rf inhomogeneity effects are negligible.

Composite Pulse Decoupling. Broadband proton decoupling is basic to the NMR of such nuclei as ^{13}C. In order to irradiate a 10-ppm ^1H spectral range, however, several watts of continuous rf power must be applied, especially on high-field spectrometers and with samples of high dielectric constant. This could give rise to sample heating, which leads to a drop in resolution and sensitivity, besides potentially causing undesired changes in the molecular species, especially where biomolecules are concerned. Inefficient decoupling, moreover, leads to unresolved residual splittings, which again result in loss of sensitivity and resolution.

Noise modulation of the carrier has been the most common technique to generate the desired 10-ppm bandwidth of proton irradiation. Other methods include squarewave phase modulation and chirp frequency modulation. The basic idea of applying a succession of inversion pulses on the protons rapidly in comparison to the inverse of the relevant coupling has of late been developed into a powerful method of efficient broadband decoupling with low rf power. Composite pulses, such as $R = 90^\circ_x\text{–}\theta_y\text{–}90^\circ_x$ ($180^\circ \leqslant \theta \leqslant 240^\circ$) may be employed as offset-insensitive (and flip angle error-tolerant) spin inversion sequences for this purpose, as discussed in the last section. An MLEV-4 cycle $RR\bar{R}\bar{R}$, where $\theta = 180^\circ$ and $\bar{R}$ has all the rf phases inverted, may be formed from R and leads to efficient decoupling over a wide band of proton frequencies when the cycle is repeated at a rate fast with respect to the coupling.

FID sampling may be performed at the end of each such cycle. Such composite pulse cycles give rise to a much wider decoupling bandwidth for a given rf field strength B_2 when compared to the other techniques; equivalently, they lead to much weaker B_2 fields i.e., lower power consumption to achieve a given decoupling bandwidth. An insight into the design of such cycles may be had by applying the principles of average Hamiltonian theory. However, an exact treatment of decoupling sequences is possible in the absence of strong couplings among the decoupled spins, by employing rotation matrices to describe the effect of pulses in the presence of offsets. Such a treatment is described in Chapter 6 in discussing the single-spin properties of multiple-pulse sequences. This approach allows exact computer simulations to evaluate cycles and their extensions for broadband decoupling, while also acting as a guide to an intuitive approach to pulse cycle design.

Supercycles may be created from cycles, leading to far greater tolerance with respect to the efficiency of the individual inversion element R. One example is MLEV-16,

$$R\,R\,\bar{R}\,\bar{R}\quad \bar{R}\,R\,R\,\bar{R}\quad \bar{R}\,\bar{R}\,R\,R\quad R\,\bar{R}\,\bar{R}\,R$$

Such supercycles, however, have a large cycle time in comparison to the digitizer dwell time for the required spectral widths, so the FID must be sampled several times during each cycle. Sidebands resulting from such subcycle sampling are attenuated considerably when the FID sampling and the decoupler cycling are desynchronized.

Extended supercycles of this kind, however, have a practical shortcoming in that their performance is quite sensitive to the setting of the quadrature phases in the sequence. Sequences that employ only 180° phase shifts, on the other hand, are far more tolerant to phase errors. In this context, cycles may be built from the basic inversion element $R = 90^\circ_x{-}180^\circ_{-x}{-}270^\circ_x$, leading to a phase-alternated rotation (PAR) of magnetization. From this basic element R, denoted $1\bar{2}3$, a wideband, alternating-phase, low-power technique for zero residual splitting, known as WALTZ-4, may be formed as: $R\ R\ \bar{R}\ \bar{R} = 1\ \bar{2}\ 3\ 1\ \bar{2}\ 3\ \bar{1}\ 2\ \bar{3}\ \bar{1}\ 2\ \bar{3}$. Expansion of this basic cycle may be performed by permuting a 90° pulse, since this has inherent self-compensation for offset effects. Thus, combination of a cyclically permuted WALTZ-4 sequence with a phase-inverted WALTZ-4 cycle leads to WALTZ-8:

$$\begin{aligned}\text{WALTZ-8} &= K\ \bar{K}\ \bar{K}\ K \\ &= \bar{2}\,3\,1\quad \bar{2}\,3\,\bar{1}\quad 2\,\bar{3}\,\bar{1}\quad 2\,\bar{3}\,1\quad 2\,\bar{3}\,\bar{1}\quad 2\,\bar{3}\,1\quad \bar{2}\,3\,1\quad \bar{2}\,3\,\bar{1} \\ &= \bar{2}\,4\,\bar{2}\,3\,\bar{1}\quad 2\,\bar{4}\,2\,\bar{3}\,1\quad 2\,\bar{4}\,2\,\bar{3}\,1\quad \bar{2}\,4\,\bar{2}\,3\,\bar{1}.\end{aligned}$$

Further expansion by a cyclic permutation of a 90° pulse from the end to the beginning of the cycle, concatenated with a matching phase-inverted cycle, leads to:

$$\begin{aligned}\text{WALTZ-16} &= Q\ \bar{Q}\ \bar{Q}\ Q \\ &= \bar{3}\,4\,\bar{2}\,3\,\bar{1}\,2\,\bar{4}\,2\,\bar{3}\quad 3\,\bar{4}\,2\,\bar{3}\,1\,\bar{2}\,4\,\bar{2}\,3\quad 3\,\bar{4}\,2\,\bar{3}\,1\,\bar{2}\,4\,\bar{2}\,3 \\ &\quad\ \bar{3}\,4\,\bar{2}\,3\,\bar{1}\,2\,\bar{4}\,2\,\bar{3}.\end{aligned}$$

The WALTZ-16 sequence is probably among the most efficient decoupling sequences that have been designed to date. It has an effective bandwidth of $2B_2$ even with resolution an order of magnitude higher than normal. A PAR-75 sequence has been proposed more recently, which has a 40% higher decoupling bandwidth than WALTZ-16.

A practical lower bound to the decoupling power that may be applied for such sequences arises because lower power leads to longer cycle times. These in turn give rise to an unacceptable sideband structure when the FID is sampled several times during the course of a cycle to achieve the required spectral width. In fact, B_2 must be maintained high enough to satisfy $\gamma_H B_2 \gg 2\pi J_{HX}$, for proton decoupling applications.

Some "Artifacts" in Decoupling Experiments. In a coupled $I_N S$ spin system, anti-phase S-spin magnetization *remains responsive* to the rf field applied on the I spins during *decoupling*. In consequence, magnetization is exchanged, via multiple-quantum coherences, between the components of the S multiplet; I-spin couplings during decoupling can also give rise to such modulation. These effects are observable whenever the S-spin system has been prepared prior to I-spin decoupling, in a state with an anti-phase magnetization component. Several heteronuclear 2D experiments that involve the creation of anti-phase magnetization, followed by decoupling and then free precession, as in the gated decoupler experiment, could exhibit such effects. In fact, any sequence where the decoupling period is not located at the beginning or the end of the evolution period t_1 could exhibit anomalous multiplets arising from this process. These effects are also responsible for the appearance of additional sidebands in heteronuclear scaling experiments involving periodic gated decoupling.

Composite Pulses with Retrograde Compensation. The inversion sequence $R^{(1)} = 2X\ 2Y\ 2X$ is very effective in compensating flip angle errors and may be expanded by:

$$R^{(m+1)} = R^{(m)}(R^{(m)}_{-\pi/2})^{-1} R^{(m)}$$

For example, $R^{(2)} = 2X\ 2Y\ 2X\ 2Y\ \overline{2X}\ 2Y\ 2X\ 2Y\ 2X$, neglecting off-resonance effects.

Correspondingly, the sequence $Q^{(1)} = 2X\ \overline{2Y}\ \overline{2X}$ acts as an inversion pulse with retrograde compensation, which inverts spins in a selected small spatial region. Expansions are now given by:

$$Q^{(m+1)} = Q^{(m)}(Q^{(m)}_{\pi/2})^{-1} Q^{(m)}_{\pi}$$

Following a nominal 90° pulse, transverse magnetization may be flipped by a Q sequence; an EXORCYCLE may be performed on the Q sequence with receiver phase alternation to achieve spatial selectivity.

Alternatively, a 27-pulse narrowband inversion sequence R with nominal 180° pulses of phases 0°, 120°, 240°, 120°, 240°, 0°, 240°, 0°, 120°, 120°, 240°, 0°, 240°, 0°, 120°, 0°, 120°, 240°, 240°, 0°, 120°, 0°, 120°, 240°, 120°, 240°, 0° may be employed, followed by a 60° read pulse. When R is phase cycled, with R_0 and R_π and the FID sum is subtracted from that following two scans with the read pulse alone, the only signal that survives arises from the region in which R inverts and the read pulse functions. Note that far less extensive phase cycling is required with this approach, compared to the strategy mentioned earlier.

Pulse sequences with such retrograde compensation may be employed with surface coils to achieve depth resolution, in effect turning to advantage the inherent rf inhomogeneity of surface coil systems. Depth selection may be varied down the object of interest simply by varying the pulsewidths or the rf power. A more detailed outline of surface coil topical imaging is given in the next section.

NMR Imaging

The possibility of spatial localization of NMR observation and its high selectivity toward a particular magnetic nucleus has led to the development of the technique of Zeugmatography. In this technique objects containing a distribution of protons are encoded with the help of special magnetic field gradients and the resulting response from these voxels (volume elements) is used to construct "proton images" using projection reconstruction techniques. This gives rise to three-dimensional images similar to x-ray or ultrasonic computer tomography. Because water is the most abundant substance in living biological systems, proton magnetic resonance imaging (MRI) has become one of the most powerful and noninvasive ways of looking at tissues, membranes, brain cells, heart, lipids, proteins, etc., both in vivo and in vitro.

Using vector gradients in mutually orthogonal directions and encoding selected points, lines, or planes it is possible to get an idea of the location and abundance of a given magnetic nucleus in a 3D object. This can be done on a whole body, organs, or perfused biological specimens. Apart from subjecting selected areas in a specimen to simple nuclear induction, one can look at the magnetization following an inversion recovery ($180°-\tau-90°$) or spin echo ($90°-\tau-180°$) pulse sequences or Carr–Purcell train. Images that are finally reconstructed will have definite contrast characteristics depending on the relaxation times T_1 and T_2 of the nuclei in different environments. There are a large number of techniques of localizing an active volume for observation, application of selected gradients during different types of excitation or acquisitionof the signal. There are also several ways of reconstructing the images. Because the current work is concerned mainly with high-resolution aspects of NMR spectroscopy we shall, in the following, deal briefly with high-resolution NMR imaging techniques. This also we do only in a cursory manner, because the field is young and expanding fast and the techniques tend to be superseded by better ones no sooner than discovered. However the bibliography given at the end of the book is expected to provide the necessary impetus for those interested.

Surface Coils and Topical Imaging. Study of live biological specimens in vivo or of perfused organs in vitro in the high-resolution mode can give information such as chemical shift and even scalar spin–spin coupling of biologically active molecules. This can give very precise clues as to the metabolic processes that take place in a live organ leading to specific medical diagnosis. The changes in the environment of a given molecule, the relative concentration of certain critical constituents in a metabolite, the dependence of chemical shift on intracellular pH, the ratio of adenosine tiphosphate (ATP) to adenosine diphosphate (ADP) etc., can lead to a finer understanding of biological as well as bioenergetic processes. Thus, the chemical shift of inorganic phosphates P_i in biospecimens acts as an internal pH meter, being critically dependent on cellular pH. Likewise, the relative concentration of phosphocreatine (PCr) and that of ATP can give precise information of the state of metabolic activity of the tissue. With the advent of molecular imaging by the use of the so-called surface coils one can look at intact specimens, such as skeletal muscle, heart, kidney, liver, and brain, in a physiologically live state. While spatial localization using "point" selection methods combined with high-resolution NMR is one way of studying high-resolution imaging, the method suffers from low sensitivity and poor shift resolution.

For ^{31}P high-resolution imaging in particular, the use of surface coils paves the way for a clearer insight into the metabolic processes. In what follows, we briefly touch upon some of these techniques employing surface coils. In conjunction with surface coil techniques for studying high-resolution NMR spectra of a selected localized area close to the surface, it is also possible in another way to select a deep localized area within the animal, by modifying the static field with superposed additional static field gradients, and yet provide high-resolution spectra leading to biochemical information. This second technique, known as "topical imaging," also provides fundamental information leading to clinical diagnosis.

In topical magnetic resonance, (TMR), using special static field gradients, a predetermined place deep within a specimen can be selected for high-resolution NMR observation. To delineate the volume from which we need high-resolution NMR information the effective homogeneous volume of the static field B_0 should be reduced so as to have maximum homogeneity at the region of interest, but rapidly varying elsewhere. The induced field at the point $B(r, \theta)$ by a profile coil with cylindrical symmetry can be given in terms of a Taylor expansion:

$$B(r, \theta) = \sum_{n=0}^{\infty} \beta_n r^n P_n(\cos\theta) \tag{1}$$

where β_n depends on coil geometry and current and $P_n(\cos\theta)$ are the associated Legendre polynomials of order n. In a coil system of maximum order $n = 4$, the field inhomogeneity ΔB across the central region corresponding to an axial extent a, given by:

$$a = \frac{3.1}{2}\left(\frac{\beta_2}{\beta_4}\right)^{1/2} \tag{2}$$

can be shown to be:

$$\Delta B = \beta_2^2/2\beta_4 \tag{3}$$

The sensitive volume obtained by adjusting β_2 and β_4 can be made to be at any point in the z axis of the field profile such that ΔB lies within 0.1 ppm in the area of interest. In adjacent areas the field changes rather rapidly, causing broadening of spectral lines from these regions (Fig. 7.2). While the resultant response contains high-resolution information from the selected volume, there is still a broad unresolved background from the remaining region, which can be subtracted by convolution difference techniques. Using this TMR technique one can study high-resolution NMR deep down in biological organs in vivo without the need for surgical procedures. An example of the TMR technique is shown in Fig. 7.3.

Surface Coils. The simplest surface coil is a single loop of rf carrier with radius r lying on the yz plane of a Cartesian coordinate system with the static field B_0 along the z axis. This coil will serve as both rf transmitter and receiver. For unit current flowing in the loop, the rf magnetic field produced at any point can be given in terms of two components B_{ax} (parallel to the axis of the surface coil) and B_{eq} (in the plane of the coil) and are given by Fig. 7.4.

$$B_{ax} = \frac{\mu}{2\pi} \frac{1}{[(r+\rho)^2 + x^2]^{1/2}} \left[K + \frac{r^2 - \rho^2 - x^2}{(r-\rho)^2 + x^2} E \right] \tag{4}$$

$$B_{eq} = \frac{\mu}{2\pi} \frac{x}{[\rho(r+\rho)^2 + x^2]^{1/2}} \left[-K + \frac{r^2 + \rho^2 + x^2}{(r-\rho)^2 + x^2} E \right] \tag{5}$$

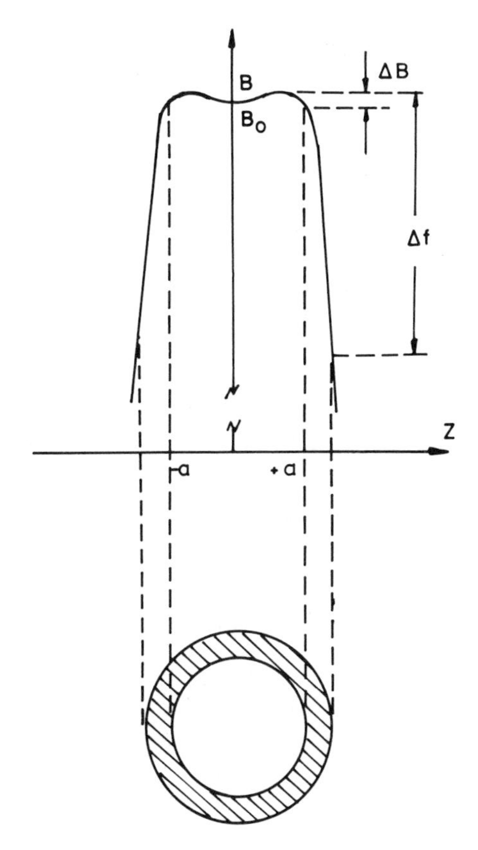

Figure 7.2. The magnetic field profile along the z axis in topical magnetic resonance. The inhomogeneity ΔB within the central region is shown in the projection below. The inner region has ΔB less than typical ^{31}P linewidths, while in the shaded regions resonances are broadened by field gradients. [Reproduced with permission. R.E. Gordon, P.E. Hanley, D. Shaw, D.G. Gadian, G.K. Radda, P. Styles, P.J. Bore and L. Chan, *Nature*, **287**, 736 (1980), copyright (c) 1980, Macmillan Journals Limited, London].

where x and ρ are components of point q parallel to the axis of the coil and perpendicular to it, r is the surface coil radius, μ is the magnetic permeability of the medium, and K and E are the complete elliptic integrals of the first and second kind. Whereas B_{ax} is always perpendicular to the coil, the B_{eq} component is given by:

$$B_{xy} = B_{eq} \sin\theta \tag{6}$$

where θ is the angle between the vectors ρ and B_0. The field generated at the point q is therefore the vector sum given by:

$$\overline{(B_1)_{xy}} = \overline{B_x} + \overline{B_{eq}} \sin\theta \tag{7}$$

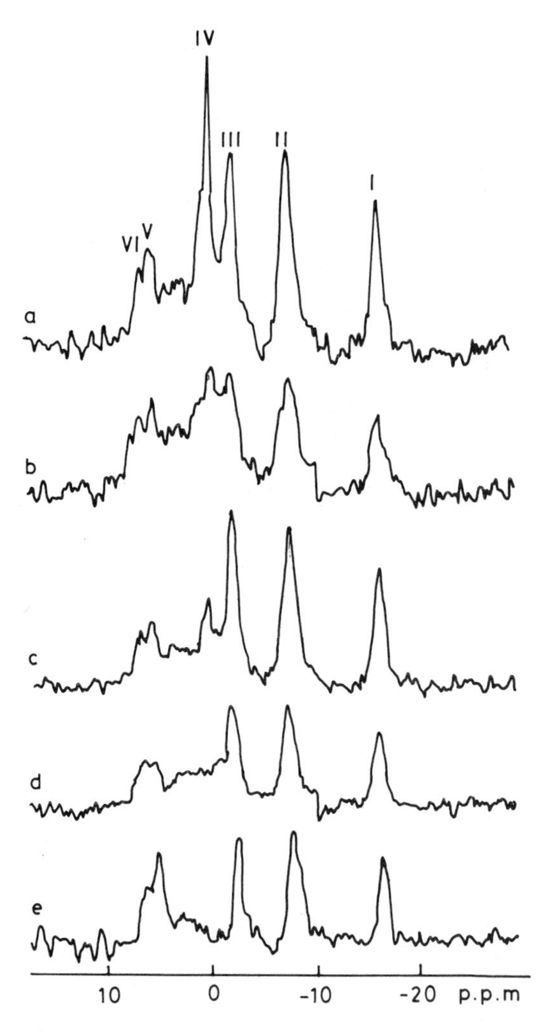

Figure 7.3. Phosphorus-31 spectrum of the liver of a live rat anesthetized and starved for 24 h. I, II, and III are from ATP; IV is from PCr, V is from Pi, and VI is from sugar phosphate (AMP and IMP). (a) spectrum in the absence of localizing fields; (b) localized with an axial extent of 20 mm; (c) same as in (a), but with 220-ms delay between pulses; (d) same as in (c), but with short pulse interval; (e) spectrum of a perfused rat liver. Convolution difference was done with line broadenings of 16 and 233 Hz. Measurement of T_1 on line II in the presence of localizing fields gave a value of 0.14 s, characteristic of ATP in liver. [Reproduced with permission R.E. Gordon, P.E. Hanley, D. Shaw, D.G. Gadian, G.K. Radda, P. Styles, P.J. Bore and L. Chan, *Nature*, **287**, 736 (1980), copyright (c) 1980, Macmillan Journals Limited, London].

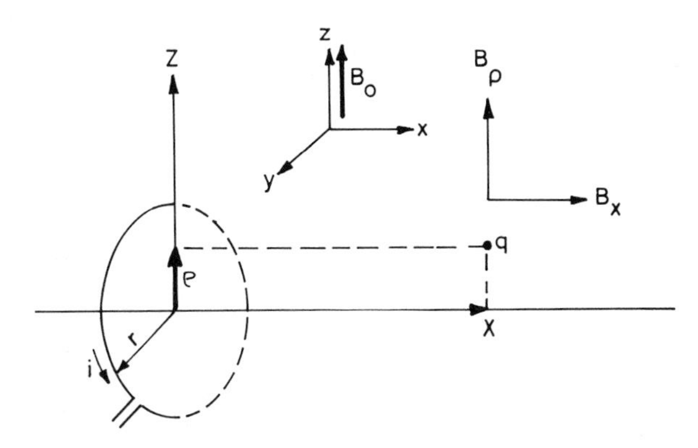

Figure 7.4. Single-loop surface coil showing the field at a point q at coordinates (ρ, x); i is the current flowing in the loop. B_ρ and B_x are the equatorial and axial components of B_1 at q.

Experimental verification of the above expression has been done for $\theta = \pi/2$, $\theta = 0$ by following the signal intensity as a function of ρ (see Fig. 7.5). It should be noted that since the equatorial field B_{xy} has a component $B_{xy} \sin\theta$ perpendicular to the B_0, the signal amplitude will be proportional to $B_{xy} \sin\theta$. The rf field is localized in a volume at the center of the coil defined roughly by the radius so that high-resolution NMR observation is possible from specific locations beneath the surface of the sample. The sine term in Eq. (7) insures that the volume element will in fact be fairly sharply defined.

In fact, experiments with phantoms do demonstrate the radial resolution of surface coils and that the active volume is confined to an area bound by the coil circumference (see Fig. 7.6).

Also variations of the phase of the rf field within the active volume which will necessarily occur should not affect the experiments in such a single-coil setup because of the principle of reciprocity. By this principle, phase changes induced at transmission will be mirrored at reception and thus cancel. This has also been verified experimentally.

Phosphorus-31 surface-coil high-resolution NMR experiments with intact rat leg muscle before and after application of a tourniquet (which produces localized ischemia) clearly show the complete loss of PCr with a concomitant increase in P_i. The pH also could be monitored using a chemical shift of P_i relative to PCr. Thus surface-coil techniques must help diagnosis of localized ischemia. Such studies can be done on brain and heart without resorting to perfusion techniques (see Fig. 7.7).

Depth Pulses. The geometry of the surface coil dictates that the nutation angle $(\gamma B_1 t_p)$ will vary through a range within the effective volume. Thus, if the rf pulse corresponds to a π nutation at the center of the coil, then at a depth of $0.75r$ (the coil radius) it will give a $\pi/2$ nutation. Because the signal intensity at any point for a pulse angle θ is given by $\theta \sin\theta$, maximum signal comes from

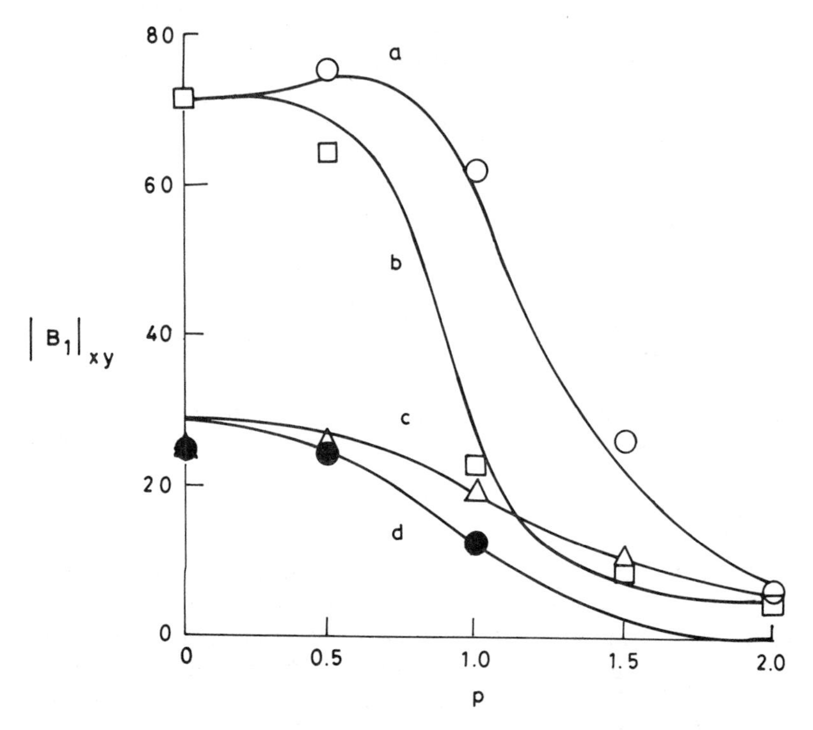

Figure 7.5. The spatial dependence of $(B_1)_{xy}$ as a function of radial coordinate ρ (in units of radius r). The solid curves are from theory. Curve (a) corresponds to $\theta = \pi/2$, $x = 0.5r$; curve (b): $\theta = 0$, $x = 0.5r$; curve (c): $\theta = \pi/2$, $x = 1.1r$; curve (d): $\theta = 0$, $x = 1.1r$ [Reproduced by permission. J.J.H. Ackerman, T.H. Grove, G.G. Wong, D.G. Gadian and G.K. Radda, *Nature*, **283**, 167 (1980), copyright (c) 1980, Macmillan Journals Limited, London].

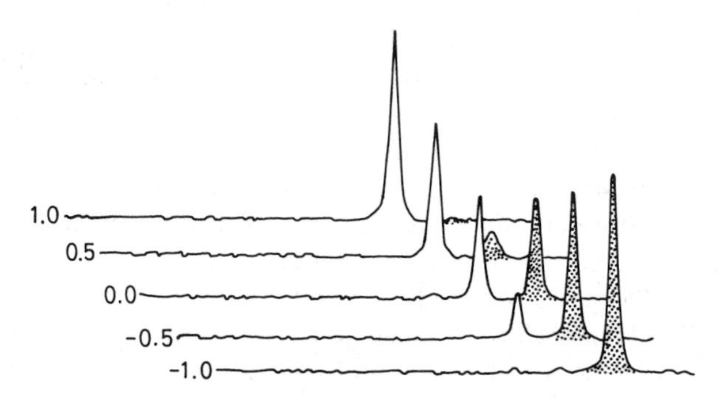

Figure 7.6. Stacked plot demonstrating the radial resolution of surface coils. Phosphorus-31 resonances from two compartments in a phantom corresponding to PCr (shaded) and P_i (unshaded). Numbers on the left correspond to distance in units of the coil radius r from the center of the coil to the partition. [Reproduced by permission. J.J.H. Ackerman, T.H. Grove, G.G. Wong, D.G. Gadian and G.K. Radda, *Nature*, **283**, 167 (1980), copyright (c) 1980, Macmillan Journals Limited, London].

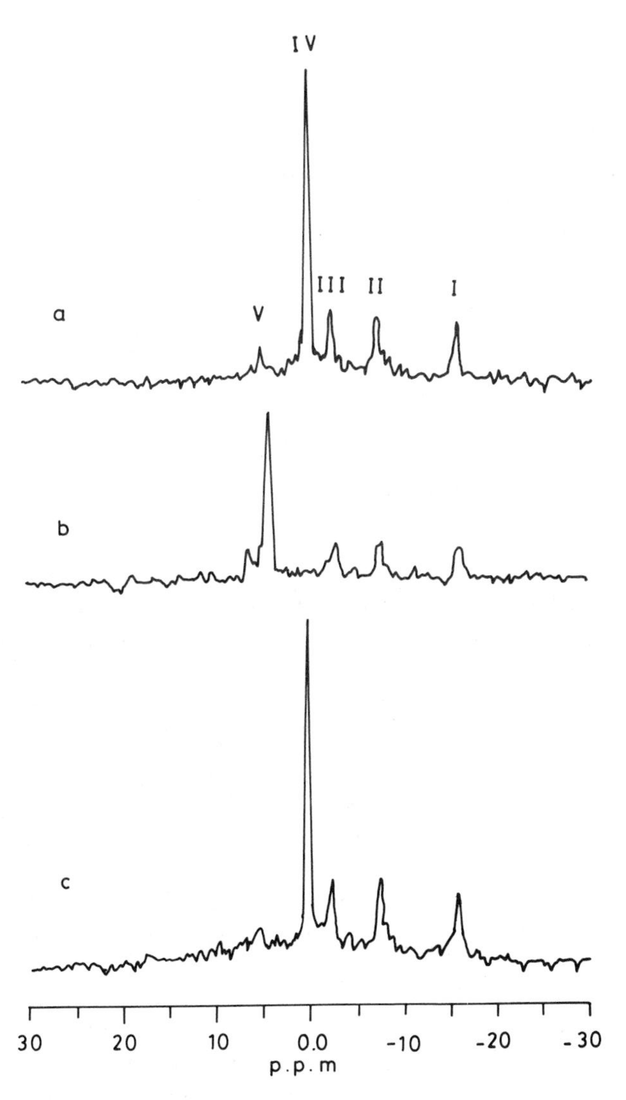

Figure 7.7 Phosphorus-31 NMR spectrum of intact rat leg muscle, using a surface coil. (a) Nonischemic muscle below the knee joint. (b) Ischemic muscle as in (a) after application of a tourniquet above the knee. (c) Spectrum of muscle above the tourniquet. The labels I to V on the peaks stand for the same as in Fig. 7.3. The dramatic reduction in IV and increase in V may be noted in (b). [Reproduced by permission. J.J.H. Ackerman, T.H. Grove, G.G. Wong, D.G. Gadian and G.K. Radda, *Nature*, **283**, 167 (1980), copyright (c) 1980, Macmillan Journals Limited, London].

a point at $0.58r$ below the surface. It is also possible to combine surface coil and TMR to obtain localized high-resolution spectra selectively in biological specimens in vivo. The main difficulty, of course, is that there is a definite likelihood of variation in the pulse angle across the sensitive volume and hence multipulse techniques, which rely on total volume excitations corresponding to $\pi/2$ or π, are likely to cause problems. One way to rectify this is to adopt the phase cycling procedures employed in 2D spectroscopy (Chapter 4 and this chapter). Thus in improving the Carr–Purcell–Meiboom–Gill sequence to account for pulse imperfections the phases of the π pulse following the $\pi/2$ pulse can be shifted through all four quadrants with inversion of receiver phase for 90° phase shifts relative to the initial $\pi/2$ excitation pulse; this can be schematically represented as:

$$\pi/2\text{–}\tau\text{–}\pi(\pm x, \pm y)\text{–}\tau\text{–acquire} \tag{8}$$

With a surface coil we write this:

$$\theta\text{–}\tau\text{–}2\theta(\pm x, \pm y)\text{–}\tau\text{–acquire} \tag{9}$$

A vector model will show that the refocused xy component of magnitude $\frac{1}{2}\sin\theta\,(1 - \cos 2\theta)$ is acquired during one cycling of phases shown above, because:

$$\tfrac{1}{2}\sin\theta\,(1 - \cos 2\theta) = \sin^3\theta \tag{10}$$

A plot of $\sin^3\theta$ versus θ shows maximum intensity at $\theta = \pi/2$ and $3\pi/2$ (negative signal), thereby selecting a region that is at a particular depth subject to a $\pi/2$ pulse. Since the delay τ has no effect on the signal intensity, one can use an excitation pulse θ, $2\theta(\pm x, \pm y)$ to select regions in the sample where the flip angle is $\pi/2$. This is termed a depth pulse. Addition of four more phase-cycled θ pulses leading to θ; $2\theta(\pm x, \pm y)$; $2\theta(\pm x, \pm y)$ leads to final signals of intensity proportional to $\sin\theta\,\frac{1}{2}(1 - \cos 2\theta)^2 = \sin^5\theta$. In general, for a pulse sequence θ; $[2\theta(\pm x, \pm y)]_n$, the acquired signal is proportional to $\sin^{2n+1}\theta$. Experimental verification using high-resolution probes shows that the intensities obtained are nearly 90% of the predicted intensities. For the specific case of surface coil the final intensity will be $\theta\sin^{2n+1}\theta$.

On exactly similar grounds it can be shown that for the depth pulse 2θ; $\theta(\pm x)$; $2\theta(\pm x, \pm y)$ or 2θ; $\theta(\pm x)$; $[2\theta(\pm x, \pm y)]_2$; the corresponding amplitude is $-\theta\cos 2\theta\sin^{2n+1}\theta$, requiring 2×4^n phase combinations. In both sequences mentioned above additional four-phase quadrature compensation can be added independently.

As far as off-resonance effects are concerned they show θ-dependent phase shifts of opposite sign depending on the offset for the normal 2θ; $\theta(\pm x)$ pulse. The 2θ; $\theta(\pm x)$; $[2\theta(\pm x, \pm y)]_2$ combined pulse over a range of ± 2000 Hz at 80 MHz corresponding to a $\pi/2$ pulse width of 9 μs for a 2-cm surface coil or 15 μs for a 4-cm surface coil showed no uncomfortable offset effects. With these depth pulse sequences it is possible to do high-resolution NMR on biological specimens. For the basic spin echo sequence one may use:

$$2\theta;\ \theta(\pm x);\ 2\theta(\pm x, \pm y)\text{–}\tau\text{–}2\theta(\pm x, \pm y)\text{–}\tau\text{–Acquire} \tag{11}$$

The depth effectiveness is unchanged by the delay τ (as long as $\tau \ll T_2$). This makes sure that chemical shift/offset, heteronuclear coupling, and $\mathbf{B}_0$ inhomogeneity all are focused. Thus, T_2 from a biological sample can be measured without worrying about the distribution of flip angles. For the measurement of T_1 using inversion recovery sequence one may use the combined pulse sequence:

$$2\theta\text{–}\tau\text{–}\theta(\pm x); \qquad [2\theta(\pm x, \pm y)]_2 \tag{12}$$

For the surface coil, where θ can take all values, the total intensity from a given volume will give a signal, when $\tau = 0$, proportional to $\theta \cos 2\theta \sin^5 \theta$, which is -1 unit for $\theta = \pi/2$. It should recover following a normal exponential $\exp(-t/T_1)$ to $\theta \sin^5 \theta$, so that the relative contribution to the total signal is $\frac{1}{2}\theta(-\cos 2\theta \sin^5 \theta + \sin^5 \theta)$. Measurement of T_1 using surface coil and the depth sequence with $\theta = \pi/2$ and $\theta = 53°$ gave values within 10% of T_1 that could be got from pure high-resolution NMR of the same specimen. Such depth pulses can therefore be used to produce high-resolution images (chemical microscopy) and can also be used to provide coupling information, for example, in a heteronuclear spin system. Thus, in a doubly tuned surface coil the following sequence will produce ^1H spectra only from those ^1H directly bonded to ^{13}C, where J is the ^{13}C–^1H single-bond scalar coupling (see Fig. 7.8). When doing double resonance, pulses on both channels should be avoided because it is difficult to match the ^1H and ^{13}C pulse angles; where ^{13}C enrichment is possible then the following sequence (Fig. 7.8) can be used for subspectral editing; thus the above sequence can be used to separate CH_n coupling networks in some systems.

Addition of every other transients in the above depth sequence gives the methylene/quaternary ($>CH_2$ and $-\overset{|}{\underset{|}{C}}-$) subspectra, while subtraction yields the methyl and methine subspectra. Thus, in spite of the inherent inhomogeneity of the rf field in surface coils we are able to do spin-echo and inversion recovery experiments just as in liquid samples. The possibility of measuring accurate T_1 even in the presence of rf and static field inhomogeneities will be particularly useful in measuring T_1 from localized regions to diagnose and study carcinogenesis. In fact these combined pulses will equally do well in ordinary high-resolution NMR instruments. Spin-echo studies with θ pulses must lead to measurement of diffusion and flow using surface coils in live intact animal organs.

Rotating-Frame Zeugmatography. It is also possible to achieve spatial localization of metabolities by ^{31}P shift-resolved imaging using a 2D variant of the depth resolution method of the previous section. In this technique, the spatially dependent B_1 field of the surface coil provides the required magnetic

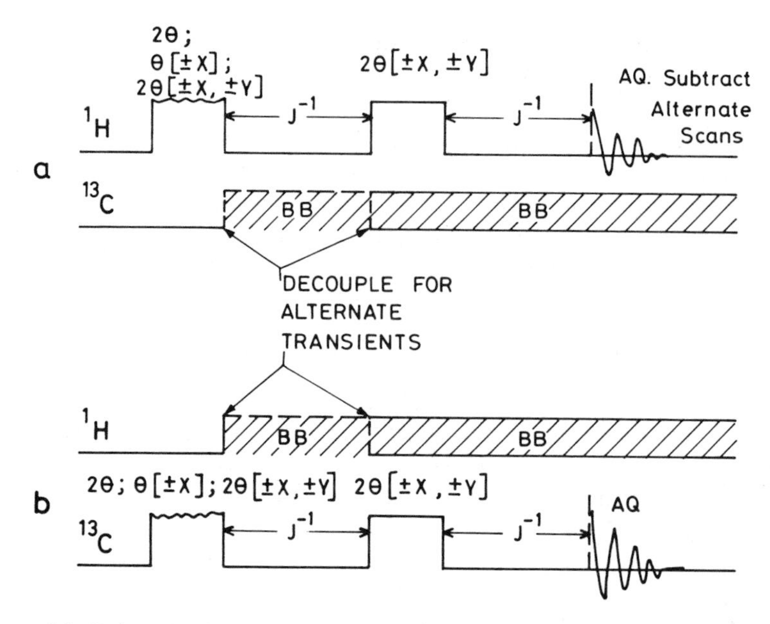

Figure 7.8. Pulse sequence for depth resolution in biological samples using surface coils. Sequence (a) will produce ^1H spectra of protons directly bonded to ^{13}C. Sequence (b) can be used for ^{13}C subspectral editing.

field gradient so that using a 2D technique it is possible to examine chemical shift information in distinct sample slices of selected spatial coordinates. The spatial rf inhomogeneity of the surface coil is used to advantage. For a given coil current the B_1 magnitude depends on the axial distance and coil geometry.

Thus the flip angle experienced by nuclei in a selected slice i will depend on its location and pulse duration t_1 being given by:

$$\theta_i = \gamma B_{1i} t_1 = \Omega_{1i} t_1 \tag{13}$$

The corresponding FID, $S(t_1, t_2)$, at the end of the excitation is:

$$S(t_1, t_2) = B_{1i} \sin(\Omega_{1i} t_1) \exp\left(i\Omega_2 t_2 - \frac{t_1}{T_2^{(1)}} - \frac{t_2}{T_2^{(2)}} \right) \tag{14}$$

where t_1 and t_2 are the familiar evolution and detection periods of a typical 2D experiment. Thus, if the exciting pulsewidth (equal to evolution time t_1) is systematically incremented, the signal amplitudes from successive FID's arising from a given slice oscillate sinusoidally as a function of t_1. The amplitude of oscillations is dependent on B_{1i} field strength, while each FID reproduces the chemical shift information. Thus, a double Fourier transformation of $S(t_1, t_2)$ gives the chemical shift along the F_2 dimension with spectral amplitudes being modulated along t_1 depending on the depth (location). The degree of spatial resolution along the axis of the surface coil was found to be

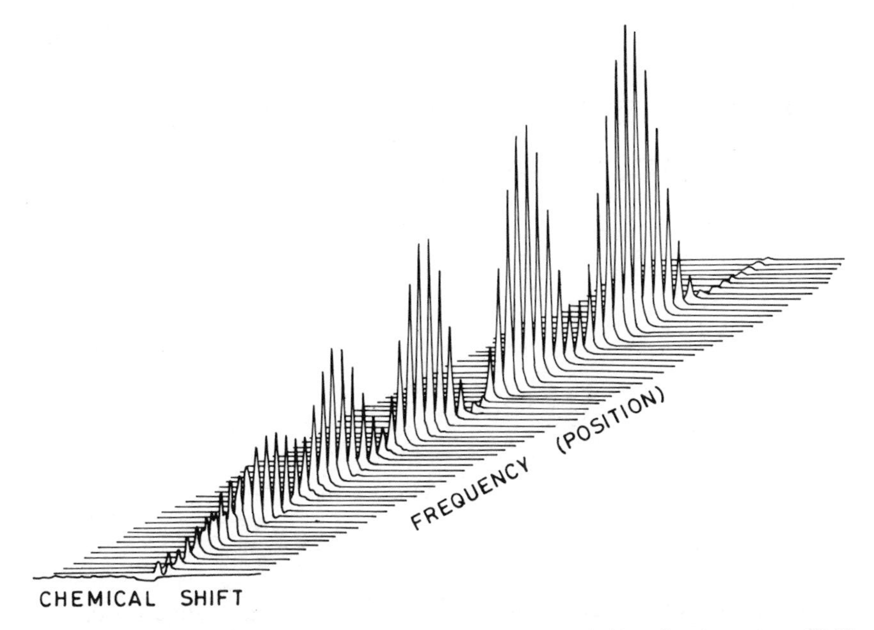

Figure 7.9. Rotating frame Zeugmatography showing spatial localization using a "0.9" surface coil. Seven capillary tubes (1.5 mm diameter) containing 20 μl of 85% H_3PO_4 spaced 4.4 mm apart along the coil axis were used as the phantom. [Reproduced by permission. M. Garwood, T. Schleich, G.B. Matson and G. Acosta, *J. Magn. Reson.*, **60**, 268 (1984), copyright 1984, Academic Press, New York].

$\Delta B_i/\Delta X_i$, where ΔB_i is the slope of the rf gradient and ΔX_i is the width of the sample slice.

It has been shown that a two-turn surface coil with the second turn 0.9 times the first and spaced 0.25 times the larger radius apart (the so-called "0.9 coil" gives nearly planar slices along the axial direction up to a distance on the order of the radius of the larger coil. The resolution is adversely affected if sufficient delay is not given between FID's because magnetization does not recover fully along the z axis. This can be avoided by providing a preparation pulse to eliminate all z components, followed by a delay to allow recovery of some z magnetization, and then applying the variable duration "evolution" pulse. It is also advisable to phase cycle the preparation pulse in quadrature. Figure 7.9 shows typical results on phantoms. Using this technique, known as rotating-frame zeugmatography, some very interesting in vivo studies have been made.

APPENDIX 1

Matrix Algebra of Spin-1/2 and Spin-1 Operators

It is frequently convenient to work with the matrix representation of spin operators in the eigenbase of the Zeeman Hamiltonian. Some results for spin-1/2 and spin-1 systems are given in this Appendix.

Eigenvectors

Eigenvectors are represented as column matrices (kets) and row matrices (bras), while operators are square matrices. The $|\alpha)$ and $|\beta)$ states for spin-1/2 are represented by:

$$|\alpha) \equiv \begin{pmatrix} 1 \\ 0 \end{pmatrix} \quad \text{and} \quad |\beta) \equiv \begin{pmatrix} 0 \\ 1 \end{pmatrix} \tag{1}$$

while, for spin 1:

$$|1) = \begin{pmatrix} 1 \\ 0 \\ 0 \end{pmatrix}, \quad |0) = \begin{pmatrix} 0 \\ 1 \\ 0 \end{pmatrix}, \quad \text{and} \quad |-1) = \begin{pmatrix} 0 \\ 0 \\ 1 \end{pmatrix} \tag{2}$$

Operators

The matrix representation of spin operators in this eigenbase may be obtained by applying the results of angular momentum theory:

$$\begin{aligned} &I_z|I,m) = m|I,m) \\ &I_+|I,m) = \sqrt{(I-m)(I+m+1)}\,|I,m+1), \quad I_+ = I_x + iI_y \\ &I_-|I,m) = \sqrt{(I+m)(I-m+1)}\,|I,m-1), \quad I_- = I_x - iI_y \\ &\mathbf{I}^2|I,m) = I(I+1)|I,m). \end{aligned} \tag{3}$$

The matrix representation of the operator $\mathbf{I}^2$ is thus $I(I + 1)$ times the unit matrix of dimension $(2I + 1)$. The other operators are given by:

$$
\begin{array}{lll}
 & \text{Spin } 1/2 & \text{Spin } 1 \\
I_z & \frac{1}{2}\begin{pmatrix} 1 & 0 \\ 0 & -1 \end{pmatrix} & \begin{pmatrix} 1 & 0 & 0 \\ 0 & 0 & 0 \\ 0 & 0 & -1 \end{pmatrix} \\
I_x & \frac{1}{2}\begin{pmatrix} 0 & 1 \\ 1 & 0 \end{pmatrix} & \frac{1}{\sqrt{2}}\begin{pmatrix} 0 & 1 & 0 \\ 1 & 0 & 1 \\ 0 & 1 & 0 \end{pmatrix} \\
I_y & \frac{i}{2}\begin{pmatrix} 0 & -1 \\ 1 & 0 \end{pmatrix} & \frac{i}{\sqrt{2}}\begin{pmatrix} 0 & -1 & 0 \\ 1 & 0 & -1 \\ 0 & 1 & 0 \end{pmatrix} \\
I_+ & \begin{pmatrix} 0 & 1 \\ 0 & 0 \end{pmatrix} & \sqrt{2}\begin{pmatrix} 0 & 1 & 0 \\ 0 & 0 & 1 \\ 0 & 0 & 0 \end{pmatrix} \\
I_- & \begin{pmatrix} 0 & 0 \\ 1 & 0 \end{pmatrix} & \sqrt{2}\begin{pmatrix} 0 & 0 & 0 \\ 1 & 0 & 0 \\ 0 & 1 & 0 \end{pmatrix} \\
\mathbf{I}^2 & \frac{3}{4}\begin{pmatrix} 1 & 0 \\ 0 & 1 \end{pmatrix} & 2\begin{pmatrix} 1 & 0 & 0 \\ 0 & 1 & 0 \\ 0 & 0 & 1 \end{pmatrix}
\end{array}
\tag{4}
$$

The element ij of these matrices represents $(i|\mathbf{O}|j), |i)$ being the ket corresponding to $m = m_i$ $(m_1 = I, m_2 = I - 1, m_3 = I - 2, \ldots)$.

The matrix representation of eigenvectors and operators in a composite spin system is obtained by working in the appropriate direct product space. Thus, for two spins 1/2,

$$
\begin{array}{ll}
|\alpha\alpha) = \begin{pmatrix} 1 \\ 0 \end{pmatrix} \otimes \begin{pmatrix} 1 \\ 0 \end{pmatrix} = \begin{pmatrix} 1 \\ 0 \\ 0 \\ 0 \end{pmatrix} & |\alpha\beta) = \begin{pmatrix} 1 \\ 0 \end{pmatrix} \otimes \begin{pmatrix} 0 \\ 1 \end{pmatrix} = \begin{pmatrix} 0 \\ 1 \\ 0 \\ 0 \end{pmatrix} \\
|\beta\alpha) = \begin{pmatrix} 0 \\ 1 \end{pmatrix} \otimes \begin{pmatrix} 1 \\ 0 \end{pmatrix} = \begin{pmatrix} 0 \\ 0 \\ 1 \\ 0 \end{pmatrix} & |\beta\beta) = \begin{pmatrix} 0 \\ 1 \end{pmatrix} \otimes \begin{pmatrix} 0 \\ 1 \end{pmatrix} = \begin{pmatrix} 0 \\ 0 \\ 0 \\ 1 \end{pmatrix}
\end{array}
\tag{5}
$$

For such an IS spin system, we have in the direct product space:

$$2I_x = \begin{matrix} & |\alpha) & |\beta) \\ (\alpha| & 0 & 1 \\ (\beta| & 1 & 0 \end{matrix} \otimes \mathbf{1}_S = \begin{pmatrix} 0 & 1 \\ 1 & 0 \end{pmatrix} \otimes \begin{pmatrix} 1 & 0 \\ 0 & 1 \end{pmatrix}$$

$$= \begin{matrix} & |\alpha\alpha) & |\alpha\beta) & |\beta\alpha) & |\beta\beta) \\ (\alpha\alpha| & 0 & 0 & 1 & 0 \\ (\alpha\beta| & 0 & 0 & 0 & 1 \\ (\beta\alpha| & 1 & 0 & 0 & 0 \\ (\beta\beta| & 0 & 1 & 0 & 0 \end{matrix} \tag{6}$$

$$2S_x = 1_I \otimes \begin{matrix} & |\alpha) & |\beta) \\ (\alpha| & 0 & 1 \\ (\beta| & 1 & 0 \end{matrix} = \begin{pmatrix} 1 & 0 \\ 0 & 1 \end{pmatrix} \otimes \begin{pmatrix} 0 & 1 \\ 1 & 0 \end{pmatrix}$$

$$= \begin{matrix} & |\alpha\alpha) & |\alpha\beta) & |\beta\alpha) & |\beta\beta) \\ (\alpha\alpha| & 0 & 1 & 0 & 0 \\ (\alpha\beta| & 1 & 0 & 0 & 0 \\ (\beta\alpha| & 0 & 0 & 0 & 1 \\ (\beta\beta| & 0 & 0 & 1 & 0 \end{matrix} \tag{7}$$

Note that the order of the spins is chosen arbitrarily; the choice made, however, must be maintained right through. The direct product of two matrices A and B is given by:

$$C = A \otimes B = \begin{bmatrix} A_{11}B & A_{12}B & \ldots\ldots & A_{1n}B \\ \vdots & & & \vdots \\ A_{n1}B & A_{n2}B & \ldots\ldots & A_{nn}B \end{bmatrix} \tag{8}$$

The direct product of an $(m \times m)$ matrix with an $(n \times n)$ matrix is thus an $(mn \times mn)$ matrix. Note that the direct product of matrices is therefore very different from matrix multiplication.

Also listed below are the matrix representations of some higher powers of spin operators. These results may be checked by usual matrix multiplication.

$$\begin{array}{lcc} & \text{Spin } 1/2 & \text{Spin } 1 \\ I_x^2 & \frac{1}{4}\begin{pmatrix} 1 & 0 \\ 0 & 1 \end{pmatrix} & \frac{1}{2}\begin{pmatrix} 1 & 0 & 1 \\ 0 & 2 & 0 \\ 1 & 0 & 1 \end{pmatrix} \\ I_y^2 & \frac{1}{4}\begin{pmatrix} 1 & 0 \\ 0 & 1 \end{pmatrix} & \frac{1}{2}\begin{pmatrix} 1 & 0 & -1 \\ 0 & 2 & 0 \\ -1 & 0 & 1 \end{pmatrix} \\ I_z^2 & \frac{1}{4}\begin{pmatrix} 1 & 0 \\ 0 & 1 \end{pmatrix} & \begin{pmatrix} 1 & 0 & 0 \\ 0 & 0 & 0 \\ 0 & 0 & 1 \end{pmatrix} \end{array} \tag{9}$$

	Spin 1/2	Spin 1
$[I_x, I_y]_+$	0	$i\begin{pmatrix} 0 & 0 & -1 \\ 0 & 0 & 0 \\ 1 & 0 & 0 \end{pmatrix}$
$[I_y, I_z]_+$	0	$\frac{i}{\sqrt{2}}\begin{pmatrix} 0 & -1 & 0 \\ 1 & 0 & 1 \\ 0 & -1 & 0 \end{pmatrix}$
$[I_z, I_x]_+$	0	$\frac{1}{\sqrt{2}}\begin{pmatrix} 0 & 1 & 0 \\ 1 & 0 & -1 \\ 0 & -1 & 0 \end{pmatrix}$

Various recursions then follow for $i = x, y, z$:

$$\begin{aligned} I_i^n &= \tfrac{1}{4} I_i^{n-2} \quad (I = 1/2, n \geqslant 2) \\ I_i^n &= I_i^{n-2} \quad (I = 1, \quad n \geqslant 3) \end{aligned} \tag{10}$$

For spin 1, the following relations also hold, with $i = x, y, z$:

$$\begin{aligned} I_i I_j I_i &= 0 \quad (i \neq j) \\ I_i I_j^2 + I_j^2 I_i &= I_i \quad (i \neq j) \end{aligned} \tag{11}$$

In general, for a spin I, we have the Cayley identity:

$$[I_z - I][I_z - (I-1)][I_z - (I-2)] \ldots [I_z + I] = 0$$

and

$$\begin{aligned} I_+^{(2I+1)} &= I_-^{(2I+1)} = 0; \\ I_+^i \left(\prod_{j=-(I-i)}^{I-i} (I_z - j) \right) I_+^i &= 0, \; i = 1, 2, \ldots, I. \end{aligned} \tag{12}$$

Finally, we give below the matrix representation of various bilinear operator products for a two-spin-1/2 situation. The results may be verified by matrix multiplication, working with the representation of the individual single-spin operators in the direct product space of the two spins 1/2.

$$I_x S_x = \frac{1}{4}\begin{bmatrix} 0 & 0 & 0 & 1 \\ 0 & 0 & 1 & 0 \\ 0 & 1 & 0 & 0 \\ 1 & 0 & 0 & 0 \end{bmatrix} \qquad I_x S_y = \frac{i}{4}\begin{bmatrix} 0 & 0 & 0 & -1 \\ 0 & 0 & 1 & 0 \\ 0 & -1 & 0 & 0 \\ 1 & 0 & 0 & 0 \end{bmatrix}$$

$$I_y S_x = \frac{i}{4}\begin{bmatrix} 0 & 0 & 0 & -1 \\ 0 & 0 & -1 & 0 \\ 0 & 1 & 0 & 0 \\ 1 & 0 & 0 & 0 \end{bmatrix} \qquad I_y S_y = \frac{1}{4}\begin{bmatrix} 0 & 0 & 0 & -1 \\ 0 & 0 & 1 & 0 \\ 0 & 1 & 0 & 0 \\ -1 & 0 & 0 & 0 \end{bmatrix}$$

$$I_zS_z = \frac{1}{4}\begin{bmatrix} 1 & 0 & 0 & 0 \\ 0 & -1 & 0 & 0 \\ 0 & 0 & -1 & 0 \\ 0 & 0 & 0 & 1 \end{bmatrix} \qquad I_zS_x = \frac{1}{4}\begin{bmatrix} 0 & 1 & 0 & 0 \\ 1 & 0 & 0 & 0 \\ 0 & 0 & 0 & -1 \\ 0 & 0 & -1 & 0 \end{bmatrix} \tag{13}$$

$$I_zS_y = \frac{i}{4}\begin{bmatrix} 0 & -1 & 0 & 0 \\ 1 & 0 & 0 & 0 \\ 0 & 0 & 0 & 1 \\ 0 & 0 & -1 & 0 \end{bmatrix} \qquad I_xS_z = \frac{1}{4}\begin{bmatrix} 0 & 0 & 1 & 0 \\ 0 & 0 & 0 & -1 \\ 1 & 0 & 0 & 0 \\ 0 & -1 & 0 & 0 \end{bmatrix}$$

$$I_yS_z = \frac{i}{4}\begin{bmatrix} 0 & 0 & -1 & 0 \\ 0 & 0 & 0 & 1 \\ 1 & 0 & 0 & 0 \\ 0 & -1 & 0 & 0 \end{bmatrix}$$

From the above, it is clear that I_xS_x, I_yS_y, I_xS_y, and I_yS_x represent zero,- and double-quantum coherences in the spin-1/2 IS system.

Operator representations for an IS system with $I = 1$, $S = 1/2$ involve (6×6) matrices. For example,

$$\sqrt{2}I_x = \begin{pmatrix} 0 & 1 & 0 \\ 1 & 0 & 1 \\ 0 & 1 & 0 \end{pmatrix} \otimes \mathbf{1}_S = \begin{pmatrix} 0 & 1 & 0 \\ 1 & 0 & 1 \\ 0 & 1 & 0 \end{pmatrix} \otimes \begin{pmatrix} 1 & 0 \\ 0 & 1 \end{pmatrix}$$

$$= \begin{array}{c} (1\alpha| \\ (1\beta| \\ (0\alpha| \\ (0\beta| \\ (-1\alpha| \\ (-1\beta| \end{array} \overset{\begin{array}{cccccc} |1\alpha) & |1\beta) & |0\alpha) & |0\beta) & |-1\alpha) & |-1\beta) \end{array}}{\begin{bmatrix} 0 & 0 & 1 & 0 & 0 & 0 \\ 0 & 0 & 0 & 1 & 0 & 0 \\ 1 & 0 & 0 & 0 & 1 & 0 \\ 0 & 1 & 0 & 0 & 0 & 1 \\ 0 & 0 & 1 & 0 & 0 & 0 \\ 0 & 0 & 0 & 1 & 0 & 0 \end{bmatrix}} \tag{14}$$

$$2S_x = \mathbf{1}_I \otimes \begin{pmatrix} 0 & 1 \\ 1 & 0 \end{pmatrix} = \begin{pmatrix} 1 & 0 & 0 \\ 0 & 1 & 0 \\ 0 & 0 & 1 \end{pmatrix} \otimes \begin{pmatrix} 0 & 1 \\ 1 & 0 \end{pmatrix} \tag{15}$$

$$= \begin{bmatrix} 0 & 1 & 0 & 0 & 0 & 0 \\ 1 & 0 & 0 & 0 & 0 & 0 \\ 0 & 0 & 0 & 1 & 0 & 0 \\ 0 & 0 & 1 & 0 & 0 & 0 \\ 0 & 0 & 0 & 0 & 0 & 1 \\ 0 & 0 & 0 & 0 & 1 & 0 \end{bmatrix} \tag{16}$$

$$\sqrt{2}I_y = i\begin{bmatrix} 0 & 0 & -1 & 0 & 0 & 0 \\ 0 & 0 & 0 & -1 & 0 & 0 \\ 1 & 0 & 0 & 0 & -1 & 0 \\ 0 & 1 & 0 & 0 & 0 & -1 \\ 0 & 0 & 1 & 0 & 0 & 0 \\ 0 & 0 & 0 & 1 & 0 & 0 \end{bmatrix} \tag{17}$$

$$2S_y = i\begin{bmatrix} 0 & -1 & 0 & 0 & 0 & 0 \\ 1 & 0 & 0 & 0 & 0 & 0 \\ 0 & 0 & 0 & -1 & 0 & 0 \\ 0 & 0 & 1 & 0 & 0 & 0 \\ 0 & 0 & 0 & 0 & 0 & -1 \\ 0 & 0 & 0 & 0 & 1 & 0 \end{bmatrix} \tag{18}$$

$$I_z = \begin{bmatrix} 1 & 0 & 0 & 0 & 0 & 0 \\ 0 & 1 & 0 & 0 & 0 & 0 \\ 0 & 0 & 0 & 0 & 0 & 0 \\ 0 & 0 & 0 & 0 & 0 & 0 \\ 0 & 0 & 0 & 0 & -1 & 0 \\ 0 & 0 & 0 & 0 & 0 & -1 \end{bmatrix} \tag{19}$$

$$2S_z = \begin{bmatrix} 1 & 0 & 0 & 0 & 0 & 0 \\ 0 & -1 & 0 & 0 & 0 & 0 \\ 0 & 0 & 1 & 0 & 0 & 0 \\ 0 & 0 & 0 & -1 & 0 & 0 \\ 0 & 0 & 0 & 0 & 1 & 0 \\ 0 & 0 & 0 & 0 & 0 & -1 \end{bmatrix} \tag{20}$$

APPENDIX 2

The Hausdorff Formula

It is frequently necessary to evaluate expressions of the form:

$$R_A = \exp(-i\mathcal{H}t)\, A \exp(i\mathcal{H}t)$$

which describe the evolution of an operator A for t seconds, under the action of a time-independent Hamiltonian $\mathcal{H}$. The Baker–Campbell–Hausdorff formula consolidates the series expansion of exponential operators to evaluate such operator evolutions:

$$\begin{aligned} R_A &\equiv \exp(-i\mathcal{H}t)\, A \exp(i\mathcal{H}t) \\ &= A - (it)[\mathcal{H}, A] + \frac{(it)^2}{2!}[\mathcal{H}, [\mathcal{H}, A]] \\ &\quad - \frac{(it)^3}{3!}[\mathcal{H}, [\mathcal{H}, [\mathcal{H}, A]]] + \ldots \end{aligned} \tag{1}$$

Note that this formula involves successive commutators with the Hamiltonian. An NMR problem has a well-defined, finite dimensional operator space associated with it and so the number of independent commutators is limited and the series expansion quickly leads to an operator recursion. It may be noted especially that it is crucial to work with the appropriate basis set of linearly independent operators, and to insure that successive commutators are expressed in this basis set, so that the operator recursions are not lost sight of. Suitable basis set operators for problems involving spin-1/2 and spin-1 systems have been discussed in Chapter 1.

We discuss below briefly some cases of interest.

CASE 1. $[\mathscr{H}, A] = 0$.

It follows that $R_A = A$ since all the commutators vanish. In other words, A is invariant to evolution under a Hamiltonian with which it commutes. For example consider a spin-1 system, with $\mathscr{H} = \frac{1}{3}(I_z^2 - I(I+1))$ and $A = [I_x, I_y]_+$.

$$
\begin{aligned}
[\mathscr{H}, A] &\sim [I_z^2, I_x I_y + I_y I_x] \\
&= [I_z^2, I_x I_y] + [I_z^2, I_y I_x] \\
&= iI_z(I_y^2 - I_x^2) + i(I_y^2 - I_x^2)I_z - iI_z(I_x^2 - I_y^2) - i(I_x^2 - I_y^2)I_z \\
&= 2i(I_z(I_y^2 - I_x^2) + (I_y^2 - I_x^2)I_z) \\
&= 2i(I_z - I_z) = 0
\end{aligned}
\tag{2}
$$

where we have employed the properties of spin-1 operators discussed in Appendix 1, and some properties of commutators to be discussed later in this Appendix. Thus, $R_A = [I_x, I_y]_+$.

CASE 2. $[\mathscr{H}, A] = \alpha A$.

In this case, $R_A = A\exp(-i\alpha t)$. As an example, consider $\mathscr{H} = \Delta I_z$ and $A = I_+ = I_x + iI_y$. Then, $[\mathscr{H}, A] = i\Delta I_y + \Delta I_x = \Delta A$. Thus, $R_A = I_+\exp(-i\Delta t)$.

CASE 3. $[\mathscr{H}, A] = B$, $[\mathscr{H}, B] = kA$, $A \neq \alpha B$.

This has been termed a problem of order 2. We find,

$$
\begin{aligned}
R_A &= \left(A + \frac{(it)^2}{2!}kA + \frac{(it)^4}{4!}k^2A + \ldots\right) - \left(itB + \frac{(it)^3}{3!}kB + \ldots\right) \\
&= A\left(1 - \frac{t^2k}{2!} + \frac{t^4k^2}{4!} - \ldots\right) - iB\left(t - \frac{t^3k}{3!} + \ldots\right) \\
&= A\cos\sqrt{k}t - \frac{1}{\sqrt{k}}iB\sin\sqrt{k}t
\end{aligned}
\tag{3}
$$

Problems of order 2 have been encountered frequently in Chapters 1 and 3 and the corresponding equations of motion, R_A, have been given there. We consider here a few representative examples.

(1) $A = I_x$, $\mathscr{H} = 2\pi J I_z S_z$.

$$
[\mathscr{H}, A] = i2\pi J I_y S_z = B; \qquad [\mathscr{H}, B] = 4\pi^2\frac{J^2}{4}I_x \text{ if } S = 1/2 = 4\pi^2\frac{J^2}{4}A \tag{4}
$$

Thus, $R_A = I_x\cos\pi Jt + 2I_yS_z\sin\pi Jt$. (5)

(2) $A = I_x$, $\mathscr{H} = \Delta I_z$.

$$
[\mathscr{H}, A] = i\Delta I_y = B; \qquad [\mathscr{H}, B] = \Delta^2 I_x = \Delta^2 A \tag{6}
$$

Thus $R_A = I_x\cos\Delta t + I_y\sin\Delta t$. (7)

(3) $A = (I_x - S_x)$; $\mathscr{H} = 2\pi J(I_y S_y + I_z S_z)$ or $2\pi \bar{J}\ \mathbf{I} \cdot \mathbf{S}$.

$$\frac{1}{2\pi J}[\mathscr{H}, A] = -iI_z S_y + iI_y S_z + iI_y S_z - iI_z S_y = 2i(I_y S_z - I_z S_y) = B$$

$$\frac{1}{2\pi J}[\mathscr{H}, B] = 2i\left(\frac{iS_x}{4} - \frac{iI_x}{4} - \frac{iI_x}{4} + \frac{iS_x}{4}\right) = (I_x - S_x) = A, \text{ if } I = S = 1/2 \tag{8}$$

Thus, $R_A = (I_x - S_x)\cos 2\pi \bar{J}t + 2(I_y S_z - I_z S_y)\sin 2\pi \bar{J}t$. (9)

Problems of order higher than two are encountered, for example, in $A_M X_N$ systems evolving under a strong coupling Hamiltonian as discussed in Chapter 1 and lead to multiple frequencies of evolution.

Commutator Algebra

We give here some of the standard results of commutator algebra that are useful in evaluating the Hausdorff formula for specific $\mathscr{H}$ and A. These expressions follow from the definition of commutators:

$$[A, B] = (AB - BA) \tag{10}$$

We find:

$$\begin{aligned} [A, B] &= -[B, A] \\ [A, kB] &= k[A, B], \ k \text{ being a constant.} \\ [AB, C] &= A[B, C] + [A, C]B \\ [A^n, B] &= \sum_{m=0}^{n-1} A^m [A, B] A^{n-m-1}, \ n \geqslant 1 \end{aligned} \tag{11}$$

APPENDIX 3

Fourier Transformation

In this Appendix we briefly discuss some important properties of the Fourier integral and related issues of sampling, and sensitivity and resolution enhancement by exponential multiplication. We also touch upon the dynamic range problem in FT NMR.

The Fourier Transform

A time domain function $f(t)$ with period $T \to \infty$ may be represented in the basis set of exponential functions $\{e^{in\omega_0 t}\}(\omega_0 = 2\pi/T)$, which goes over into $\{e^{i\omega t}\}$ as the period $T \to \infty$. We thus have:

$$f(t) = \frac{1}{2\pi} \int_{-\infty}^{\infty} F(\omega) e^{i\omega t} \, d\omega \tag{1}$$

the inverse Fourier transform (FT) of $F(\omega)$, with

$$F(\omega) = \int_{-\infty}^{\infty} f(t) e^{-i\omega t} \, dt \tag{2}$$

the Fourier Transform of $f(t)$. Note that these prescriptions for $f(t)$ and its Fourier transform $F(\omega)$ correspond to an exponential Fourier analysis of a periodic function, in the limit of the period $T \to \infty$. We denote an FT pair by $f(t) \leftrightarrow F(\omega)$. $F(\omega)$ is in general a complex function:

$$F(\omega) = |F(\omega)| e^{i\theta(\omega)} \tag{3}$$

If $f(t)$ is a real function of t, the complex conjugate of $F(\omega)$, i.e., $F^*(\omega) = F(-\omega)$. Thus

$$F(-\omega) = |F(\omega)| e^{-i\theta(\omega)} \tag{4}$$

The magnitude spectrum $|F(\omega)|$ is an even function of ω, while the phase spectrum $\theta(\omega)$ is odd. Several properties of the FT follow from the definition, considering the FT pair $f(t) \leftrightarrow F(\omega)$.

1. Symmetry

$$F(t) \leftrightarrow 2\pi f(-\omega)$$

If $f(t)$ is an even function, we have $F(t) \leftrightarrow 2\pi f(\omega)$

2. Linearity

$$a_1 f_1(t) + a_2 f_2(t) \leftrightarrow a_1 F_1(\omega) + a_2 F_2(\omega)$$

for arbitrary constants a_1 and a_2.

3. Scaling

$$f(at) \leftrightarrow \frac{1}{|a|} F\left(\frac{\omega}{a}\right)$$

for a real constant a.

4. Frequency shifting

$$f(t)e^{i\omega_0 t} \leftrightarrow F(\omega - \omega_0)$$

5. Time shifting

$$f(t - t_0) \leftrightarrow F(\omega)e^{-i\omega t_0}$$

6. Time differentiation and integration

$$\frac{df}{dt} \leftrightarrow i\omega F(\omega)$$

if the FT of df/dt exists; and

$$\int_{-\infty}^{t} f(\tau)\, d\tau \leftrightarrow \frac{1}{i\omega} F(\omega)$$

if $F(\omega)/\omega$ is bounded at $\omega = 0$.

7. Frequency differentiation

$$-itf(t) \leftrightarrow \frac{dF}{d\omega}$$

and

$$(-it)^n f(t) \leftrightarrow \frac{d^n F}{d\omega^n}$$

8. Convolution

$$f_1(t) * f_2(t) \equiv \int_{-\infty}^{\infty} f_1(\tau) f_2(t - \tau)\, d\tau \leftrightarrow F_1(\omega) F_2(\omega)$$

Also,

$$f_1(t)f_2(t) \leftrightarrow \frac{1}{2\pi}\int_{-\infty}^{\infty} F_1(v)F_2(\omega - v)\,dv$$
$$\equiv \frac{1}{2\pi}[F_1(\omega) * F_2(\omega)] \tag{5}$$

Convolutions obey the commutative, associative, and distributive laws.

We list below some useful FT pairs:

$f(t)$	$F(\omega)$
1	$2\pi\delta(\omega)$
$\delta(t)$	1
$u(t)$	$\pi\delta(\omega) + (1/i\omega)$
$e^{-at}u(t)$	$(a + i\omega)^{-1}$
$\cos\omega_0 t$	$\pi[\delta(\omega - \omega_0) + \delta(\omega + \omega_0)]$
$\sin\omega_0 t$	$i\pi[\delta(\omega + \omega_0) - \delta(\omega - \omega_0)$
$\cos\omega_0 t\, u(t)$	$\frac{\pi}{2}[\delta(\omega - \omega_0) + \delta(\omega + \omega_0)] + (i\omega/(\omega_0^2 - \omega^2))$
$\sin\omega_0 t\, u(t)$	$(\omega_0/(\omega_0^2 - \omega^2)) + \frac{\pi}{2i}[\delta(\omega - \omega_0) - \delta(\omega + \omega_0)]$
$e^{-at}\sin\omega_0 t\, u(t)$	$\omega_0/[(a + i\omega)^2 + \omega_0^2]$
$G_\tau(t)$	$\tau Sa(\omega\tau/2)$
$(\omega_0/2\pi)Sa(\omega_0 t/2)$	$G_{\omega_0}(\omega)$
$e^{-t^2}/2\sigma^2$	$\sigma\sqrt{2\pi}\, e^{-\sigma^2\omega^2/2}$
$\delta_T(t) \equiv \sum_{n=-\infty}^{\infty} \delta(t - nT)$	$\omega_0\delta_{\omega_0}(\omega) \equiv \omega_0 \sum_{n=-\infty}^{\infty} \delta(\omega - n\omega_0)\left(\omega_0 = \frac{2\pi}{T}\right)$
$e^{-a\lvert t\rvert}$	$2a(a^2 + \omega^2)$

Notes

1. The step function

$$u(t) = 1, \quad t > 0$$
$$= 0, \quad t < 0$$

2. The Dirac delta function

$$\delta(t) = \lim_{a \to 0}\frac{d}{dt}(u_a(t))$$
$$= \lim_{a \to 0}\frac{1}{a}[u(t) - u(t - a)],$$

where $u_a(t)$ is a step function with rise time $t = a$.

$$\int_{-\infty}^{\infty} \delta(t)\,dt = 1$$

3. The sampling function

$$Sa\left(\frac{\omega\tau}{2}\right) = \sin(\omega\tau/2)/(\omega\tau/2)$$

4. The gate function

$$G_\tau(t) = 1, \quad |t| < \tau/2$$
$$= 0, \quad |t| > \tau/2$$

It is of particular interest for FT NMR to consider the time-domain functions

$$S_1(t) = u(t)e^{-at}\sin\omega_0 t \text{ and } S_2(t) = u(t)e^{-at}\cos\omega_0 t, \tag{6}$$

which represent a general FID ($a \geqslant 0$). We find from the definitions that their Fourier transforms are given by:

$$S_1(t) \leftrightarrow \int_{-\infty}^{\infty} u(t)e^{-at}\sin\omega_0 t e^{-i\omega t}\,dt$$
$$= \frac{1}{2}\left[\frac{(\omega+\omega_0)}{a^2+(\omega+\omega_0)^2} - \frac{(\omega-\omega_0)}{a^2+(\omega-\omega_0)^2}\right] \tag{7}$$
$$- \frac{i}{2}\left[\frac{a}{a^2+(\omega-\omega_0)^2} - \frac{a}{a^2+(\omega+\omega_0)^2}\right]$$

The real part, therefore, corresponds to a Lorentzian dispersion signal at ω_0, whereas the imaginary part is a Lorentzian absorption at ω_0 (full width at half height is equal to a/π Hertz), ignoring the responses at $-\omega_0$. On the other hand:

$$S_2(t) \leftrightarrow \int_{-\infty}^{\infty} u(t)e^{-at}\cos\omega_0 t e^{-i\omega t}\,dt$$
$$= \frac{1}{2}\left[\frac{a}{a^2+(\omega-\omega_0)^2} + \frac{a}{a^2+(\omega+\omega_0)^2}\right] \tag{8}$$
$$- \frac{i}{2}\left[\frac{(\omega-\omega_0)}{a^2+(\omega-\omega_0)^2} + \frac{(\omega+\omega_0)}{a^2+(\omega+\omega_0)^2}\right]$$

In this case, the real part corresponds to an absorption at ω_0, and the imaginary part corresponds to a dispersion.

These results have been employed throughout the book in discussions of lineshapes. In the above expressions, $a = (T_2^*)^{-1}$.

Resolution enhancement may be achieved by multiplying $S_1(t)$ and $S_2(t)$ by $\exp(bt)$ ($b \geqslant 0$), leading to a linewidth of $(a - b)/\pi$ Hz; sensitivity is reduced in this process because the later, noisy portions of the FID are weighted in preference to the initial portion, which has the maximum signal content.

Sensitivity enhancement, on the other hand, is achieved by multiplying $S_1(t)$ and $S_2(t)$ by $\exp(-bt)$, ($b \geqslant 0$), leading to a preferential weighting of the signal-rich initial portion of the FID relative to the noisy FID tail. From the expressions for the FT of $S_1(t)$ and $S_2(t)$, however, it is clear that in this process resolution is lost, the resulting linewidth being $(a + b)/\pi$ Hz.

Sampling

In actual practice, most FT NMR spectrometers perform a discrete Fourier transform of the FID to retrieve the frequency spectrum. To this end, the FID, which is an analog voltage waveform, is sampled at regular intervals and digitized for computer handling. The sampling frequency must be sufficiently high to reproduce the frequency spectrum with fidelity. In fact for a band-limited signal with no frequency components above ν_m Hz, the sampling frequency should be $\nu_s \geqslant 2\nu_m$ to produce the correct spectrum. This result of "Nyquist sampling theorem" may be proved as follows for a function $f(t)$ which is sampled every T seconds, generating $f(t)\delta_T(t)$, where:

$$\delta_T(t) = \sum_{n=-\infty}^{\infty} \delta(t - nT) \tag{9}$$

For $f(t) \leftrightarrow F(\omega)$, we have:

$$f(t)\delta_T(t) \leftrightarrow F(\omega) * \omega_s\delta_{\omega_s}(\omega) \tag{10}$$

Employing the frequency shift property, we find:

$$F(\omega) * \delta_{\omega_s}(\omega) = F(\omega) * \sum_n \delta(\omega - n\omega_s)$$

$$= \sum_n F(\omega) * \delta(\omega - n\omega_s) \leftrightarrow \frac{1}{2\pi} \sum_n f(t) e^{in\omega_s t} \tag{11}$$

$$\Rightarrow F(\omega) * \delta_{\omega_s}(\omega) = \frac{1}{2\pi} \sum_n F(\omega - n\omega_s)$$

$$\Rightarrow f(t)\delta_T(t) \leftrightarrow \frac{1}{T} \sum_{n=-\infty}^{\infty} F(\omega - n\omega_s), \tag{12}$$

which is the function $F(\omega)$ repeating itself every ω_s radians per second; this periodic repetition occurs without overlap as long as $\omega_s \geqslant 2\omega_m$. This implies that the sampling interval (also called dwell time)

$$T = \frac{2\pi}{\omega_s} = \frac{1}{\nu_s} \leqslant \frac{1}{2\nu_m} \tag{13}$$

If this condition is violated, the overlap of successive $F(\omega - n\omega_s)$ leads to a frequency $\nu_s + \Delta\nu$ being represented in the spectrum as $\nu_s - \Delta\nu$, which is known as aliasing. Aliased signals in FT NMR spectra may be identified readily because they occur with wrong phases which cannot be corrected by the zero- and first-order phase correction routines.

Dynamic Range Problem

The detection of very weak signals in the presence of very strong ones is a problem that is often encountered in practical NMR, e.g., in the study of the ^1H NMR of biomolecules in D_2O. The small percentage of HDO even in a highly deuterated sample of D_2O can cause considerable problems. In CW NMR, the problem may sometimes be avoided by not exciting the HDO signal at all; in pulsed methods of excitation however, this requires special approaches, such as tailored excitation, described in Chapter 2.

In the presence of very strong resonances, weak resonances in an FID may be detected only if the digitizer resolution and computer word length are sufficiently high. In the frequency domain, single-scan proton spectra can have a dynamic range as high as $\pm 40{,}000:2$ with a 12-bit digitizer (analog to digital converter, or ADC). The dynamic range increases further with a 16-bit digitizer and can be as high as $\pm 280{,}000:2$. In order to utilize the available dynamic range, however, the computer word length should be sufficiently high. For instance, the dynamic range of $\pm 40{,}000:2$ cannot be availed if the computer word length is only 16 bits (± 32 K).

APPENDIX 4

Dipolar Relaxation

The magnetic interaction between nuclear dipoles was the subject of Chapter 6, where the line broadening induced by it in solids was discussed, together with means to counter it. Some consequences of cross-relaxation arising from dipole–dipole interactions in liquids and solids were also described in Chapter 2 and Chapter 6, in the context of sensitivity enhancement by incoherent polarization transfer. In this Appendix, we summarize some basic results for relaxation under dipolar coupling.

The Hamiltonian describing dipolar interaction between two spins I and S is given by:

$$\mathscr{H}_d = \frac{\gamma_I \gamma_S h^2}{4\pi^2}\left[\frac{\mathbf{I}\cdot\mathbf{S}}{r^3} - \frac{3(\mathbf{I}\cdot\mathbf{r})(\mathbf{S}\cdot\mathbf{r})}{r^5}\right] \tag{1}$$

$$\begin{aligned}
&= \frac{\gamma_I \gamma_S}{r^3}\left(\frac{h}{2\pi}\right)^2 \Big[I_z S_z + \frac{1}{2}(I^+ S^- + I^- S^+) - 3(\sin^2\theta\cos^2\varphi I_x S_x \\
&\quad + \sin^2\theta\sin^2\varphi I_y S_y + \cos^2\theta I_z S_z \\
&\quad + \sin^2\theta\cos\varphi\sin\varphi(I_x S_y + I_y S_x) \\
&\quad + \sin\theta\cos\theta\sin\varphi(I_x S_z + I_z S_x) \\
&\quad + \sin\theta\cos\theta\cos\varphi(I_y S_z + I_z S_y))\Big]
\end{aligned}$$

$$= \left(\frac{\gamma_I \gamma_S h^2}{4\pi^2 r^3}\right)(A + B + C + D + E + F) \tag{2}$$

with

$$
\begin{aligned}
A &= I_z S_z (1 - 3\cos^2\theta) \\
B &= -\tfrac{1}{4}(I^+ S^- + I^- S^+)(1 - 3\cos^2\theta) \\
C &= -\tfrac{3}{2}(I^+ S_z + I_z S^+)\sin\theta\cos\theta\, e^{-i\varphi} \\
D &= -\tfrac{3}{2}(I^- S_z + I_z S^-)\sin\theta\cos\theta\, e^{i\varphi} \\
E &= -\tfrac{3}{4} I^+ S^+ \sin^2\theta\, e^{-2i\varphi} \\
F &= -\tfrac{3}{4} I^- S^- \sin^2\theta\, e^{2i\varphi}
\end{aligned}
\tag{3}
$$

The structure of this Hamiltonian insures zero-, single-, and double-quantum connectivities. We express $\mathscr{H}_d$ as a product of a spin operator part and a function of spatial coordinates:

$$\mathscr{H}_d(t) = \mathbf{O}_d f(t)$$

where random motions lead to the time dependence in $f(t)$. Employing the standard results of time-dependent perturbation theory to compute the probability of relaxation transitions induced by $\mathscr{H}_d(t)$ between a pair of levels $|i)$ and $|j)$ of the two-spin system, we have, assuming for the fluctuations in $\mathscr{H}_d$ a stationary random process:

$$
\begin{aligned}
W_{ij} &= \frac{1}{T}\frac{4\pi^2}{h^2}|\mathbf{O}_{ij}|^2 \int_0^T dt \int_{-t}^{T-t} \overline{f(t+\tau)f(t)}\, e^{-2\pi i(E_i - E_j)\tau/h}\, d\tau \\
&= \frac{1}{T}\frac{4\pi^2}{h^2}|\mathbf{O}_{ij}|^2 \int_0^T dt \int_{-t}^{T-t} G(\tau) e^{-2\pi i(E_i - E_j)\tau/h}\, d\tau \\
&= \frac{4\pi^2}{h^2}|\mathbf{O}_{ij}|^2 \int_{-\infty}^{\infty} G(\tau) e^{-2\pi i(E_i - E_j)\tau/h}\, d\tau \\
&= \frac{4\pi^2}{h^2}|\mathbf{O}_{ij}|^2 J(\omega_{ij})
\end{aligned}
\tag{4}
$$

where $G(\tau)$ is an autocorrelation function and is generally a rapidly decaying function of τ. The transition probability between levels i and j under random fluctuations in $\mathscr{H}_d(t)$ is thus proportional to the spectral density $J(\omega)$ of the fluctuations at $\omega_{ij} = (E_i - E_j)$; $J(\omega)$ is the Fourier transform of the correlation function:

$$G(\tau) = \overline{f(t)f(t+\tau)} \tag{5}$$

the statistical avarage of $f(t)f(t+\tau)$.

Three distinct probabilities are to be calculated for zero-, single-, and double-quantum processes, as indicated above. The probabilities are all products of the square of the appropriate matrix element, with the function of spatial coordinates averaged suitably. We find for $I = S = 1/2$:

$$
\begin{aligned}
W_0 &\sim \tfrac{1}{16}\overline{(1 - 3\cos^2\theta)^2}\, J(\omega_{23}) = \tfrac{1}{16} \times \tfrac{4}{5} J(\omega_{23}) \\
W_{1I} &\sim \tfrac{9}{16}\overline{\sin^2\theta\cos^2\theta}\, J(\omega_{12}) = \tfrac{9}{16} \times \tfrac{2}{15} J(\omega_{12}) \\
W_2 &\sim \tfrac{9}{16}\overline{\sin^4\theta}\, J(\omega_{14}) = \tfrac{9}{16} \times \tfrac{8}{15} J(\omega_{14})
\end{aligned}
\tag{6}
$$

with the spatial coordinates isotropically averaged over the unit sphere. When dealing with white spectral conditions for $J(\omega)$, i.e., in the extreme narrowing limit, we therefore have: $W_0 : 2W_1 : W_2 = 1 : 3 : 6$. For homonuclear interactions in the "black" spectral limit, on the other hand $W_1 = W_2 = 0$, and W_0 alone is nonzero. In general therefore the spectral density at $\omega = \omega_I$, ω_S, $|\omega_I - \omega_S|$ and $(\omega_I + \omega_S)$ are all relevant parameters contributing to relaxation under dipolar coupling; for T_2 relaxation $J(0)$ contributes additionally via the $A(t)$ term of $\mathscr{H}_d(t)$.

In many practical situations the correlation function $G(\tau)$ may be characterized with a single time constant which is the correlation time τ_c:

$$
G(\tau) \sim \exp(-|\tau|/\tau_c) \tag{7}
$$

$J(\omega)$ now takes the form:

$$
\begin{aligned}
J(\omega) &= \int_{-\infty}^{\infty} \exp(-|\tau|/\tau_c)\exp(-i\omega\tau)\, d\tau \\
&= \frac{2\tau_c}{1 + \omega^2\tau_c^2}
\end{aligned}
\tag{8}
$$

A plot of $J(\omega)$ versus ω is given in Fig. A.4.1 for three situations: $\omega_0 < 1/\tau_c$, $\omega_0 \sim 1/\tau_c$ and $\omega_0 > 1/\tau_c$, ω_0 being the Larmor frequency.

The nuclear Overhauser effect may be derived at once from the relaxation and energy level diagram (Fig. A.4.2) for two spins 1/2, employing the above discussion. Notice that observing I spins while saturating S, W_{1I} contri-

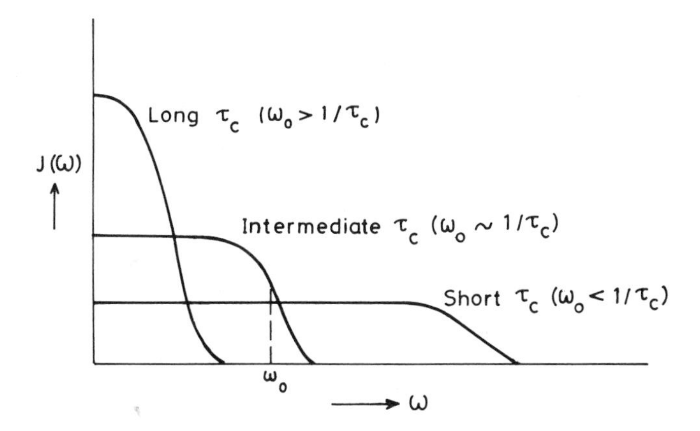

Figure A.4.1. Spectral density $J(\omega)$ versus frequency ω.

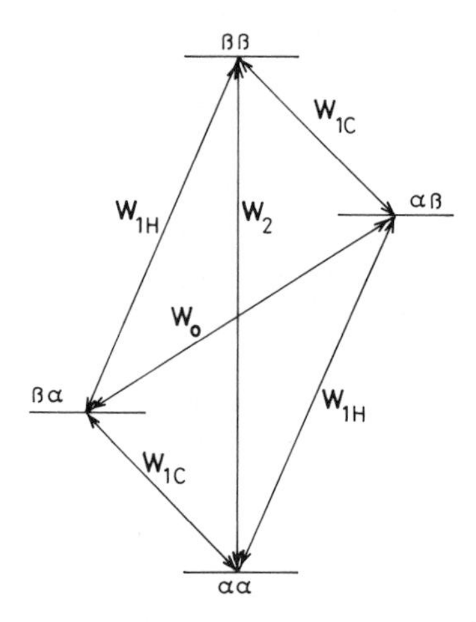

Figure A.4.2. Energy level diagram and relaxation probabilities in a system of two spins 1/2.

butes to establishment of Boltzmann populations, while W_2 and W_0 lead to nonequilibrium populations of opposite sign. This leads at once to the enhancement factor:

$$\frac{\langle I_z \rangle - \langle I_0 \rangle}{\langle I_0 \rangle} = \eta = \frac{(W_2 - W_0)}{(W_0 + 2W_{1I} + W_2)} \frac{\gamma_S}{\gamma_I} \tag{9}$$

In the white spectral limit, therefore,

$$\frac{\langle I_z \rangle - \langle I_0 \rangle}{\langle I_0 \rangle} = \frac{1}{2} \frac{\gamma_S}{\gamma_I} \tag{10}$$

When $I = {}^{13}C$ and $S = {}^{1}H$, for example, upon saturating the proton resonances, the dipole coupled ${}^{13}C$'s attain a polarization which is three times the Boltzmann equilibrium quantity. A negative NOE (~ 5) results for ${}^{15}N$ dipolar coupled to protons ($I = {}^{15}N$, $S = {}^{1}H$), since ${}^{15}N$ has a negative magnetogyric ratio.

In practical situations, the dipolar interaction between spins may or may not be the dominant mechanism of relaxation; in case other mechanisms such as spin–rotation interaction, chemical shift anisotropy, etc., contribute significantly to the establishment of Boltzmann equilibrium, observed nuclear Overhauser enhancements will be below the maximum derived above:

$$\frac{\langle I_z \rangle - \langle I_0 \rangle}{\langle I_0 \rangle} = \frac{\gamma_S}{\gamma_I} \frac{(W_2 - W_0)}{(W_0 + 2W_{1I} + W_2 + W^0)} \tag{11}$$

where W^0 is the transition probability under other relaxation mechanisms.

In the homonuclear situation, we find:

$$\frac{\langle I_z \rangle - \langle I_0 \rangle}{\langle I_0 \rangle} = \tfrac{1}{2} \text{ in the white spectral limit} \qquad (12)$$
$$= -1 \text{ in the ``black'' spectral limit}$$

It is interesting to note that in the isotropic phase, negative NOE's can result for a pair of nuclei with the same sign of γ, if τ_c is above a threshold value, *provided* the larger γ is not more than 2.38 times the smaller γ. This again corresponds to the situation that $W_0 > W_2$, i.e., $J(\omega_{23}) \gg J(\omega_{14})$.

APPENDIX 5

Magnus Expansion and the Average Hamiltonian Theory

The time evolution of the density matrix of the system under the influence of the internal Hamiltonians is governed by the von Neumann equation:

$$\dot{\sigma}(t) = -i[\mathscr{H}, \sigma(t)] \tag{1}$$

When the Hamiltonian is time independent then the above equation integrates to:

$$\sigma(t) = U(t)\sigma(0)U^{-1}(t)$$

with:

$$U(t) = \exp(-i\mathscr{H}t) \tag{2}$$

However, when the Hamiltonian is not constant throughout the time interval of interest but can be assumed to change with a proviso that when the total interval is divided into n subintervals of duration $\tau_1, \tau_2, \ldots, \tau_n$, the corresponding Hamiltonians $\mathscr{H}_k (k = 1, \ldots, n)$ are constant in the kth interval we can write $U(t)$ as:

$$U(t) = \exp(-i\mathscr{H}_n\tau_n)\exp(-i\mathscr{H}_{n-1}\tau_{n-1})\ldots\exp(-i\mathscr{H}_1\tau_1) \tag{3}$$

provided that all the $\mathscr{H}_k$'s commute.

If, on the other hand, the Hamiltonian varies continuously we have to express $U(t)$ in integral form as:

$$U(t) = T\exp\left(-i\int_0^t \mathscr{H}(t')\,dt'\right) \tag{4}$$

where T is the Dyson time ordering operator, which orders the operators of higher time arguments in the expanded exponential to the left.

Magnus expansion is the tool used to express the operator $U(t_c)$, (this replaces $U(t)$, when dealing with cyclic pulse sequences such as WAHUHA-4,

MREV-8, see Chapter 6) in the form of $\exp(-iFt_c)$. The form of F is readily found when we expand $U(t_c)$ in a power series (under conditions of actions corresponding to infinitesimal rotations):

$$\begin{aligned} U(t_c) = &\left[1 + (-i\mathscr{H}_n\tau_n) + \frac{(-i\mathscr{H}_n\tau_n)^2}{2!} + \cdots\right] \times \cdots \\ &\left[1 + (-i\mathscr{H}_k\tau_k) + \frac{(-i\mathscr{H}_k\tau_k)^2}{2!} + \cdots\right] \times \cdots \\ &\left[1 + (-i\mathscr{H}_1\tau_1) + \frac{(-i\mathscr{H}_1\tau_1)^2}{2!} + \cdots\right] \times \cdots \end{aligned} \tag{5}$$

If the time intervals are so divided as to make $\mathscr{H}_k\tau_k$ a small quantity of first order we can rewrite the above equation as:

$$\begin{aligned} U(t_c) = &\ 1 + (-i)[\mathscr{H}_1\tau_1 + \mathscr{H}_2\tau_2 + \cdots + \mathscr{H}_n\tau_n] \\ &+ \frac{(-i)^2}{2!}[\mathscr{H}_1^2\tau_1^2 + \mathscr{H}_2^2\tau_2^2 + \cdots + \mathscr{H}_n^2\tau_n^2 \\ &+ 2(\mathscr{H}_2\mathscr{H}_1\tau_2\tau_1 + \mathscr{H}_3\mathscr{H}_2\tau_3\tau_2 + \cdots \\ &+ \mathscr{H}_n\mathscr{H}_{n-1}\tau_n\tau_{n-1})] + \frac{(-i)^3}{3!}[\mathscr{H}_1^3\tau_1^3 + \mathscr{H}_2^3\tau_2^3 \\ &+ \cdots 3(\mathscr{H}_2\mathscr{H}_1{}^2\tau_2\tau_1^2 + \mathscr{H}_2^2\mathscr{H}_1\tau_2^2\tau_1 + \cdots) \\ &+ 6(\mathscr{H}_3\mathscr{H}_2\mathscr{H}_1\tau_3\tau_2\tau_1 + \mathscr{H}_4\mathscr{H}_3\mathscr{H}_2\tau_4\tau_3\tau_2 + \cdots) \\ &\cdots + \cdots] \end{aligned} \tag{6}$$

The time ordering operator makes sure that all operators are arranged in ascending order from right to left.

We said we wanted to express $U(t_c)$ as $\exp(-iFt_c)$. We now make the assumption that F can be written as a sum of Hamiltonians of different order of magnitude (similar to perturbation order); i.e.,

$$F = \bar{\mathscr{H}} + \bar{\mathscr{H}}^{(1)} + \bar{\mathscr{H}}^{(2)} + \cdots \tag{7}$$

This leads to:

$$\begin{aligned} \exp(-iFt_c) &= \exp(-it_c(\bar{\mathscr{H}} + \bar{\mathscr{H}}^{(1)} + \bar{\mathscr{H}}^{(2)} + \cdots)) \\ &= 1 + (-it_c)(\bar{\mathscr{H}} + \bar{\mathscr{H}}^{(1)} + \bar{\mathscr{H}}^{(2)} + \cdots) \\ &\quad + \frac{(-it_c)^2}{2!}[(\bar{\mathscr{H}})^2 + \bar{\mathscr{H}}\bar{\mathscr{H}}^{(1)} + \bar{\mathscr{H}}^{(1)}\bar{\mathscr{H}} + (\bar{\mathscr{H}}^{(1)})^2 \\ &\quad + \bar{\mathscr{H}}\bar{\mathscr{H}}^{(2)} + \bar{\mathscr{H}}^{(2)}\bar{\mathscr{H}} + \bar{\mathscr{H}}^{(2)}\bar{\mathscr{H}}^{(1)} + \cdots] \\ &\quad + \frac{(-it_c)^3}{3!}[(\bar{\mathscr{H}})^3 + (\bar{\mathscr{H}})^2\bar{\mathscr{H}}^{(1)} + \bar{\mathscr{H}}^{(1)}(\bar{\mathscr{H}})^2 \\ &\quad + \bar{\mathscr{H}}\bar{\mathscr{H}}^{(1)}\bar{\mathscr{H}} + (\bar{\mathscr{H}}^{(1)})^3 + \cdots] \end{aligned} \tag{8}$$

Grouping terms that are correct to a given order, the above equation can be rewritten:

$$\begin{aligned}\exp(-iFt_c) = 1 - it_c\bar{\mathscr{H}} - it_c\bar{\mathscr{H}}^{(1)} + \frac{(-it_c)^2}{2!}(\bar{\mathscr{H}})^2 \\ - it_c\bar{\mathscr{H}}^{(2)} + \frac{(-it_c)^2}{2!}[\bar{\mathscr{H}}\bar{\mathscr{H}}^{(1)} + \bar{\mathscr{H}}^{(1)}\bar{\mathscr{H}}] + \cdots\end{aligned} \tag{9}$$

we shall now equate the terms in the above equation to those given in the expansion of $U(t_c)$, leading to:

$$\begin{aligned}\bar{\mathscr{H}} &= 1/t_c(\mathscr{H}_1\tau_1 + \mathscr{H}_2\tau_2 + \mathscr{H}_3\tau_3 + \cdots) \\ \bar{\mathscr{H}}^{(1)} &= (-i/(2t_c))([\mathscr{H}_2, \mathscr{H}_1]\tau_2\tau_1 + [\mathscr{H}_3, \mathscr{H}_1]\tau_3\tau_1 \\ &\quad + [\mathscr{H}_3, \mathscr{H}_2]\tau_3\tau_2 + \cdots) \text{ etc.}\end{aligned} \tag{10}$$

It is obvious that $t_c\bar{\mathscr{H}}$ must be a quantity of first order, $t_c\bar{\mathscr{H}}^{(1)}$, of second order, etc.

As an example, if the $\mathscr{H}$ takes only two values during a cycle such that during 0 to $t_c/2$, $\mathscr{H}_1\tau_1 = A$ and during $t_c/2$ to t_c, $\mathscr{H}_2\tau_2 = B$, then Magnus expansion leads to:

$$\begin{aligned}t_c\bar{\mathscr{H}} &= A + B \\ t_c\bar{\mathscr{H}}^{(1)} &= (-i/2)[B, A] \\ t_c\bar{\mathscr{H}}^{(2)} &= (1/12)\{[B, [B, A]] + [[B, A], A]\}\end{aligned} \tag{11}$$

From which it follows that, for noncommuting operators A and B,

$$\begin{aligned}\exp(-iB)\exp(-iA) = \exp\{-i(A + B) - 1/2[B, A] \\ + (i/12)([B, [B, A]] + [[B, A], A]) + \cdots\}\end{aligned} \tag{12}$$

which is the well known Baker–Campbell–Hausdorff formula.

In the limit $\tau_k \to 0$:

$$\bar{\mathscr{H}} = (1/t_c)\int_0^{t_c} \mathscr{H}(t)\,dt \tag{13}$$

where $\bar{\mathscr{H}}$ is the *average Hamiltonian* correct to first order. Likewise,

$$\bar{\mathscr{H}}^{(1)} = (-i/2t_c)\int_0^{t_c} dt_2 \int_0^{t_2} dt_1\,[\mathscr{H}(t_2), \mathscr{H}(t_1)] \tag{14}$$

For a symmetrical cycle of time-dependent Hamiltonians such that:

$$\begin{aligned}\bar{\mathscr{H}}(t) &= \bar{\mathscr{H}}(t_c - t) \\ U^{-1}(t_c) &= \exp(i\mathscr{H}_1\tau_1)\exp(i\mathscr{H}_2\tau_2)\cdots\exp(i\mathscr{H}_n\tau_n) \\ &= \exp(-i\mathscr{H}_n\tau_n)\exp(-i\mathscr{H}_{n-1}\tau_{n-1})\cdots\exp(-i\mathscr{H}_1\tau_1) \\ &= U(-t_c)\end{aligned} \tag{15}$$

For such cycles it is straightforward to show that:

$$\bar{\mathscr{H}}^{(k)} = (-1)^k \bar{\mathscr{H}}^{(k)} \tag{16}$$

and hence:

$$\bar{\mathscr{H}}^{(k)} = 0$$

for k odd. If, on the other hand, we consider antisymmetrical cycles corresponding to the equality:

$$\bar{\mathscr{H}}(t) = -\bar{\mathscr{H}}(t - t_c)$$

then:

$$U(t_c) = \exp(i\mathscr{H}_1\tau_1)\exp(i\mathscr{H}_2\tau_2)\cdots \times \cdots \exp(i\mathscr{H}_{n-1}\tau_{n-1})\exp(i\mathscr{H}_n\tau_n)$$
$$= U^{-1}(t_c)$$

i.e.,

$$\exp(-iF_a t_c) = \exp(iF_a t_c)$$

Therefore:

$$F_a = 0 \tag{17}$$

Applying the average Hamiltonian theory to the Carr–Purcell sequence corresponding to 90°_y–τ–180°_y–2τ–$180^\circ_y 2\tau$, we can see that the sequence is cyclic with cycle time 4τ. It is used mainly to suppress the applied field inhomogeneity term:

$$\begin{aligned}\mathscr{H}_{\text{inh}} &= -\gamma \sum_k \Delta B_{0k} I_{kz} \\ &= -\sum_k \Delta\omega_k I_{kz}\end{aligned} \tag{18}$$

where ΔB_{0k} is the deviation of the static field from its mean value at the kth nucleus. In the interaction representation I_{kz} toggles between I_{kz} and $-I_{kz}$ so that for integer cycle times chemical shift evolution and field inhomogeneities are removed. Under δ pulse approximation the Hamiltonian commutes at all times and hence no contribution from $\bar{\mathscr{H}}^{(n)}$ $(n > 1)$ exists. For finite pulse width t_w,

$$\bar{\mathscr{H}}^{(1)} = \frac{1}{\pi t_w}(1 - (t_w/t_c)) \sum_k (\Delta\omega_k)^2 I_{kz} \neq 0 \tag{19}$$

For the phase alternated sequence $-\tau$–θ_{-y}–2τ–θ_y–$\tau-$ first-order average Hamiltonian is given by:

$$\bar{\mathscr{H}} = (1/2)(\mathscr{H}_x + \mathscr{H}_z) \tag{20}$$

when $\theta = \pi/2$. For dipolar interaction:

$$\bar{\mathscr{H}}_{\text{D}} = (1/2)(\mathscr{H}_{\text{D}x} + \mathscr{H}_{\text{D}z}) = -(1/2)\mathscr{H}_{\text{D}y} \tag{21}$$

and for chemical shift interaction:

$$\bar{\mathscr{H}}_{\mathrm{CS}} = \frac{1}{2}\sum_k (\delta_k + \Delta)(-I_{kx} + I_{kz})$$

$$= \frac{1}{\sqrt{2}}\sum_k (\delta_k + \Delta) I^{(k)}_{\langle 101 \rangle}$$

where:

$$I^{(k)}_{\langle 101 \rangle} = \frac{1}{\sqrt{2}}(-I_{kx} + I_{kz}) \tag{22}$$

with the chemical shifts being scaled by a factor $(\sqrt{2})^{-1}$. The cycle is symmetrical and hence all $\bar{\mathscr{H}}^{(k)}$, k = odd vanishes. The leading even-order correction term is given by:

$$\bar{\mathscr{H}}^{(2)} = -\frac{\tau^2}{12}\sum_k (\delta_k + \Delta)^3 (2I_{kz} - I_{kx}) \tag{23}$$

When the pulses are θ pulses ($\theta \neq \pi/2$),

$$\bar{\mathscr{H}} = \frac{1}{2}\sum_k [I_{kz}(1 + \cos\theta) - I_{kx}\sin\theta] \tag{24}$$

and the corresponding scale factor can be shown to be $|\cos(\theta/2)|$. The phase-alternated pulse sequence is used to scale down chemical shifts uniformly.

APPENDIX 6

Tensor Representation of Spin Hamiltonians

All interactions that we come across in magnetic resonance spectroscopy—dipolar coupling, chemical shift, spin–spin coupling, quadrupole coupling, etc.,—can be written in compact form using the tensor notation in terms of irreducible spherical tensor operators because the spin-dependent and coordinate-dependent parts transform as second-rank tensors.

The spin operators in the spherical basis and the various spherical harmonics are very similar and it is possible to construct tensor operator equivalents by a systematic substitution of position variables by the corresponding spin components using the equivalent operator formalism. The position operators do commute with each other, while the spin operators need not and to avoid any complication out of this we follow Rose's definition of a tensor operator. The mth component of an lth rank tensor is given by:

$$T_{l,m}(\mathbf{I}) = (\mathbf{I} \cdot \nabla)^l r^l Y_{l,m}(r) \tag{1}$$

where the operator $(\mathbf{I} \cdot \nabla)^1$ operates on the position coordinates of the spherical harmonic $Y_{l,m}$ and systematically replaces them by the appropriate spin component. The spherical harmonics are given by:

$$Y_{l,m}(\theta, \varphi) = (-1)^{(m+|m|)/2} \left[\frac{(2l+1)(l-|m|)!}{4\pi(l+|m|)!} \right]^{1/2} \times P_{l,m}(\cos\theta) \exp(im\varphi) \tag{2}$$

where $P_{l,m}(\cos\theta)$ are the associated Legendre polynomials. Equation (1) automatically gives for first-rank tensor elements:

$$T_{1,1} = -(3/4)I_+/\sqrt{2}$$
$$T_{1,0} = (3/4)I_0 \tag{3}$$
$$T_{1,-1} = (3/4)I_-/\sqrt{2}$$

where $I_0 = I_z$, $I_+ = I_x + iI_y$, and $I_- = I_x - iI_y$.

The higher rank tensors can be generated from the first-rank tensors by the recursion relation:

$$T_{l,m} = \left(\frac{4\pi}{3}\right)^{1/2}\left[\frac{l!(2l+1)!!}{4\pi}\right]^{1/2} \times \sum_{\mu} C(1, l-1, l; -\mu, \mu+m) \times T_{l-1,\mu+m}T_{1,-\mu} \tag{4}$$

where $C(\ldots)$ are the Clebsch–Gordon coefficients. Thus $T_{2,0}$ is given by:

$$T_{2,0} = \frac{4\pi}{3}\left(\frac{15}{2\pi}\right)^{1/2}[C(1,1,2;-1,1)T_{1,1}T_{1,-1} + C(1,1,2;0,0)T_{1,0}T_{1,0} + C(1,1,2;1,-1)T_{1,-1}T_{1,1}] \tag{5}$$

The second-rank tensors are thus derived to be:

$$T_{2,0} = \sqrt{\frac{5}{4\pi}}(3I_0^2 - \mathbf{I}^2) \tag{6}$$

$$T_{2,\pm1} = \pm\sqrt{\frac{15}{8\pi}}(I_\pm I_0 \pm I_0 I_\pm)$$
$$T_{2,\pm2} = \sqrt{\frac{15}{8\pi}}I_\pm^2 \tag{7}$$

The nine components T_{ij} of a Cartesian second-rank tensor $\{T\}$ can be decomposed into a scalar:

$$T_0 = \tfrac{1}{3}\mathrm{Tr}\{\mathbf{T}\} \tag{8}$$

and an antisymmetric first-rank tensor:

$$\mathbf{T}_1 : T_{ij}^1 = (1/2)(T_{ij} - T_{ji}) \tag{9}$$

of which there will be three components with zero trace and a traceless second-rank tensor;

$$\mathbf{T}_2 : T_{ij}^2 = \tfrac{1}{2}(T_{ij} + T_{ji}) - \tfrac{1}{3}\mathrm{Tr}\{\mathbf{T}\} \tag{10}$$

having five components. Thus any of the nine elements can be given by:

$$T_{ij} \doteq \tfrac{1}{3}\mathrm{Tr}\{\mathbf{T}\} + T_{ij}^1 + T_{ij}^2 \tag{11}$$

Just as rotation in coordinate space can be represented by Euler rotation matrices $R(\alpha\beta\gamma)$ such that:

$$T'_{kq} = R(\alpha\beta\gamma) T_{kq} R^{-1}(\alpha\beta\gamma) \tag{12}$$

the corresponding irreducible spherical tensor components are governed for coordinate space rotation by the Wigner rotation matrices $\mathscr{D}^k_{pq}(\alpha\beta\gamma)$ (see Table A.6.1). For second-rank tensors one need consider only irreducible spherical tensor operators up to order 2.

Hamiltonians can be expressed as the product of two irreducible tensors $\mathbf{A}_k$ and $\mathbf{T}_k$ with components A_{kq}, T_{kp} so that:

$$\mathbf{A}_k \cdot \mathbf{T}_k = \sum_{q=-k}^{+k} (-1)^q A_{kq} T_{k-q} = \sum_{q=-k}^{+k} (-1)^q A_{k-q} T_{kq} \tag{13}$$

Any Hamiltonian involving spin interactions can be expressed as:

$$\mathscr{H} = \mathbf{X} \cdot \mathbf{A} \cdot \mathbf{Y} = \sum_{i,j} A_{ij} X_i Y_j \tag{14}$$

where $\mathbf{X}$ and $\mathbf{Y}$ are vectors and $\mathbf{A}$ is a 3×3 matrix. In terms of irreducible spherical tensor operators,

$$\mathscr{H} = \sum_{k=0}^{2} \sum_{q=-k}^{+k} (-1)^q A_{kq} T_{k-q} \tag{15}$$

with:

$$T_{ij} = Y_i X_j$$

and:

$$\begin{aligned}
T_{00} &= -\frac{1}{3}(T_{xx} + T_{yy} + T_{zz}) \\
T_{10} &= -\frac{i}{\sqrt{2}}(T_{xy} - T_{yx}) \\
T_{1\pm1} &= -\frac{1}{2}(T_{zx} - T_{xz} \pm i(T_{zy} - T_{yz})) \\
T_{20} &= \frac{1}{\sqrt{6}}(3T_{zz} - (T_{xx} + T_{yy} + T_{zz})) \\
T_{2\pm1} &= \mp\frac{1}{2}(T_{xz} + T_{zx} \pm i(T_{yz} + T_{zy})) \\
T_{2\pm2} &= \frac{1}{2}(T_{xx} - T_{yy} \pm i(T_{xy} + T_{yx}))
\end{aligned} \tag{16}$$

The corresponding elements of the $\mathbf{A}$ tensor are given by replacing T by spin operators. For symmetric tensors,

$$\mathscr{H} = A_{00} T_{00} + A_{20} T_{20} - (A_{2,-1} T_{21} + A_{21} T_{2-1}) \tag{17}$$

Table A.6.1. Wigner Rotation Matrix $D^2_{m'm}(\alpha, \beta, \gamma)$

m' \ m	2	1	0	−1	−2
2	$\frac{(1+\cos\beta)^2}{4} e^{2i(\alpha+\gamma)}$	$\frac{1+\cos\beta}{2} \sin\beta e^{i(2\gamma+\alpha)}$	$\sqrt{\frac{3}{8}} \sin^2\beta e^{2i\gamma}$	$\frac{1-\cos\beta}{2} \sin\beta e^{i(2\gamma-\alpha)}$	$\frac{(1-\cos\beta)^2}{4} e^{2i(\gamma-\alpha)}$
1	$-\frac{1+\cos\beta}{2} \sin\beta e^{i(2\alpha+\gamma)}$	$\left[\cos^2\beta - \frac{1-\cos\beta}{2}\right] e^{i(\alpha+\gamma)}$	$\sqrt{\frac{3}{8}} \sin 2\beta e^{i\gamma}$	$\left[\frac{1+\cos\beta}{2} - \cos^2\beta\right] e^{i(\gamma-\alpha)}$	$\frac{1-\cos\beta}{2} \sin\beta e^{i(\gamma-2\alpha)}$
0	$\sqrt{\frac{3}{8}} \sin^2\beta e^{2i\alpha}$	$-\sqrt{\frac{3}{8}} \sin 2\beta e^{i\alpha}$	$\frac{3\cos^2\beta - 1}{2}$	$\sqrt{\frac{3}{8}} \sin 2\beta e^{-i\alpha}$	$\sqrt{\frac{3}{8}} \sin^2\beta e^{-2i\alpha}$
−1	$-\frac{1-\cos\beta}{2} \sin\beta e^{i(2\alpha-\gamma)}$	$\left[\frac{1+\cos\beta}{2} - \cos^2\beta\right] e^{i(\alpha-\gamma)}$	$-\sqrt{\frac{3}{8}} \sin^2\beta e^{-i\gamma}$	$\left[\cos^2\beta - \frac{1-\cos\beta}{2}\right] e^{-i(\alpha+\gamma)}$	$\frac{1+\cos\beta}{2} \sin\beta e^{-i(2\alpha+\gamma)}$
−2	$\frac{(1-\cos\beta)^2}{4} e^{2i(\alpha-\gamma)}$	$-\frac{1-\cos\beta}{2} \sin\beta e^{i(\alpha-2\gamma)}$	$\sqrt{\frac{3}{8}} \sin^2\beta e^{-2i\gamma}$	$-\frac{1+\cos\beta}{2} \sin\beta e^{-i(\alpha+2\gamma)}$	$\frac{(1+\cos\beta)^2}{4} e^{-2i(\alpha+\gamma)}$

with:

$$A_{2\pm1} = \mp (A_{xz} \pm iA_{yz})$$

For dipolar (D) and quadrupolar (Q) interaction the corresponding elements are:

$$\begin{aligned}
A_{00} &= -\frac{1}{3}\operatorname{Tr}\mathrm{D} \\
A_{10} &= A_{1\pm1} = 0 \\
A_{20} &= (3/\sqrt{6})\mathrm{D}_{zz} \\
A_{2\pm1} &= \mp(\mathrm{D}_{xz} \pm i\mathrm{D}_{yz}) \\
A_{2\pm2} &= \frac{1}{2}(\mathrm{D}_{xx} - \mathrm{D}_{yy} \pm 2i\mathrm{D}_{xy})
\end{aligned} \tag{18}$$

and:

$$\begin{aligned}
T_{20} &= (1/\sqrt{6})(3I_zS_z - \mathbf{I}\cdot\mathbf{S}) \\
T_{2\pm1} &= \mp\frac{1}{2}(I_zS_\pm + I_\pm S_z) \\
T_{2\pm2} &= (1/2)I_\pm S_\pm
\end{aligned} \tag{19}$$

Table A.6.2 gives the various interactions in terms of spin operators corresponding to the $T_{l,m}$'s.

Table A.6.3 summarizes the transformation properties under spin operator space rotation (cf., Chapter 6, multiple-pulse line-narrowing sequences) of operators that are not rejected by truncation (i.e., that correspond to first-order in perturbation).

As can be seen I-spin homonuclear bilinear spin operators and single spin

Table A.6.2. T_{lm}'s Corresponding to Various Interactions

Interaction	λ	$I_\alpha A^\lambda_\beta$	T_{00}	T_{20}	$T_{2\pm1}$	$T_{2\pm2}$
Shielding	CS	$I^i_\alpha B_\beta$	$I^i_0 B_0$	$\sqrt{\frac{2}{3}}I^i_0 B_0$	$\frac{1}{\sqrt{2}}I^i_\pm B_0$	0
Spin rotation	SR	$I^i_\alpha J_\beta$	$\mathbf{I}^i\cdot\mathbf{J}$	$\frac{1}{\sqrt{6}}(3I^i_0J_0 - \mathbf{I}^i\cdot\mathbf{J})$	$\frac{1}{\sqrt{2}}(I^i_{\pm1}J_0 + I^i_0J_{\pm1})$	$I^i_{\pm1}J_{\pm1}$
Quadrupole	Q	$I^i_\alpha I^i_\beta$	$(\mathbf{I}^i)^2$	$\frac{1}{\sqrt{6}}(3(I^i_0)^2 - (\mathbf{I}^i)^2)$	$\frac{1}{\sqrt{2}}(I^i_{\pm1}I^i_0 + I^i_0I^i_{\pm1})$	$(I^i_{\pm1})^2$
Dipole indirect Spin–spin	D J	$I^i_\alpha I^k_\beta$	$\mathbf{I}^i\cdot\mathbf{I}^k$	$\frac{1}{\sqrt{6}}(3I^i_0I^k_0 - \mathbf{I}^i\cdot\mathbf{I}^k)$	$\frac{1}{\sqrt{2}}(I^i_{\pm1}I^k_0 + I^i_0I^k_{\pm1})$	$I^i_{\pm1}I^k_{\pm1}$

Table A.6.3. Transformation of T_{lm}'s in I-spin Space

Interaction	T_{l0}^{λ}	After truncation	Transforms in I-spin space according to: $l =$	$m =$
Shielding	T_{00}^{CS}	$I_0^i B_0$	1	0
	T_{20}^{CS}	$\sqrt{\frac{2}{3}} I_0^i B_0$	1	0
Quadrupolar	T_{20}^{Q}	$\frac{1}{\sqrt{6}}(3(I_0^i)^2 - (\mathbf{I}^i)^2)$	2	0
Direct and indirect between $\mathbf{I}^i$ and $\mathbf{I}^k$	T_{00}^{J}	$\mathbf{I}^i \cdot \mathbf{I}^k$	0	0
	$T_{20}^{D,J}$	$\frac{1}{\sqrt{6}}(3I_0^i I_0^k - \mathbf{I}_0^i \cdot \mathbf{I}_0^k)$	2	0
Direct and indirect between $\mathbf{I}^i$ and $\mathbf{S}^k$	T_{00}^{J}	$I_0^i S_0^k$	1	0
	$T_{20}^{D,J}$	$\sqrt{\frac{2}{3}} I_0^i S_0^k$	1	0

quadrupolar operators are identical in their transformation properties in both full space and I spin subspace. However, operators that are linear in I space transform differently in full space and I-spin subspace and this aids in selective averaging by multiple-pulse sequences.

Selected Bibliography

Some Fundamental Papers

1. F. Bloch, *Phys. Rev.* **70**, 460 (1946).
2. F. Bloch, W.W. Hansen, and M. Packard, *Phys. Rev.* **70**, 474 (1946).
3. N. Bloembergen, E.M. Purcell, and R.V. Pound, *Phys. Rev.* **73**, 679 (1948).
4. E.L. Hahn, *Phys. Rev.* **80**, 580 (1950).
5. J.H. Van Vleck, *Phys. Rev.* **74**, 1168 (1948).
6. E.L. Hahn and D.E. Maxwell, *Phys. Rev.* **88**, 1070 (1952).
7. N.F. Ramsay, *Phys. Rev.* **78**, 699 (1950).
8. H.S. Gutowsky, D.W. McCall, and C.P. Slichter, *J. Chem. Phys.* **21**, 279 (1953).
9. I. Solomon, *Phys. Rev.* **99**, 559 (1955).
10. H.Y. Carr and E.M. Purcell, *Phys. Rev.* **94**, 630 (1954).
11. U. Fano, *Rev. Mod. Phys.* **29**, 74 (1957).
12. E.R. Andrew, A. Bradbury, and R.G. Eades, *Arch. Sci. Geneva*, **11**, 223 (1958).
13. I.J. Lowe and R.E. Norberg, *Phys. Rev.* **107**, 46 (1957).
14. A. Abragam and W.G. Proctor, *Phys. Rev.* **109**, 1441 (1958).
15. C.P. Slichter and W.C. Holton, *Phys. Rev.* **122**, 1701 (1961).
16. U. Haeberlen and J.S. Waugh, *Phys. Rev.* **175**, 453 (1968).
17. F.M. Lurie and C.P. Slichter, *Phys. Rev.* **123**, A1108 (1964).
18. S.R. Hartmann and E.L. Hahn, *Phys. Rev.* **128**, 2042 (1962).
19. A. Pines, M.G. Gibby and J.S. Waugh, *J. Chem. Phys.* **56**, 1776 (1972).
20. W.P. Aue, E. Bartholdi and R.R. Ernst, *J. Chem. Phys.* **64**, 2229 (1976).
21. R. Freeman and G.A. Morris, *Bull Magn. Reson.* **1**, 5 (1979).
22. K.J. Packer and K.M. Wright, *Mol. Phys.* **50**, 797 (1983).
23. O.W. Sørensen, G. Eich, M.H. Levitt, G. Bodenhausen, and R.R. Ernst, *Prog. NMR Spectroscopy*, **16**, 163 (1983).
24. G. Bodenhausen, *Prog. NMR Spectroscopy*, **14**, 137 (1981).
25. D.P. Weitekamp, *Adv. Magn. Resonance*, **11**, 111 (1983).

Books and Monographs

1. E.R. Andrew: *Nuclear Magnetic Resonance*, Cambridge University Press, Cambridge, 1955.
2. A. Abragam: *The Principles of Nuclear Magnetism*, Clarendon Press, Oxford, 1961.
3. J.A. Pople, W.G. Schneider, and H.J. Bernstein: *High Resolution Nuclear Magnetic Resonance*, McGraw-Hill, New York, 1957.
4. M.E. Rose: *Elementary Theory of Angular Momentum*, Wiley, New York, 1957.
5. E.P. Wigner: *Group Theory and its Applications to the Quantum Mechanics of Atomic Spectra*, Academic Press, New York, 1959.
6. C.P. Slichter: *Principles of Magnetic Resonance*, Springer-Verlag, New York, 1980.
7. F.W. Wehrli and T. Wirthlin: *Interpretation of Carbon-13 NMR Spectra*, Heyden and Son Ltd., London, 1976.
8. T.C. Farrar and E.D. Becker: *Pulse and Fourier Transform NMR*, Academic Press New York, 1983.
9. U. Haeberlen: *High Resolution NMR in Solids—Selective Averaging*, Academic Press, New York, 1976.
10. M. Goldman: *Spin Temperature and Nuclear Magnetic Resonance in Solids*, Oxford University Press, Oxford, 1970.
11. M. Mehring: *Principles of High Resolution NMR in Solids*, Springer-Verlag, New York, 1983.
12. Ad Bax: *Two Dimensional Nuclear Magnetic Resonance in Liquids*, Delft University Press, D. Reidel Publishing Co. Dordrecht, 1982.
13. D. Shaw, *Fourier Transform NMR Spectroscopy*: Elsevier, Amsterdam, 1976.
14. P. Mansfield and P.G. Morris: *NMR Imaging in Biomedicine*, Academic Press, New York, 1982.
15. K. Roth: *NMR Tomography and Spectroscopy in Medicine, an Introduction*, Springer-Verlag, New York, 1984.

Coherence Transfer

1. O. W. Sørensen and R.R. Ernst, *J. Magn. Reson.* **51**, 477 (1983).
2. H. Bildsøe, S. Dønstrup, H.J. Jakobsen, and O.W. Sørensen, *J. Magn. Reson.* **53**, 154 (1983), and **55**, 347 (1983).
3. M.R. Bendall, D.T. Pegg, J.R. Wesener, and H. Gunther, *J. Magn. Reson.* **59**, 223 (1984).
4. G.C. Chingas, A.N. Garroway, W.B. Moniz, and R.D. Bertrand, *J. Amer. Chem. Soc.* **102**, 2526 (1980).
5. G.C. Chingas, A.N. Garroway, R.D. Bertrand, and W.B. Moniz, *J. Magn. Reson.* **35**, 283 (1979).
6. D.B. Zax, A. Bielecki, K.W. Zilm and A. Pines, *Chem. Phys. Lett.* **106**, 550 (1984).
7. O.W. Sørensen, R. Freeman, T. Frenkiel, T.H. Mareci, and R. Schuck, *J. Magn. Reson.* **46**, 180 (1982).
8. P.J. Hore, E.R.P. Zinderweg, K. Nicolay, K. Dijkstra, and R. Kaptein, *J. Amer. Chem. Soc.* **104**, 4286 (1982).
9. O.W. Sørensen, M.H. Levitt, and R.R. Ernst, *J. Magn. Reson.* **55**, 104 (1983).
10. P.J. Hore, R.M. Scheek, and R. Kaptein, *J. Magn. Reson.* **52**, 339 (1983).
11. O.W. Sørensen, M.H. Levitt, and R.R. Ernst, *J. Magn. Reson.* **55**, 104 (1983).
12. M.H. Levitt and R.R. Ernst, *Chem. Phys. Lett.* **100**, 119 (1983).
13. O.W. Sørensen, M. Rance, and R.R. Ernst, *J. Magn. Reson*, **56**, 527 (1984).
14. O.W. Sørensen, U.B. Sørensen, and H.J. Jakobsen, *J. Magn. Reson.* **59**, 332 (1984).

15. N. Chandrakumar and S. Subramanian, *J. Magn. Reson.* **62**, 346 (1985).
16. N. Chandrakumar, *J. Magn. Reson.* **63**, 202 (1985).
17. N. Chandrakumar, G.V. Visalakshi, D. Ramaswamy, and S. Subramanian, *J. Magn. Reson.* **67**, 307 (1986).

Two-Dimensional NMR in Liquids

1. D.L. Turner and R. Freeman, *J. Magn. Reson.* **29**, 587 (1978).
2. K. Nagayama, P. Bachmann, K. Wüthrich, and R.R. Ernst, *J. Magn. Reson.* **31**, 133 (1978).
3. H. Kessler, H. Oschkinot, O.W. Sørensen, H. Kogler, and R.R. Ernst, *J. Magn. Reson.* **55**, 329 (1983).
4. A.A. Maudsley, L. Müller, and R.R. Ernst, *J. Magn. Reson.* **28**, 463 (1977).
5. G.A. Morris, *J. Magn. Reson.* **44**, 277 (1981).
6. M.H. Levitt, O.W. Sørensen, and R.R. Ernst, *Chem. Phys. Lett.* **94**, 540 (1983).
7. A. Bax and R. Freeman, *J. Magn. Reson.* **44**, 542 (1981).
8. M. Rance. O.W. Sørensen, G. Bodenhausen, G. Wagner, R.R. Ernst, and K. Wüthrich, *Biochem. Biophys. Res. Comm.* **117**, 479 (1983).
9. Anil Kumar, R.V. Hosur, and K. Chandrasekhar, *J. Magn. Reson.* **60**, 143 (1984).
10. G. Eich, G. Bodenhausen, and R.R. Ernst, *J. Amer. Chem. Soc.* **104**, 3731 (1982).
11. A. Bax, *J. Magn. Reson.* **53**, 149, (1983).
12. H. Kogler, O.W. Sørensen, G. Bodenhausen, and R.R. Ernst, *J. Magn. Reson.* **55**, 157 (1983).
13. L. Braunschweiler and R.R. Ernst, *J. Magn. Reson.* **53**, 521 (1983).
14. J. Jeener, B.H. Meier, P. Bachmann, and R.R. Ernst, *J. Chem. Phys.* **71**, 4546 (1979).
15. S. Macura and R.R. Ernst, *Mol. Phys.* **41**, 95 (1980).
16. S. Macura, K. Wüthrich, and R.R. Ernst, *J. Magn. Reson.* **46**, 269 (1982).
17. C.A.G. Haasnoot, F.J.M. van de Ven, and C.W. Hilbers, *J. Magn. Reson.* **56**, 343 (1984).
18. G. Bodenhausen, G. Wagner, M. Rance, O.W. Sørensen, K. Wüthrich, and R.R. Ernst, *J. Magn. Reson.* **59**, 542 (1984).
19. G. Bodenhausen and R.R. Ernst, *J. Magn. Reson.* **45**, 367 (1981).
20. N. Kurihara, O. Kamo, M. Umeda, K. Sato, K. Hyakuna, and K. Nagayama, *J. Magn. Reson.* **65**, 405 (1985).
21. K. Nagayama, *Analytical Sciences*, **1**, 95 (1985).

Multiple-Quantum Spectroscopy

1. T.H. Mareci and R. Freeman, *J. Magn. Reson.* **48**, 158 (1982).
2. L. Braunschweiler, G. Bodenhausen, and R.R. Ernst, *Mol. Phys.* **48**, 535 (1983).
3. T.H. Mareci and R. Freeman, *J. Magn. Reson.* **51**, 531 (1983).
4. Y.S. Yen and D.P. Weitekamp, *J. Magn. Reson.* **47**, 476 (1982).
5. L. Müller, *J. Magn. Reson.* **59**, 326 (1984).
6. G. Drobny, A Pines, S. Sinton, D.P. Weitekamp, and D. Wemmer, *Symp. Faraday Soc.* **13**, 49 (1979).
7. D.P. Weitekamp, *Adv. Magn. Reson.* **11**, 111 (1983).
8. D.P. Weitekamp, J.R. Garbow, J.B. Murdoch, and A. Pines, *J. Amer. Chem. Soc.* **103**, 3578 (1981).
9. W.S. Warren, S. Sinton, D.P. Weitekamp, and A. Pines, *Phys. Rev. Lett.* **43**, 1791 (1979).

10. A.N. Garroway, J. Baum, M.G. Munowitz, and A. Pines, *J. Magn. Reson.* **60**, 337 (1984).
11. S. Vega, T.N. Shattuck, and A. Pines, *Phys. Rev. Lett.* **37**, 43 (1976).
12. S. Vega, T.N. Shattuck, and A. Pines, *Phys. Rev.* **A22**, 638 (1980).
13. P. Brunner, M. Reinhold, and R.R. Ernst, *J. Chem. Phys.* **73**, 1086 (1980).

High Resolution NMR of Solids

1. J. Schaefer, E.O. Stejskal, and R. Buchdahl, *Macromolecules*, **10**, 384 (1977).
2. J. Herzfeld and A.E. Berger, *J. Chem. Phys.* **73**, 6024 (1980).
3. D. Ellett, U. Haeberlen, and J.S. Wuagh, *J. Amer. Chem. Soc.* **92**, 411 (1970).
4. U. Haeberlen and J.S. Waugh, *Phys. Rev.* **175**, 453 (1968).
5. W.K. Rhim, D.D. Elleman and R.W. Vaughan, *J. Chem. Phys.* **59**, 3740 (1973), and **58**, 1772 (1973).
6. W.K. Rhim, D.D. Elleman, L.B. Schreiber, and R.W. Vaughan, *J. Chem. Phys.* **60**, 4595 (1974).
7. N. Chandrakumar, D. Ramaswamy, and S. Subramanian, *J. Magn. Reson.* **43**, 345 (1983).
8. A. Pines. M.G. Gibby, and J.S. Wuagh, *J. Chem. Phys.* **56**, 1776 (1972).
9. E.O. Stejskal, J. Schaefer, and J.S. Wuagh, *J. Magn. Reson.* **59**, 569 (1973).
10. W.T. Dixon, *J. Chem. Phys.* **77**, 1806 (1982).
11. R.G. Griffin, G. Bodenhausen, R.H. Haberkorn, T.H. Huang, M. Munowitz, R. Osredkar, D.J. Ruben, R.E. Stark, and H. van Willigan, *Phil, Trans. Royal. Soc.* **A299**, 547 (1981).
12. R.K. Hester, J.L. Ackerman, B.L. Neff, and J.S. Waugh, *Phys. Rev. Lett.* **36**, 1081 (1976).
13. M. Linder, A. Höhener, and R.R. Ernst, *J. Chem. Phys.* **73**, 4959 (1980).
14. W.P. Aue. D.J. Ruben, and R.G. Griffin, *J. Magn. Reson.* **43**, 472 (1981).
15. P. Caravatti, G. Bodenhausen, and R.R. Ernst, *Chem. Phys. Lett.* **89**, 363 (1982).
16. P. Caravatti, L. Braunschweiler, and R.R. Ernst, *Chem. Phys. Lett.* **100**, 305 (1983).
17. D.P. Weitekamp, A. Bielecki, D. Zax, K. Zilm, and A. Pines, *Phys. Rev. Lett.* **50**, 1807 (1983).
18. D.B. Zax, A. Bielecki, K.W. Zilm, and A. Pines, *Chem. Phys. Lett.* **106**, 550 (1984).
19. T. Terao, T. Fujii, T. Onodera, and A. Saika, *Chem. Phys. Lett.* **107**, 145 (1984).

Composite Pulses and Imaging

1. M.H. Levitt, R. Freeman, and T. Frankiel in "*Advances in Magnetic Resonance*" (J.S. Waugh, Ed.), Vol. 11, Academic Press, New York, 1983.
2. R. Tycko, H.M. Cho, E. Schneider, and A. Pines, *J. Magn. Reson.* **61**, 60 (1985).
3. R.E. Gordon, P.E. Hanley, D. Shaw, D.G. Gadian, G.K. Radda, P. Styles, P.J. Bore, and L. Chan, *Nature*, **287**, 736 (1980).
4. J.J.H. Ackerman, T.H. Grove, G.G. Wong, D.G. Gadian, and G.K. Radda, *Nature*, **283**, 167 (1980).
5. M.R. Bendall, R.E. Gordon, *J. Magn. Reson*, **53**, 365 (1983).
6. M. Garwood, T. Schleich, G.B. Matson, and G. Acosta, *J. Magn. Reson.* **60**, 268 (1984).

Index